GAMMA-RAY BURSTS
3rd Huntsville Symposium

GAMMA-RAY BURSTS

3rd Huntsville Symposium

Huntsville, AL   October 1995
PART ONE

EDITORS
Chryssa Kouveliotou
USRA/Marshall Space Flight Center

Michael F. Briggs
University of Alabama, Huntsville

Gerald J. Fishman
NASA/Marshall Space Flight Center

AIP CONFERENCE
PROCEEDINGS 384

American Institute of Physics   Woodbury, New York

Authorization to photocopy items for internal or personal use, beyond the free copying permitted under the 1978 U.S. Copyright Law (see statement below), is granted by the American Institute of Physics for users registered with the Copyright Clearance Center (CCC) Transactional Reporting Service, provided that the base fee of $6.00 per copy is paid directly to CCC, 222 Rosewood Drive, Danvers, MA 01923. For those organizations that have been granted a photocopy license by CCC, a separate system of payment has been arranged. The fee code for users of the Transactional Reporting Service is: 1-56396-603-4/ 96 /$6.00.

© 1996 American Institute of Physics

Individual readers of this volume and nonprofit libraries, acting for them, are permitted to make fair use of the material in it, such as copying an article for use in teaching or research. Permission is granted to quote from this volume in scientific work with the customary acknowledgment of the source. To reprint a figure, table, or other excerpt requires the consent of one of the original authors and notification to AIP. Republication or systematic or multiple reproduction of any material in this volume is permitted only under license from AIP. Address inquiries to Office of Rights and Permissions, 500 Sunnyside Boulevard, Woodbury, NY 11797-2999; phone 516-576-2268; fax: 516-576-2499; e-mail: rights@aip.org.

L.C. Catalog Card No. 96-79458
ISBN 1-56396-685-9 (set)
ISBN 1-56396-603-4 (Part One)
ISBN 1-56396-684-0 (Part Two)
DOE CONF- 9510332

Printed in the United States of America

Contents

Preface xix

PART ONE

TEMPORAL STUDIES

What Do the Statistics of Temporal Signatures Tell Us about GRBs? 3
    I. G. Mitrofanov

Canonical Timescales in GRBs—1995 13
    J. P. Norris

Morphological Study of Short Gamma-Ray Bursts 23
    P. N. Bhat, G. J. Fishman, C. A. Meegan, R. B. Wilson, and W. S. Paciesas

Brightness-Independent Measurements of GRB Durations 28
    J. T. Bonnell, J. P. Norris, R. J. Nemiroff, and J. D. Scargle

Study of the Very Short Time Structure in Three BATSE Events
Using TTE Data 33
    D. B. Cline, S. Otwinowski, and D. A. Sanders

Comparison of the Properties of Short and Long Gamma-Ray Bursts
Observed with PHEBUS 37
    J.-P. Dezalay, J. P. Lestrade, C. Barat, R. Talon, O. Terekhov, R. Sunyaev, and A. Kuznetsov

Correlations between Duration, Hardness, and Intensity in GRBs 42
    C. Kouveliotou, T. Koshut, M. S. Briggs, G. N. Pendleton, C. A. Meegan, G. J. Fishman, and J. P. Lestrade

Time Profiles and Pulse Structure of Bright, Long Gamma-Ray
Bursts Using BATSE TTS Data 47
    A. Lee, E. Bloom, and J. Scargle

OSSE Limits on Pre- and Post-Burst Emission from GRB 950421 52
    S. M. Matz, J. E. Grove, W. N. Johnson, and G. H. Share

Quantifying Variability of GRB Light Curves Using Multifractal Analysis ... 57
    D. C. Meredith, J. M. Ryan, and C. A. Young

"Equal Flux" Averaging of BATSE Gamma-Ray Burst Time Profiles 62
    I. Mitrofanov, A. Pozanenko, A. Chernenko, M. S. Briggs, R. D. Preece, W. S. Paciesas, and G. J. Fishman

Peak-to-Peak Dilation of Gamma-Ray Bursts 67
    J. Neubauer and B. E. Schaefer

Calibration of Tests for Time Dilation in GRB Pulse Structures 72
    J. P. Norris, R. J. Nemiroff, J. T. Bonnell, and J. D. Scargle

Test for Time Dilation of Intervals between Pulse Structures in GRBs 77
    J. P. Norris, J. T. Bonnell, R. J. Nemiroff, and J. D. Scargle

Flux-Duration Correlations and Cosmological Time Dilation 82
    V. Petrosian and T. T. Lee

Intrinsic Dependence of Gamma-Ray Burst Durations on Energy 87
    G. Richardson, T. Koshut, W. Paciesas, and C. Kouveliotou

Profiles of Gamma-Ray Bursts and Their Component Pulses 91
  J. D. Scargle, J. P. Norris, R. J. Nemiroff, and J. T. Bonnell
How Exponential are FREDs? 96
  B. E. Schaefer and S. E. Dyson
**Hardness Ratio versus Duration for PVO Compared to BATSE
and PHEBUS** 101
  I. A. Smith, A. Crider, E. P. Liang, B. D. Dunne, E. E. Fenimore,
  and H. Li
Origin of the Gamma-Ray Burst Duration Bimodality 106
  V. C. Wang
**A Multifractal Study of the Scaling Property of Gamma-Ray Burst
Time Profiles** 111
  Y. Yan, J. P. Lestrade, J.-P. Dezalay, M. Adams, and B. Stark
**A Compact Representation of GRB Time Series Using Multiscale
Edge Detection** 116
  C. A. Young, D. C. Meredith, and J. M. Ryan

SPECTRAL STUDIES

Gamma-Ray Burst Continua: A Review 123
  D. L. Band
Low-Energy Spectral Features in GRBs 133
  M. S. Briggs
The Statistics of the Search for Absorption Lines in Burst Spectra 143
  D. Band, L. Ford, J. Matteson, M. S. Briggs, W. Paciesas,
  G. Pendleton, R. Preece, D. Palmer, and B. Teegarden
Testing the Compton Attenuation Theory of Gamma-Ray Burst Spectra 148
  J. J. Brainerd, R. D. Preece, M. S. Briggs, G. N. Pendleton,
  and W. S. Paciesas
A Comprehensive Search for Low-Energy Lines in BATSE GRBs 153
  M. S. Briggs, D. L. Band, R. D. Preece, G. N. Pendleton, W. S. Paciesas,
  L. Ford, and J. L. Matteson
EGRET TASC Observations of the Bright April 25, 1995, GRB 158
  J. R. Catelli, B. L. Dingus, and E. J. Schneid
Extrapolations of Gamma-Ray Burst Spectra to the Optical-UV Band 162
  L. Ford and D. Band
Short Time Scale Characteristics of Gamma-Ray Burst Continua 167
  L. Ford, D. Band, and J. Matteson
BATSE SD Observations of Hercules X-1 172
  P. E. Freeman, D. Q. Lamb, R. B. Wilson, M. S. Briggs, W. S. Paciesas,
  R. D. Preece, D. L. Band, and J. L. Matteson
Fitting of Combined Ulysses/COMPTEL GRB Spectra 177
  J. Greiner, T. Aigner, M. Sommer, O. R. Williams, R. M. Kippen,
  K. Hurley, M. Boër, and M. Niel
**Effects of Compton Scattering on BATSE Gamma-Ray Burst
Spectral Analysis** 182
  X.-M. Hua and R. E. Lingenfelter

Cyclotron Line Formation in a Relativistic Outflow 187
   M. Isenberg, D. Q. Lamb, and J. C. L. Wang
The Spectral Evolution of Gamma-Ray Burst Pulses 192
   V. E. Kargatis and E. P. Liang
COMPTEL Measurements of MeV Gamma-Ray Burst Spectra 197
   R. M. Kippen, J. Ryan, A. Connors, M. McConnell, C. Winkler,
   L. O. Hanlon, V. Schönfelder, J. Greiner, M. Varendorff, W. Collmar,
   W. Hermsen, and L. Kuiper
Exponential Decay of Gamma-Ray Burst Spectral Break Energy
with Photon Fluence 202
   E. P. Liang and V. Kargatis
The νF_ν Peak Energy Distributions of Gamma-Ray Bursts Observed
by BATSE 204
   R. S. Mallozzi, W. S. Paciesas, G. N. Pendleton, M. S. Briggs,
   R. D. Preece, C. A. Meegan, and G. J. Fishman
Statistical Analysis of BATSE Gamma-Ray Burst Spectra 209
   I. Mitrofanov, A. Chernenko, A. Pozanenko, M. S. Briggs, R. D. Preece,
   W. S. Paciesas, and G. J. Fishman
Performance Evaluation of the BATSE Spectroscopy Detectors 213
   W. S. Paciesas, M. S. Briggs, R. B. Wilson, R. D. Preece, J. L. Matteson,
   and D. L. Band
High-Resolution GRB Spectra from the Transient Gamma-Ray
Spectrometer (TGRS) 218
   D. M. Palmer, H. Seifert, B. J. Teegarden, N. Gehrels, T. L. Cline,
   R. Ramaty, K. Hurley, R. Pehl, N. Madden, and A. Owens
Recent Gamma-Ray Burst Continuum Observations and Their Implications
for Line Features in Gamma-Ray Bursts 223
   G. N. Pendleton, M. S. Briggs, R. D. Preece, R. S. Mallozzi,
   W. S. Paciesas, G. J. Fishman, C. A. Meegan, and C. Kouveliotou
Detailed Spectral Analysis Revealing Two Distinct Classes of Pulses
in Gamma-Ray Bursts 228
   G. N. Pendleton, W. S. Paciesas, R. D. Preece, M. S. Briggs, C. Kouveliotou,
   and C. A. Meegan
Are There MeV Gamma-Ray Bursts? 233
   T. Piran and R. Narayan
BATSE Observations of GRB Spectra at Low Energies 238
   R. D. Preece, M. S. Briggs, G. N. Pendleton, W. S. Paciesas, J. L. Matteson,
   D. L. Band, and C. A. Meegan
Time-Resolved Spectroscopy with the BATSE Large Area Detectors:
High-Energy Behavior 243
   R. D. Preece, M. S. Briggs, G. N. Pendleton, W. S. Paciesas, D. L. Band,
   L. A. Ford, and C. Kouveliotou
Burst Spectra over a Wide Energy Range 248
   B. E. Schaefer, D. Palmer, C. E. Fichtel, B. L. Dingus, E. J. Schneid,
   R. M. Kippen, C. Winkler, L. Hanlon, and V. Schönfelder

TASC Measurements/Upper Limits for Energetic Gamma-Ray Bursts
within the EGRET Field of View 253
    E. J. Schneid, D. L. Bertsch, J. R. Catelli, B. L. Dingus, J. A. Esposito,
    C. E. Fichtel, R. C. Hartman, S. D. Hunter, G. Kanbach, D. A. Kniffen,
    Y. C. Lin, H. A. Mayer-Hasselwander, P. F. Michelson, C. von Montigny,
    R. Mukherjee, P. L. Nolan, P. Sreekumar, and D. J. Thompson

The Detector Response Matrices of the Transient Gamma-Ray
Spectrometer (TGRS) 258
    H. Seifert, T. L. Cline, N. Gehrels, B. Mitra, D. M. Palmer, R. Ramaty,
    B. J. Teegarden, K. Hurley, N. Madden, R. Pehl, and A. Owens

Cosmological Gamma-Ray Bursts and the Cosmic Ray Flux
above 10^{19} eV 263
    E. Waxman

GLOBAL STUDIES

Gamma-Ray Burst Models and the Angular Distribution of 3B 271
    D. H. Hartmann

Implications of the Observed Angular Distribution of Gamma-Ray
Bursts for Galactic and Cosmological Models 281
    D. Q. Lamb

The BATSE 3B Catalog 291
    C. A. Meegan, G. N. Pendleton, M. S. Briggs, C. Kouveliotou,
    T. M. Koshut, J. P. Lestrade, W. S. Paciesas, M. L. McCollough,
    J. J. Brainerd, J. M. Horack, J. Hakkila, W. Henze, R. D. Preece,
    R. S. Mallozzi, and G. J. Fishman

Gross Spectral Differences between Bright and Very Bright
Gamma-Ray Bursts 301
    J.-L. Atteia, C. Barat, M. Boër, J.-P. Dezalay, M. Niel, R. Talon,
    G. Vedrenne, K. Hurley, M. Sommer, R. Sunyaev, A. Kuznetsov,
    and O. Terekhov

Luminosity Function and Cosmological Evolution of Gamma-Ray Bursts ... 306
    W. J. Azzam and V. Petrosian

Repeater Models and Sky Coverage 311
    D. L. Band

Analysis of the Space Distribution of the Gamma-Ray Bursts
in the BATSE 3B Catalog 316
    B. M. Belli

The Corrected Log N–Log Fluence Distribution of Cosmological
Gamma-Ray Bursts 321
    J. S. Bloom, E. E. Fenimore, and J. in 't Zand

On the Distribution of BATSE Gamma-Ray Bursts 326
    M. Borumand and W. Kluźniak

Comparison of Two Burst Repetition Tests 330
    J. J. Brainerd

Testing the Dipole and Quadrupole Moments of Galactic Models 335
    M. S. Briggs, W. S. Paciesas, G. N. Pendleton, C. A. Meegan,
    G. J. Fishman, J. M. Horack, C. Kouveliotou, D. H. Hartmann,
    and J. Hakkila

Detection of Gamma-Ray Bursts from Andromeda 340
    T. Bulik, P. S. Coppi, and D. Q. Lamb

Constraints on the Galactic Corona Models of Gamma-Ray Bursts
from the 3B Catalogue 345
    T. Bulik and D. Q. Lamb

Search for Repeating Classical Bursts with the Interplanetary Network 350
    T. L. Cline, K. C. Hurley, M. Boër, M. Niel, M. Sommer, C. Kouveliotou,
    G. J. Fishman, and C. A. Meegan

A Simultaneous Spectral Invariant Analysis of the GRB Count
Distribution and Time Dilation 353
    E. Cohen and T. Piran

Constraints on Galactic Corona Models of Gamma-Ray Bursts
Imposed by Andromeda and the BATSE 3B Catalog 358
    P. Coppi, T. Bulik, and D. Q. Lamb

A Wide-Ranging Search for Correlations among Burst Properties 363
    S. E. Dyson and B. E. Schaefer

Continuing Search of the EGRET Data for High-Energy Gamma-Ray
Microsecond Bursts 368
    C. E. Fichtel, D. L. Bertsch, R. C. Hartman, S. D. Hunter,
    C. von Montigny, D. J. Thompson, B. L. Dingus, J. A. Esposito,
    R. Mukherjee, P. Sreekumar, G. Kanbach, H. A. Mayer-Hasselwander,
    D. A. Kniffen, Y. C. Lin, P. F. Michelson, P. L. Nolan, L. McDonald,
    and E. J. Schneid

The GRB Rate at High Photon Energies 373
    B. Funk, K. Mannheim, and D. Hartmann

GRBs Observed with WATCH and BATSE (3B Catalogue) 378
    J. Gorosabel, A. J. Castro-Tirado, N. Lund, S. Brandt, O. Terekhov,
    and R. Sunyaev

Analysis of the Systematic Errors in the Positions of BATSE
Catalog Bursts 382
    C. Graziani and D. Q. Lamb

Constraints on the Luminosities of Cosmological Gamma-Ray Bursts 387
    J. Hakkila, C. A. Meegan, J. M. Horack, G. N. Pendleton, M. S. Briggs,
    R. S. Mallozzi, T. M. Koshut, R. D. Preece, and W. S. Paciesas

Repetition/Clustering in the BATSE 3B Catalog 392
    J. Hakkila, C. A. Meegan, M. S. Briggs, D. H. Hartmann, G. N. Pendleton,
    and J. M. Horack

Searsch for Supergalactic Anisotropies in the 3B Catalog 397
    D. H. Hartmann, M. S. Briggs, and K. Mannheim

BATSE Detection Biases against "Slow Rising" Gamma-Ray Bursts 402
    J. C. Higdon and R. E. Lingenfelter

Reconciling GRB Time Dilation Measurements to the Brightness
Distribution in Standard Cosmology 407
    J. M. Horack, R. S. Mallozzi, and T. M. Koshut

Analytic Constraints on Gamma-Ray Burst Luminosity Functions 412
    J. M. Horack, C. A. Meegan, J. Hakkila, and A. G. Emslie

The Compatibility of Friedmann Models with Observed Properties
of GRBs and a Large Hubble Constant 417
    J. M. Horack, A. G. Emslie, T. M. Koshut, R. S. Mallozzi,
    and C. A. Meegan

The Ulysses Supplement to the BATSE 3B Catalog 422
    K. Hurley, T. Cline, G. J. Fishman, C. Kouveliotou, C. Meegan,
    M. Sommer, M. Boër, and M. Niel

BACODINE/3rd Interplanetary Network Burst Localization 427
    K. Hurley, S. Barthelmy, P. Butterworth, T. Cline, M. Sommer,
    M. Boër, M. Niel, C. Kouveliotou, G. J. Fishman, and C. Meegan

Threshold Effects in GRB Brightness Distributions 431
    J. J. M. in 't Zand and E. E. Fenimore

The Angular Distribution of COMPTEL Gamma-Ray Bursts 436
    R. M. Kippen, J. Ryan, A. Connors, M. McConnell, D. H. Hartmann,
    C. Winkler, L. O. Hanlon, V. Schönfelder, J. Greiner, M. Varendorff,
    W. Collmar, W. Hermsen, and L. Kuiper

A Search for Untriggered Events in the BATSE Data Base 441
    J. M. Kommers, W. H. G. Lewin, J. van Paradijs, C. Kouveliotou,
    G. J. Fishman, and M. S. Briggs

Consistency of GRB Durations and Spectra with Standard Cosmology 446
    T. Koshut, R. Mallozzi, J. Horack, W. Paciesas, C. Kouveliotou,
    and R. Rutledge

Clusters of Cosmic Gamma-Ray Bursts as a Manifestation
of the Local Sources 451
    A. V. Kuznetsov

GRB Brightness Ratio Distribution Analysis 455
    J. G. Laros

GRB Localizations from BATSE, Mars Observer, and
Ulysses Observations 459
    J. G. Laros, W. V. Boynton, K. C. Hurley, C. Kouveliotou, G. J. Fishman,
    C. A. Meegan, T. L. Cline, D. M. Palmer, R. D. Starr, J. I. Trombka,
    M. Boër, M. Niel, M. Sommer, and A. E. Metzger

Threshold Effects on Gamma-Ray Burst Distributions 462
    T. T. Lee and V. Petrosian

Expected Gamma-Ray Burst Excess from M31 Based on Halo Models 467
    H. Li and E. Liang

Constraints from 3B & PVO on the Halo Beaming Model 472
    H. Li and R. Duncan

Likelihood Analysis of GRB Repetition 477
    S. Luo, T. Loredo, and I. Wasserman

Gamma-Ray Burst Redshift Constraints from BATSE Spectral Data 482
    R. S. Mallozzi, G. N. Pendleton, and W. S. Paciesas

A New Gravitational Lens Search for Gamma-Ray Bursts 487
    G. F. Marani, R. J. Nemiroff, J. P. Norris, and J. T. Bonnell

Progress with the Konus-W Gamma-Ray Burst Spectrometer
on GGS-Wind 492
    E. P. Mazets, R. L. Aptekar, D. D. Frederiks, S. V. Golenetskii,
    V. N. Ilynskii, M. M. Terekhov, T. L. Cline, P. S. Butterworth,
    and D. E. Stilwell

Geocenter Angle Distribution of the 3B Catalog 497
    M. L. McCollough, C. A. Meegan, and G. N. Pendleton

Constraints on the Distribution of Neutron Star Birth Velocities
from the Properties of Rotation-Powered Pulsars 502
    L. Munoz-Franco, D. Q. Lamb, and T. Bulik

Time-Dilation, Log N–Log P, and Cosmology 507
    R. J. Nemiroff, J. P. Norris, J. T. Bonnell, and J. D. Scargle

A Constraint on the Distance Scale to Cosmological Gamma-Ray Bursts 512
    J. M. Quashnock

Constraints on the Gamma-Ray Burst Luminosity-Duration Relationship
in the Galactic Scenario 517
    R. E. Rutledge, W. H. G. Lewin, J. Hakkila, G. Pendleton,
    J. P. Lestrade, C. Kouveliotou, and C. Meegan

Search for Constraints on the Gamma-Ray Burst Peak Flux-Distance
Relation in the Cosmological Scenario 522
    R. E. Rutledge, W. H. G. Lewin, J. Hakkila, T. Koshut, G. Pendleton,
    J. P. Lestrade, C. Kouveliotou, J. Horack, and C. Meegan

The Correlation between Gamma-Ray Burst Duration and Peak Flux 527
    R. E. Rutledge, W. H. G. Lewin, G. Pendleton, J. P. Lestrade,
    C. Kouveliotou, and C. Meegan

A New Distance Scale for GRBs: The Local Group Halo 532
    B. E. Schaefer

Contribution of Galactic Arm Sources to the 3B Catalog 536
    I. A. Smith

Spherical Harmonic Analysis of the Angular Distribution of GRBs 540
    M. Tegmark, D. H. Hartmann, M. S. Briggs, and C. A. Meegan

DMSP Satellite Detections of Gamma-Ray Bursts 545
    J. Terrell, P. Lee, R. W. Klebesadel, and J. W. Griffee

Gamma-Ray Burst Repetition and BATSE 3B Position Uncertainties 550
    V. C. Wang and R. E. Lingenfelter

A Search for Micro Cosmic Gamma-Ray Bursts in BATSE One
Second Continuous Data 555
    C. A. Young, M. B. Arndt, D. A. Biesecker, and J. M. Ryan

PART TWO

COUNTERPART SEARCHES

Searches for Gamma-Ray Burst Counterparts: Current Status and
Future Prospects 565
    F. J. Vrba

Searching for Prompt GRB Counterparts at 74 MHz 575
    R. J. Balsano, S. E. Thorsett, W. A. Coles, P. S. Ray, J. Rhodes,
    B. J. Rickett, S. Barthelmy, P. Butterworth, T. Cline, N. Gehrels,
    G. J. Fishman, C. Kouveliotou, and C. A. Meegan

Progress with the Real-Time GRB Coordinates Distribution Network (BACODINE) 580
    S. D. Barthelmy, P. S. Butterworth, T. L. Cline, N. Gehrels,
    G. J. Fishman, C. Kouveliotou, C. Meegan, and K. Hurley

Search for UHE Counterparts of Gamma-Ray Bursts 585
    P. N. Bhat, K. Sivaprasad, B. S. Acharya, P. R. Vishwanath,
    and M. V. S. Rao

Deep Infrared Imaging of the Putative X-Ray Counterpart to GRB 920501 589
    O. Blaes, R. Antonucci, K. Hurley, and T. Hurt

The TAROT Project: An Optical Glance at GRBs 594
    M. Boër, J. L. Atteia, C. Barat, M. Niel, J. F. Olive, C. Chevalier,
    S. Ilovaisky, and H. Pedersen

Search for Ultra High Energy (UHE) Gamma-Ray Counterparts of BATSE 3B Catalog Events 598
    M. Catanese, M. Chantell, C. E. Covault, J. W. Cronin, B. E. Fick,
    L. F. Fortson, J. W. Fowler, K. G. Gibbs, M. A. K. Glasmacher, K. D. Green,
    D. B. Kieda, J. Matthews, B. J. Newport, D. Nitz, R. A. Ong, D. Sinclair,
    and J. C. VanderVelde

Searches for TeV Counterparts to Classical Gamma-Ray Bursts 603
    V. Connaughton, C. W. Akerlof, S. Biller, J. Buckley, D. A. Carter-Lewis,
    M. Catanese, M. F. Cawley, D. J. Fegan, J. Finley, J. Gaidos, A. M. Hillas,
    R. C. Lamb, R. Lessard, J. McEnery, G. Mohanty, N. A. Porter, J. Quinn,
    H. J. Rose, F. Samuelson, M. S. Schubnell, G. Sembroski, R. Srinivasan,
    T. C. Weekes, C. Wilson, J. Zweerink, S. Barthelmy, T. Cline, N. Gehrels,
    G. J. Fishman, C. Kouveliotou, and C. Meegan

Looking for the Source of ~Hour-Long Soft X-Ray Emission following GRB 780506 607
    A. Connors and M. McConnell

Status of the High Energy GRB Counterpart Search with the HEGRA Experiment 612
    B. Funk, H. Krawczynski, L. Padilla, S. Barthelmy, P. Butterworth,
    T. Cline, N. Gehrels, G. J. Fishman, C. Kouveliotou, C. Meegan,
    and the HEGRA Collaboration

Limits on Prompt Radio Emission from GRB 950430 and GRB 950706 at 151 MHz 617
    D. A. Green, C. A.-C. Dessenne, P. J. Warner, D. J. Titterington,
    E. M. Waldram, S. D. Barthelmy, P. S. Butterworth, T. L. Cline, N. Gehrels,
    D. M. Palmer, G. J. Fishman, C. Kouveliotou, and C. A. Meegan

Simultaneous Optical/Gamma-Ray Observations of GRBs 622
    J. Greiner, W. Wenzel, R. Hudec, P. Spurný, J. Florián, J. Ziener,
    E. I. Moskalenko, A. V. Barabanov, N. S. Chernych, K. Birkle, N. Bade,
    S. B. Tritton, T. Ichikawa, C. Kouveliotou, G. J. Fishman, C. A. Meegan,
    W. S. Paciesas, and R. B. Wilson

Rapid Follow-Up ROSAT Observation of GRB 940301 627
    J. Greiner, N. Bade, K. Hurley, R. M. Kippen, and J. Laros

Results from Searching Astronomical Catalogues for the Counterparts of Gamma-Ray Bursts 632
    T. E. Harrison, W. R. Webber, and B. J. McNamara

A 23 GHz Survey of GRB Error Boxes 637
    J. N. Hewitt, C. A. Katz, S. D. Barthelmy, W. H. Baumgartner, T. L. Cline, B. E. Corey, G. J. Fishman, N. Gehrels, K. C. Hurley, C. Kouveliotou, C. A. Meegan, C. B. Moore, R. E. Rutledge, and C. S. Trotter

Why Do We Not See Simultaneous and Fading Optical Counterparts to GRB? 642
    R. Hudec

Searches for Optical Counterparts to GRBs: EON and AIO 646
    R. Hudec, J. Soldán, P. Spurný, J. Florián, P. Štěpán, M. Tichý, J. Tichá, L. Vyskočil, W. Wenzel, S. Barthelmy, T. Cline, N. Gehrels, G. J. Fishman, C. Meegan, C. Kouveliotou, and A. Mutafov

The STARE Project: First Light 651
    C. A. Katz, J. N. Hewitt, C. B. Moore, J. D. Ellithorpe, B. Rabii, S. D. Barthelmy, T. L. Cline, N. Gehrels, G. J. Fishman, C. Kouveliotou, and C. Meegan

Search for TeV Counterparts of BATSE Gamma-Ray Bursts with the HEGRA Air Shower Arrays 656
    H. Krawczynski, J. Prahl, D. Schmele, B. Funk, L. Padilla, S. Barthelmy, P. Butterworth, T. Cline, N. Gehrels, G. J. Fishman, C. Kouveliotou, C. Meegan, and the HEGRA Collaboration

Searches for Optical Counterparts of BATSE Gamma-Ray Bursts with the Explosive Transient Camera 661
    H. A. Krimm, R. K. Vanderspek, and G. R. Ricker

Luminous Galaxies near Gamma-Ray Burst Positions 666
    S. B. Larson, I. S. McLean, and E. E. Becklin

Results from GROCSE I: A Real-Time Search for Gamma-Ray Burst Optical Counterparts 671
    B. Lee, C. Akerlof, E. Ables, R. M. Bionta, L. Ott, H.-S. Park, E. Parker, S. Barthelmy, P. Butterworth, T. Cline, N. Gehrels, G. J. Fishman, C. Kouveliotou, C. Meegan, and D. Ferguson

Results from the USNO Quiescent Optical Counterpart Search of IPN[3] GRB and Optical Transient Localizations 676
    C. B. Luginbuhl, F. J. Vrba, R. Hudec, D. H. Hartmann, and K. Hurley

Ground-Based Gamma-Ray Burst Follow-Up Efforts: The First Three Years of the BATSE/COMPTEL/NMSU Gamma-Ray Burst Rapid Response Network 680
    B. J. McNamara, T. E. Harrison, J. Ryan, R. M. Kippen, G. J. Fishman, C. Kouveliotou, and C. A. Meegan

X-Ray Observations of Gamma-Ray Burst Counterparts with ASCA 685
    T. Murakami, R. Shibata, A. Yoshida, K. Hurley, P. Li, C. Kouveliotou, and G. J. Fishman

GRB Optical Search in the Wide-Field Plate Database and in the Flare Stars Database 690
    A. S. Mutafov, A. I. Makarieva, M. K. Tsvetkov, K. P. Tsvetkova, R. Hudec, and K. Hurley

Hubble Space Telescope Observations of Four Gamma-Ray Burst Error Boxes 695
    B. E. Schaefer, T. L. Cline, and K. Hurley

The Deepest Optical Investigation of the GRB 790613 Error Box 697
    V. V. Sokolov, V. G. Kurt, S. V. Zharykov, A. I. Kopylov, and A. V. Berezin

A Continued Search for Transient Events in the COBE DMR Database Simultaneous with Cosmic Gamma-Ray Bursts 702
    J. G. Stacy, P. D. Jackson, Tj. R. Bontekoe, and C. Winkler

MODELS

Theory of Gamma-Ray Bursts 709
    S. E. Woosley

An Alternative View of Gravitational Lensing of GRBs 719
    J. A. Ball

Relativistic Expansions in Gamma-Ray Bursts: Constraints from Photon-Photon Pair Production 724
    M. G. Baring and A. K. Harding

Short Gamma-Ray Bursts and Primordial Black Hole (PBH) Evaporation 729
    D. B. Cline

Gamma-Ray Bursts from the Interaction of Degenerate Disks with Fast Neutron Stars (Part I) 734
    S. A. Colgate and H. Li

Gamma-Ray Bursts from the Interaction of Degenerate Disks with Fast Neutron Stars (Part II) 739
    S. A. Colgate and H. Li

Gamma-Ray Bursts from Comet–Antimatter Comet Collisions in the Oort Cloud 744
    C. D. Dermer

The Time Evolution of GRB Spectra by a Precessing Lighthouse Gamma Jet 749
    D. Fargion and A. Salis

Precessing Gamma Jets in the Extended and Evaporating Galactic Halo as the Sources of GRBs 754
    D. Fargion and A. Salis

Long and Short GRBs 759
    J. I. Katz and L. M. Canel

Prospects of Fluorescent Nuclear Line Searches in GRB Afterglows 764
    E. P. Liang and G. J. Mathews

General Relativistic Simulation of Close Neutron Star Binaries: Implications for Cosmological Gamma-Ray Bursts 768
    G. J. Mathews and J. R. Wilson

Gamma-Ray Bursts from Internal Shocks in a Relativistic Wind:
Temporal and Spectral Properties 772
    R. Mochkovitch and Y. Fuchs
Coalescing Neutron Stars as Gamma-Ray Bursters? 777
    M. Ruffert, H.-T. Janka, and G. Schäfer
Hydrodynamic Time Scales and Temporal Structure of GRBs 782
    R. Sari and T. Piran
A Shock Emission Model for Gamma-Ray Bursts 787
    M. Tavani
Spectral Models of GRBs 792
    M. Tavani
A Cosmological Test for Shock-Powered GRBs 797
    M. Tavani
Gamma-Ray Emission from Compact Relativistic MHD Winds 802
    C. Thompson

INSTRUMENTATION AND TECHNIQUES

The Prompt Acquisition of Gamma-Ray Optical Counterparts
with the Bradford Robotic Telescope 809
    J. E. F. Baruch, C. Bennett, M. J. Cox, and R. Davis
The Burst Observer and Optical Transient Exploring System (BOOTES) ... 814
    A. J. Castro-Tirado, R. Hudec, and J. Soldán
High Spectral Resolution Studies of Gamma-Ray Bursts on New Missions .. 819
    U. D. Desai, M. H. Acuna, T. L. Cline, B. R. Dennis, L. E. Orwig,
    J. I. Trombka, and R. D. Starr
BASIS: A GRB Mission Concept 824
    N. Gehrels, B. Teegarden, L. Barbier, T. Cline, A. Parsons, J. Tueller,
    S. Barthelmy, D. Palmer, J. Krizmanic, E. Fenimore, G. J. Fishman,
    C. Kouveliotou, K. Hurley, W. Paciesas, J. van Paradijs, S. Woosley,
    M. Leventhal, D. McCammon, W. Sanders, and B. Schaefer
Gamma-Ray Burst Studies with the Energetic X-Ray Imaging Survey
Telescope (EXIST) 829
    F. A. Harrison, J. E. Grindlay, N. Gehrels, C. J. Hailey, W. A. Mahoney,
    T. A. Prince, B. D. Ramsey, P. Ubertini, G. K. Skinner, and M. C. Weisskopf
Burstman: A Portable GRB Detector for Really Long Voyages 834
    K. Hurley, J. H. Primbsch, P. Berg, K. Ziock, I. Mitrofanov,
    D. Anfimov, A. Chernenko, V. Dolidze, V. Loznikov, A. Pozanenko,
    A. Tonshev, D. Ushakov, T. Cline, R. Baker, D. Stilwell, D. Sheppard,
    and N. Madden
Can a Strong GRB Affect the Natural Jovian Radio Emission? 837
    S. Klose
ALLEGRO: A New Approach to Gamma-Ray Bursts 841
    S. M. Matz, D. A. Grabelsky, W. R. Purcell, M. P. Ulmer, G. N. Pendleton,
    J. M. Cordes, J. P. Finley, W. A. Wheaton, and R. B. Wilson

Recent Results from the BATSE/OSSE Rapid Burst Response 846
    S. M. Matz, C. A. Meegan, G. J. Fishman, J. E. Grove, W. N. Johnson,
    and G. H. Share

Using BATSE to Measure Gamma-Ray Burst Polarization 851
    M. McConnell, D. Forrest, W. T. Vestrand, and M. Finger

EREBUS: An Experiment to REveal the BUrster Sites 856
    C. A. Meegan, G. J. Fishman, B. A. Harmon, J. M. Horack, R. B. Wilson,
    J. J. Brainerd, M. S. Briggs, W. S. Paciesas, G. N. Pendleton,
    C. Kouveliotou, and J. Hakkila

Gamma-Ray Burst Optical Counterpart Search Experiment (GROCSE) 861
    H.-S. Park, E. Ables, R. M. Bionta, L. Ott, E. Parker, C. Akerlof, B. Lee,
    S. Wallace, S. Barthelmy, P. Butterworth, T. Cline, N. Gehrels, G. J. Fishman,
    C. Kouveliotou, C. Meegan, and D. Ferguson

BART-Burst Alert Robotic Telescope 866
    J. Soldán, R. Hudec, and M. Němček

Gamma-Ray Burst Monitoring with the Hard X-Ray Detector Onboard the ASTRO-E Mission 870
    A. Yoshida, H. Ezawa, Y. Fukazawa, M. Hirayama, E. Idesawa, H. Ikeda,
    Y. Ishisaki, N. Iyomoto, T. Kamae, J. Kataoka, H. Kaneda, H. Kubo,
    K. Makishima, K. Matsushita, K. Matsuzaki, T. Mizuno, T. Murakami,
    K. Nagata, S. Nakamae, M. Nomachi, H. Obayashi, T. Otsuka, H. Ozawa,
    Y. Saito, M. Sugizaki, T. Takahashi, T. Tamura, M. Tashiro, N. Tsuchida,
    and K. Tsukada

Special Evening Session

The BATSE Burst Location Algorithm 877
    G. N. Pendleton, M. S. Briggs, and C. A. Meegan

SOFT GAMMA REPEATERS

Are the Soft Gamma Repeaters a Motley Crew? 889
    K. Hurley

Astrophysics of the Soft Gamma Repeaters 897
    C. Thompson

Radiative Opacities and Photosphere Models for Soft Gamma Repeaters 907
    V. G. Bezchastnov, G. G. Pavlov, Yu. A. Shibanov, and V. E. Zavlin

SGR 0525−66 Only 2.6 Days After 912
    A. J. Castro-Tirado

Is SGR 1900+14=GRS 1915+105? 916
    A. J. Castro-Tirado

Spectral Studies of Magnetic Photon Splitting in the March 5 Event and SGR 1806−20 921
    H.-K. Chang, K. Chen, E. E. Fenimore, and C. Ho

A Search for March 5th−Like Bursts in the PVO Database 926
    A. Crider and E. E. Fenimore

A Periodic Variable X-Ray Counterpart to SGR 0525−66? 931
    R. Danner, J. Trümper, and S. R. Kulkarni
Search for an Extended X-Ray Counterpart of SGR 1900+14 936
    J. Greiner
A Photon Splitting Cascade Model of Soft Gamma Repeaters 941
    A. K. Harding and M. G. Baring
Infrared Observations of SGR 1900+14 946
    W. A. Mahoney, P. Durouchoux, T. N. Gautier, J. C. Ling, P. Wallyn,
    and Wm. A. Wheaton
Rosat Observations of Supernova Remnant N49 951
    D. Marsden, R. E. Rothschild, R. E. Lingenfelter, and R. C. Puetter
Compton Scattering Effects in the Spectra of Soft Gamma Repeaters 956
    M. C. Miller and T. Bulik
Observations of Soft Gamma Repeaters: SGR 1806−20
and SGR 0525−66 961
    T. Murakami, R. Shibata, A. Yoshida, N. Kawai, I. Hayashi,
    S. R. Kulkarni, C. Kouveliotou, and E. E. Fenimore
Infrared, Submillimeter, and Millimeter Observations of the Soft
Gamma Repeaters 966
    I. A. Smith, K. Hurley, A. S. B. Schultz, J. van Paradijs,
    L. B. F. M. Waters, L. M. Chernin, R. Joyce, F. J. Vrba, D. Hartmann,
    C. Kouveliotou, P. Durouchoux, P. Wallyn, and S. Corbel

MISCELLANEOUS

A Diamond Jubilee Debate 973
    J. T. Bonnell, R. J. Nemiroff, and C. J. Graziani
A Brief History of the Discovery of Cosmic Gamma-Ray Bursts 977
    J. T. Bonnell and R. W. Klebesadel
An Interactive Gamma-Ray Burst Educational Text for the
World-Wide-Web 981
    J. M. Horack, S. Rizvi, and L. Friend
A Gamma-Ray Burst Bibliography, 1973−1995 985
    K. Hurley
Temporal and Spectral Characteristics of Terrestrial Gamma Flashes 990
    R. J. Nemiroff, J. T. Bonnell, and J. P. Norris

Conference Participants 995
Author Index 1003

Preface

The Third Huntsville Gamma-Ray Burst Symposium was held October 25–27, 1995, in Huntsville, Alabama, USA. It had the largest attendance to date of any workshop on the subject. There were 187 registered attendees, most of whom attended the full three days of talks and poster presentations.

The majority of the papers were in one of three distinct areas:

- detailed analyses of recent data on bursts
- searches for burst counterparts
- descriptions of new observational initiatives in the field.

Theoretical models and interpretations were rather sparse compared to the above categories, a real departure from past gamma-ray burst conferences. Also, many theorists appeared to be more uncertain of their favorite burst models and mechanisms than at previous conferences; they defended them less intensely. It is clear that models for gamma-ray bursts are more primitive than any other observed phenomenon in all of astronomy.

About two months prior to the conference, the Third BATSE Gamma-Ray Burst catalog was released, containing 1122 bursts with improved location accuracy over previous catalogs. The conference thus became a forum for presentations and discussions of data from this catalog as well as subsequent analyses. To facilitate future analyses, a special evening session was devoted to a description of the BATSE burst location methodology and possible sources of error.

The search for cosmological signatures from gamma-ray bursts continues; the results are still debated hotly. A new analysis reported evidence for redshift in the spectral continua of gamma-ray bursts consistent with that expected with cosmological distances for bursts. Other papers dealt with tests for consistency in different cosmological signatures and the use of burst data to constrain cosmological parameters.

Since the last conference, the notion of high-velocity neutron stars populating an extended galactic halo has gained impetus. Bursts from such neutron stars might be compatible with the observed isotropy of bursts. Even so, it is apparent that the allowed parameter space consistent with such halo models is decreasing as more bursts are observed and the isotropy remains.

In other developments, the BATSE-BACODINE system has become fully operational, distributing near-realtime burst locations to observers worldwide. Comprehensive searches for optical and radio counterparts were reported using data from this network in conjunction with many ground-based facilities covering virtually all wavelength regions. In all cases, the bottom line was that no counterpart was seen, but the ubiquitous upper limits were given.

In these volumes, we have assembled the papers, both from talks and poster presentations, into sections of similar general topics. These topical section titles are nearly the same as those of past conferences and workshops on gamma-ray bursts.

We were gratified to see numerous new and young researchers in the meeting, even at the graduate level. It strengthened our conviction that the field is extremely active, with new "blood" joining in and imaginative, new ideas for more experiments and missions to provide much-needed, better observational data. Everyone loves a mystery and this is one of the biggest and best that astronomy has to offer.

The organizers are grateful to Ms. Susan Benefield and other members of the staff of the USRA Astronomy Program in Huntsville for their superb assistance with the conference and with these proceedings.

Chryssa Kouveliotou
Michael F. Briggs
Gerald J. Fishman

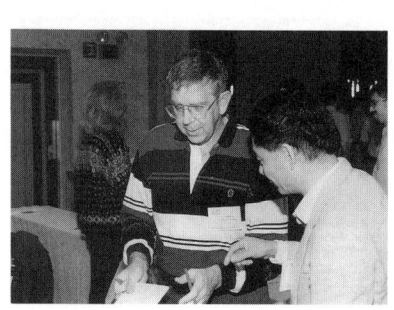

TEMPORAL STUDIES

What do the statistics of temporal signatures tell us about GRBs?

I.G.Mitrofanov

Institute for Space Research
Profsojuznaja 84/32 117810 Moscow Russia

Due to very successful measurements by the Compton/BATSE instrument, a large volume of GRB data has now become available. These data allow a *new statistics* of GRBs focused on the average temporal signatures of fluxes and on spectral variability. We apply these methods to the BATSE GRB data, and ask some new questions. These questions, when properly answered, may lead to a basic concept of the GRB phenomena.

I. WHAT DO WE NEED TO KNOW ABOUT GRBS?

At present we have two conflicting paradigms for gamma-ray bursts (GRBs). The first associates GRBs with sporadic outbursting activity of neutron stars in an extended galactic halo. The second associates them with collisions of compact objects at cosmological distances.

The observational data do not support unambiguously either of these paradigms, and some recent results even contradict both. New approaches are necessary and new methods should be used to look for features of the phenomenon, which probably might add these key questions:

Are GRBs a single phenomenon, or can several morphological groups be resolved among them? Astronomers know very well how similar stars and quasars can be, when their optical images are compared, and how different they are in origin. Do we observe two different groups of star-like and quasar-like GRBs? or do we deal with a single phenomenon associated with one astronomical population? To establish the paradigm(s) of GRBs, this question has to be clarified with the highest priority.

Are GRBs a time-reversible phenomenon? GRBs are commonly known to have very complicated time histories, which are thought to originate from the random light curves of emitters during the outbursts. Physical processes which drive an outburst might either have time-reversible evolution, like impacts of comet fragments with a star, or time-irreversible one, like stellar flares.

Are temporal signatures of bright and dim GRBs similar or different? The concept of *standard candles* is tested by this question. If bright and dim events are different only due to the distances of the emitters, a single morphological group of sources could be identified as a homogeneous astronomical population. In the opposite case, either some spread of intrinsic properties should

© 1996 American Institute of Physics

be postulated, or some evolution-like variance of these properties should be assumed with increasing distances, or some more complicated inhomogeneity could be expected for these emitters.

Are cosmological effects of redshift and time-dilation really observed for GRBs? This question directly tests the cosmological paradigm for GRBs. Because of drastically different time histories and spectra of individual events, one has no chance to perform a direct comparison of individual GRBs to reveal these effects, and therefore statistical methods of burst averaging should be used to study them.

Is a model of an extended galactic halo compatible with the observed temporal and spectral signatures of GRBs? If the sources of GRBs in the extended galactic halo have a standard power, then bright and dim events should not have any differences, excepting their apparent intensities. Therefore, any difference in temporal or spectral signatures, provided such were found, would point out real differences in the emitters in the halo.

We introduce below the main temporal parameters used in our study and address the above five questions in detail.

A. Fluence-based time parameters We define t_{50} as the time it takes to accumulate from 25% to 75% of the total fluence of a burst. Similarly, we define t_{90} as the time to accumulate from 5% to 95% of the fluence (4). Fluence-based time parameters are appropriate, when one needs to establish the full duration of emission. However, these time parameters t_{50} and t_{90} depend on the particular shapes of a GRB time history: while for single pulse bursts they represent the pulse width, for multi-pulse events with long *valleys* between the pulses these parameters reflect more the pulse separations rather than the pulse widths.

Moreover, for some bursts with particular time histories small changes in the profiles could drastically change the values of t_{50} or t_{90}. For example, if a burst has two pulses about ~ 1 sec long with a background valley ~ 100 sec between them, the value of t_{50} should *jump* between ~ 1 sec and ~ 100 sec depending on whether the weaker pulse contributes $< 25\%$ or $> 25\%$ to the total fluence. For such events, the fluence-based time parameters could be drastically changed by small statistical fluctuations of counts. One study of the dependence of t_{50} and t_{90} on the burst profile has appeared (8).

B. Profile-based time parameters We propose two complementary profile-based parameters to measure the *total pulse duration*, t_{PD}, and the *total valley duration*, t_{VD}, of a GRB at the some flux level. To compare time histories of GRBs with different intensities, one could use a dimensionless flux $f_* = F/F_{max}$, where F_{max} is the flux at the principal peak. By defining t_{PD} and t_{VD} using the dimensionless flux f_*, the parameters t_{PD} and t_{VD} can be used to compare the total durations of the pulses and valleys of events of differing intensities (20).

For each particular event one might also estimate a number of pulse intersections n at the level f_*. For GRBs with multi-pulse time histories $n > 1$, the average durations of the pulses and the valleys could be defined, as $\bar{t}_{PD} = t_{PD}/n$ and $\bar{t}_{VD} = t_{VD}/(n-1)$, respectively.

Provided that the flux level f_* of an intersection could be selected well above the level of background fluctuations (even for dim events), these profile based temporal parameters will have negligible systematic dependence on burst intensity. The average values of these parameters would properly represent the general temporal signatures of the selected GRBs.

C. Temporal parameters of the average curves of emissivity and hardness ratios Temporal widths may also be attributed to the *average curve of emissivity* (ACE), which can be created for any selected sub-set of GRBs (17). For all GRBs in the selected set, using a particular temporal resolution, all the profiles should be normalized to $f_{max} = 1$ and aligned at moments t_{max} of their principal peaks. The average in the time-bins δt_i of the aligned time scale provides the average emissivity levels $\langle f_i \rangle$ of all of the contributing events (14,15). The time width of the ACE above the level of dimensionless flux f_* is defined as

$$t_{ACE} = \sum_{i, \text{where} f_i > f_*} \langle f_i \rangle \cdot \delta t_i. \qquad (1)$$

The widths t_{ACE} can be estimated separately for the rise fronts and the back slopes, which are defined as the portions of the time histories before and after the times t_{max} of the principal peaks.

It was found that the *average curve of hardness ratio* (ACHR) has a broad maximum at the rise front of the ACE peak, and decreases with decreasing average flux in the back slope (see Fig. 1).

D. Average auto-correlation curves The auto-correlation function (ACF) of GRB time profiles helps to determine the typical time scales of the internal correlations of their histories (27). When ACFs are averaged for a selected subset of GRBs, one peak curve represents an auto-correlation of all

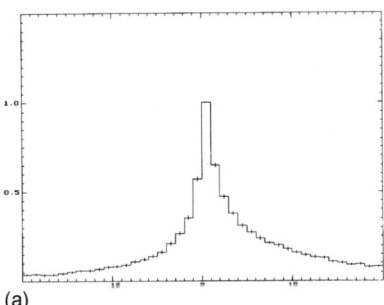
(a)

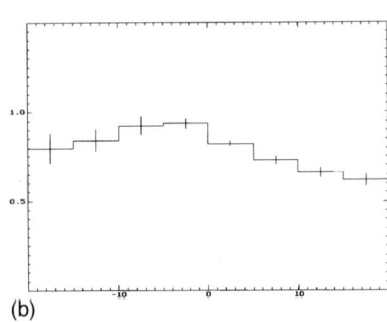
(b)

FIG. 1. The ACE (a) and ACHR (b) at the 1024 ms time scale and for discriminator channels 2+3 are presented for 628 GRBs of the 3B catalog (12)

time histories taken into account. The time width of the peak of an average ACF can be easily measured, and a general temporal parameter t_{ACF} can be introduced, e.g., full width at $e^{-0.5}$ of the maximum level.

The average ACF can be considered to be a self-similar function. The systematic error was shown to be less or equal to 10% (27): if time histories of contributing events were stretched by some factor, the average ACF curve would be stretched by the same factor. Therefore, the average ACF can certainly be used as an appropriate signature to study auto-correlated variability for different subsets of GRBs, to study autocorrelation time scales in different energy ranges, etc.

When general signatures of GRBs temporal variability are studied, the complementary methods of peak alignment and auto-correlation function should be used. The latter seems to be less sensitive to systematic effects caused by differing S/N, while the former method is thought to be more appropriate to reproduce the actual variations of emissivity during outbursts.

II. ARE GRBS A TIME-SYMMETRIC AND SINGLE-MODE?

The main result of GRB duration statistics is their bimodality–the existence of two classes, short and long (3,4). The classes are also associated with differences in spectral hardness and number/intensity statistics (1,24,5).

The peak alignment method permits the comparison of the emissivity along the rise fronts of the ACE with that of the back slopes. Using 1024 ms time data, the time-widths t_{ACE} for the rise front and the back slope of the ACE for 338 GRBs of the 2B catalog (11) are 2.33 ± 0.04 s and 4.21 ± 0.05 s, respectively (17). Therefore, the rise of the ACE is about 1.8 times faster than the decay. The same difference is seen in the ACE profile of 628 GRBs of the 3B catalog (Figure 1).

On the other hand, when the ACE is created using the much shorter averaging time scale of 64 ms, the peak becomes more symmetrical (16). It is possible that the non-symmetrical shape of ACE at 1024 ms resolution represents the long-time evolution of GRBs, while the ACE at 64 ms resolution represents the evolution of the main peak of each GRBs. Thus it appears that the main peaks of GRBs are more time symmetric than overall GRB profiles.

The difference between the rise and decay of GRBs is also clearly seen in the ACHR shape. The ACHR shows a broad maximum extending about 10 seconds before the rapid rise of the main peak, while during the decay the average hardness ratio decreases along with the average flux (Figure 1).

III. ARE TEMPORAL SIGNATURES OF BRIGHT AND DIM GRBS SIMILAR OR DIFFERENT?

As mentioned above, when one studies the morphological unity of GRBs, a comparison between subsets of bright and dim events has key importance. If classical GRBs are a mixture of bursts from two different astronomical populations, one would expect that they would contribute different proportions to bright and dim subsets. Therefore, the comparison of bright and dim subsets might provide evidence of types of GRBs.

On the other hand, if classical GRBs are emitted by a single astronomical population, one has to check for possible differences between close and distant emitters, which might result either from cosmological time-dilation and redshift, or from an intrinsic non-homogeneity of outbursting sources. Again, the average signatures of bright and dim subsets might resolve these differences.

Before comparing bright and dim sub-sets of GRBs, one has to exclude as much as possible brightness dependent systematic effects and to properly define the intensity measure used to separate bursts into bright and dim subsets. Classical bursts are known to have very different durations, from several milliseconds up to several hundreds of seconds.

Originally peak flux was proposed as the proper parameter to measure burst intensity (10). The key question is the proper estimation of this quantity. Firstly, for bursts with similar light curves, the time profiles of dim events would have relatively larger statistical fluctuations than bright events. Secondly, searching for the principal peak of flat-topped bursts, one would systematically tend to select positive fluctuations. Thirdly, spectral deconvolution should be done to transform counts into photons, and events with similar counts might have different peaks in photons, and *vise versa*.

To overcome these difficulties, peak fluxes F_{max} in units of photons cm^{-2} s^{-1} at the 1024 ms time scale, as presented in the BATSE catalogs (7,11,12), were used as the intensity measure. These fluxes were used to distinguish bright, medium and dim events.

To compare bright and dim GRBs, several pairs of bright and dim subsets have been formed. When the bursts are divided into two groups with no intervening gap, both bright and dim subsets include events of medium intensities. If a brightness dependent stretching exists, these medium GRBs would weaken the difference between the ACE profiles of bright and dim GRBs. For a stronger test, datasets were used with a broad gap between the bright and dim sets.

No significant difference was found between the ACE curves for these sets. In particular, sets of 208 bright and 204 dim GRBs of the 3B catalog were selected via $F_{max} > 1.4$ photons cm^{-2} s^{-1} and $F_{max} < 0.6$ photons cm^{-2} s^{-1}, respectively. ACE curves for these subsets for discriminator channels 2+3 are presented in Figure 2. The curves are very similar. For events from the 2B catalog (11) the total widths of the ACE above the $f_* = 0.1$ level are $t_{ACE}^{bright} = 6.64 \pm 0.10$ s and $t_{ACE}^{dim} = 6.57 \pm 0.09$ s, respectively (17).

This result is in contradiction with the conclusion (21) that the average curve for bright events is significantly narrower than that for dim events. This contradiction could result either from different criteria used to select the subsets of bright and dim GRBs, or from differences in the averaging procedures.

This discrepancy has been investigated (17) using the same groups of events which were selected as bright and dim subsets in accordance with wavelet-

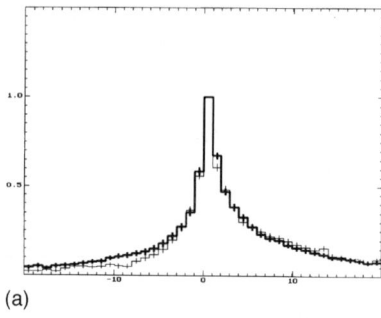 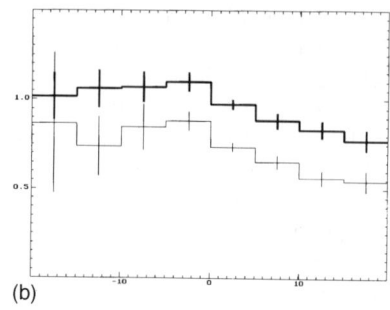

FIG. 2. The ACE (a) and ACHR (b) at the 1024 ms time scale and for discriminator channels 2+3 are presented for subsets of 209 bright (thick line) and 204 dim (thin line) events of the 3B catalog (12)

thresholding criteria (21). Both averaging procedures give qualitatively similar ACE profiles when they are applied to the same data sets. Additionally, it was found that the events of bright and dim subsets selected by the wavelet-thresholding criteria (21) are not complete, and the difference between the average time histories might be due to some systematic effects (Figure 3). Further study is necessary to resolve this discrepancy, but now the definite statement can be made that subsets of bright and dim GRBs selected by peak flux criteria have non-distinguishable ACE curves both for 1024 ms and 64 ms time scales (16,17).

On the other hand, a real difference between subsets of bright and dim GRBs was found using average spectral signatures. It is an effect of hardness/intensity correlation. It was found first in the PHOBOS/APEX database, when the average energy spectra were compared at peaks of bright and dim events (13). The effect of hardness/intensity correlation was seen also in early samples of 126 (23) and 205 (15) BATSE GRBs. The average curves of hardness evolution for subsets of bright and dim BATSE GRBs are shown in Figure 2. Both curves have similar shapes, but the ACHR for the bright subset goes well above the curve for the dim subset.

For five intensity groups of BATSE GRBs selected by the peak flux criteria, the average energy of the peak of the νF_ν photon spectra E_p has been estimated. A well-pronounced correlation was found between the average peak flux $F_{\max}$ of these groups and the average peak energies E_p of these intensity groups (9). The averaged photon spectra of bright and dim subsets of BATSE GRBs in 16 energy channels are also significantly different, with much harder spectra for the group of bright events (19).

So, a GRB hardness/brightness correlation is well established and generally accepted.

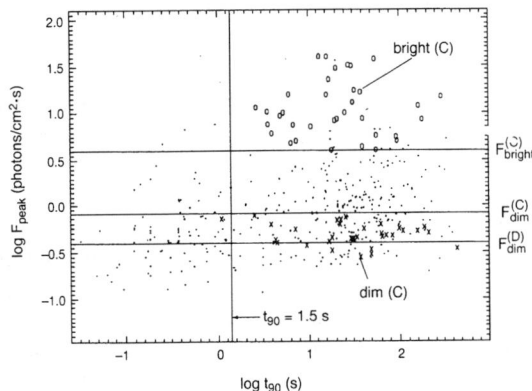

FIG. 3. Peak flux as a function of duration for all events of the 2B catalog (11). The axes are logarithmic. The vertical line separates the events with $t_{90} < 1.5$ s that were excluded from the analysis in (21). Sets of bright and dim events selected in (21) are indicated as "O" and "X", respectively. The horizontal lines represent the photon flux thresholds on the 1024 ms time scale which separate sets of bright (C), dim (C) and dimmest (D) events (17).

IV. ARE COSMOLOGICAL EFFECTS OF REDSHIFT AND TIME-DILATION SEEN IN GRBS?

The effect of the hardness/intensity correlation of GRBs looks very similar to a cosmological red-shift, provided that bright and dim events are identified with emitters at small and large cosmological red-shifts, Z_{br} and Z_{dim}, respectively. The relative redshift factor between observed spectral features of bright and dim events, which have the same energy $E^{(0)}$ in co-moving frames, corresponds to

$$E^{(0)}_{dim}/E^{(0)}_{bright} = Y(Z_{br}, Z_{dim}) = (1+Z_{dim})/(1+Z_{br}). \qquad (2)$$

If the observed correspondence between average photon fluxes and average spectral peak energies for the five intensity groups of GRBs are converted to redshift factors, the largest factor is found between the average energies of the brightest group No 5 and the dimmest group No 1: $Y_{redshift}(Z_5, Z_1) = 1.62 - 2.22$ (9). For standard cosmological models, the redshift factors correspond to $Z_{br} = 0.08 - 0.12$ and $Z_{dim} = 0.75 - 1.49$.

These values qualitatively agree with the previous estimations from the time-dilation effect (21), which has been claimed for the average light curve of the dimmest GRBs in comparsion to the average curve of bright events. Originally, the time-dilation factor was estimated as $Y_{dilation} \sim 2.25$, but more recent calculations (22) led to the smaller value $Y_{dilation} \sim 1.5$. An auto-correlation function approach has also been used to test for cosmological effects. If time-dilation of dim events exists, the ACF should be stretched by the same factor as the time histories. The value of time-stretch factor derived from comparison of ACF curves for bright and dim subsets is ~ 1.3 (6).

As mentioned above, when the subsets are selected by the peak flux criteria, no stretching was found between the subsets of bright and dim events. If cosmological redshifting occurs, when bright and dim events are observed in the same energy band, the photons were emitted in different energy bands. Therefore, one has to take into account that cosmological time-dilation of dim events might be compensated by the intrinsic narrowing of their light curves at the higher energies where the photons were emitted in the co-moving frame. Therefore, a special test had to be done which took into account the dependence of GRBs time histories on photon energy (17). Two subsets of BATSE GRBs were used: 71 brightest events with $V/V_{max} < 0.5$ and 53 dimmest events with $V/V_{max} > 0.6$. The ACE curve for the bright subset was transformed in accordance with energy redshift and time-dilation to obtain the best fit to the ACE of the dimmest set. The resulting best-fit value of the cosmological factor was $Y_* = 0.92 \pm 0.2$, which indicates that average curve for dim events is insignificantly narrower (not broader!) than the bright one.

On the other hand, a complementary test based upon the hardness/intensity correlation has been used to estimate cosmological redshift factors (9). For each intensity group $i = 1-5$, three time-width parameters $t_{ACE}(E_{1,2,3})$ were calculated by the peak alignment method (18). The dependence of the width of the ACE on photon energy for the brightest group $i = 5$ was interpolated by the power law

$$t^{(5)}_{ACE}(E) = 5.25 \cdot (175\,\text{keV}/E)^{0.27}. \tag{3}$$

If the redshift factors $Y_{i,5}$ have a cosmological origin, the time-widths for the ACEs of intensity groups $i = 1-4$ should be as follows

$$t^{(i)}_{ACE}(E) = 5.25 \cdot Y_{i,5} \cdot (175\,\text{keV}/E\ Y_{i,5})^{0.27}. \tag{4}$$

Comparing the model values (eq. 4) and the actual values of the time-widths of the ACEs for the intensity groups $i = 1-4$, very large χ^2_2 values are obtained. For the dimmest and dim groups $i = 1$ and 2, χ^2_2 is 299.0 and 185.0, respectively (18).

V. IS THE GALACTIC HALO MODEL COMPATIBLE WITH THE OBSERVED TEMPORAL AND SPECTRAL SIGNATURES OF GRBS?

The isotropic and inhomogeneous distribution of GRBs could be explained by their origin in an extended galactic halo (2,25,26). One might assume that the sources are not *standard candles* and have variable peak emissivities P distributed between P_{max} and P_{min} according to a power law $W(P) \sim (P/P_{min})^\beta$.

In this case, GRBs with similar observed peak fluxes f_{max} would belong to the same intensity group, but they might be emitted from very different distances in the halo. For events with similar intensities, the ratio of distances to the most distant and the closest sources is $(P_{max}/P_{min})^{1/2}$. Therefore, different intensity groups of GRBs could consist of different proportions of emitters with high and low emissivities.

The perfect similarity between the average temporal parameters of bright and dim GRBs means that all the emitters in the halo have the same average light curves, independent of the magnitude of their intrinsic power.

On the other hand, to explain the hardness/intensity correlation of GRBs, one should assume (18) that the energy E_p correlates with emissivity, as $E_p(P) = E_p^{(max)} \cdot (P/P_{max})^\delta$. Different contributions of strong and weak emitters in the same intensity group lead to different average spectral hardness for these groups. The models of extended galactic halo with parameters $P_{min}/P_{max} = 0.10 - 0.03$, $E_p^{(max)} = 1900 - 2300$ keV and $\delta = 0.9 - 1.8$ were shown to explain the effect of hardness/intensity correlation of GRBs.

VI. CONCLUSIONS

The following conclusions can be drawn about the general properties of GRBs based on their average temporal signatures and spectral variability:

1) *Two morphological groups are resolved among GRBs.* At the present time the conclusion can be drawn that GRBs have a double-peak distribution either over fluence-based or profile-based temporal parameters. However, the main question remains to be solved, is the bimodality of GRBs due to two different phenomena with different astronomical origins, or is it due to an internal bimodality of the emission properties of sources belonging to the same astronomical population.

2) *GRBs are not a time-reversible phenomenon.* The rise fronts and back slopes of GRBs, defined with respect to the time t_{max} of the principal peak, are not time-symetric. Generally, the rise fronts are faster than the back slopes, and the rise fronts are associated with much harder radiation. One might conclude that the origin of GRBs has to be associated with non-reversible physical process(es).

3) *Temporal signatures of bright and dim GRBs are similar.* Subsets of bright and dim GRBs selected by peak flux criteria were shown to have indistinguishable average emissivity curves both at 1024 ms and 64 ms time scales. On the other hand, a difference between bright and dim events was found in their average spectral signatures, namely a hardness/intensity correlation. The average curves of hardness evolution of bright and dim subsets have similar shapes, but the curve for bright events is well above the similar one for the dim subset.

4) *Cosmological effects of spectral redshift and light curves time-stretching are not seen for GRBs.* When both redshift and time-stretch transformations of bright GRBs were used to fit the average temporal and spectral signatures of dim events, no evidence for cosmological effects was found. Either the cosmological model with standard emitters in co-moving frames should be excluded, or very special internal evolution of emitters should be postulated

which compensates the cosmological stretching by an internal narrowing of time histories.

5) *Models of extended galactic halo are comparable with observed temporal and spectral signatures of GRBs.* The perfect similarity of average light curves for bright and dim GRBs points out that all emitters in the halo, both close and distant, have the same time scale of bursts. To explain the hardness/intensity correlation, the spectral hardness of radiated photons has to be postulated to correlate with intrinsic emissivity of emitters.

Acknowledgments. I thank very much Drs. Jerry Fishman, Michael Briggs, Bill Paciesas and Rob Preece for fruitful cooperation which made this paper possible and Local Organizing Committee for kind and warm hospitality. I thank also Dr. Anton Chernenko, who helped to prepare this paper.

REFERENCES

1. B.M. Belli, Astrophys. Sp. Sci. **231**, 43 (1995).
2. T. Bulik and D.Q. Lamb, Astrophys. Sp. Sci. **231**, 373 (1995).
3. J.-P. Dezalay, et al., in Gamma Ray Bursts, AIP Conference Proceedings **265**, eds. W.S. Paciesas and G.J. Fishman, 304 (1991).
4. C. Kouveliotou, et al., ApJ Letters **413**, L101 (1993).
5. C. Kouveliotou, these proceedings (1996).
6. E.E. Fenimore, these proceedings (1996).
7. G.J. Fishman, et al., ApJ Suppl. **92**, 229 (1994).
8. T. Koshut, et al., ApJ **463**, 570 (1996).
9. R.S. Mallozzi, et al., ApJ, in press (1995).
10. E.P. Mazets, et al., Ioffe Physical-Technical Institute Preprint No. 686 (1980).
11. C. A. Meegan, et al., electronic catalog, GROSSC (1995).
12. C. A. Meegan, et al., ApJ Suppl., in press (1996).
13. I.G. Mitrofanov, et al., in Gamma-Ray Bursts: Observations, Analysis and Theories, eds. C. Ho, R.I. Epstein and E.E. Fenimore (CUP), 203 (1992).
14. I.G. Mitrofanov, et al., in Compton Gamma-Ray Observatory, Eds. M. Friedlander, N. Gehrels & D. Macomb, AIP Conference Proceedings **280**, 761 (1993).
15. I.G. Mitrofanov, et al., in Gamma-Ray Bursts,Eds. G.J. Fishman, J.J. Brainerd & K. Hurley, AIP Conference Proceedings **307**, 187 (1994).
16. I.G. Mitrofanov, et al., Astrophys. Sp. Sci. **231**, 103 (1995).
17. I.G. Mitrofanov, et al., ApJ, in press (1996).
18. I.G. Mitrofanov, M. L. Litvak and A. M. Chernenko, ApJ, in press (1996).
19. I.G. Mitrofanov, A.M. Chernenko, et al., these proceedings (1996).
20. I.G. Mitrofanov, A.S. Pozanenko, et al., in preparation (1996).
21. J.P.Norris, et al., ApJ **424**, 540 (1994).
22. J.P. Norris, these proceedings (1996).
23. W.S. Paciesas, et al., in Gamma Ray Bursts, Eds. W.S. Paciesas and G.J. Fishman, AIP Conference Proceedings **265**, 190 (1992).
24. G. Pizzichini, Adv. Space Res. **15**, 5 (1995).
25. P. Podsiadlowski, M.J. Rees and M. Ruderman, MNRAS **273**, 755 (1995).
26. I.S. Shklowskii and I.G. Mitrofanov, MNRAS **212**, 545 (1985).
27. J.J. in't Zand and E.E. Fenimore, submitted.

Canonical Timescales In GRBs–1995

Jay P. Norris

NASA/Goddard Space Flight Center, Greenbelt, MD 20771

Understanding the bimodal duration distribution (dynamic range > 10^4) of γ-ray bursts is central to determining if the phenomenon is in fact a singular one. A unifying concept, beyond isotropy and inhomogeneity of the two groups separately, is that bursts consist of pulses, organized in time and energy: wider pulses are more asymmetric, their centroids are shifted to later times at lower energies, and shorter, more symmetric pulses tend to be spectrally harder. Long bursts tend to have many pulses while short bursts usually have few, relatively narrow pulses. Two factors, viewing angle and beaming, may account for pulse asymmetry and the large dynamic range (~ 200) in pulse widths.

A cosmological time-dilation signature, with an expected dynamic range of order two, would be difficult to measure against these large intrinsic variations and low signal-to-noise levels of dimmer bursts. Some statistics (T_{90}, pulse intervals) are particularly sensitive to brightness bias, noise, and apparently minor variations in definition. Also, spectral redshift would move narrower, high-energy emission from dim bursts into the band of observation, constituting a countering effect to time dilation. With analyses restricted to bursts longer than ~ 2 s, tests for time dilation that are constructed to be free of brightness bias have yielded time-dilation factors ~ 2–3, for pulse structures, intervals between structures, and durations.

GRB BIMODAL DURATION DISTRIBUTION

Beyond the isotropy and departure from homogeneity of γ-ray bursts (GRBs), their bimodal duration distribution (1) may be the most defining feature. If we only understood why this phenomenon (or phenomena) has a dynamic range of almost five decades in event duration, we might begin to understand GRBs. We do know a few things in connection with the bimodal appearance, enough perhaps to speculate that it reflects a unified phenomenon: Long and short bursts, on either side of the "valley" at ~ 2 s, are separately isotropically distributed (2). There is evidence that both groups are undernumerous at low peak fluxes, compared to the Euclidean expectation in a homogeneously filled space (3), although the best measurement of peak flux for very short bursts is probably yet to be realized. That is, if one truly believes peak flux to be an indicator of distance with some fidelity, then time-tagged event data (2-μs resolution) should be employed to measure the peak fluxes of short bursts, since pulses in short bursts can be considerably shorter

© 1996 American Institute of Physics

than the shortest (64-ms) timescale on which peak fluxes are tabulated in the BATSE 3B catalog.

Other indicators of kinship are found when the time profiles are examined in detail. Both short and long bursts consist of pulses, that are organized in time and energy, as discussed in the "Pulse Paradigm" section below (4,5). In terms of spectral softening and asymmetry, pulses in short bursts appear to be carbon copies of pulses in long bursts, except that they are, on average, compressed by a factor of ~ 20. A possible difference between the two groups is that long bursts tend to have many pulses, whereas short bursts often have just a few major pulses structures (6). If the GRB phenomenon is singular, then "telescoping" distributions – in pulse width, pulse interval, and number of pulses per burst – may explain the valley and bimodal appearance (6,7). Short bursts do tend to be spectrally harder (8). But in general, bursts tend to soften as they progress, as demonstrated convincingly by Ford et al. (9). Thus, the softer event-averaged spectra of long bursts may be a matter of the radiation transfer of later pulses being affected by prior burst history.

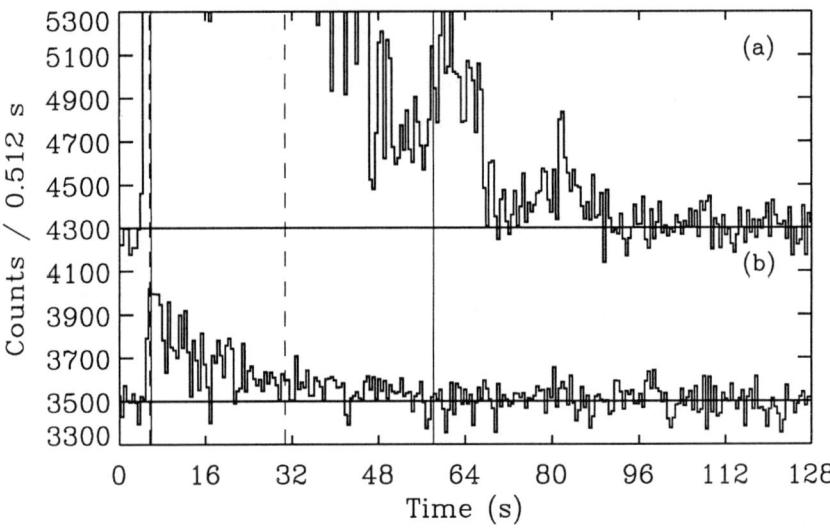

FIG. 1. BATSE burst # 678 with (a) original peak intensity, and with (b) peak intensity reduced to 1400 counts s^{-1} and variance from the background interval of a dim burst added. The t_5 and t_{95} points are shown as solid (a) and dashed (b) lines, determined from 4σ threshold above background on timescales up to 16 s.

Establishing duration measurements in a brightness-independent manner is desirable. As illustrated in Figure 1, one can either have accurate durations for bright bursts, or estimations for all bursts (to some threshold) which are relatively free of brightness bias. Figure 1 depicts BATSE trigger # 678 with (a) original peak intensity, and with (b) peak intensity reduced to 1400 counts s^{-1} and variance from the background interval of a dim burst added.

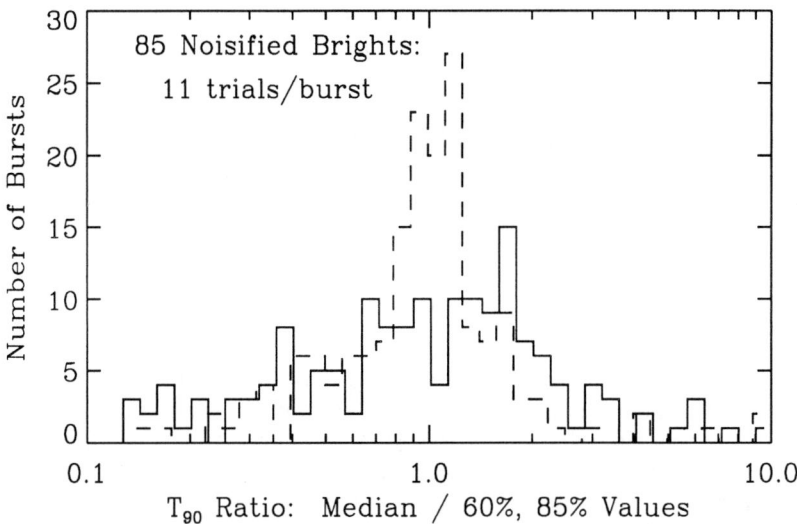

FIG. 2. T_{90} estimations for 85 long, bright bursts, 11 realizations per burst, with peak intensity and signal-to-noise equalized to dim burst level. Solid and dashed histograms are median value divided by 1st and 11th, or by 3rd and 9th ranked values, respectively. Note logarithmic scale on abscissa.

The t_0 and t_{100} points are estimated by seeking the first and last fluctuation, respectively, to exceed the 4-σ level above background. For the original peak intensity profile, due to the structure near bin $T = 60$ s, T_{90} is almost twice as long as that for the dimmed and noise-equalized profile. Since for the 3B durations, t_0 and t_{100} points were determined by eye and at original peak intensity and signal-to-noise (s/n) levels, one may conclude that durations of bright bursts are fairly accurately estimated, but that those of dim bursts are underestimated. Since backgrounds are necessarily fitted to the original time profiles, there will be a tendency, regardless of succeeding steps in a duration measurement procedure, to declare low-intensity portions of the burst to be background intervals. This problem exacerbates the underestimation of dim burst durations, and is particularly difficult to circumvent since it is desirable to specify the background near the burst, thereby more accurately fitting any curvature.

The T_{90} statistic is particularly vulnerable to statistical fluctuations, since the t_5 and t_{95} points are necessarily near the t_0 and t_{100} points, which are ill-defined. Figure 2 illustrates T_{90} estimations for 85 long ($T_{90} > 2$ s) bright bursts (peak flux [256 ms] > 4.6 photons cm^{-2} s^{-1}). Eleven realizations were performed per burst, with peak intensity and s/n equalized to dim burst levels as described for Figure 1. Solid and dashed histograms are for the median value divided by 1st and 11th, or by 3rd and 9th ranked values ($\sim 85\%$ and $\sim 60\%$ confidence, respectively). The spread in T_{90} (note logarithmic scale on abscissa) is attributable in many bursts to low-intensity tails or small outlier structures, which are (de)accentuated by fluctuations present in one run but not in another, and which contribute little to the total counts, but considerably to the total event duration. One must conclude that T_{90} is not a terribly robust and useful statistic. Obviously, T_{50} is influenced by the determination of the t_0 and t_{100} points, but less dramatically so. Notice that minor variations in t_0 and t_{100} will have a much larger relative effect on duration estimates for short bursts.

TIME DILATION

In the early 1980s, upon discerning the quasi-isotropy of the KONUS 11/12 burst localizations, Upendra Desai suggested to me to search for the signature of cosmological time dilation in the KONUS sample (duration distributions were constructed and nothing interesting was found). In the BATSE era, Paczynski (10) and Piran (11) made quantitative predictions concerning time dilation in GRB time profiles, based on the BATSE isotropy picture, integral number-intensity relation, standard cosmology and simple source population assumptions (nonevolving luminosity [L_{GRB}] and space density [η_{GRB}] functions). Because these assumptions have practically never been found to obtain for other cosmological populations (12,13), we should not be surprised to find in the end that L_{GRB} and/or η_{GRB} evolve with cosmic time (14,15) or that any presently observed time stretching effects anti-correlated with GRB brightness are partly or purely special relativistic manifestations (16).

The timescales on which cosmological time dilation would be observed in GRBs and corresponding methods used to search for the signature are summarized in Table 1. Each timescale and method has its peculiar drawbacks, related to brightness bias; to the necessity of determining and applying a correction for spectral redshift of temporal structure; or to the heterogeneous nature of burst profiles. It is difficult designing sensitive tests to measure a cosmological time-dilation factor, TDF = $[1+z_{dim}]/[1+z_{brt}]$, which may be of order 2–3 (10,11,19,21), when the dynamic range in durations is $> 10^4$, and the dynamic range of pulse-structure widths is $\sim 10^2$.

Assuming that the intrinsic burst process does not evolve with cosmic time, then all timescales in the table, except one, would necessarily be required to manifest equal TDFs after correction for redshift effects. The exception (since η_{GRB} is not guaranteed a constant value) is the timescale between bursts, probed by the number-intensity relation. Several studies (17–21), which take into account brightness-bias effects and energy-dependent narrowing effects arising from redshift, have found mutually consistent results for long bursts, thus supporting the cosmological time-dilation interpretation. For short bursts, several difficulties – smaller sample with requisite temporal resolution, greater dispersion in relative duration from noise, difficulty of precise measurement of peak flux for pulses shorter than 64 ms – combine to make bias-free analyses more problematic. Recent automated approaches attempt to address these complications (22).

TABLE 1. Timescales for Cosmic Time Dilation in Long Bursts

Phenomenon	Timescale	Method	Primary Complications
Redshift	$\sim$ 300 keV/c	peak in $\nu F(\nu)$ [a]	spectral shape, bandpass
Pulse widths	$\sim$ 0.2 – 1 s	pulse fitting	ID'ing true pulses
Pulse Intervals	$\sim$ 0.3 – 3 s	peak finding	ID'ing true intervals
"Burst Core"	$\sim$ 2 – 4 s	ACF	width correction
	$\sim$ 8 – 16 s	peak align	ID'ing peak, width corr.
Durations	2 – 600 s	T_{50}, T_{90}	brightness bias
Burst Intervals	$\sim$ 1 day	log(N)-log(P)	η_{GRB}, L_{GRB}

[a]ref: (23)

As implied by Figure 1, a primary problem with using event durations to measure time dilation is that virtually all of a bright BATSE burst is easily apprehended by eye, whereas for a relatively dim BATSE burst ($\sim$ 20 – 100 times less intense), low-intensity structures are lost in the statistical fluctuations. By constructing two duration distributions for long, bright bursts, with original peak intensity and with peak intensity equalized to that of dim bursts, and then applying a uniform threshold to define t_0 and t_{100} points, we can roughly estimate the degree of brightness bias that must be inherent in the 3B durations. Figure 3 illustrates these two distributions for the T_{90} measure. The distribution with equalized peak intensity is shifted to lower values relative to that for original peak intensity by factors of 1.7 (Gaussian fits), 1.4 (K-S test), or 2.8 (average of ratio of log[duration]). The corresponding factors for T_{50} are 1.3, 1.2, and 1.7, respectively. These brightness-bias factors will tend to obscure a time-dilation effect of order two. Therefore, careful consideration should be given to nullifying brightness-bias effects when estimating durations.

Two methods have been used to measure the width of the "burst core" – the region near burst peak intensity – often comprising several pulses. These are the peak-alignment method (17,24) and the auto-correlation function (ACF) (19,25). Such methods are widely appreciated to be common-sense approaches to search for time dilation in GRB profiles. Both approaches utilize the region near maximum count rate. The peak-alignment method is a linear sum of peak-normalized profiles, but finding "the" peak in dim bursts and accounting for noise bias at the peak can be problematic. The ACF is a nonlinear combination of the peak region with itself. The peak-finding problem is much ameliorated, but the ACF width is affected by noise in dim bursts. In both methods, the noise problems are addressable by equalization of the s/n levels of intensity groups that are being compared (19).

But, concomitant with time dilation is redshift: the temporal structure associated with each energy range is shifted to lower energies in the observer's frame of reference. Since GRB temporal structures are narrower at higher energy, this redshift-dependent narrowing of temporal structure competes with cosmological time dilation. Moreover, in "stack and average" approaches, *dilation of intervals between pulses* is included, and this diminishes the original

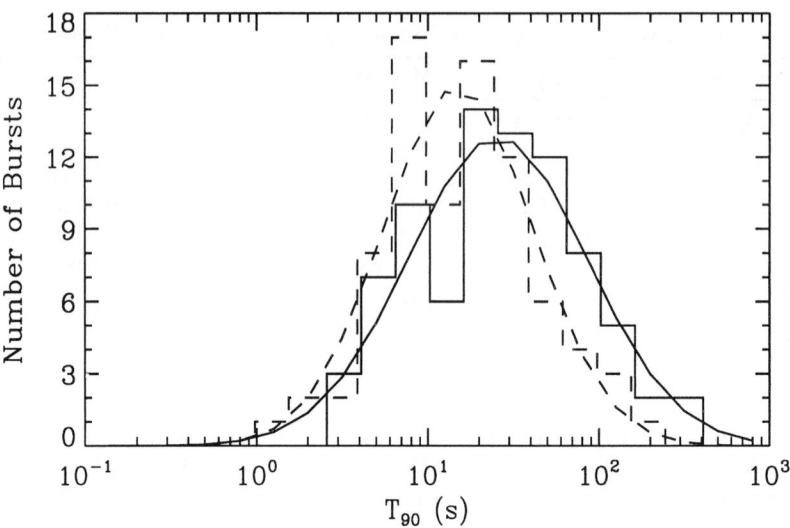

FIG. 3. T_{90} duration distributions and Gaussian fits for long, bright bursts, with original peak intensity (solid) and with peak intensity equalized to that of dim bursts (dashed). The distribution for dimmed bursts is shifted to lower values, illustrating the brightness bias against "seeing" the entirety of dim bursts.

effect one wishes to measure, dilation of temporal structure. Thus, width corrections must be applied in order to compare with expectations of any cosmological modelling. The combined correction for redshift and interval dilation is largest for the ACF statistic, and practically as large for the peak-alignment method (19).

Figure 4 is a schematic illustration of these two effects combined. The top panel represents a burst at zero redshift. The tallest pulse has been located to be used in some "burst core" measure of time dilation. The bottom panel is the same burst coming from a redshift of unity, TDF = 2. The solid curve shows the time-dilation effect. Note that there is more space between pulses. The dashed curve includes the degree of pulse narrowing for a redshift of approximately unity. Combined, interval dilation and pulse narrowing constitute a large width correction to the observed time-dilation measure. Dilation of intervals between pulses was not properly taken into account in previous treatments (17,25). An estimation is made of the combined correction, using 16-channel data, by redshifting and time-dilating profiles of bright bursts, yielding correction factors of ~ 1.4 (peak-align) and ~ 1.6 (ACF), assuming for example, an actual TDF of ~ 2.5 between bright and dim bursts (19). Both approaches yield consistent corrected TDFs, but only constrain the allowable TDF to lie within the range 2–3.

Note that since bursts are shorter at higher energy (9), a correction analogous to that required to compensate for pulse narrowing is necessary for

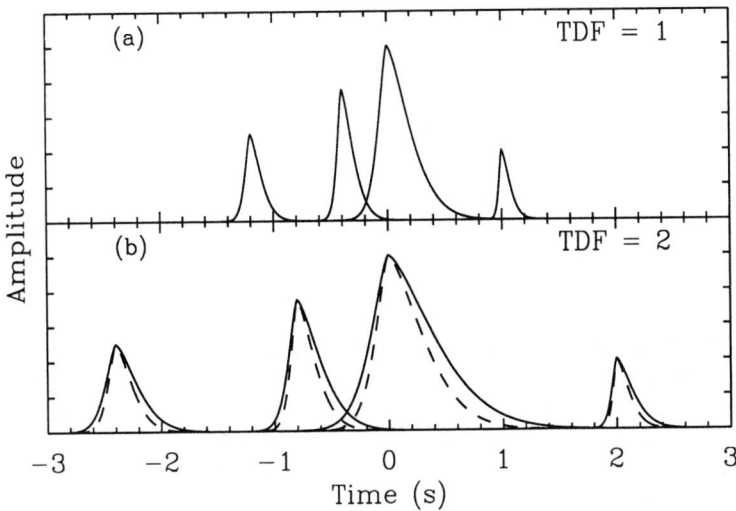

FIG. 4. Schematic of dilation of pulses and intervals, and energy-dependent narrowing of pulse structure. (a) Burst at zero redshift. (b) Same burst from redshift of unity (TDF = 2); solid curve shows time-dilation effect on pulses and intervals; dashed curve includes pulse narrowing for redshift of ~ 1. Combined, interval dilation and pulse narrowing necessitate large time-dilation correction for methods which average time profiles or ACFs.

duration measures of time dilation (21). If the true TDF were 2 (or 3), then the correction factor is of order 1.10 (or 1.25). Thus the time-dilation measure for durations is less affected by redshift than those for the "burst core."

For one potential measure of time dilation, [3B fluence]/[3B peak flux], three adjustments would be necessary to correct for weighting of the counts by the redshifted photon energy (in fluence but not flux), narrowing of pulse structure, and shorter durations at higher energy. In fact, this measure would yield uncorrected TDFs *less than unity* for dim bursts. In addition, systematic errors in fluence for dim bursts may arise in the process of background fitting.

One might think that measurements of intervals between pulse structures would be most free of systematics and energy-dependent effects, and thus would yield an observed TDF closest to the actual TDF. But some consideration shows that related problems will plague interval measures of time dilation: Intervals between pulses in bright bursts are much more clearly delineated than in dim bursts. As a time profile is dilated, additional, shorter intervals can appear near the limit of resolution of the data. Also, narrower temporal structures at higher energy redshifted into the observation band will result in deeper minima between peaks, more readily sensed by a thresholding algorithm. The latter two effects will tend to result in a smaller observed TDF as two new, shorter intervals arise from bifurcation of one interval.

Procedures have been devised in attempts to overcome these problems

(20,26). Rendering all burst profiles to the same s/n level helps defeat the brightness-bias problem. Wavelet- (or Fourier-) denoising of profiles (34) may be performed to diminish the frequency of insignificant peaks. Approximate calibration of systematic effects arising from time dilation and redshift can be achieved by stretching and redshifting bright bursts at original resolution, and performing measurements on all bursts at several times the original resolution. Some attempts have been made to devise good automated tests of interval dilation, but these methods are still somewhat immature. Distributions of pulse widths and intervals were measured by Davis using a semi-automated approach (27). Scargle has developed a fully automated pulse-fitting algorithm that should allow model variations to be tested for short bursts (22).

If the time dilation observed in bursts is cosmological, then *intervals between bursts* would also be time-dilated, and this would be reflected in the number-intensity distribution since the observable is a "density rate" (e.g., bursts yr^{-1} Gpc^{-3}) – observed volumes and time intervals at high redshift are larger than in the comoving frame of the burst. These General Relativistic effects are correctly taken into account in some (14,25,28,29), but not all, recent literature. However, the possibilities of cosmological source-density and/or luminosity evolution, or nonstandard cosmology, introduce considerable latitude into the modelling of redshift parameter space, enough so to make problematic – or at least questionable – the derivation of useful constraints vis-a-vis inclusion of time-dilation measurements. Also, some treatments so far have (inconsistently) analyzed results of time-dilation measurements for long bursts (> 2 s) and the number-intensity distribution for all bursts. Very interestingly, it may be that a GRB luminosity function is implied by the fact that dim, soft-spectrum events appear to follow a number-intensity relation rather close to a -3/2 power law (30,31,1). This could have important ramifications for many aspects of GRB analyses and modelling.

PULSE PARADIGM

In the hard X-ray/low energy γ-ray portion of the spectrum – now well-mapped by BATSE – GRBs appear to be composed of pulses which are organized in time and energy. This observation is important because pulses must reflect the "atoms" of the emission process, the details upon which we must ultimately rely to understand much of the high-energy physics, even if bursts are eventually observed in other wavebands. A global characteristic to be explained by physical modelling is that the envelopes of burst time profiles tend toward asymmetry (longer decays) (32). This tendency appears to extend to shorter timescales within a burst (33), that is, to pulses.

Figure 15 of ref. (5) schematically illustrates the range of pulse shapes that are found in relative isolation, where we do not have to worry too much about blended, overlapping shapes distorting the statistics. Shown are the approximate extremes of pulse shapes and spectral dependences of pulses that have been fitted in bright bursts longer than ~ 2 s. The range of pulse behavior is seen to go from {symmetric, narrow, with negligible lag between energy bands, and tending to be spectrally harder} to {asymmetric with longer decays, wider, low energy lagging high energy, and spectrally softer}. Similar patterns are found in a preliminary study of pulses in short bursts using time-tagged event data, but with the pulse timescale reduced by a factor of $\sim$ 20 compared to that for long bursts. Pulses in short bursts are more often completely separated from their neighbors (frequently a short burst appears to comprise only a few pulses), and perhaps for this reason the pulse paradigm appears cleaner for them (see Figure 2 of ref. (4)). Automated pulse fitting and denoising approaches may clarify this picture (22,34,35).

Some bursts clearly exhibit the pulse paradigm throughout the whole event. For example, trigger # 2083 consists of two main pulses, both wide, with longer decays than rises, and with the pulse centroids shifted to later times at low energy relative to high energy. Trigger # 678 (shown in Figure 1, but truncated and with insufficient resolution!) has a large number of spiky pulses, each one (as far as can be disentangled) symmetric down to a timescale of 16 ms with no shift in centroid with energy (33). Thorough analyses need to be conducted to determine to what degree this pattern prevails in bursts.

The pulse asymmetry/centroid lag pattern extends across a range in pulse widths from $\sim$ 10 ms to $\sim$ 1 s. It would seem that a variation in boost factor, Γ, cannot entirely account for the range of pulse widths, since then a comparable range in spectra would result. But a combination of dynamic range in Γ and variation in some geometrical factor, e.g., viewing angle, may account for the range of pulse widths. In such a scenario, viewing angle would be related to degree of pulse asymmetry, which is correlated with pulse width in long bursts (for physical modelling of pulses, see refs. (36–39)). This would also be consistent with the observation that short bursts are, on average, spectrally harder than long bursts (8), and with the slight trend for more symmetric, narrower pulses to be harder (5). The observed spectral hardness of a pulse, however, must also be a function of intrinsic beaming factor and position within a burst (radiation transfer dependent on preceding pulses) since we know that bursts tend to soften as they progress (9).

In conclusion, I wish to point out that, once again at a GRB conference (last time, Stanford 1984), the temporal domain was given first billing. Clearly this is more in tribute to BATSE's capabilities for recording bursts' time histories, allowing us to examine more closely their chaotic nature, rather than to our ability to understand the phenomenon. But let us not forget that the burst physics is written in large proportion in their temporal evolution.

Acknowledgements: It is a pleasure to acknowledge innumerable enlightening conversations with with many GRB aficionados, including especially Jerry Bonnell, Ed Fenimore, Robert Nemiroff, and Jeffrey Scargle.

REFERENCES

1. C. Kouveliotou, et al., these proceedings.
2. M. S. Briggs, et al., ApJ **459**, 40 (1996).
3. C.A. Meegan, these proceedings.

4. J.P. Norris, Ap Space Sci **231**, 95 (1995).
5. J.P. Norris, et al., ApJ **459**, in press (1996).
6. J.P. Norris, et al., in *Gamma-Ray Bursts 1993*, (AIP **307**: New York), 172 (1994).
7. V. Wang, these proceedings.
8. C. Kouveliotou, et al., ApJ **413**, L101 (1993).
9. L.A. Ford, et al., ApJ **439**, 307 (1995).
10. B. Paczynski, Nature **355**, 521 (1992).
11. T. Piran, ApJ **389**, L45 (1992).
12. V. Trimble, in *Gamma-Ray Bursts 1993*, (AIP **307**: New York), 717 (1994).
13. C. Hazard, in *The Space Distribution of Quasars*, (ASP: San Francisco), 170 (1991).
14. J.M. Horack, et al., these proceedings.
15. J.M. Horack, R.S. Mallozzi, T. Koshut, ApJ in press (1996).
16. J.J. Brainerd, ApJ **428**, L1 (1994).
17. J.P. Norris, et al., ApJ **424**, 540 (1994).
18. J.P. Norris, et al., ApJ **439**, 542 (1995).
19. J.P. Norris, et al., these proceedings (peak-align, ACF).
20. J.P. Norris, et al., these proceedings (peak intervals).
21. J.T. Bonnell, et al., these proceedings.
22. J.D. Scargle, et al., these proceedings.
23. R.S. Mallozzi, et al., ApJ **454**, 597 (1995); and these proceedings (1996).
24. I.G. Mitrofanov, Ap Space Sci **231**, 103 (1995).
25. E.E. Fenimore and J.S. Bloom, ApJ **453**, 25 (1995).
26. J. Neubauer and B. Schaefer, these proceedings.
27. S.P. Davis, *Ph.D. thesis*, The Catholic University of America (1995).
28. E. Cohen and T. Piran, ApJ (1995).
29. R.J. Nemiroff, et al., these proceedings.
30. G. Pizzichini, talk given at COSPAR, Hamburg, (1994); and these proceedings.
31. B.M. Belli, Ap Space Sci **231**, 43 (1995).
32. B. Link, R.I. Epstein, and W.C. Priedhorsky, ApJ **408**, L81 (1993).
33. R.J. Nemiroff, et al., ApJ **423**, 432 (1994).
34. C.A. Young, D.C. Meredith, and J.M. Ryan, these proceedings.
35. A. Lee, E. Bloom, and J.D. Scargle, these proceedings.
36. E.E. Fenimore, C.D. Madras, and Nayakshin, preprint (1996).
37. R. Mochkovitch, V. Maitia, and R. Marques, Ap Space Sci **231**, 441 (1995).
38. N.J. Shaviv, Ap Space Sci **231**, 445 (1995).
39. R. Sari and T. Piran, these proceedings.

Morphological Study of Short Gamma Ray Bursts

P. N. Bhat*, G. J. Fishman[†],
C.A. Meegan[†], R.B. Wilson[†] and W. S. Paciesas**

*Tata Institute of Fundamental Research,
Bombay 400 005, India.
[†]ES-66, Space Science Laboratory
NASA/Marshall Space Flight Center, Huntsville, AL 35812.
**Department of Physics, University of Alabama in Huntsville
Huntsville, AL 35899.

Gamma Ray Bursts (GRB) of duration less than about 2 s, detected by the Burst and Transient Source Experiment (BATSE) on board the Compton Gamma Ray Observatory have been selected for temporal analysis. These bursts constitute nearly 25% of the total and presumably form a separate class. Several parameters to describe the complexity and rapidity based on the burst temporal structure are derived and their dependence on other temporal and spectral properties are explored. A parameter is derived for each burst to characterize its spectral evolution based on its light curves in 4 energy channels. Bursts detected during April 1991 and March 1993 have been analysed yielding a sample size of 51 bursts. It has been found that the burst complexity is independent of its spectral content. The spectral evolution of short bursts is same as that of longer bursts. Also a systematic search for a coherent emission of γ- rays in short bursts yielded a negative result.

INTRODUCTION

GRB durations vary greatly, spanning at least 5 orders of magnitude. The median burst duration is about 10–15 s. Burst duration is a detector and distance dependent parameter (1). However there are reports of a break in the burst duration distribution at $\sim$ 600 ms based on the bursts detected by SIGNE (2,3). Based on 66 bursts detected by the PHEBUS experiment on GRANAT, Dezalay et al. (4) find that short rise times and hard spectra are common features of bursts lasting $<$ 2 s and hence may form a separate class of GRB's. More recently BATSE data also show a clear bimodal duration distribution and that on the average shorter bursts have a significantly higher hardness ratio compared to longer bursts (5). Attempts (6) to search for possible correlations of more spectral properties with temporal structures of longer bursts were not successful.

© 1996 American Institute of Physics

In this paper we address the following questions:
(a) Is there any evidence for a coherent emission of γ-rays in a burst? (b) Is there a spectral evolution during short bursts? If so how does it differ from that of longer bursts? (c) Are any of the spectral characteristics related to any of the temporal features of the burst?

OBSERVATIONS

BATSE large area detectors are described elsewhere (7). Among the various data types available we use the time tagged event (TTE) data which have the highest time resolution (2 μs) available in any GRB experiment so far. TTE data have 4 channel spectral information; the approximate photon energies at the channel boundaries are: 25, 50, 100 and $>$ 300 keV respectively.

DEFINITIONS

The spectral characteristics are defined by:
(a) Hardness Ratio (HR): HR is defined as the ratio of the number of signal photons above 100 keV to that in the range 25–100 keV. Both the numerator and the denominator are summed over the triggered detectors in a burst. The mean value of HR for short bursts is 0.96.

(b) Spectral Evolution parameter: It has been noted by Bhat et al. (1994) (8) that in a majority of the longer, selected sample of GRB's the higher energy photons arrive earlier than the lower energy photons while a smaller number of GRBs show evidence for the contrary. It has been found that a similar behaviour is true for the shorter bursts too. A spectral evolution parameter is derived for each burst based on the above property. Centroids of burst light curves have been derived for 4 different energy channels. Because of the spectral evolution, the centroids shift to earlier times with increasing photon energy. The slope of the straight line fit to the centroids as function of logarithm of mean photon energy is defined as the spectral evolution parameter. The mean photon energy for each channel is computed as the geometric mean of the channel boundaries. Fig. 1 shows a frequency distribution of this parameter. The mean value of the spectral evolution parameter is -7.32×10^{-3}.

The temporal characteristics are defined by:

(a) Burst Complexity Index, CI: This is a measure of the total number of statistically significant peaks in a burst time-history. The number of peaks are the number of zero-crossings of the first differential of the burst time history (expressed in units of σ). The peaks are counted both during the burst and background. A low pass filter is applied successively to burst and background to remove the statistical noise. The number of peaks during the burst while the background shows 0 or 1 peak is called the CI. Fig. 2 shows a frequency distribution of the CI for 51 short GRB's in our sample. The mean value

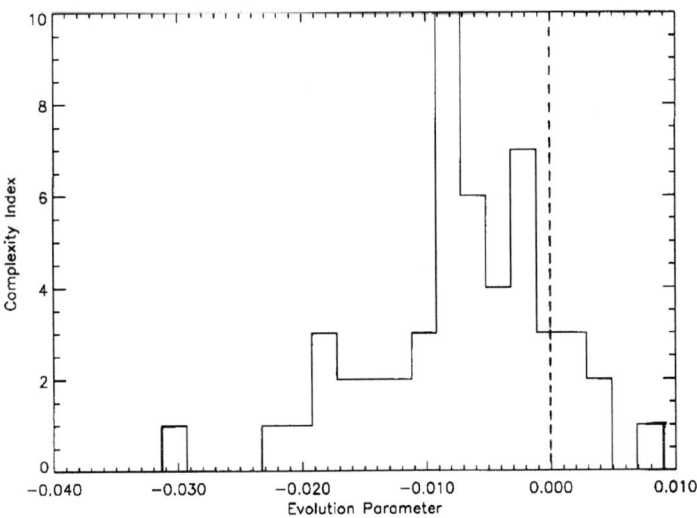

FIG. 1. The frequency distribution of the spectral evolution parameter showing the diversity of spectral evolution in short bursts.

of the complexity index is 2.9 showing that on the average bursts have three peaks.

(b) Rapidity Index, RI: Defined as the ratio of the variances of the first differential of the burst and background. The signal during the burst interval is expressed in units of standard deviation. This is a measure of the sharpness of the peaks in the burst time history. RI is a function of photon energy, generally increases with the average photon energy. CI and RI are found to be uncorrelated. Fig. 3 shows a frequency distribution of this parameter. The bursts with large RI correspond to those with fine time structures (9).

RESULTS

Using the statistical distribution of event arrival time differences in pairs of bright detectors we find no evidence for coherent emission of γ-ray photons (on time scales as short as 10 μs) in any of the bursts analysed, a result consistent with a similar search carried out on longer, stronger bursts (10).

The asymmetry of the distribution shown in Fig. 1 shows that the majority of bursts show spectral evolution even at ms/sub-ms time scales (11). A lesser number of bursts either do not show any spectral evolution or they evolve such that the harder photons arrive later than the softer photons. The diversity of spectral evolution among short bursts is same as that for longer bursts (10).

The CI is seen to be independent of HR, consistent with earlier results (12).

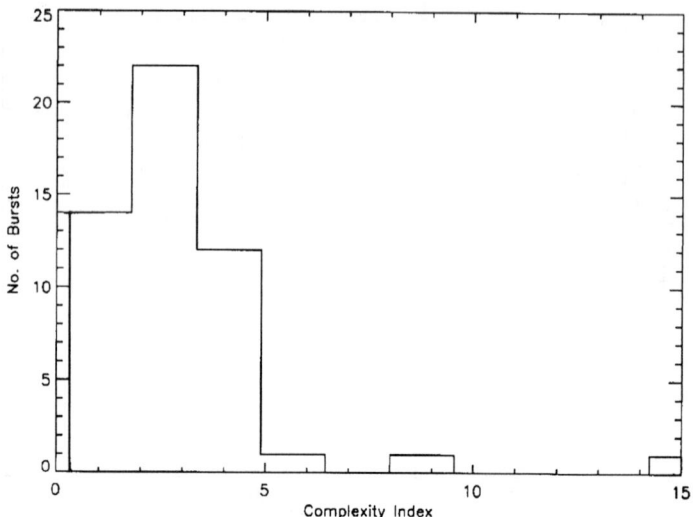

FIG. 2. The frequency distribution of the complexity index parameter.

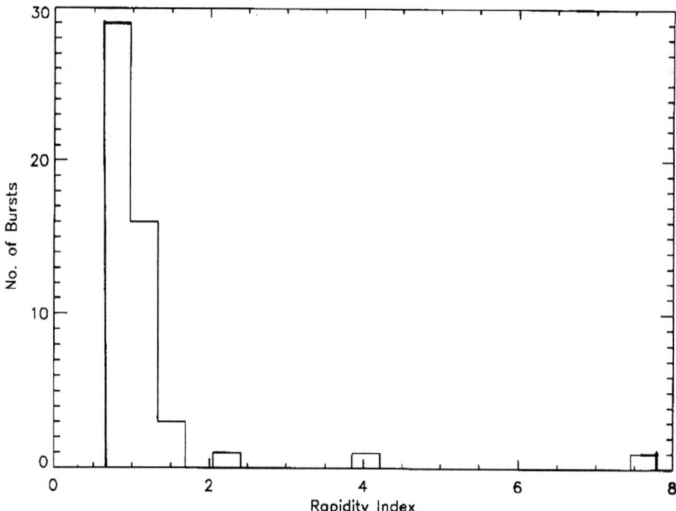

FIG. 3. The frequency distribution of the rapidity index parameter.

REFERENCES

1. E.P. Mazets and S.V. Golenetskii, Astronomia **32**, 16 (1987).
2. E.P. Mazets and S.V. Golenetskii Ap. & Space Sci. **75**, 47 (1981).
3. C. Barat, et al., ApJ **285**, 791 (1984).
4. J-P. Dezalay, et al., in Gamma Ray Bursts, eds. W.S. Paciesas and G.J. Fishman, AIP Conf. Proc. **265**, 304 (AIP, New York, 1992).
5. C. Kouveliotou, et al., ApJ Letters **413**, L101 (1994).
6. E. Jourdain, Ph. D. Thesis, C.E.S.R., Paul Sabatier Univ., Tolouse, France (1990).
7. G.J. Fishman, et al., Proc. GRO Science Workshop; Greenbelt, ed. W.N. Johnson, 2-39 (1989).
8. P. N. Bhat, et al., ApJ **426**, 604 (1994).
9. P. N. Bhat, et al., in Compton Gamma Ray Observatory, eds. M. Friedlander, N. Gehrels and D. J. Macomb, AIP Conf. Proc. **280**, 953 (AIP, New York, 1993).
10. D. Band, et al., ApJ **413**, 281 (1993).
11. B.E. Schaefer, et al., ApJ **404**, 673 (1992).
12. J.P. Lestrade, et al., in Compton Gamma Ray Observatory, eds. M. Friedlander, N. Gehrels and D. J. Macomb, AIP Conf. Proc. **280**, 969 (AIP, New York, 1993).

Brightness-Independent Measurements of GRB Durations

J.T. Bonnell,[1,2] J.P. Norris,[1] R.J. Nemiroff,[3] and J.D. Scargle[4]

[1] *NASA/Goddard Space Flight Center, Greenbelt, MD 20771*
[2] *Universities Space Research Association*
[3] *George Mason University, Fairfax, VA 22030*
[4] *NASA/Ames Research Center, Moffett Field, CA 94035*

> We present measurements of T_{90} and T_{50} durations of BATSE gamma-ray bursts in the 3B catalog, based on an automated algorithm designed to reduce brightness-bias effects by scaling burst peak intensities and variances to those of the dimmest burst in the sample. A multi-scale threshold is applied to estimate the beginning and end of burst activity. T_{90} and T_{50} measures indicate an observed time-dilation factor of order 2 between average durations of bright and dim bursts. We also estimate the dependence of duration measure on energy band, and find that an increase of order 10–25% would be required if the actual cosmological time-dilation factor were 2–3.

INTRODUCTION

Time dilation of gamma-ray bursts (GRBs) was first reported by Norris *et al.* (1) and has since been confirmed (2–4) for GRBs in the BATSE 2B and 3B catalogs. These reports find that bright burst time profiles are stretched with respect to dim bursts by approximately a factor 2. If the burst sources are at cosmological distances, the time-dilation factor (TDF) could have a simple interpretation, namely, $TDF = (1+z)_{\text{dim}}/(1+z)_{\text{bright}}$.

This time dilation effect is observed using several different brightness-bias free measurement techniques which are sensitive to stretching on different timescales (4). Perhaps the simplest conceptual measure of stretching is burst duration. The T_{90} (T_{50}) duration measure, which we adopt here, is defined as the interval between times when 5% and 95% (25% and 75%) of the burst counts above background are accumulated (5). Although simple in concept, unbiased duration measurements of this type are difficult to realize: A bias-free algorithm must be designed to estimate burst onset and cessation in a manner that is *independent of burst brightness*. This task might be easy if burst temporal development were somehow predictable; but the intrinsic variations in burst profile shape and the large dynamic range of burst durations make for an interesting problem. The burst durations reported here are estimated by an automated algorithm designed to address these issues. After normalization of all bursts to a fiducial peak count level and equalization of signal-to-noise (S/N), the algorithm applies a predetermined threshold on multiple timescales to the background-subtracted burst profile to estimate onset (t_0) and cessation (t_{100}) times. Background fitting is accomplished by visual examination of the eight-detector and four-channel data.

MEASUREMENT PROCEDURE

Data Preparation. Count rate data from all BATSE 3B bursts with T_{90} duration estimates > 2 seconds were included in this analysis. Bursts with peak count rate (>25 keV) less than 1400 counts s^{-1} were considered too weak to derive reliable duration measurements. After discarding bursts with data gaps, 510 bursts remained. Burst profiles were prepared by combining all routinely available four-channel data – DISCLA, PREB, and DISCSC data types – for the triggered detectors. The resulting burst record typically included about 120 s before and up to about 400 s after burst trigger. Whenever necessary, the data-stream start and stop times were adjusted to include all obvious and/or reported (3B) pre- and post-cursor burst emission.

Background Fitting. Because duration measures require the consideration of burst emission as near as possible to background levels, background estimation is a crucial issue. The degree of background variation in the BATSE data requires that appropriate background regions be chosen from times adjacent to the apparent burst. Simple automated procedures can accomplish this only approximately and were used to generate a first guess at reasonable background regions. Backgrounds were fitted using typically second order (occasionally higher order) polynomials. In a second pass, background estimation was refined by careful examination of DISCLA data from all eight BATSE detectors and the four energy channels of the prepared burst record. The data were uniformly binned to 1024 ms and plotted in both counts and residuals (sigmas). Two operators simultaneously viewed the data for the eight detectors and the four channels of the summed, triggered detectors on two separate CRTs. The text from the 3B Catalog Comments Table was also displayed and reviewed. Overlayed on the data stream were initial background fits along with symbols marking the times when 5% and 95% of the burst counts had accumulated based on the 3B Catalog values. This dual viewing approach proved to greatly enhance the operators' ability to distinguish between burst and non-burst emission, particularly near background levels, allowing background intervals to be chosen with confidence. We noticed, however, that since low-level structure in dim bursts is in practice indistinguishable from non-Poisson background variations, there is a tendency to fit backgrounds in dim bursts closer to, or even including part of, the burst interval.

Automated Duration Measurements. T_{90}s and T_{50}s were determined from the background-subtracted data (> 25 keV) using an automated procedure which applies a threshold criterion on multiple timescales to estimate the t_0 and t_{100} times, respectively. The primary goal of this "Auto-Thresholding" algorithm is to generate duration measurements which are relatively free of brightness-bias effects. After determining backgrounds as described above, the remaining procedure is automated, thus allowing rapid estimation of burst durations without human intervention.

To eliminate brightness bias, the background-subtracted records were scaled to the peak intensity of the dimmest burst in the sample (1400 counts s^{-1}). Necessarily, low-intensity structures, easily detectable in the original data for bright bursts, will become statistically less significant and more difficult to detect. This effect is illustrated in Figure 1 of ref. (6) for the bright burst trigger 678; as shown there, measurements of bright burst durations based on identification of statistically significant fluctuations will be shorter when the uniformly scaled data is compared to the burst at original intensity. In addition, to address the problem of non-Poisson variance, the partial reduction of variance (more-so the brighter the burst) incurred in the intensity scaling procedure was nullified via equalization of S/N levels of all bursts by adding the requisite variance from background regions of dim bursts, according to a procedure similar to that described in (1).

The t_0 and t_{100} times were then determined by a multi-scale thresholding procedure, which searches for a 4σ positive fluctuation on multiple timescales, ranging in dyadic steps from $\approx$ 0.5 s to 16 s, and assigns t_0 and t_{100} based on the first and last time bins identified, respectively. Thus, both tall and narrow as well as broad and short 4σ fluctuations trigger the threshold. Such a prescription appears necessary to sense both fast rising structures and long, low decays that are prevalent in GRBs. Background regions as well as regions where non-burst emission was identified were excluded from the t_0 and t_{100} search. The final T_{90} and T_{50} duration estimates adopted for each burst are the median result of 11 peak intensity and S/N equalization trials.

RESULTS AND CONCLUSION

The total sample was rank ordered by peak flux and divided into 6 subsamples with $\approx$ 85 bursts per group. Duration distributions were constructed for each group. To search for a trend of peak flux vs. duration, we fitted Gaussian centroids to each distribution. We also stretched the unbinned durations of the bright group to find best agreement with those of the five dimmer groups, using the Kolmogorov-Smirnov (K-S) test. These two measures yielded comparable TDF values. In Figures 1 and 2 the TDFs for five dimmer groups relative to that of the brightest group are plotted for the T_{90} and T_{50} measures, respectively. The central values (most probable stretch factors) and 1σ error bars shown in the figures are based on confidence determinations via the K-S test. The central values and probability of unity stretch factor (in parentheses) are listed in Table 1. The two dimmest burst groups have TDFs relative to the bright group of about 2. These results are statistically consistent with previous time-dilation factors derived for comparable *peak-intensity* (3), rather than *peak-flux*, groupings. This analysis differs from the earlier work in that all usable bursts with durations > 2 s and peak intensities above the 1400 counts s^{-1} level were analyzed. Thus a trend of larger TDF with lower peak flux appears to emerge, as found on other timescales in GRBs (4).

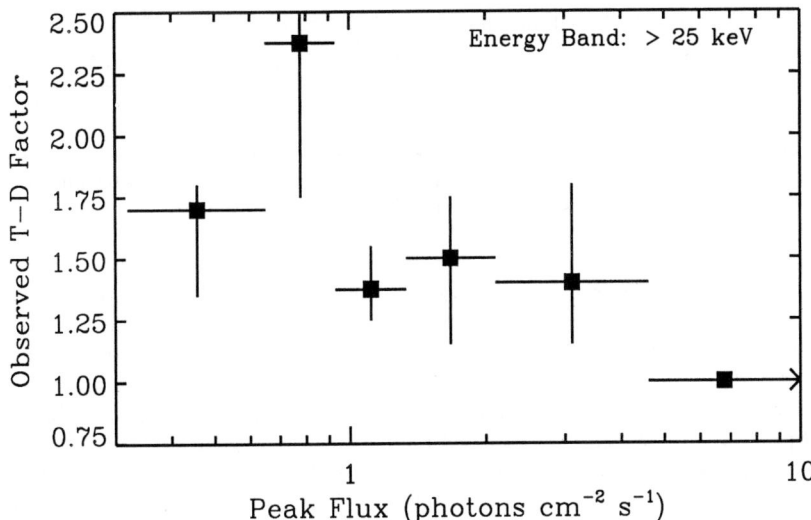

FIG. 1. The T_{90} TDF for each brightness group relative to the brightest group. Central values and 1σ error bars are based on confidence determinations via the K-S test. The two dimmest burst groups have relative TDFs of about 2.

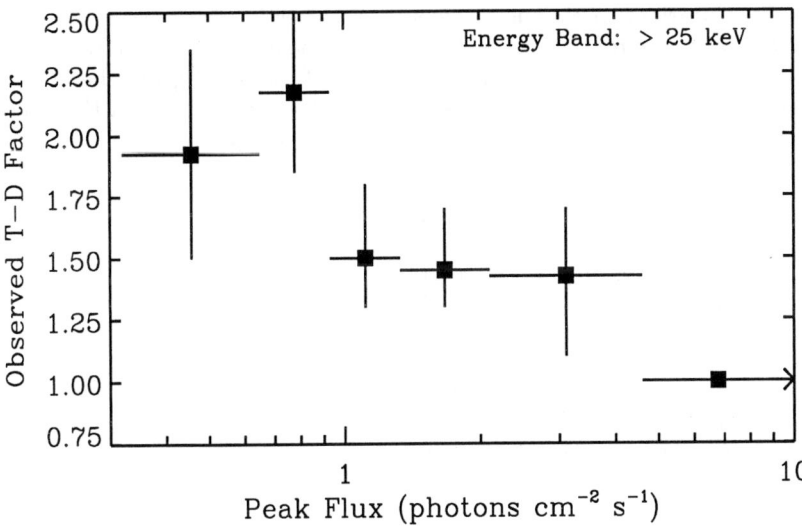

FIG. 2. Similar to Figure 1; the T_{50} TDF for each brightness group relative to the brightest group.

TABLE 1. Duration Time-Dilation Factors vs 3B Peak Flux

Peak Flux[a]	2.10	1.33	0.93	0.65	0.32
T_{90} TDF	1.40 (0.095)	1.50 (0.038)	1.38 (0.053)	2.38 (0.00001)	1.70 (0.0019)
T_{50} TDF	1.42 (0.10)	1.45 (0.016)	1.50 (0.074)	2.18 (0.0001)	1.93 (0.0034)

[a]Lower boundary in ph cm^{-2} s^{-1}; boundary for brightest group is 4.60 ph cm^{-2} s^{-1}. Values in parentheses are probabilities for unity stretch factor.

Energy Correction. Ford et al. (7) demonstrated that bursts are intrinsically shorter at higher energies as burst spectra tend to soften with time. If GRBs are at cosmological distances, a self-consistent determination requires that the *observed* TDF be corrected for this energy dependence to determine the *actual* TDF, since time profiles of dim, presumably more distant bursts would be redshifted into the observed bandpass from higher energies, whereas in the burst rest frames, the profiles would be intrinsically longer (at lower energy). Norris et al. (4) discuss analogous corrections for other measures of TDFs in GRBs. We have coarsely quantified the dependence of duration on energy by comparing T_{90} and T_{50} estimations for bright bursts based on summed counts in the 25–110 keV band (channels 1+2) to those in the 55–320 keV (channels 2+3) and 110–320 keV (channel 3 alone) bands. For both T_{90}s and T_{50}s, the average durations for channels 2+3 must be multiplied by ~ 1.1 to best match (via K-S test) the durations measured for channels 1+2. For durations in channel 3 alone, the correction factor is ~ 1.3. Unfortunately, the appropriate correction factor depends on the unknown actual TDF. Because the uncertainty in *observed* TDF is large, given the current sample size, duration measurements by themselves only constrain the *actual* $TDF, (1+z)_{\mathrm{dim}}/(1+z)_{\mathrm{bright}}$, to lie in the range ~ 2–3.

REFERENCES

1. J.P. Norris, et al., ApJ **424**, 540 (1994).
2. E.E. Fenimore and J.S. Bloom, ApJ **453**, 25 (1995).
3. J.P. Norris, et al., ApJ **439**, 542 (1995).
4. J.P. Norris, et al., these proceedings.
5. C. Kouveliotou, et al., ApJ **413**, L101 (1993).
6. J.P. Norris, these proceedings.
7. L.A. Ford, et al., ApJ **439**, 307 (1995).

Study of the Very Short Time Structure in Three BATSE Events Using TTE Data

David B. Cline, Stanislas Otwinowski and David A. Sanders

University of California at Los Angeles
Center for Advanced Accelerators, Department of Physics and Astronomy
Los Angeles, CA 90095-1547

The short time structure (possibly shot noise) might reveal much about the spatial nature of the GRB source. In the case of GRBs from Primordial Black Holes (PBHs), the initial explosion region is $\sim 10^{-13}$ cm in size, much smaller than any other known compact object. The BATSE team has already reported on one interesting event. We report here on the study of the TTE data from three BATSE events that may be PBH candidates. The latest results will be presented at the meeting.

INTRODUCTION

This paper presents a preliminary analysis of the time structure of three short duration Gamma Ray Burst (GRB) events. The data analyzed were Time Tagged Event (TTE) data from the BATSE 2B Catalog (1). By studying the time structure of short duration GRBs one can look for a periodicity and high frequency structure that would indicate an object of small size.

It is hoped that this analysis will indicate that there is an object with a size much smaller than any known compact object. If so, this would be consistent with the model of an explosion from the evaporation of a Primordial Black Hole (2,3). In this model the size of the initial explosion would be of the order of 10^{-13} cm.

ANALYSIS

The analysis presented in this paper was performed using the following procedures. The TTE data type was sorted by the energy of the γ-ray. The γ-rays were divided into four energy bins; $20 < E < 50$ keV, $50 < E < 100$ keV, $100 < E < 300$ keV and $E > 300$ keV.

The structure of the GRB may be shown by plotting the number of γ-rays versus time for time-bins of 1 millisecond width. An example for one energy range, $100 < E < 300$ keV, is shown in Figure 1 for the first second of the burst. This figure shows the plot of the number of γ-rays versus time for the three TTE events; burst trigger numbers are 432, 480 and 512. It should

© 1996 American Institute of Physics

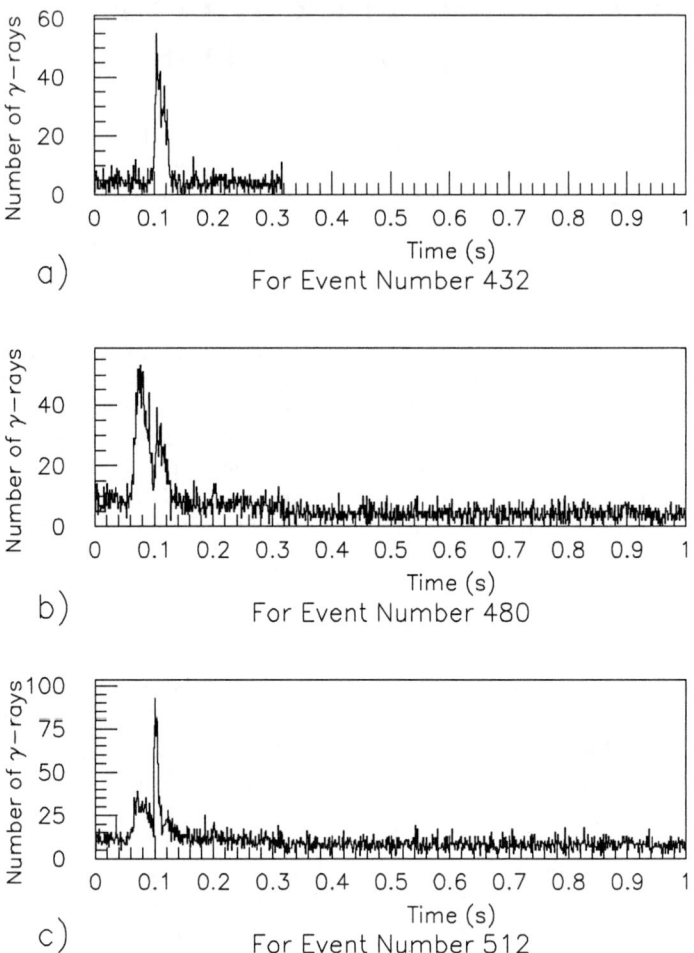

FIG. 1. The number of γ-rays versus time for burst trigger numbers a) 432, b) 480 and c) 512; for the energy range $100 < E < 300$ keV.

be noted that for burst number 432 the remaining data were overwritten by burst number 433.

To examine the fine time structure of short duration Gamma Ray Bursts, the Power Spectral Density (PSD) is calculated for both the peak and the background of the event. This is done by performing a discrete Fast Fourier

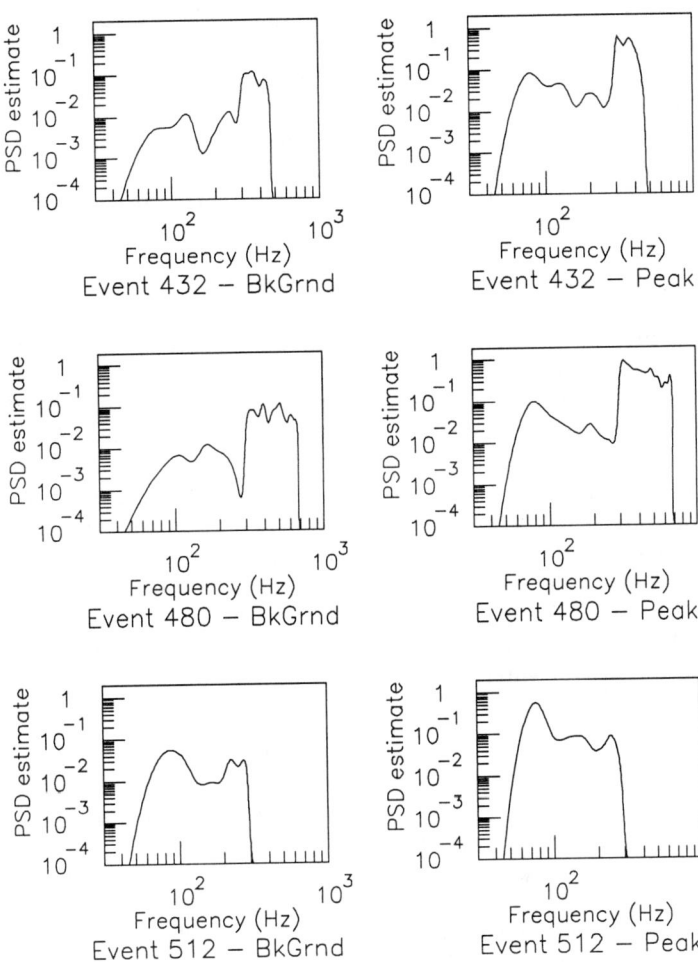

FIG. 2. The Power Spectral Density versus frequency of the background and the peak for burst trigger numbers a) 432, b) 480 and c) 512 for the energy range $100 < E < 300$ keV.

Transform (FFT) of the distribution of counts versus time from an array of selected time bins. The peak selection contains counts from the full time duration of the peak. The background selection contains the counts from an equal time duration.

The Fourier Transform takes a Real function of time and transforms it into a Complex function of frequency, $F(t) \to C(\omega)$. Therefore, to examine the Power Spectral Density, the normalized magnitude of $C(\omega)$ is plotted versus the frequency, using the method of Belli (4-6). The normalization, N, is the integration of the background counts for same time duration as the given peak. Thus, $|C(\omega)|/N$ is plotted versus the $f = \omega 2\pi$ for both the peak and the background. Due to the space available only a sample of the Fourier Analyses, for one energy range, is shown. These plots, shown in Figure 2, are the Fourier Analysis of the time distributions plotted in Figure 1.

FUTURE PLANS

The future plans entail a more detailed examination of the three events presented in this paper followed by the examination of more short duration TTE events. This will include careful examination of effects of different binning in an attempt to use the smallest possible time bins. As part of this examination, generation of Monte Carlo simulations of the background radiation will be performed. As part of the background, examinations a few longer duration bursts will be examined. An additional test that may be used is an auto-correlation function to examine periodicity of multiple spikes in a peak.

We wish to thank the following people for their aid in obtaining and analyzing the data presented in this paper: Gerald Fishman at NASA's Marshal Space Flight Center and both Jay Norris and Jerry Bonnell at NASA's Goddard Space Flight Center.

REFERENCES

1. C. A. Meegan, et al., electronic catalog available from grossc.gsfc.nasa.gov (1994).
2. S. W. Hawking, Nature **248**, 30 (1974).
3. S. W. Hawking, Commun. Math. Phys. **43**, 199 (1975).
4. B. M. Belli, in Gamma-Ray Bursts, eds. G.J. Fishman, J.J. Brainerd and K. Hurley, AIP Conf. Proc. **307**, 192 (AIP, New York,1993).
5. B. M. Belli, Astron. Astrophys. Suppl. Ser. **97**, 63 (1993).
6. B. M. Belli, ApJ **393**, 266 (1992).

Comparison of the Properties of Short and Long Gamma-Ray Bursts Observed with Phebus

J-P. Dezalay[1,2], J-P. Lestrade[1], C. Barat[2], R. Talon[2],
O. Terekhov[3], R. Sunyaev[3], and A. Kuznetsov[3]

[1] *Department of Physics and Astronomy*
Mississippi State University, MS39762, USA
[2] *Centre d'Etude Spatiale des Rayonnements (CNRS/UPS)*
BP 4346, 31029 Toulouse Cedex, France
[3] *IKI, Russian Academy of Sciences*
Profsoyuznaya 84/34, 117810 Moscow, Russia

> The observation of subclasses within the classical Gamma-Ray Burst population is the first clue that the bursts may be produced by a small number of distinct mechanisms, sources, or initial conditions. Events in one subclass are characterized by a short duration (less than 2 seconds) and, on average, harder spectra compared to longer bursts. Here, we present a statistical study of GRB properties as observed with the PHEBUS experiment such as intensity, duration, spectral hardness, and the shape of their time profiles. We show which burst characteristics are common to both the long and short populations and which are not. Finally, we compare our results with BATSE observations. Specifically, we discuss the possible causes for the striking difference observed in the hardness-duration diagrams for the two experiments.

THE TWO SUBCLASSES OF GRBS

The search for classes in the classical GRB population is made difficult by the very large diversity in their time profiles and spectra. The evidence that subclasses exist within this sample comes from the hardness-duration correlation (1-3) and the bimodality of the duration distribution (2). The evidence shows that events with a duration less than 2 s constitute a homogeneous sample and therefore have to be considered separately when studying the statistical properties of GRBs. It is also important to search for common and specific characteristics between subclasses as a function of energy in order to determine what makes them different. Data from the Phebus experiment are useful for subclass comparison at high energies (100 keV–100 MeV) with a high temporal and spectral resolution. One other major advantage of the Phebus data set is the clear separation of the two subclasses in a Duration-Hardness diagram in the 120 keV–7 MeV energy range.

© 1996 American Institute of Physics

PROPERTIES OF PHEBUS BURSTS

Phebus is a French-Russian experiment on board the GRANAT satellite. It has been operating since December 1989. Its 6 independent cylindrical BGO detectors (7.8 by 12 cm) anticoincident with a plastic shield are positioned on the vehicle in order to cover 4π sr. The data recorded in burst mode for each detector consist of: a time profile in the ~ 120keV $- 1.6$MeV energy range with a temporal resolution varying from 8 ms to 32 ms, 176 time-to-spill spectra, and 640 spectra integrated over 1 s. In the following, we compare the statistical distributions of the two populations for several quantities: the duration, the intensity, the hardness, and a new estimator of the shape of the time profile.

Between 1 December 1989, and 30 September 1994, the Phebus experiment has observed 174 cosmic GRBs out of 500 triggers. More than 80% of these bursts have been observed with other instruments. The other triggers are due to solar flares, particle precipitations, and atmospheric events. Of the 174 GRBs, 131 fall into the long (> 2 s) category and 43 are short.

Durations and Intensities

The durations have been estimated using the same T90 algorithm used for the BATSE data set (2). The distribution of T90 durations is quite similar to that obtained using the *total duration* algorithm (3). The logarithmic mean for the entire population is 7.79 ± 1.08 s, while it is $D_S = 0.55 \pm 0.08$ s for the sample of short bursts, and $D_L = 18.56 \pm 1.67$ s for the long ones. The ratio of durations of the two subclasses is therefore $D_L/D_S \approx 30$ for Phebus compared to ≈ 80 for BATSE. The average duration of short events is relatively high compared to the BATSE value of 0.33 ± 0.21 s due to the absence of triggers on the 64-ms time scale.

The intensity distribution (C_{max}/C_{min}) is a crucial parameter to test whether or not the two subclasses have similar spatial distributions. This comparison has to be done over a wide energy range since the spectrum of short events is significantly harder. The intensities are calculated using the C_{max} on $1/4$ s in the $150-1000$keV energy range. The $1/4$ s time scale is used to trigger the experiment on shortest events so the bias due to the binning size is minimized. A Kolmogorov-Smirnov test yields a probability of 0.18 that the two C_{max} distributions are drawn from different parent distributions. This result confirms earlier studies made with BATSE data (4) that short and long events have comparable intensity distributions.

We have also tested the homogeneity of the radial distribution of our two samples using the V/V_{max} test (5). Among the 165 events with measurable values of V/V_{max}, 37 are short while 128 are long. The average values for the test are $\langle V/V_{=max}\rangle = 0.40 \pm 0.022$ for the whole population, $\langle V/V_{max}\rangle = 0.46 \pm 0.047$ for the short bursts, and $\langle V/V_{max}\rangle = 0.39 \pm 0.029$ for the long bursts. The $\langle V/V_{max}\rangle$ values for the two subclasses are compatible within the 2σ error bars.

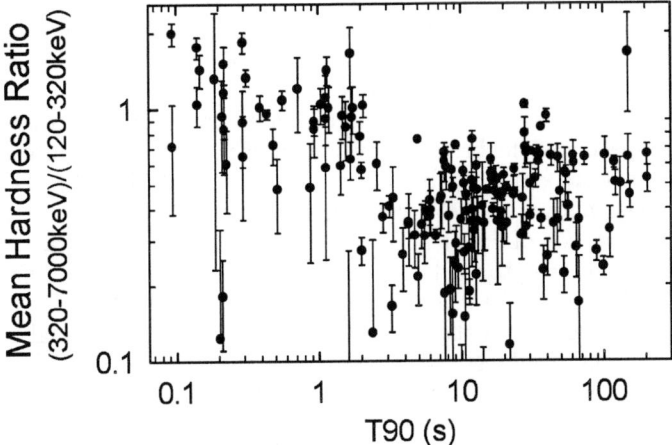

FIG. 1. Hardness-Duration diagram for 174 GRBs observed with PHEBUS.

Spectral Hardness

Using Phebus data, we have shown previously that the spectral hardness distributions are significantly different for the two subclasses (1,3). Our spectral hardness measure is the mean hardness ratio (HR). It is calculated from the spectrum integrated over the total duration of the event. The energy ranges in the ratio are 320 − 7000keV over 120 − 320keV. Figure 1 shows the HR as a function of T90 for the 174 events detected with Phebus. The separation between the average hardness of short bursts and long bursts is significant at a 6σ confidence level. This difference between the spectral hardness is still significant at the $> 4\sigma$-level, even if we use other algorithms either for the hardness ratio (e.g., Peak hardness ratio (3)) or duration (T50 or FWHM duration).

Even though the Phebus HR distribution is not bimodal and the duration distribution is only weakly bimodal (3), surprisingly, the two subclasses are well separated in the HR-T90 plane as compared with an identical diagram obtained using BATSE data. (In a BATSE Hardness-Duration diagram the HR is roughly anticorrelated with T90 (2)). This difference could be explained by selection effects which prevent the detection with PHEBUS of short and soft events or by the fact that our HR is calculated on brighter events and with different energy ranges as compared to BATSE. The answer to this question can be found using the set of common events detected by the two instruments and will be discussed in a future paper.

Total output energy

One important question concerning these two classes of bursts is whether or not the total energy output is comparable between an average short burst and an average long burst. That is, does the increased hardness of the short bursts compensate for the shorter duration resulting in a similar total luminosity? In this case the total energy released in the two kinds of bursts would be the same, if their typical distances are comparable.

In order to test this hypothesis, we have calculated two sum spectra by adding the average spectra of the brightest events for each of the two samples. These two sum spectra are significantly different up to several MeV (3). The ratio of total energies (above 120 keV) contained in the two sum spectra is $E_{Long}/E_{Short} \approx 9$. The use of count spectra instead of photon spectra has little influence on this ratio due to the large peak efficiency of the Phebus BGO detectors ($> 50\%$ at 2 MeV). The same ratio calculated from the BATSE 3B Catalog (6) using the fluences in the $50 - 300$keV energy range is equal to ≈ 18. Therefore, for the brightest GRBs, although the proportion of high energy photons is greater in short events, there is still an order of magnitude more energy emitted during longer softer events.

Shape of the time profile

Todate, duration seems to be the only temporal characteristic that is different between the two subclasses. However, careful observation of the Phebus time profiles gives the impression that the short-event light curve is more "square" on average. This not only means that the typical rise times and decay times are less but also that the counts are more uniformly distributed over the duration of the event. In order to quantify this impression, we introduce a parameter which is a measure of the degree of "squareness" of a time profile.

The squareness parameter (SP) is defined as the total time (expressed in percentage of the duration of the burst) for which the count rate is at least 50% of the maximum count rate in the event. This parameter is calculated using the distribution of count rates between the beginning and the end of the event (including time spent at background level) as determined by T90. The temporal resolution of the data used to compute this parameter is 32 ms. Among the 163 bursts available for this study, 42 have a duration less than 2 s and only 6 have a duration less than 200 ms (the SP is strongly affected by the bin size for these events). Figure 2 displays the mean hardness ratio as a function of this squareness parameter. Although there is a significant separation in this diagram between the two subclasses, the SP does not increase the difference observed with T90.

DISCUSSION

This paper presents several statistical properties of the two major subclasses of gamma-ray bursts observed by PHEBUS. We confirm that the intensity distributions for the short and long bursts are similar. This result added to the isotropy of the angular distributions for both suggests a comparable

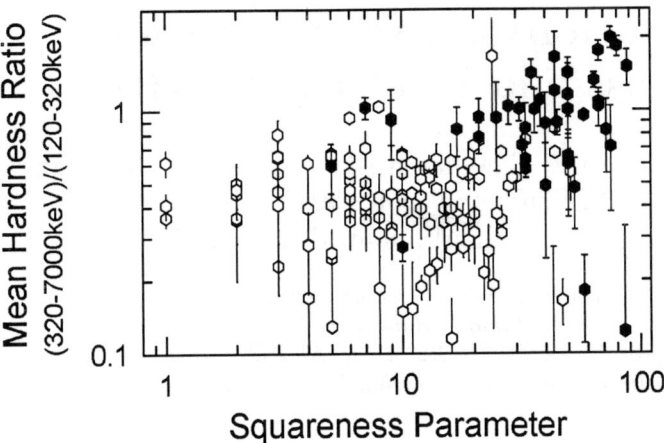

FIG. 2. Mean Hardness Ratio as a function of the squareness of the time profile. Short events (T90≤ 2 s, closed) and long events (T90> 2 s, open)

distance scale for the two populations. Under this assumption, we find that the longer (softer) bursts emit an order of magnitude more energy than the shorter (harder) bursts.

Finally, we show, using T90 as the measure of duration, that the two subclasses observed with PHEBUS are distinctly separated in the hardness-duration diagram. This separation is in striking contrast to the BATSE result which shows a more contiguous distribution in hardness-duration space. The BATSE subclasses are more clearly separated using the bimodality of the duration histogram. On the other hand, the Phebus subclasses are more easily delineated in the Hardness-Duration Diagram. It is interesting to note that with these two different approaches of selecting the events and with two completly different experiments we find the same $\sim 2\ s$ boundary (see however (7) for an alternative method of selecting both samples from the Hardness-Duration Diagram). Further analysis of the characteristics of the two populations is required to explain the differences between the two experiments.

REFERENCES

1. J-P. Dezalay, et al., in Gamma-Ray Bursts, eds. W.S. Paciesas and G.J. Fishman, AIP Conf. Proc. **265**, 304 (AIP, New York, 1992).
2. C. Kouveliotou, et al., ApJ **413**, L101 (1993).
3. J-P. Dezalay, et al., Astroph. and Spa. Sci. **231**, 115 (1995).
4. S. Mao, R. Narayan, and T. Piran, ApJ **420**, 171 (1994).
5. M. Schmidt, J.C. Higdon and G. Hueter, ApJ **329**, L85 (1988).
6. C.A. Meegan, et al., submitted to ApJ, (1995).
7. B.M. Belli, Astroph. and Spa. Sci. **231**, 43 (1995).

Correlations between Duration, Hardness and Intensity in GRBs

Chryssa Kouveliotou*,††, Thomas Koshut*,††, Michael S. Briggs†,
Geoffrey N. Pendleton†, Charles A. Meegan††,
Gerald J. Fishman††, John P. Lestrade$

*Universities Space Research Association
†University of Alabama in Huntsville, AL 35899, USA
††ES-84,NASA/MSFC, Huntsville, AL 35812, USA
$Mississippi State University, MS 39762,USA

We present here the T_{90} and T_{50} distributions for different trigger time scales from the 3B catalog and confirm the bimodality established before with the 1B data. We show that the Hardness Ratio/duration anti-correlation is present in the 3B data set and we explore Hardness and Intensity correlations by studying various burst subsets. We discuss the impact of these correlations within the framework of a cosmological GRB origin hypothesis.

INTRODUCTION

There are very few non-controversial (established) results associated with gamma-ray bursts (GRBs). The majority of these results is related to the GRB temporal structures: we know that their morphologies have defied classification efforts for almost three decades (1), that their profiles are asymmetric (2,3), and that their light-curves are chaotic with no evidence of periodicities beyond the statistically expected outliers (4). Finally, the first clear evidence for subclass separation in the GRB population has come from their durations (5). Here, I confirm and expand on earlier BATSE temporal results (5), present further analyses taking also into account spectral data and speculate on the consequences of these latest findings.

Time domain - GRB duration bimodality

We will use here T_{90} as a measure of burst durations. T_{90} is the time interval in which the integrated counts from the burst increase from 5% to 95% of the total counts. We have computed the T_{90} values only for events that had no data gaps during the burst. Out of a total of 1122 GRBs in the 3B catalog, only 834 bursts had computable durations. Further, to (i) reduce the effect of an instrumental cutoff on short bursts and (ii) to study a sample selected with uniform criteria, we have plotted in Figure 1 only the GRBs that triggered

© 1996 American Institute of Physics

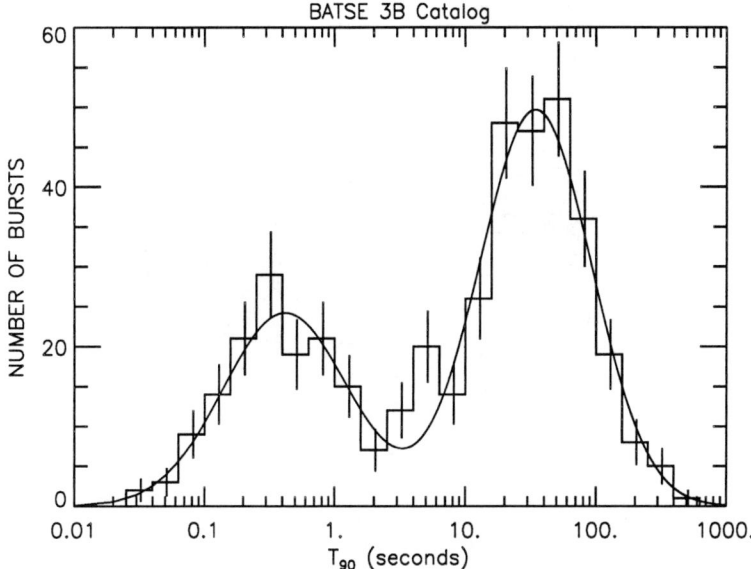

FIG. 1. T_{90} distribution for 427 bursts from the 3B catalog. The solid line represents a fit of two log-normal Gaussians.

on the 64 ms timescale. The ensuing T_{90} distribution of 427 GRBs is then fitted with two log-normal Gaussians, weighted by the uncertainties of each histogram bin. We clearly see the bimodality of the distribution, with the two modes centered at 0.42±0.05 s (short bursts) and 34.4±2.3 s (long bursts); the minimum occurs between 2 and 4 s.

Hardness Ratio - Duration Diagram

We define as hardness ratio (HR) the ratio of the total counts during the T_{90} interval of a GRB between 100–300 keV to the total counts between 50–100 keV. This will be abbreviated here as HR32, indicating that the two energy ranges correspond to BATSE discriminator channels 3 and 2, respectively. In the past (5) we have shown that there is a trend for short events to be harder, while long ones are predominantly softer. The same trend was found earlier with data from the Phebus experiment (6), albeit not related to the duration bimodality. Figure 2 shows the HR32-T_{90} diagram for 795 events from the 3B catalog, for which we could calculate their T_{90} values irrespective of trigger time scales *and* their peak fluxes (see also next section). In the figure, we distinguish two clusters of events and a downward trend in the HRs from short to longer events. The average values of HR32 for GRBs below and above 2 seconds are 1.48±0.05 and 0.90±0.02, respectively. The dash-dotted lines in Fig. 2 correspond to these averages.

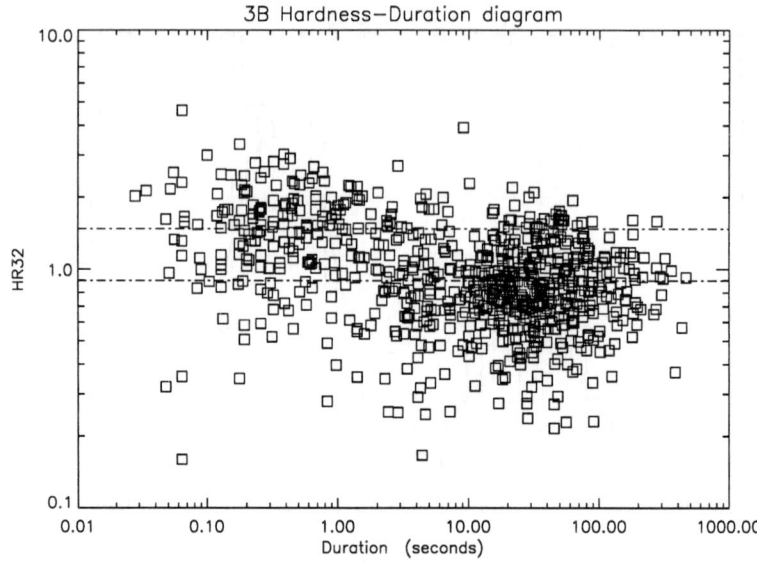

FIG. 2. HR32 vs T_{90} diagram for 795 bursts from the 3B catalog. The dash-dotted lines represent average values of the HR32s for the short (top) and the long events (bottom), with the separator set at 2 seconds.

DURATION-HARDNESS-INTENSITY CORRELATIONS

Pizzichini (7,8) and Belli (9,10) suggested selecting horizontal strips of the HR vs T_{90} diagram and comparing the $\log N$-$\log P$ diagrams of each strip, i.e., to compare the radial distribution of the bursts in each hardness region assuming the same overall duration distribution. Their results are intriguing, but so far without an obvious physical interpretation; the $\log N$-$\log P$ diagrams varied from near homogeneity to strong deviations from it. Here we attempt to select our subsets based on a strict physical selection criterion. Our reasoning is as follows:

Had there been two different GRB subclasses separated by their durations and hardnesses (short+hard, long+soft), the "purest" members of each class would be those located (i) above HR32 = 1.5 and (ii) below HR32 = 0.9. In other words, we select the hardest and the softest of the events, in the eventual hope that this way we have the least contamination of one class with the other.

Figure 3 shows the $\log N$ - $\log P$ diagram of the two sets thus selected. An additional completeness criterion of 0.8 photons/cm^2 s is set for the peak intensity (over 64 ms) of the events. The final numbers of events plotted are 135 hard (thick solid line) and 281 soft (thick dotted line) GRBs. The dashed line shows the $-3/2$ power-law corresponding to a homogeneous distribution in Euclidean space. The thin solid and dotted lines correspond to upper limits

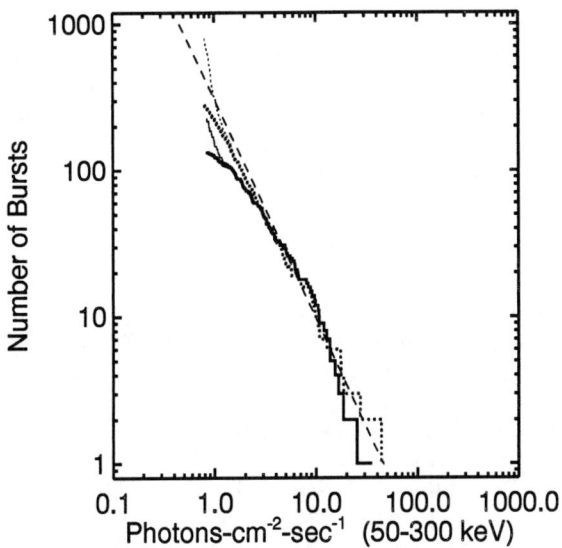

FIG. 3. Log$N$-log$P$ diagram for 135 hard (solid line) and 379 soft (dotted line) GRBs. The dashed line shows a $-3/2$ power-law.

imposed by corrections for non-uniform sky-coverage. Figure 3 indicates that there is a significant difference in the radial density distribution between hard and soft events: the latter appear clearly more homogeneous.

Another way of expressing these results is by the computation of the average $V/V_{\max}$ for these subsets. The available sample sizes are further reduced if we select by the 64 ms time scale: we now have 96 hard and 170 soft GRBs with $\langle V/V_{\max}\rangle$ of 0.338±0.029 and 0.467±0.022, respectively. These values differ by 3.5σ from one another; the hard events are 5.6σ away from homogeneity and the soft ones are consistent with homogeneity (1.5σ away). If we calculate $V/V_{\max}$ irrespective of trigger time scale, the average is essentially unchanged for the hard events but it changes dramatically for the soft events: it now is 0.354±0.017, which brings them 8.6σ away from homogeneity.

CONCLUSIONS

Although these are preliminary results, we believe we are setting the ground for the physical selection criteria for GRB subsets: *one should select both on duration and on spectral properties* in order to isolate different GRB subclasses. A more detailed analysis corroborating these qualitative results, albeit using a different selection criteria, is presented elsewhere in these proceedings (11). Here we draw the following conclusions: assuming the two subsets (hard+short and soft+long) have the same spatial distribution (i.e., they belong to the same population of objects), then soft and long GRBs are intrinsically fainter in the 64 ms timescale. In other words, we sampled farther the distribution of the hard events but we are still undersampled for the soft ones. Since these results are highly dependent on the timescale used for the $\langle V/V_{\max}\rangle$ calculation, we believe they reflect mainly the instrumental sensitivity on short timescales and not fundamental GRB properties. We are currently studying these effects in detail.

REFERENCES

1. C. Kouveliotou, et al., Ann.N.Y. Acad. Sci. **759**, 411 (1995).
2. R. Nemiroff, et al., in Gamma-ray Bursts, Eds. G.J Fishman, J.J. Brainerd & K. Hurley, AIP Conf. Proc. **307**, (AIP, New York), 237 (1994).
3. I. Mitrofanov, et al., in Gamma-ray Bursts, Eds. G.J Fishman, J.J. Brainerd & K. Hurley, AIP Conf. Proc. **307**, (AIP, New York), 187 (1994).
4. C. Kouveliotou, et al., in Gamma-ray Bursts, Eds. W.S. Paciesas & G.J Fishman, AIP Conf. Proc. **265**, (AIP, New York), 299 (1992).
5. C. Kouveliotou, et al., ApJ **413**, L101 (1993).
6. J.-P. Dezalay, et al., in Gamma-ray Bursts, Eds. W.S. Paciesas & G.J Fishman, AIP Conf. Proc. **265**, (AIP, New York), 304 (1992).
7. G. Pizzichini, in Proceedings of the XXX COSPAR General Assembly, Hamburg, Germany (1994).
8. G. Pizzichini, these proceedings (1996).
9. B.M. Belli, Astroph. Sp. Sci. **231**, 43 (1995).
10. B.M. Belli, these proceedings (1996).
11. G.N. Pendleton, et al., these proceedings (1996).

Time Profiles and Pulse Structure of Bright, Long Gamma-Ray Bursts Using BATSE TTS Data

Andrew Lee*, Elliott Bloom* and Jeffrey Scargle°

*Stanford Linear Accelerator Center, Stanford University, Stanford, CA 94309
°NASA / Ames Research Center, Moffett Field, CA 94035

> The time profiles of many gamma-ray bursts observed with BATSE consist of distinct pulses, which offer the possibility of characterizing the temporal structure of these bursts using a relatively small set of pulse-shape parameters. This pulse analysis has previously been performed on some bright, long bursts using binned data, and on some short bursts using BATSE Time-Tagged Event (TTE) data. The BATSE Time-to-Spill (TTS) burst data record the times required to accumulate a fixed number of photons, giving variable time resolution. The spill times recorded in the TTS data behave as a gamma distribution. We have developed an interactive pulse-fitting program using the pulse model of Norris et al. and a maximum-likelihood fitting algorithm to the gamma distribution of the spill times. We then used this program to analyze a number of bright, long bursts for which TTS data are available. We present statistical information on the attributes of pulses comprising these bursts.

BATSE TIME-TO-SPILL DATA

The BATSE Time-to-Spill (TTS) burst data record the time intervals to accumulate a fixed number of photons, usually 64 photons, in each of four energy channels. Relatively little analysis has been done with the TTS data, because they are less convenient to use with standard algorithms than binned data or time-tagged event (TTE) data. However, TTS data offer variable time resolution, ranging from under 50 ms at low background rates to under 1 ms in the peaks of the brightest bursts. In contrast, the finest time resolution available for binned data is 16 ms for the medium energy resolution (MER) data, and then only for the first 33 seconds after the burst trigger. The TTS data usually allow the complete time profiles of bright, long bursts—up to 16,384 spill events (over 10^6 photons) for each channel—to be stored in the limited memory on board the CGRO. This is unlike the TTE data, which are limited to 32,768 photons in all four energy channels combined (see Fig. 1). For short bursts, the TTE data have the advantages of finer time resolution, and of containing data from before the burst trigger time. Some of the shortest bursts are nearly over by the time burst trigger conditions have been met.

© 1996 American Institute of Physics

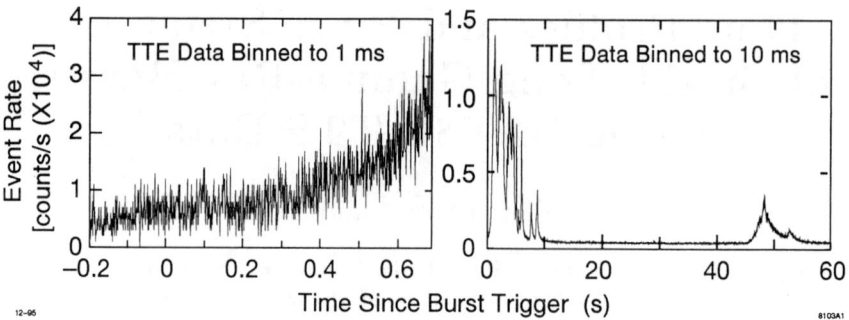

FIG. 1. BATSE Trigger Number 143, a bright, long burst.

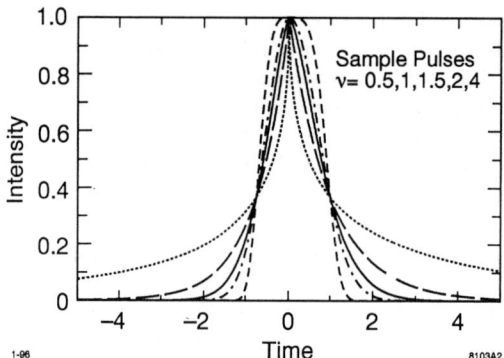

FIG. 2. Sample Pulses, $\nu = 0.5, 1.0, 1.5, 2.0, 4.0$.

The spill times recorded in the TTS data behave as a gamma distribution, for which the probability of observing a spill time t is

$$P(t) = \frac{t^{N-1} R^N e^{-t}}{\Gamma(N)},$$

where N is the number of events per spill and R is the individual event rate. This probability distribution is closely related to the Poisson distribution. For large N, the gamma distribution approaches a normal distribution, while for $N = 1$, it is the exponential distribution for individual time-tagged events.

THE PULSE MODEL

We have used the phenomenological pulse model of Norris et al. to fit burst time profiles. In this model (see Fig. 2), each pulse is described by five parameters with the functional form

$$I(t) = A \exp\left(-\left|\frac{t - t_{\max}}{\sigma_{r,d}}\right|^\nu\right),$$

where $t_{\max}$ is the time at which the pulse attains its maximum, σ_r and σ_d are the rise and decay times, respectively, A is the pulse amplitude, and ν

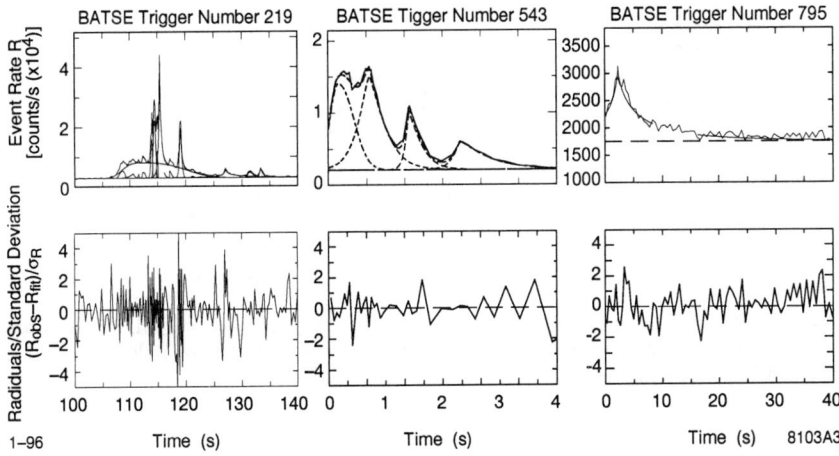

FIG. 3. Gamma-Ray Burst Time Profiles, Energy Channel 3, 100-300 keV.

(the "peakedness") gives the sharpness or smoothness of the pulse at its peak. For $\nu = 1$, the rise and decay are both simple exponentials, and for $\nu = 2$, the rise and decay are Gaussian. Pulses can, and frequently do, overlap.

PULSE-FITTED BURSTS

We have developed an interactive pulse-fitting program, written in IDL. When fitting pulses to a gamma-ray burst time profile, the user sets the initial pulse parameters, along with the initial parameters for a background with constant slope, graphically. The fitting routine uses a version of the standard IDL routine CURVEFIT, modified to perform a maximum-likelihood fit for the gamma distribution that the TTS spill times follow, rather than the usual χ^2 fit. The algorithm used by this routine is the Levenberg-Marquardt gradient-expansion method.

Each spill time in the TTS data file gives the inverse rate at the time of the spill. The program displays the burst time profile in one window, and a second window normally displays the pulse-fit residuals; that is, the difference between the observed and fitted rates for both the recorded individual spill times and after smoothing. We show only the smoothed data. Instead of the pulse fit residuals, we show the residuals divided by the standard deviation of the data (see Fig. 3).

BURST AND PULSE CHARACTERISTICS

To date, we have fitted pulses to 109 gamma-ray bursts from the BATSE 2B catalog in energy channel 3, which covers approximately 100–300 keV, for a total of 756 pulses. Of these pulses, we considered to be statistically significant only those with amplitudes that differed from zero by at least three standard deviations, as reported by the fitting routine. Sixteen pulses failed this test,

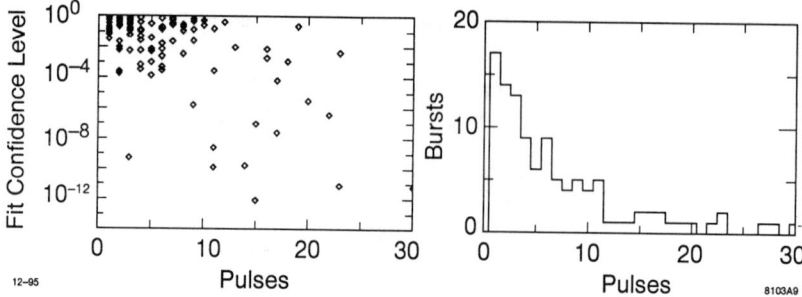

FIG. 4. (left) Fits to complex bursts requiring large numbers of pulses are more likely to have poorer confidence levels. (right) Average number of pulses per burst is 6.9 ± 6.5; the median is 5.

and only the remaining 740 pulses were used to obtain the statistics shown. The results are consistent with those found by Norris et al., fitting all four energy channels to binned BATSE data for 40 gamma-ray bursts. (See Figs. 4, 5, and 6.)

CONCLUSIONS

This phenomenological pulse model gives an accurate and compact representation of the time profiles of many of the simpler gamma-ray bursts, allowing statistical studies of their characteristics. Some issues arise for more complex bursts that must be fit using many pulses. One issue is the uniqueness of the pulse decomposition when there is large overlap between different pulses. Another is the low confidence levels of many of the fits. It would be possible to obtain fits with higher confidence levels by simply adding more pulses, but this eventually defeats the goal of a compact representation for the burst time profile. In addition, as smaller pulses are added, the statistical significance of the fit parameters decreases, eventually becoming statistically insignificant.

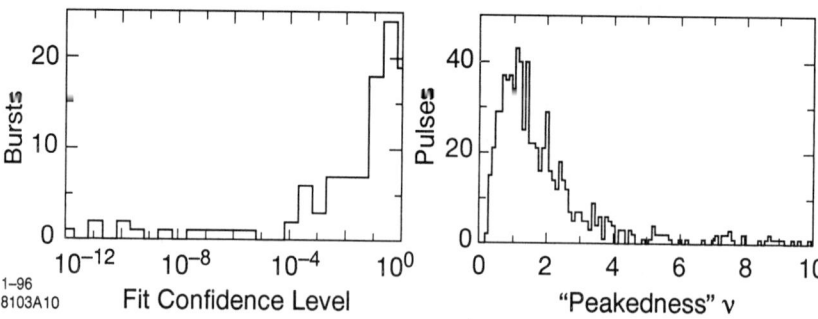

FIG. 5. Average value of "peakedness" parameter ν: 2.0 ± 1.9 Median: 1.4.

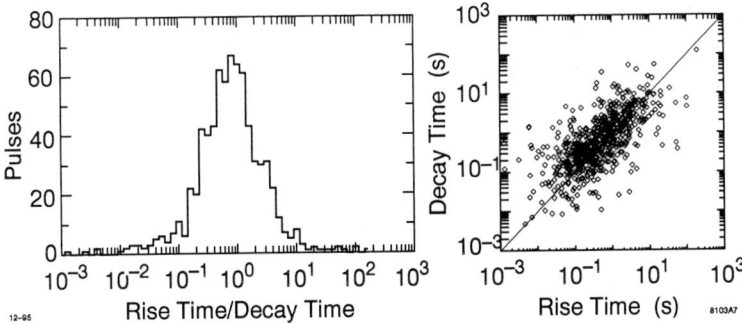

FIG. 6. The ratio of rise to decay times covered a very broad range. The geometric mean of the ratio is 0.74 with a one standard deviation range of 0.19 - 2.9, and the median is 0.77. (The usual arithmetic mean of the ratio is 2.2 ± 8.3, and of the inverse ratio is 5.5 ± 34.)

FUTURE WORK

We plan to complete this pulse analysis for all bright, long BATSE bursts for which TTS data are available, and for all four energy channels, comparing the characteristics of the different energy channels. We will also analyze the pulse-fit residuals to see if they differ from white noise, which would indicate temporal behavior not represented in this pulse model. We will also examine the use of other pulse models.

Acknowledgments: This work was supported in part by Department of Energy contract DE-AC03-76SF00515. Jeffrey Scargle acknowledges support from the NASA Astrophysics Data Program.

REFERENCES

1. U.D. Desai, Astrophys. Space Sci. **75**, 15 (1981).
2. G.J. Fishman, et al., in Proc. Gamma Ray Observatory Science Workshop, ed. W.N. Johnson, 2 (Greenbelt, MD: NASA/GSFC, 1989).
3. B. Link, R.I. Epstein, and W.C. Priedhorsky, ApJ **408**, L81 (1993).
4. C.A. Meegan, et al., BATSE Flight Software User's Manual, (Huntsville, AL: NASA/MSFC, 1991).
5. C.A. Meegan, et al., "The Second BATSE Burst Catalog," (Compton Observatory Science Support Center, 1994).
6. J.P. Norris, et al., in Gamma-Ray Bursts, AIP Conf. proc. **265**, eds. W.S. Paciesas and G.J. Fishman, 294 (AIP, New York, 1992).
7. J.P. Norris, et al., in Gamma-Ray Bursts, AIP Conf. proc. **265**, eds. W.S. Paciesas and G.J. Fishman, 172 (AIP, New York, 1992).
8. J.P. Norris, et al., submitted, (1996).

OSSE Limits on Pre- and Post-Burst Emission from GRB950421

S. M. Matz*, J. E. Grove[†], W. N. Johnson[†], and G. H. Share[†]

*Dept. of Physics and Astronomy, Northwestern University, Evanston, IL
60208-2900
[†]Naval Research Lab, Washington, DC 20375

> The search for gamma-ray burst counterparts is one of the most important fields of current investigation. Measurements of pre- and post-burst emission can provide crucial constraints on emission processes and physics. OSSE can make deep searches for precursors, recurrent burst emission, and fading or persistent counterparts from 50 keV to 10 MeV for gamma-ray bursts serendipitously in or near the field of view at the time of the burst. We present limits from one such event which was well-localized by COMPTEL/IPN, GRB950421, in terms of absolute flux, and as a fraction of the measured burst flux.

INTRODUCTION

There is no consensus on the actual nature of the gamma-ray burst sources or their energy production and radiation mechanisms. Part of the difficulty is that bursts exist as detectable sources for typically only 10–100 s. It is akin to trying to unravel the detailed physics of supernovae from the neutrino burst alone, with no knowledge of the precursor star, the SN light curve, or the ejecta/SNR.

Observation of an X-ray or gamma-ray counterpart before or after the burst could contribute significantly to identifying the nature of the source and to understanding the physics of the burst. Evidence already exists for precursors and extended emission in soft X-rays (1,2) and delayed (up to 90 min) high energy emission (3,4) in a few events. This indicates that particle acceleration or energy release is not strictly confined to the main impulsive burst. In addition, a number of theoretical calculations predict both line and continuum emission extending after the end of the primary burst. For example, in the model of Woosley (5), accretion of stellar debris may continue for hours after the burst, producing more, and harder, total emission than the initial burst.

The OSSE FOV is relatively small, but we have still observed, by chance, a number of gamma-ray bursts in our central detectors during the 5 years of the GRO mission. These events are particularly valuable because they allow us to place limits on both pre- and post-burst emission. When the burst itself is observed, these limits can be presented as fractions of the burst flux, which is physically meaningful and independent of uncertainties due to position errors.

© 1996 American Institute of Physics

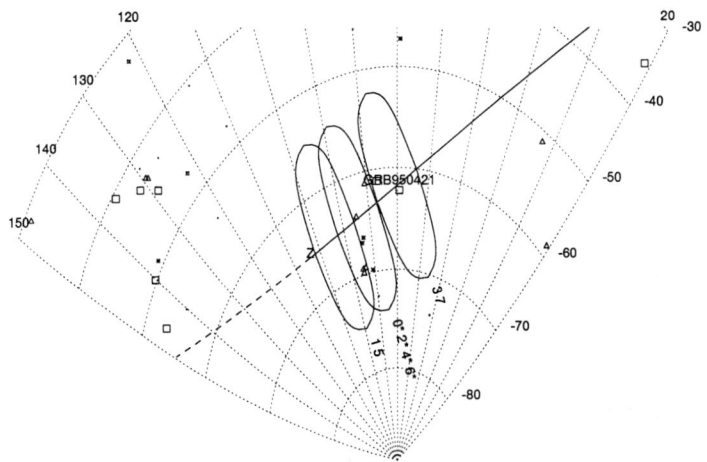

FIG. 1. OSSE chopping strategy for the observation of the source position for GRB950421. Coordinates are RA and Dec.

Some OSSE observations of FOV events have been described in earlier papers (6,7).

Here we present results from OSSE observations of GRB950421. The advantages of these OSSE observations are that they are: 1) prompt — no gap between burst and observations, 2) available before as well as after the event, 3) in the only energy range where bursters are known to emit, and 4) the most sensitive ever in this energy range. Results from GRB940301 have been previously published (8). Results of prompt slewing observations of burst locations are discussed elsewhere (9).

OSSE INSTRUMENT ON GRO

The OSSE instrument (11) on GRO consists of 4 large ($\sim$ 500 cm^2 area each) cylindrical NaI/CsI phoswich detectors, each surrounded by an active NaI shield. Tungsten collimators limit the field-of-view of the detectors to $3.8° \times 11.4°$ (FWHM). The detectors can be rotated through 192° about an axis which is parallel to the long direction of the collimator. The usual observation mode is to alternate source and background pointings with each detector every 2 minutes. The normal spectral energy range of the OSSE detectors is 50 keV to 10 MeV.

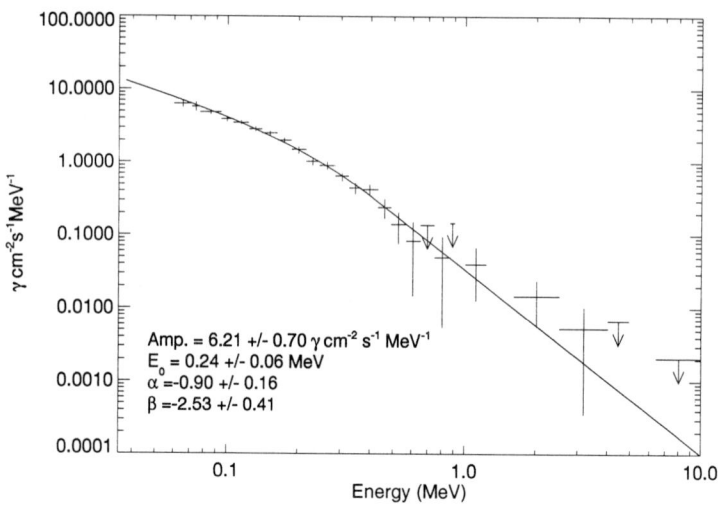

FIG. 2. Burst spectrum and best-fit Band function (10).

OSSE OBSERVATIONS OF GRB950421

GRB950421 (BATSE trigger 3516) is an example of a recent event which occurred in the OSSE field-of-view. It was seen by OSSE in its central detectors to > 300 keV. The combined COMPTEL/IPN position (RA: 69.4, Dec: −62.2) is near the center of the FOV ($\sim$ 90% response) for a background pointing of an LMC X–3 observation being conducted at the time of the burst (Figure 1). A modified pointing table designed to improve response to the preliminary COMPTEL position was manually uploaded at trigger +2.5 hr. After $\sim$ 7 hr OSSE returned to its scheduled observation. In all, the source position was observed by OSSE from trigger−10 days to trigger+4 days.

Figure 2 shows the OSSE spectrum accumulated over the 16.4 s interval which includes the burst. Overplotted is the best-fit Band function (10). Note that uncertainty in the burst location produces some additional systematic uncertainty in the unfolding and fitting which is not reflected in the errors on the data points or fitted parameters.

GRB950421: PRE- AND POST-BURST LIMITS

Figure 3 shows the 3σ upper limits derived for average flux from the nominal position of GRB950421, integrated over a number of time scales, before and after the burst. The source position is occulted by the Earth each orbit, so we have chosen integration times corresponding to unocculted periods: from source rise to just before the burst (15 min duration), just after the burst to source set (2–21 min after the burst), and the next orbit (77–112 min post-burst). In addition, two longer integrations include all the data before and after the burst. Also shown is the measured flux from the burst itself, averaged over the 16.4 s spectral accumulation time. These limits are significantly better than those derived for GRB940301 (8) because the burst location is nearer the center of the OSSE FOV. As with the burst spectrum, there is some additional uncertainty in the absolute flux limits since these depend on the position assumed for the source. The upper limits as a fraction of the averaged burst flux are listed in the Table 1; these limits are not sensitive to the burst location.

The results can also be stated as limits on the *total fluence* in the integration interval vs. the total fluence in the burst (Table 2). This is a more direct way to compare total energy output in the different phases. These results imply that the total fluence in the lower energy ranges immediately before and after the burst must be less than 1.3 times the total burst fluence. The upper limit in the next orbit is still roughly comparable to the total burst fluence.

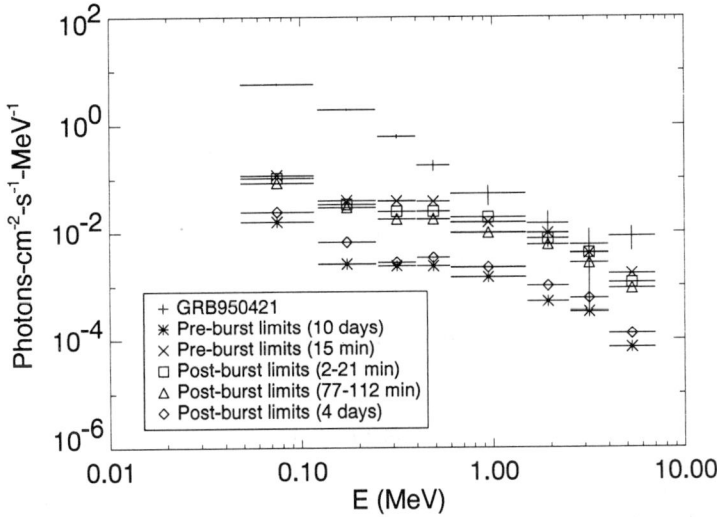

FIG. 3. Burst spectrum and 3σ upper limits on average flux for GRB950421 on a number of time scales, based on the OSSE observations.

CONCLUSIONS

OSSE has observed a number of gamma-ray bursts in or near our field of view. For GRB950421 we have derived limits on absolute flux from the source location both before and after the event. We have also given limits as a fraction of the measured burst flux. At low energies (< 250 keV), the

TABLE 1. GRB950421: Limits as a Fraction of Burst Flux

Energy Range (MeV)	Averaged GRB Flux (16.4 s) ($\gamma/cm^2/s/MeV$)	(3σ Upper limits)/(GRB flux)			
		Pre-GRB (15 min)	Post-GRB (2–21 min)	Post-GRB (77–112 min)	Post-GRB (4 days)
0.05–0.12	5.7 ± 0.2	0.02	0.02	0.015	0.004
0.12–0.25	2.0 ± 0.1	0.02	0.02	0.015	0.003
0.25–0.40	0.61 ± 0.05	0.07	0.04	0.029	0.005
0.40–0.60	0.18 ± 0.04	0.21	0.14	0.098	0.019
0.60–1.5	0.05 ± 0.02	0.30	0.36	0.19	0.043

TABLE 2. GRB950421: Limits as a Fraction of Total Burst Fluence

Energy Range (MeV)	(3σ Upper limits)/(GRB fluence)			
	Pre-GRB (15 min)	Post-GRB (2–21 min)	Post-GRB (77–112 min)	Post-GRB (4 days)
0.05–0.12	1.3	1.3	1.8	90.
0.12–0.25	1.3	1.2	1.8	72.
0.25–0.40	4.0	2.9	3.5	96.
0.40–0.60	13.3	9.8	11.8	400.
0.60–1.5	18.8	26.1	22.5	910.

integrated flux immediately (15–20 min) before and after the burst must be less than 1.3 times the total fluence in the ~ 10 s of the burst.

REFERENCES

1. A. Yoshida, et al., PASJ **41**, 509 (1989).
2. T. Murakami, et al., Nature **350**, 592 (1991).
3. M. Sommer, et al., ApJ **422**, L63 (1994).
4. K. Hurley, et al., Nature **372**, 652 (1994).
5. S. E. Woosley, ApJ **405**, 273 (1993).
6. G. H. Share, et al., in Gamma-Ray Bursts, eds. G. J. Fishman, J. J. Brainerd, and K. Hurley, AIP Conf. Proc. **307**, 283 (AIP, New York, 1994).
7. L. Hanlon, et al., A&A **296**, 41 (1995).
8. S. M. Matz, et al., Ap&SS **231**, 123 (1995).
9. S. M. Matz, et al., Ap&SS **231**, 127 (1995).
10. D. Band, et al., ApJ **413**, 281 (1993).
11. W. N. Johnson, et al., ApJS **86**, 693 (1993).

Quantifying Variability of GRB Light Curves Using Multifractal Analysis

D. C. Meredith, J. M Ryan, and C. A. Young

Physics Department, University of New Hampshire, Durham, NH 03824-3568

Gamma Ray Burst light curves have strikingly different morphologies, ranging from a smooth single pulse to a complex series of overlapping pulses. We present multifractal analysis as a method for quantifying the variability in their morphology. Multifractal analysis was chosen because it measures the structure on different time scales which is typical of many GRBs. We provide an estimate of the error on this variability measure and look for correlations with other GRB characteristics including intensity and duration. This new measure may help to classify bursts, as well as identify repeaters and gravitationally lensed partners.

INTRODUCTION

One striking difference between different GRB time series is their morphology. Some bursts are very simple (e.g., the FRED's: Fast Rise, Exponential Decay) while others show great complexity. The variability, therefore, is as much a part of the character of a burst as is its peak intensity or duration, and a robust objective measure of that variability is needed.

There has been previous work on variability. The work of Graziani and Lamb (1) which uses the ratio of $C_{max}^{64}/C_{max}^{1024}$ as a measure of variability has been criticized for having an intrinsic correlation with burst length (2,3). Lestrade (4) uses a measure based on numbers of runs up and runs down as a measure of variability. The multifractal analysis provides a qualitatively different measure of variability.

FUNDAMENTALS OF MULTIFRACTALS

Since multifractal analysis is not part of the standard astrophysicist's tool box, we present a short description of this method. To begin, fractals are sets that show detail on every scale. The canonical example is the coastline of Alabama which is very rough whether you look at it on a kilometer scale or a meter scale. Similarly, multifractals are measures on sets which show detail on every scale. (In our case the set is the total time interval of the burst, and the measure is the count rate.) A quick look at some of the more complicated GRB light curves shows that they have many time scales and this suggests that they may be amenable to multifractal analysis.

© 1996 American Institute of Physics

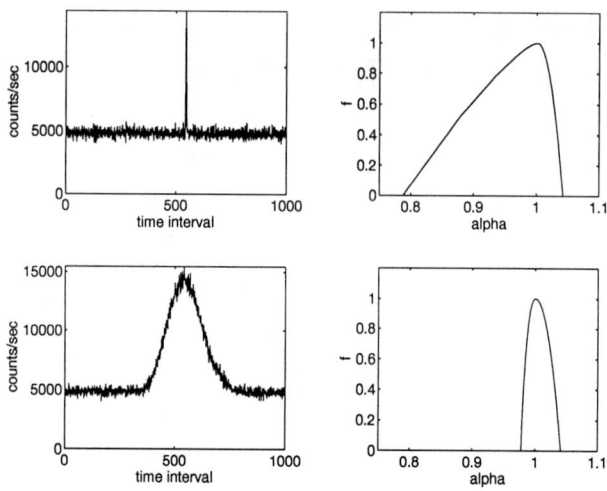

FIG. 1. In the left hand column are two lognormal curves with added Poisson noise and widths that differ by a factor of twenty. In the right hand column are the associated $f(\alpha)$ curves. The narrow peak gives a much wider $f(\alpha)$ curve, and is therefore judged more variable.

In order to investigate these different scales, the whole set (total time interval) is broken into smaller and smaller pieces, which we will call bins. The parameter α tells how a particular measure scales with bin size: measure=(bin size)$^\alpha$ (Full mathematical details are given in the review paper by Mandelbrot (5).) Therefore α values greater than one indicate that the measure decreases quickly as bin size decreases, while those values less than one indicate that the measure decreases slowly as bin size decreases (bin sizes are always less than one, α is always greater than zero). In other words, small values are associated with the large, narrow peaks; large values are associated with the small, narrow valleys.

We then introduce a new variable f which is a function of α. (see Figure 1, left hand column). The value of $f(\alpha_0)$ gives the fractal dimension of the underlying set with α in the interval $[\alpha_0, \alpha_0 + d\alpha]$. For GRBs, the underlying set is the time line, hence the maximum value of f is one. The wider the $f(\alpha)$ curve, the greater the variability of the data because of the greater range of α values. Smooth curves, on the other hand, generate $f(\alpha)$ curves made of the single point $(1,1)$. This agrees with our intuition for the variability of GRBs: (ignoring noise for the moment) the smooth FRED bursts are not variable while complex bursts with many narrow spikes are. Given that the width measures the variability, we will take α_{min} as the single measure of variability; the smaller the value of α_{min}, the larger the variability.

DETAILS OF MULTIFRACTAL ANALYSIS

We used the canonical method of Chhabra et al. (6) to calculate the $f(\alpha)$ curve for GRBs. This method begins by calculating new normalized measures:

$$\mu(q,\epsilon)_i \equiv \frac{p_i(\epsilon)^q}{\sum_j p(\epsilon)_j^q} \quad (1)$$

where $p_i(\epsilon)$ is the fraction of the measure in the i'th bin of size ϵ. For our application, the largest value of ϵ is the total time interval of the burst, the smallest value is 64 msec (the BATSE DISCSC data type time resolution). From this variable μ new quantities are made:

$$g(q,\epsilon) \equiv \sum_i \mu(q,\epsilon)_i \log(\mu(q,\epsilon)_i) \qquad \beta(q,\epsilon) \equiv \sum_i \mu(q,\epsilon)_i \log(p(\epsilon)_i) \quad (2)$$

A least squares fit to a straight line (with equal weights) is done for both of the above variables as a function of $\log(\epsilon)$. The slope of each fit is denoted as $f(q)$ and $\alpha(q)$, and we have a parametric representation of $f(\alpha)$. It is nontrivial to see that these quantities match the description given for them in the introduction.

In order to complete our analysis, we need an estimate of the error in the $f(\alpha)$ curve. We assume that the majority of the error in the data is from counting statistics. However, we cannot easily propagate these errors to get errors on f and α. To understand the difficulty, note that the data points to which we are fitting a straight line are not independent, hence the usual error analysis for least squares fitting does not apply. Instead, we run the multifractal analysis on the burst four times: once with the original data, and three times with three different renoisings of the data (that is, we take the data and add Poisson noise to them). The purpose of this is to simulate taking the data four different times. We take the range in α values as an error estimate on the α value of the true data.

MULTIFRACTAL ANALYSIS OF GRBS

In order to understand our results better, we ran the analysis on a set of noisy lognormal curves, since lognormals have the general shape of a GRB pulse. We found that $f(\alpha)$ was independent of pulse asymmetry, number of pulses, and weakly dependent on pulse height (i.e., the slope near the top of the curve was smaller for larger bursts), but strongly dependent on pulse width (see Figure 1).

We then calculated $f(\alpha)$ curves for the strongest 84 bursts in triggers 105 – 999. We looked for correlations of α_{min} with other parameters, and found it to be uncorrelated with duration, location, but correlated with total peak flux (see Figure 2). We see that both the lower left and upper left corners of Figure 2 are empty. The lower left corner corresponds to bursts with high variability and low peak flux. We believe that this correlation is real: a burst

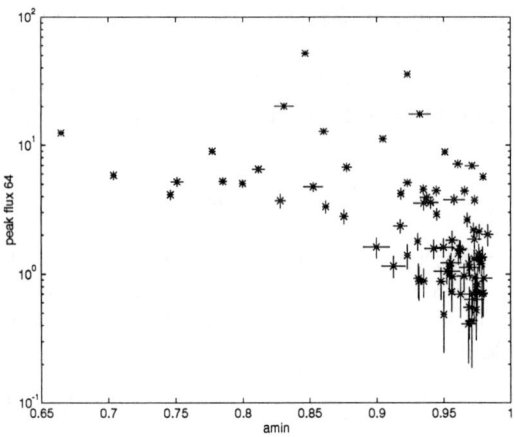

FIG. 2. Plot of variability vs. peak fluence.

with low peak flux has no large narrow spikes, and is therefore not variable. We also see no bursts with high peak flux and high variability. We suspect that this may be simply a matter of low statistics: The upper left hand corner of the plot is probing the tails of both the distribution in peak flux and in variability, and is therefore not expected to have many counts.

We show in Figure 3 the results for four very different GRB (triggers 219, 160, 249, and 332, 100−300 keV energy range) to test if the multifractal results agree with our intuitions. Trigger 332 is a FRED (fast rise, exponential decay) and is the smoothest to the eye; it also has the narrowest $f(\alpha)$ curve. Triggers 160 and 219 appear equally variable (according to the multifractal analysis), even though 160 has more spikes. This is consistent with the test runs that show that only narrowness of spikes, not number, increases the variability. Trigger 249 is judged less variable since its tallest pulses are much wider than in 219 or 160.

CONCLUSION

We have presented here a new measure of variability which is most sensitive to the presence of narrow pulses. This makes it distinctly different than runs up and runs down of Lestrade, or the measure of Graziani and Lamb. We are able to estimate the error of this measure by simulating different data runs. This measure does not correlate with position, duration, but does correlate with peak flux since bursts with small peak fluxes are lacking in narrow pulses. Our future work will look at more bursts, and investigate using this measure of variability for searching for repeaters and lensed partners.

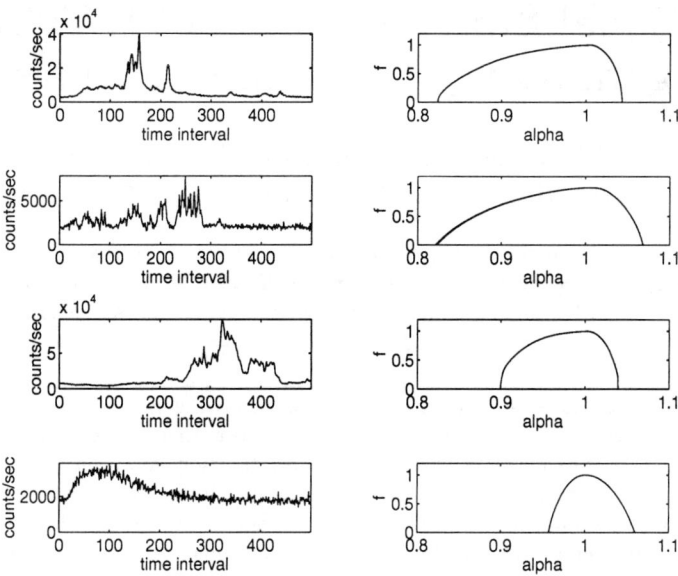

FIG. 3. The left hand column shows the light curve between the 100–300 keV energy range of burst triggers 219, 160, 249, and 332; the right hand curve shows the $f(\alpha)$ curve for each event.

Acknowledgements. This work was supported through NASA contract NAS5-26645 and through NASA Compton GRO Guest Investigation under Grant NAG5-2731.

REFERENCES

1. C. Graziani and D. Q. Lamb, in *Gamma-Ray Bursts Second Workshop*, edited by G. Fishman, J. J. Brainerd, and K. Hurley (AIP, **307** New York) 227 (1993).
2. R. Rutledge and W. H. G. Lewin, in *Gamma-Ray Bursts Second Workshop*, edited by G. Fishman, J. J. Brainerd, and K. Hurley (AIP, **307** New York) 232 (1993).
3. C. Meegan and C. Kouveliotou, in *Gamma-Ray Bursts Second Workshop*, edited by G. Fishman, J. J. Brainerd, and K. Hurley (AIP, **307** New York) 232 (1993).
4. J. P. Lestrade, in *Gamma-Ray Bursts Second Workshop*, edited by G. Fishman, J. J. Brainerd, and K. Hurley (AIP, **307** New York) 212 (1993).
5. B. Mandelbrot, *PAGEOPH* **131**, 5 (1989).
6. A. B. Chhabra, C. Meneveau, R. Jensen, and K. R. Sreenivasan, *Phys. Rev. A* **40**, 5284 (1989).

"Equal flux" averaging of BATSE gamma-ray burst time profiles

I. Mitrofanov[1], A. Pozanenko[1], A. Chernenko[1],
M. S. Briggs[2], R. D. Preece[2], W. S. Paciesas[2], G. J. Fishman[3]

[1] *Space Research Institute (IKI), Profsoyznaya 84/32, 117810 Moscow, Russia,*
[2] *Dept. of Physics, University of Alabama in Huntsville, Huntsville, AL 35899,*
[3] *NASA Marshall Space Flight Center, Huntsville, AL 35812*

We introduce new parameters for characterizing gamma ray burst time profiles. The method is based on estimating the average parameters of gamma-ray bursts (e.g., pulse duration, duration between pulses, hardness ratio) for corresponding flux intervals. The profiles of all bursts are normalized to unity at their count rate peaks, then the parameters are calculated for each particular dimensionless flux interval. In this paper we discuss application of the method to bursts of the 3B Catalog.

INTRODUCTION

The time profiles of gamma-ray bursts show the most statistically significant information about the bursts. However GRB time profiles are complex and and have resisted classification. Therefore statistical analysis of profiles is an attractive approach to investigate the time/structure signatures of bursts. The bimodal duration distribution (2) is almost the sole general time signature of bursts. On the other hand, comparison of the profiles of different burst sets may help distinguish groups of bursts. The method of peak aligned averaging have been used for analysis of GRB temporal properties (3). Parameterization of pulses was used to investigate burst profiles and separate pulse asymmetry was found (1,5).

Here we use a non-parametric method called "equal flux" averaging and use it for time/structure evaluation. While the peak alignment method (4) estimates the average flux for successive time bins around the peak, this method estimates the average time duration for corresponding flux intervals. The method can be used for the calculation of the distribution of burst parameters at a particular flux interval. We use this method also to compare bright and dim gamma-ray bursts.

DEFINITIONS

The flux of any burst varies between its highest peak and zero at the background level. If a time profile is normalized to unity at the main peak, one

© 1996 American Institute of Physics

may compare time histories of all bursts at the same level of dimensionless flux f, where $f \in [0, 1]$. For any fixed f_* one can find all times during which the burst flux lies above this level and we define the Pulse Duration t_{PD} as the total duration of all episodes above f_*. We define the Equivalent Pulse Width t_{EPW} as the integral under the normalized profile for all times that the normalized flux exceeds the level f_*. The number of intersection of the time profile with the level $f = f_*$ is twice the number of peaks (n) above f_*. The total time between all peaks at the level f_* is the Valleys Duration t_{VD}.

For every burst we calculate these parameters as well as the mean pulse duration t_{PD}/n and the mean valley duration (or mean interval between peaks) $t_{VD}/(n-1)$. Some of the bursts may have only one peak, in which case the last ratio is not defined.

DATA ANALYSIS AND RESULTS

In present analysis we use two datasets. The first set consists of 64 ms Large Area Detector data (DISCSC + PREB data types) for 273 GRBs of the 2B catalog, while the second set consists of 1024 ms Large Area Detector data (DISCLA) for 625 bursts of the 3B catalog with $T_{90} > 1$ s. Bursts with data gaps were excluded from both sets. In both sets we use time profile in channels 2+3 ($\sim 50 - 300$ keV).

The choice of the value of f_* is somewhat arbitrary. However, the smaller f_*, the more that parameter values will be subject to statistical uncertainties due to background fluctuations. The most susceptible parameters are t_{VD} and n. It is evident also that parameters derived from 64 ms data (versus 1024 ms data) are more to statistical uncertainties. By varying the strength of the data profile we have found that this bias is less than 15% at $f_*=0.4$ for t_{VD} and n. For 1024 ms data the bias at the level $f_* = 0.4$ is less than 5% for all parameters.

The distributions of t_{PD} and t_{EPW} at the level $f_* = 0.4$, derived using 64 ms data, are presented in Figure 1 and Figure 2, respectively. The left side of each distribution has a sharp cutoff due to the temporal resolution (64 ms) of the data used. All single peak events with durations less than 64 ms fall in the left-most bin of the distributions.

It was found that the T_{90} distribution is bimodal with a minimum near 2 s (2). It is evident from Figures 1, 2 that right peak of the t_{PD} and t_{EPW} distributions consists of GRBs with $T_{90} > 2$ s while the left peak consist mainly of GRB with $T_{90} < 2$ s.

Therefore, the bimodality of the distributions is present for both t_{PD} and t_{EPW} parameters. The T_{90}/short mode mainly consists of the single-pulse bursts. The T_{90}/long mode contains single-pulse events as well as multi-pulse bursts.

Figure 3 presents the distribution of the mean of the intervals between peaks $t_{VD}/(n-1)$ for 64 ms data. Because single peak bursts have no valley this figure includes only multi-peak events ($n > 1$), and therefore it represents the T_{90}/long mode mainly. The distribution of $t_{VD}/(n-1)$ presented in the

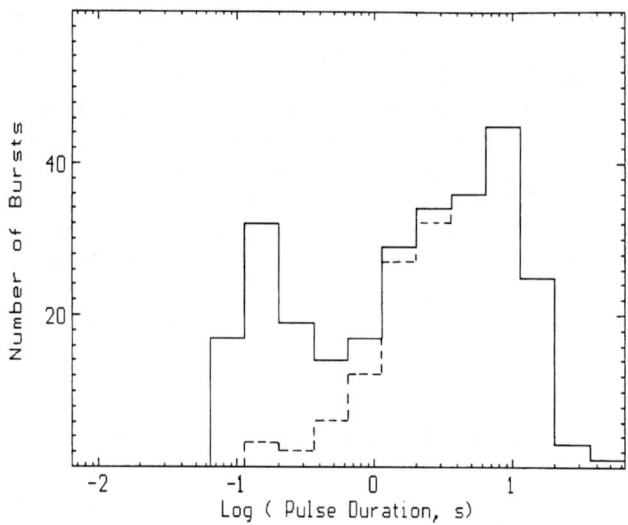

FIG. 1. Distribution of GRB Pulse Duration t_{PD} based on 64 ms data. Solid line: all GRBs, Dashed line: GRBs with $T_{90} > 2$ s.

Figure 3 is flat or unimodal rather than bimodal.

We subdivided each set of bursts (64 ms and 1024 ms data types) into two groups: bright and dim bursts. Each group consists of approximately an equal number of bursts, and the separation is based on peak flux measured at the same time scale as the data. The logarithmic means of t_{PD}, t_{EPW} and $t_{VD}/(n-1)$ are shown in Tables 1 and 2.

There is no significant difference in the mean values of the parameters of bright and dim bursts for all of the cases presented in the Tables.

TABLE 1. Logarithmic means of t_{PD}, t_{EPW} and $t_{VD}/(n-1)$ at $f_* = 0.4$ for 64 ms data type, seconds

Parameter	bright bursts	dim bursts
t_{PD}	$1.10^{+0.19}_{-0.17}$ [a]	$1.55^{+0.27}_{-0.23}$
t_{EPW}	$0.74^{+0.11}_{-0.09}$	$1.10^{+0.18}_{-0.16}$
$t_{VD}/(n-1)$	$0.58^{+0.11}_{-0.09}$	$0.66^{+0.13}_{-0.11}$

[a] Estimation of errors based on a sample variance.

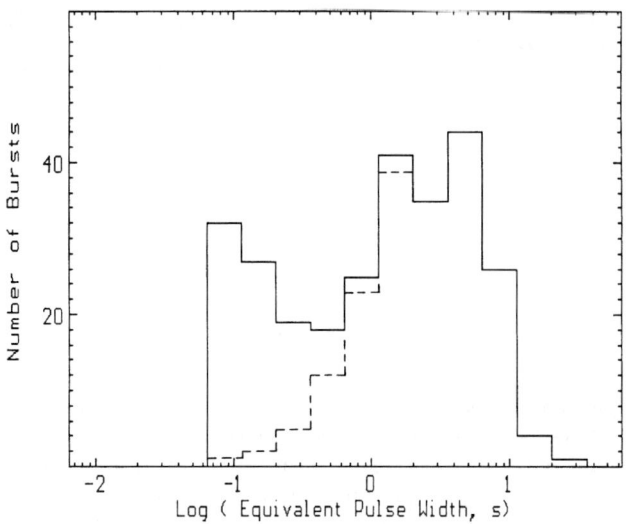

FIG. 2. Distribution of GRB Equivalent Pulse Width $t_{\rm EPW}$ based on 64 ms data. Solid line: all GRBs, Dashed line: GRBs with $T_{90} > 2$ s.

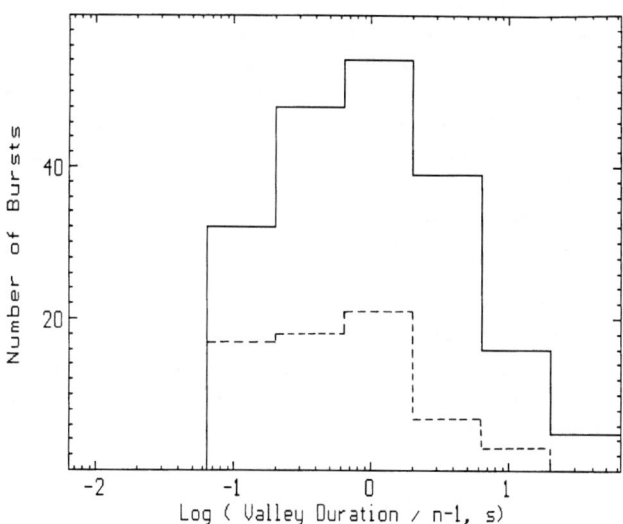

FIG. 3. The distribution of $\log t_{\rm VD}/(n-1)$ based on 64 ms data and $f_{\sharp} = 0.4$. Solid Line: all GRBs, Dashed line: bright GRBs (peak flux > 2 phot cm^{-2} s^{-1} on the 64 ms time scale)

TABLE 2. Logarithmic means of $t_{\rm PD}$, $t_{\rm EPW}$ and $t_{\rm VD}/(n-1)$ at $f_* = 0.4$ for 1024 ms data type, seconds

Parameter	bright bursts	dim bursts
$t_{\rm PD}$	$5.89^{+0.42}_{-0.39}$	$6.92^{+0.32}_{-0.31}$
$t_{\rm EPW}$	$4.26^{+0.31}_{-0.29}$	$4.90^{+0.23}_{-0.22}$
$t_{\rm VD}/(n-1)$	$4.57^{+0.56}_{-0.50}$	$4.79^{+0.46}_{-0.42}$

DISCUSSION

We introduced the Equivalent Pulse Width, Valleys Duration and Number of apparent peaks as measures of burst time/structure. Investigation of these parameters can reveal intrinsic features of bursts on a particular time scale. Along with the peak alignment method, these parameters can be used to search for GRB time signatures.

Similarly to the bimodality of the T_{90} distribution (2), $t_{\rm EPW}$ and $t_{\rm PD}$ manifest bimodal distributions. On the other hand, the distribution of $t_{\rm VD}/(n-1)$ does not show any feature and corresponds to a flat distribution between 64 ms and 6 s. Each burst consists of some number of peaks $n > 0$. If a characteristic peak duration existed, then values of $t_{\rm PD}$ and $t_{\rm EPW}$ would be multiples of such a characteristic time.

One might assume that the bimodality of T_{90}, $t_{\rm EPW}$ and $t_{\rm PD}$ distributions represents the existence of two physical groups of bursts with different time profiles. One group consists of a few (or single) pulse events with narrow pulses, which are responsible for the left mode of the bimodal distributions of T_{90}, $t_{\rm EPW}$ and $t_{\rm PD}$. The other group contains both single-pulse and multi-peak events, and average pulses are much broader for this group. It populates the right peak of bimodal distributions. The difference in time profiles of these two modes will be studied in more detailes elsewhere (6).

Comparison of bright and dim subsets reveal no significant difference between subsets. Therefore estimations of $t_{\rm PD}$ and $t_{\rm PD}$ indicate that the present data of the BATSE experiment is consistent with no, or very small stretching between bright and dim bursts. However, the mean values of bright bursts are systematically less than the mean values of dim bursts, which needs to be studied with more data sets.

REFERENCES

1. S. P. Davis, J. P. Norris, in AIP Conf. Proc. **307**, (New York: AIP), 182 (1994).
2. C. Kouveliotou, et al., ApJ **413**, L101 (1993).
3. I. G. Mitrofanov, et al., in AIP Conf. Proc. **307**, (New York: AIP), 187 (1994).
4. I. G. Mitrofanov, these Proceedings (1996).
5. J. P. Norris, et al., submitted to ApJ (1995).
6. I. G. Mitrofanov, et al., in preparation (1996).

Peak-to-Peak Dilation of Gamma Ray Bursts

Jeff Neubauer and Bradley E. Schaefer

Yale University, Physics Department, JWG 463, New Haven CT 06520-8121

If Gamma Ray Bursts are at cosmological distances, then the faint bursts must have light curves which are stretched in time when compared to bright bursts. Such a time dilation has been reported by Norris and coworkers, where the relative stretching of pulse durations is roughly a factor of two between bright and faint burst. Nevertheless, models have been presented which can account for this peak-duration dilation within galactic models. A solution to this ambiguity can be had by looking for dilation of different measures of time scale, as these would not be subject to the same effects. One such time scale is the peak-to-peak time. Cosmological models can only be true if the peak-to-peak dilation agrees exactly with the peak-duration dilation. This is a sharp test.

In this paper, we report on our study of peak-to-peak dilation of 465 multipeaked bursts from the BATSE 2B catalog. We used a wide variety of peak selection criteria and peak-to-peak time scale definitions. We find that many tests can find dilations with a factor of two, yet that many other tests can prove that there is no dilation. That is, by selecting which test to adopt, we can make peak-to-peak dilation either appear or disappear at high levels of significance. We conclude that the measurement of dilation has subtle systematic errors of unknown nature, and these can only be recognized by detailed Monte Carlo analyses.

INTRODUCTION

One of the principal questions concerning Gamma Ray Bursts (GRBs) is their typical distances. The GRB community is divided as to whether bursters form a halo around our Milky Way galaxy or are at cosmological distances. A distinct prediction of the cosmological models is that the light curve shapes of the distant bursts must be stretched in time when compared to nearby bursts.

Norris et al. (1) have studied the peak durations of GRBs and find a time stretching factor of ~2 between the bright and faint bursts. Such dilation would presumably place the faint bursts at distances with cosmological z of around 1 (but see (2)). Such distance scales are in agreement with those deduced from the LogN-LogP curve as well as that required by the available energy from neutron stars. At face value, this discovery appears to prove the cosmological origin of GRBs. Nevertheless, two types of challenges have been presented for this conclusion.

The first challenge is that Mitrofanov et al. (3) have not been able to reproduce the claimed dilation. In fact, their results are consistent with no dilation at all. This discrepancy is disturbing. At the Hunstville meeting, both sides agreed that the discrepancy depends on the details of the selection for which bursts are included in the bright and faint classes. Until this detail is resolved, the existence of dilation in GRBs must remain undetermined.

The second challenge is that explanations for dilation have been advanced within the framework of galactic models (4). Here, a beaming model can produce a brightness/duration dependence in a natural manner for any volume limited distribution of bursters. Such an explanation leaves the resulting apparent dilation as a free parameter, so the match between Norris' dilation factor and that expected for cosmological models can only be called a coincidence. This particular beaming model only works for the durations of individual pulses.

A sharp test of these two challenges can be made by searching for peak-to-peak dilation in GRBs. Within cosmological models, all time scales in GRB light curves must be dilated by the same factor. Within the galactic beaming model, the peak-to-peak time is governed by some 'clock' within the source itself, and the selection effects on this time scale are likely to be greatly different from those of the peak-duration. Thus, cosmological models require that the peak-to-peak dilation agree exactly with the peak-duration dilation and that this dilation be a factor of roughly two between the brights and the faints, while galactic models expect that the dilations for peak-to-peak and peak-duration should be different. As such, the peak-to-peak dilation can solve the GRB distance problem.

OBSERVATIONS

We have used the bursts in the BATSE 2B catalog, as presented on the CDROM version available from the Compton Observatory Science Support Center. For our first study, we have used the DISCLA data type from the BATSE Large Area Detectors. This provides 64 ms time bins, but only after the BATSE trigger. To avoid any biases based on this asymmetric binning, we will not consider peaks that occur before the highest peak. We also 'noisified' each light curve as prescribed by (1) to avoid any biases related to peak flux. We summed BATSE discriminator channels 1, 2, and 3 so as to get the best signal-to-noise.

We used one procedure (with choices for several input parameters) for identifying the time and peak flux of peaks within a given light curve. This procedure operated on the 'noisified' light curve, where the fiducial peak flux level was itself an input parameter. The first task was to identify local maxima and minima. Adjacent peaks were then combined together when the individual peaks were found to be indistinct or insignificant. Peaks were distinct if separated by a valley whose depth was at least some fraction (typically 33%) of the highest peak flux below the lowest of the peaks. Peaks were considered significant if they exceeded the background level by some fraction (typically 33%) of the highest peak flux. We explored a variety of parameters for the fiducial flux, the distinctiveness criterion, and the significance criterion.

For each choice of parameters, we derived the times and fluxes for all peaks in the bursts of the BATSE 2B catalog. Detailed checking of our program results with light curve plots by eye, demonstrated that our algorithm was indeed selecting peaks much as a human might. About half of the BATSE 2B bursts had only one peak, and these were subsequently not used in the analysis. A typical number of remaining bursts was 465.

We used three definitions for converting these derived peak times into a peak-to-peak time. Our first method was to simply use all times between successive peaks. Thus, a burst with N peaks would provide N-1 peak-to-peak times, all of which have equal weight. In this case, the times from multipeaked bursts will dominate the statistics. Our second method was to average all the times between successive peaks for each individual burst. Thus, a burst with N peaks and a time of T between the first and last peaks will yield a peak-to-peak time of T/N. Our third method was to consider only the time between the highest peak and the second highest peak. Keep in mind that we do not consider peaks before the highest peak.

Each burst was placed into a brightness bin. The brightness was taken as the peak flux with 256 ms time bins as tabulated by the BATSE Team. We created four bins, each with one quarter of the bursts. This utilization of all bursts is better than just creating two classes (brights and faints) because more bursts are used, brightness gaps are avoided, and bins are sufficiently narrow to avoid dilution of any cosmological effect.

The peak-to-peak time for each bin was identified as the median value. (Similar results were obtained when all the logarithms of the times were averaged together.) The rms scatter of all logarithmic values within each bin was also calculated. The uncertainty in the time for each bin was taken to be the rms scatter divided by the square root of the number of times in each bin. (The scatter resulting from Poisson errors in the photon statistics are always much smaller.) This procedure results in the median peak-to-peak times (τ) for GRBs in four brightness bins centered around a median flux (F).

RESULTS

A logarithmic plot of τ as a function of F should show any dilation present. Cosmological models require that the faint end be roughly a factor of two longer than the bright end, which is 0.3 on a log plot. Galactic models are likely to yield a different slope, with a zero slope being the expected value in the absence of selection effects.

The table above lists the logarithm of the median peak to peak times (in seconds) for each of the four brightness bins. The logarithm of the median brightness within each bin is 4.35, 4.01, 3.77, and 3.61 from brightest to faintest. The three lines are for peak-to-peak times as defined by our three methods.

Peak-to-Peak definition	Bright Bin	2nd Bin	3rd Bin	Faint Bin
Between successive peaks	0.68 ± 0.07	0.72 ± 0.08	0.81 ± 0.07	1.03 ± 0.07
Average	0.90 ± 0.08	0.89 ± 0.11	0.83 ± 0.07	1.07 ± 0.08
Second highest to highest	1.11 ± 0.08	1.18 ± 0.09	1.20 ± 0.08	1.21 ± 0.08

These values can be tested for consistency with no dilation by means of a chi-square test. The chi-squares for the three definitions are 14.6, 5.3, and 1.0 for three degrees of freedom. The first case is inconsistent with no dilation to a high degree of significance, while the other two cases are easily consistent with no dilation. In the first case, the apparent dilation is by a factor of 2.2 (0.35 in the logarithm) just as expected for cosmological models.

Similar discrepancies were found based on alternative choices of the peak identification criteria. That is, the variation of parameters within reasonable ranges can make dilation appear or disappear. However, it may be significant that we have never found a significant case of 'anti-dilation'. The cases we examined had apparent dilation factors ranging uniformly from 1.0 (no dilation) to around 2.5 (cosmological dilation).

DISCUSSION

The evidential existence of peak-to-peak dilation is extremely sensitive to fine choices of peak indentification criteria and the definition of the peak-to-peak time. In such a case, we conclude that the measurement of the peak-to-peak time is being dominated by some unknown systematic effects. Until these effects are identified and controlled, the peak-to-peak dilation measurement cannot fulfill its potential for solving the GRB distance scale.

The peak-duration dilation question is currently faced with a similar problem, in that slight variations of definitions can make or break cosmological dilation. So we conclude that the peak-duration measures must have similar systematic effects which preclude any current decision on the GRB distance scale.

How can these deadlocks be broken? The only means that we know of, is to test the procedures with simulated data where the dilation factor has been inserted by hand. Such a test would have to be performed in a Monte Carlo sense for many randomly created simulated data bases. These simulated data bases should be constructed from bright bursts which are dimmed and dilated and redshifted in accordance with some model. The entire procedure (as applied to the real BATSE data) should then be applied to the simulated data. The resulting dilation factor can then be compared with the known answer. When this is repeated for many procedures and parameter choices, it should become clear which techniques reliably reproduce the input answer. Then these reliable techniques should be applied to the actual BATSE data.

In summary, the peak-to-peak dilation of GRB light curves has a strong potential of solving the GRB distance scale question. Nevertheless, current measures give ambiguous results, and definite results are likely to come only after extensive Monte Carlo studies.

REFERENCES

1. J.P. Norris, et al., ApJ **424**, 540 (1994).
2. E.E. Fenimore, and J.S. Bloom, ApJ in press (1995).
3. I.M. Mitrofanov, et al., in Gamma Ray Bursts, eds. G. J. Fishman, J. J. Brainerd, and K. Hurley, AIP Conf. Proc. **307**, 187 (AIP, New York, 1994).
4. J.J. Brainerd, ApJL **428**, L1 (1995).

Calibration of Tests for Time Dilation in GRB Pulse Structures

J.P. Norris,[1] R.J. Nemiroff,[2] J.T. Bonnell,[3] and J.D. Scargle[4]

[1] *NASA/Goddard Space Flight Center, Greenbelt, MD 20771*
[2] *George Mason University, Fairfax, VA 22030*
[3] *Universities Space Research Association*
[4] *NASA/Ames Research Center, Moffett Field, CA 94035*

> Two tests for cosmological time dilation in γ-ray bursts—the peak alignment and auto-correlation statistics—involve averaging information near the times of peak intensity. Both tests require width corrections, assuming cosmological origin for bursts, since narrower temporal structure from higher energy would be redshifted into the band of observation, and since intervals between pulse structures are included in the averaging procedures. We analyze long (> 2 s) BATSE bursts and estimate total width corrections for trial time-dilation factors (TDF $= [1+z_{\rm dim}]/[1+z_{\rm brt}]$) by time-dilating and redshifting bright bursts. Both tests reveal significant trends of increasing TDF with decreasing peak flux, but neither provides sufficient discriminatory power to distinguish between actual TDFs in the range 2–3.

TWO WIDTH CORRECTIONS TO OBSERVED TIME DILATION

If γ-ray bursts (GRB) are at cosmological distances, then the temporal structure associated with each energy range at the source is shifted to lower energies in the observer's frame of reference. Since GRB pulse structures are narrower at higher energy, this redshift-dependent *narrowing* would compete with cosmological time dilation. Note that both redshift-dependent narrowing and time dilation have mathematical analogues in special relativistic (SR) beaming models—blue-shifting of the radiation, and time contraction of the temporal structure. So far, there is no observational data which affords a way to distinguish between SR and cosmology in GRBs. We discuss the problem in terms of the cosmological hypothesis, keeping in mind that both cases have the same temporal mensuration problems.

Each measure of time dilation in GRBs requires a different width correction for narrowing of temporal structure, but all such corrections operate in the same sense: The *actual* time-dilation factor (TDF) is decreased by the redshift effect, so that the *observed* TDF, between bright and dim groups of bursts, is smaller, and a function of temporal structure in both groups,

$$TDF_{\rm obs} = F[\, TDF_{\rm act},\ \Lambda(TDF_{\rm act} \times E_c),\ \Lambda(E_c)\,] \quad (1)$$

© 1996 American Institute of Physics

where E_c is the energy band of observation, and $TDF_{act} \times E_c$ is the band, relative to z_{brt} of bright bursts, from whence the temporal structure was redshifted (1–3). Of course, $TDF_{act} = [1+z_{dim}]/[1+z_{brt}]$) is what we are after. The insidious part is the dependence of Λ, a generic width statistic, on TDF_{act}. Thus the energy-dependent width correction for redshift is like a ratio of widths of temporal structure, but it is not a simple ratio for all time dilation measures, for the following reason: For measures which utilize information from a portion of the time profile that contains any intervals or valleys between pulse structures, time dilation of these regions interjacent to peaks "subtracts" from time dilation of regions of emission, since interval regions and emission regions are not segregated. Thus when these portions of time profiles are averaged—as in Peak Aligned profile (PA) and Auto-Correlation Function (ACF) measures—regions of emission and intervals between pulses are thrown together in the averaging process. The resulting situation is illustrated in Figure 4 of ref (4), in these proceedings.

For the PA and ACF statistics we have calibrated the resulting diminution of time dilation that would arise for the combined effects of redshift of temporal structure and averaging of regions that contain intervals. Note that correction for the latter effect, previously not addressed (1,3), is not required for time-dilation measures which rely upon distributions of a measured parameter, such as distributions of pulse widths, intervals between pulses, or burst durations. Instead, these measures have simple width-correction ratios, that can be obtained from the distributions in two relevant energy bands.

CALIBRATION OF WIDTH CORRECTIONS

The PA and ACF tests for time dilation have been described before (1,3,5). The basic procedures are: divide bursts into groups based on some measure of brightness; and within a brightness group, average the profiles with their highest intensity peaks in registration, or average the profiles' ACFs. The "common-sense" appeals of the PA test are that it operates entirely in the time domain, and makes use of the most intense part of a burst. The efficacy of the ACF test is that it probes short timescales, to the limit of temporal resolution, without the need to worry about finding the exact location of peak intensity in dim bursts. Also, the ACF has a well-defined correction for co-added noise at zero lag. For both tests the main problem is that the width corrections are appreciable, resulting in less than satisfactory discriminatory power in the TDF_{act} range $\sim$ 2–3, given the present sample variance.

Data preparation. We use BATSE DISCSC data summed over channels 1 and 2 ($\sim$ 27–115 keV) to construct average peak-aligned profiles and ACFs for bursts longer than $\sim$ 2 s and with peak intensities higher than 1400 counts s^{-1}. The bursts are divided into six brightness groups, $\sim$ 85 bursts per group, according to their BATSE 3B peak fluxes determined at 256 ms. This timescale compromises between 64 ms, where noisier estimates are obtained for dim bursts, and 1024 ms, which integrates over pulse widths (pulses in long bursts having FWHM $\sim$ 100–500 ms, dependent on energy band (6)). Quadratic (infrequently, higher order) backgrounds are fitted and subtracted. For the PA test the time profiles are rendered to 512-ms resolution; the original 64-ms resolution is preserved for the ACF test. To approximately nullify brightness bias, the comparison of average profiles or ACFs between different brightness groups (described below) is performed with the signal-to-noise (s/n) levels of the individual time profiles of the bright group equalized to the s/n levels of the profiles of the other groups. For each of the five brightness groups below the brightest, ten such noisy realizations per bright burst are computed, with the new peak intensity chosen randomly from among the peak intensities of the bursts of each respective group. Thus, about $850 \times 5 = 4250$ noisy realizations of bright bursts are created.

Estimating observed time dilation. For -8 to $+16$ (512-ms) bins of the peak of the average PA profiles of the bright bursts, the intensity levels and corresponding profile widths for a dimmer group are found, and width ratios computed for the 24 bins (N_{PA}). A similar procedure is followed for the ACFs, for ± 60 64-ms lag bins (N_{ACF}). However, since pulse widths (FWHM) in individual bursts are ~ 500 ms, only $N_{\text{indep}} \approx 12$ and 4 independent ratio estimates result for PA and ACF procedures, respectively. We estimate means, standard and sample errors by a bootstrap procedure (7). The ~ 85 burst profiles (or ACFs) in each brightness group (850 noisy realizations for the brightest group) are considered the "parent population" from which 85 profiles are drawn randomly with replacement. For each brightness group the random selection is repeated 500 times. For each run the average profile is computed, and the width ratios computed as described above. The $500 \times N_{\text{PA}}$ (or N_{ACF}) width ratios are rank ordered. The resulting 50th, 15nd, and 84th percentile levels are taken as the median width ratio—box symbols in Figures 1 and 2—negative and positive 1σ standard errors, respectively. The sample errors plotted are these 1σ errors reduced by the factor $\sqrt{N_{\text{indep}}}$.

For the PA and ACF measures, the *observed* TDFs range up to ~ 1.75 and ~ 1.45, respectively, for the dimmest group relative to the brightest. But the *actual* TDFs would be larger in both the cosmological and SR beaming hypotheses, by width-correction factors we estimate as follows.

Estimating width corrections, interval dilation+redshift effects. The expected time-dilation "signal", *sans* redshift effect, is easily simulated by merely stretching the profiles (using the original 64-ms data) of the bright group by factors of 2.0 and 3.0, and comparing with the unstretched profiles, but now using 16-channel MER data. For *actual* TDFs of 2.0 and 3.0, the *recovered* TDFs for the PA measure are ~ 1.7 (85%) and ~ 2.55 (85%), respectively (circle symbols, Figure 1). Similarly for the ACF measure, the recovered TDFs are ~ 1.6 (80%) and ~ 2.0 (66%), respectively (circle symbols, Figure 2). Incomplete recovery of the input TDF is attributable to inclusion of regions which contain stretched intervals as well as stretched pulse structures.

Recall that we analyzed the six brightness groups in the 25–115 keV band (box symbols, Figures 1 and 2). The additional effect of redshift of temporal structure is simulated by using the 16-channel data for bright bursts: corresponding approximately to TDF=1, Σ chans: 2–6, ~ 22–100 keV; TDF=2,

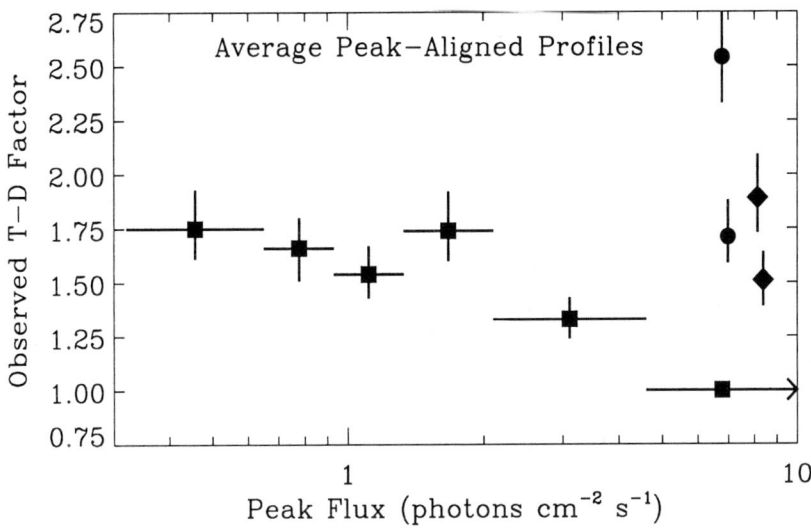

FIG. 1. Box symbols indicate observed time-dilation factor vs. BATSE 3B peak flux (256 ms), determined from average peak-aligned profiles. Errors estimated by bootstrap method. On right side, upper four points (without horizontal bars) are the bright burst sample time-dilated (circle symbols) and then redshifted as well (diamond symbols), by factors of 2 and 3 (lower and upper symbols, respectively). Approximately 85 bursts per group.

Σ chans: 4–8 + half of 9, $\sim$ 41–200 keV; and TDF=3, Σ chans: half of 5 + 6–10, $\sim$ 65–315 keV. The combined effect of stretching and redshifting profiles of bright bursts is indicated by diamond symbols in Figures 1 and 2. Narrower structure redshifted into the band of observation further reduces the observed TDF. For actual TDFs and redshifts of 2.0 and 3.0, the recovered TDFs are now $\sim$ 1.5 (75%) and $\sim$ 1.9 (63%), respectively, for the PA measure; the corresponding values for the ACF measure are $\sim$ 1.4 (70%) and $\sim$ 1.6 (54%), respectively. As can be seen by comparing the pairs of diamond symbols (Figures 1 and 2) with the observed TDF determinations for the six brightness groups, the uncertainties are such that TDF_{act} is only constrained to the range $\sim$ 2–3 for the dimmer groups via both the PA and ACF measures.

In conclusion, on the short (64 ms – few s) and intermediate (1–20 s) timescales probed by the PA and ACF tests, observed TDFs, relative to bright bursts, range up to $\sim$ 1.45 (ACF) and $\sim$ 1.75 (PA). From calibrations using the bright sample we conclude that, for the same input time-dilation and redshift factor, the ACF *is expected* to yield smaller TDF_{obs}. In fact, actual cosmological TDF's would be somewhat larger than TDF_{obs}: Two effects, redshift of narrower structure into the band of observation, and inclusion of stretched intervals, result in smaller observed time-dilation factors. The second effect was not appropriately simulated in previous estimates which used ratios of average pulses (1) or ratios of average ACFs (3) in different energy bands of

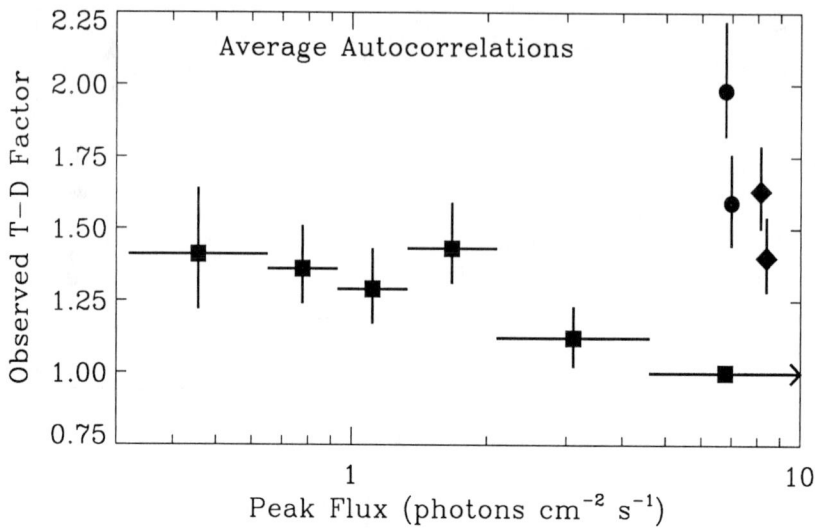

FIG. 2. Similar to Figure 1, except determined from the ACFs of bursts in six brightness groups. Observed TDFs (boxes) are lower than for peak-aligned profiles, but calibrations obtained by time-dilating (circles) and then redshifting (diamonds) bright bursts by factors of 2 and 3 are lower as well. Relatively larger error bars result than for PA measure since fewer independent time bins were used.

bright bursts. The width corrections are more pronounced at higher TDFs, such that with present uncertainties, both the PA and ACF measures only constrain $TDF_{act} = [1+z_{dim}]/[1+z_{brt}]$ to lie in the range 2–3.

REFERENCES

1. J.P. Norris, et al., ApJ **424**, 540 (1994).
2. J.P. Norris, Ap Space Sci **231**, 95 (1995).
3. E.E. Fenimore and J.S. Bloom, ApJ **453**, 25 (1995).
4. J.P. Norris, these proceedings.
5. I.G. Mitrofanov, Ap Space Sci **231**, 103 (1995).
6. J.P. Norris, et al., ApJ **459**, in press (1996).
7. B. Efron and R.J. Tibshirani, *An Introduction to the Bootstrap*, (New York: Chapman and Hall) (1993).

Test for Time Dilation of Intervals Between Pulse Structures in GRBs

J.P. Norris,[1] J.T. Bonnell,[2] R.J. Nemiroff,[3] and J.D. Scargle[4]

[1] *NASA/Goddard Space Flight Center, Greenbelt, MD 20771*
[2] *Universities Space Research Association*
[3] *George Mason University, Fairfax, VA 22030*
[4] *NASA/Ames Research Center, Moffett Field, CA 94035*

> If γ-ray bursts are at cosmological distances, then not only their constituent pulses but also the intervals between pulses should be time-dilated. Unlike time-dilation measures of pulse emission, intervals would appear to require negligible correction for redshift of narrower temporal structure from higher energy into the band of observation. However, stretching of pulse intervals is inherently difficult to measure without incurring a timescale-dependent bias since, as time profiles are stretched, more structure can appear near the limit of resolution. This problem is compounded in dimmer bursts because identification of significant structures becomes more problematic. We attempt to minimize brightness bias by equalizing signal-to-noise (s/n) level of all bursts. We analyze wavelet-denoised burst profiles binned to several resolutions, identifying significant fluctuations between pulse structures and interjacent valleys. When bursts are ranked by peak flux, an interval time-dilation signature is evident, but its magnitude and significance are dependent upon temporal resolution and s/n level.

EXPECTED SYSTEMATICS

It might naively be thought that time dilation of intervals between peaks or pulses in γ-ray bursts (GRB) should be free of the energy-dependent effects that plague measures of time dilation of temporal structures (1,2). However, at least two effects are expected to give rise to systematic biases that make attempts to determine the actual measure of time dilation of pulse intervals difficult:

(1) Structure appearing at limit of temporal resolution. For the present purpose, define "interval between pulses" to be the interval between two discernible peaks of emission, desired significance being adjustable. The average width of GRB pulses in long ($T_{90} > 2$ s) bursts is $\sim$ 100–500 ms, dependent on energy band. However, there is a large dispersion in pulse width (3). Since the timescale for intervals between pulses is also of this order, there is often a high degree of pulse overlap. Consequently, for the 64-ms resolution data which we employ, time dilating a burst profile by a factor $S \sim 2$ will result in the appearance of newly resolved structure at the shortest resolved timescale. Some

intervals between peaks which were not resolved in the unstretched burst will then become discernible, with the result that some intervals in the original profile divide into two shorter ones. The average pulse interval in a stretched burst will therefore not be S times longer than in the unstretched burst, but somewhat less than S.

(2) Redshift of narrower pulses from higher energy (deeper valleys). This is the same effect which diminishes the observed measure of time dilation in pulse widths: Since pulses are narrower at higher energy, the dimmer bursts – presumably suffering more redshift – will have narrower pulses redshifted into the band of observation. The effect on pulse-interval measures is that valleys between pulses will be deeper and more significant in the redshifted bursts since there will be less pulse overlap. This effect will result in additional (otherwise time-dilated) intervals being bifurcated, and therefore shortened. Also, some new valleys will appear that were not present in the non-redshifted burst profile.

Measures of pulse-interval dilation. A time-dilation measure for intervals which appears relatively unbiased is the average (or median) interval between pulses or peaks, per burst (see also the definitions in ref (4), these proceedings). Alternatively, all intervals found within a given brightness group might be weighted equally (5), but this would tend to weight longer bursts more heavily. A measure like event duration is the interval between first and last significant peaks. A more complex formulation might take into account the significance (e.g., depth of interjacent valley) of an interval. How such definitions are to be corrected (assuming cosmological hypothesis) for redshift and resolution effects should be estimated by performing simulations. In this paper we report results only for a test of the time-dilation effect between pulse intervals, and leave the understanding of corrections for a more detailed study.

PROCEDURE

Data preparation. BATSE bursts in the 3B catalog with measured $T_{90} > 2$ s (6) above a peak-intensity threshold form the sample. The threshold is either 2400 or 1400 counts s^{-1}, with the sample divided into 5 or 6 groups ($\sim$ 85 bursts per group), respectively, according to BATSE 3B peak flux (256-ms timescale). BATSE DISCSC data (64-ms resolution) summed over channels 1–4 (> 25 keV) are used; quadratic (infrequently, higher order) backgrounds are fitted and subtracted. Burst profiles are prepared by rendering their signal-to-noise (s/n) levels equal to that of the burst with the lowest peak intensity in the sample, according to a procedure discussed in ref (1). This step renders variances and peak intensities approximately equal. The prepared profiles are then run through a Haar wavelet-denoiser to remove insignificant ($< 2\sigma$) fluctuations on all timescales. Without denoising, identification of some valleys would often be compromised by insignificant fluctuations.

Interval identification. The profiles are searched for occurrences of two peaks separated by a valley, requiring a significance of the intensity difference of at least 4σ between the lower peak and the valley. By requiring a highly sig-

TABLE 1. Interval Time-Dilation Factors vs 3B Peak Flux, Resolution

	\multicolumn{5}{c}{Peak Flux (ph cm^{-2} s^{-1}) [a]}				
	2.10	1.33	0.93	0.65	0.32
		Threshold: 1400 cts s^{-1}			
64 ms	1.55 (0.38)	2.85 (0.035)	1.25 (0.87)	3.25 (0.018)	2.30 (0.054)
128 ms	1.65 (0.085)	1.72 (0.072)	1.75 (0.13)	2.35 (0.0008)	2.42 (0.004)
256 ms	1.28 (0.36)	1.35 (0.59)	1.25 (0.77)	1.98 (0.013)	1.50 (0.14)
512 ms	1.15 (0.45)	1.35 (0.25)	1.60 (0.093)	2.18 (0.016)	2.38 (0.0013)
		Threshold: 2400 cts s^{-1}			
64 ms	0.85 (0.67)	1.18 (0.32)	1.28 (0.20)	1.72 (0.029)	
128 ms	1.30 (0.22)	1.75 (0.032)	1.40 (0.29)	2.15 (0.008)	
256 ms	1.05 (0.60)	1.55 (0.037)	1.40 (0.10)	2.20 (0.0016)	
512 ms	1.05 (0.82)	1.30 (0.21)	1.22 (0.62)	2.20 (0.0002)	

[a] Lower peak-flux boundary for 5 brightness groups; boundary for brightest group: 4.60 ph cm^{-2} s^{-1}. Values in parentheses are probabilities for stretch factor of unity.

nificant interval, we are essentially identifying intervals between major pulse structures, rather than individual pulses, thereby (hopefully) ameliorating some systematic effects described in the previous section. The interval search is performed for the prepared profiles binned to 64-ms, 128-ms, 256-ms, and 512-ms resolutions. Each binning timescale was analyzed separately.

RESULTS

We adopt the first measure of interval time-dilation described above, the median interval per burst. We then form distributions of median intervals for each brightness group. By stretching the distribution of intervals for the brightest group on a grid of trial time-dilation factors and performing a Kolmogorov-Smirnov (K-S) test for degree of agreement between the interval distribution of the brightest group and those of the five dimmer groups, we estimate observed time-dilation factors and associated errors.

For each binning timescale and the two peak intensity thresholds, Table 1 lists the measured interval time-dilation factors (TDF), and the probability (in parentheses) of agreement given a stretch factor of unity, of the five dimmer groups relative to the brightest group. More significant determinations result more often for the higher peak intensity threshold, presumably because a higher s/n level is realized in the noise equalization procedure. However, the higher threshold necessarily cannot examine the dimmer bursts. For all timescales a trend is evident of longer median intervals towards lower peak flux.

Significances of disagreement between the interval distributions of dimmest (or second dimmest) and brightest groups range from $\sim 2\sigma$ to 3.5σ, with longer timescales tending to be more significant. A partial explanation for this must be that for coarser binning (higher counts per bin), a larger number of bursts survive to contribute to the distribution: the numbers of occurrences of bursts with 2 or more peaks with a $> 4\sigma$ valley in between, increases as the timescale increases. For 1400 counts s^{-1} threshold, the number of such occurrences per group increases from ~ 24 (64 ms) to 38 (256-512 ms); for 2400 counts s^{-1} threshold, the number of contributing bursts is approximately constant with timescale, ~ 50 occurrences.

Two examples of the trend are illustrated in Figures 1 and 2. The first case is for the 1400 counts s^{-1} threshold on the longest timescale analyzed, 512 ms; the second case shows results for the 2400 counts s^{-1} threshold for 128 ms resolution. Both figures illustrate the more conservative result (in terms of significance) for their respective timescales obtained for the bright group relative to dimmest (or second dimmest) group, as can be seen by comparing probabilities for unity stretch factor for the two thresholds in Table 1.

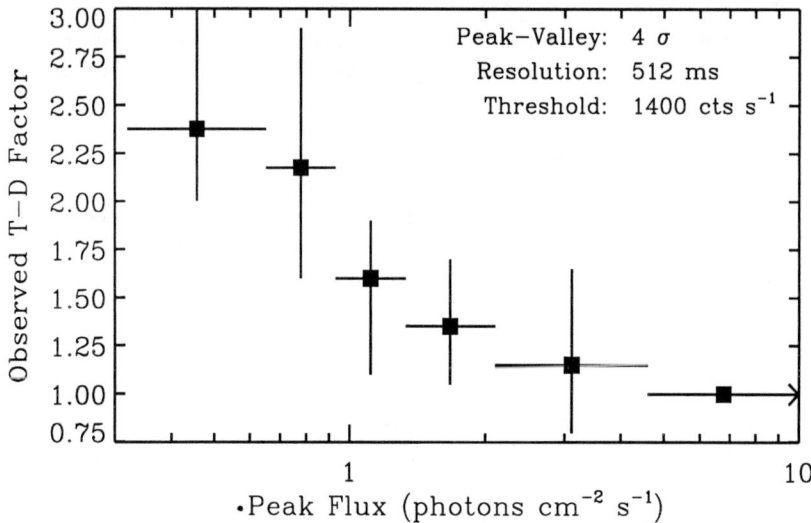

FIG. 1. Observed interval time-dilation factor vs. BATSE 3B peak flux, for 1400 counts s^{-1} threshold for profiles rendered to 512 ms resolution. Central values and 1σ uncertainties determined via K-S test, by stretching distributions of intervals for bright burst group and comparing with distributions of dimmer groups.

CONCLUSIONS

When bursts are grouped by BATSE peak flux, we find a relative time-dilation effect for intervals between pulse structures, at a significance level of $\sim 2.5\sigma$, between brightest and dimmest / next to dimmest burst groups. This *observed* time-dilation factor is of order 2. Actual time-dilation factors would probably be somewhat larger: Two effects – appearance of new structure at the limit of resolution as bursts are stretched, and narrowing of pulses with

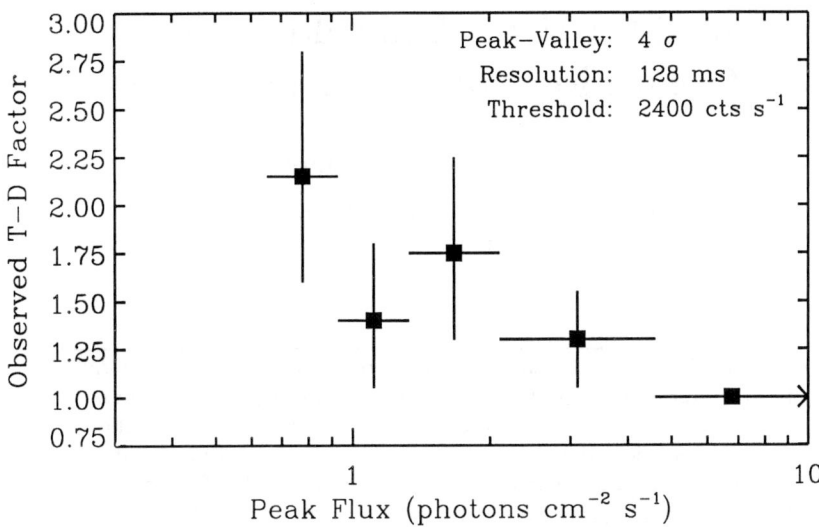

FIG. 2. Observed interval time-dilation factor vs. BATSE 3B peak flux, for 2400 counts s^{-1} threshold for profiles rendered to 128 ms resolution.

higher energy, thus better defining valleys between pulse structures – probably result in the observed time-dilation factor being smaller than the actual value. As this is an exploratory study in need of robust simulations to calibrate these effects, we conclude that this result tentatively and qualitatively confirms the result of Davis (5), in which intervals between pulse structures were measured using a pulse-fitting approach.

REFERENCES

1. J.P. Norris, et al., ApJ **424**, 540 (1994).
2. J.P. Norris, Ap Space Sci **231**, 95 (1995).
3. J.P. Norris, et al., ApJ **459**, in press (1996).
4. J. Neubauer and B. Schaefer, these proceedings.
5. S.P. Davis, *Ph.D. thesis*, The Catholic University of America (1995).
6. J.T. Bonnell, et al., these proceedings.

Flux-Duration Correlations and Cosmological Time Dilation

Vahé Petrosian and Theodore T. Lee

Center for Space Science and Astrophysics
Stanford University
Stanford, California 94305

> We perform several nonparametric correlation tests between brightness and duration of bursts in the BATSE 3B catalog to search for evidence of cosmological time dilation. These tests account for the effects of data truncation due to threshold effects in both limiting brightness and limiting duration, enabling us to utilize up to 46% of all triggered bursts in our analysis. Previous tests have been limited to $\sim 20\%$ of all triggered bursts. We find no significant evidence for correlation between various measures of peak intensity and duration, but there is evidence for a positive correlation between fluence and duration, which is in the opposite sense of the correlation expected from time dilation.

INTRODUCTION

The angular distribution of gamma-ray bursts is isotropic (1), while the $\log N$-$\log S$ distribution flattens for weaker bursts. Such a distribution would be expected if bursts were of cosmological origin with a typical redshift of about unity (2, 3) and would imply (4, 5) that their light curves should be stretched due to cosmological time dilation. However, the expected time dilation would be a factor of a few while the burst durations cover a large dynamic range. Therefore, a time dilation effect can only be detected statistically. Norris et al. (6, 7) found evidence that brighter bursts had shorter durations than dimmer ones, consistent with a time dilation factor of about 2. However, there is disagreement about the reality or magnitude of the observed time dilation. For example, Mitrofanov et al. (8) find no evidence for any time dilation and Fenimore et al. (9) obtain a much smaller time dilation factor (1.3 instead of 2) from a larger set of data.

Clearly, despite the numerous works published on the subject, time dilation of gamma-ray bursts remains controversial. Here we present new results on this topic which differ from previous works in two important ways. We include a larger fraction of the detected bursts and use several different measures of burst strength and duration. Previous studies of burst time dilation have been limited to bright, long duration bursts.

© 1996 American Institute of Physics

ANALYSIS METHOD

The problem of searching for a time dilation (or a redshift-distance relation) amounts to searching for a correlation between two variables, one of which, the burst duration, is a measure of the redshift, and the other, some measure of burst brightness, is a measure of the distance. Due to the detection biases against short and weak bursts, the practice in previous dilation studies has been to confine them to relatively bright, long bursts and utilize only about 20% of the total number of triggered bursts.

However, it is possible to extend such a test to a much larger sample if the observational selection criteria or data truncations are well defined, because there exist methods to test for correlations in the presence of such truncations. A simple test which is easily applied to burst data is the t_w test developed by Efron & Petrosian (10) and described in detail in several of our previous publications (11–14). Briefly, the test relies on the concept of the *associated set* of points for each data point which is then assigned a rank amongst the points in this set. The value of t_w gives the probability $P(t_w) = \mathrm{erfc}(|t_w|/\sqrt{2})$ that the data were drawn from an uncorrelated distribution, so that a large and positive (negative) value of t_w would imply significant correlation (anticorrelation).

CHOICE OF VARIABLES

All previous studies have used the peak photon count rate or the peak flux as a measure of the distance. These assume a nearly standard candle peak photon luminosity, which seems not very likely considering the large dispersion in the durations and pulse shapes of GRBs. It is more likely that the total energy (or total number of photons) emitted by a burst is a standard candle, so that the energy or photon fluence will be a better measure of the distance to the bursts. We carry out our tests using both fluxes and fluences.

The BATSE catalog provides several quantities for this task. The most tractable quantity is the peak photon count rate $\bar{C}_P$ averaged over the three trigger times $\Delta t = 64, 256$, and 1024 ms. The burst selection criterion is that $\bar{C}_P \geq \bar{C}_{\lim}$, where $\bar{C}_{\lim}$ is the threshold value (for each Δt). A more appropriate quantity, which is directly related to the peak photon luminosity F_P, is the peak photon flux $\bar{f}_P$ (averaged over Δt). However, because of the dependence of Δt on redshift and the bias against detection of short duration bursts (duration $T < \Delta t$) a more accurate measure is (12) $f_P = \bar{f}_P(1 + \Delta t/T)$. If F_P is a standard candle (independent of T) then $f_P = F_P/4\pi d_L^2$, would be a measure of the appropriate luminosity distance d_L. On the other hand, if the total energy ε or the total number of photons has a small dispersion, then the appropriate measure of distance found in the BATSE catalog is the fluence $\mathcal{F} = \varepsilon/4\pi d_L^2$.

As described in Lee & Petrosian (11) the threshold on all of these quantities is obtained as

$$\mathcal{X}_{\lim} = \mathcal{X} \bar{C}_{\lim}/\bar{C}_P, \tag{1}$$

where $\mathcal{X}$ stands for $\bar{f}_\mathrm{P}$, f_P, or $\mathcal{F}$.

To test the cosmological hypothesis we now need a measure of redshift. Because of the complex and varied burst pulse shapes, it is not clear what measure of the time structure, or which of the several available time scales associated with the pulse profiles, would be a reliable measure of redshift. In addition to using the durations T_{50} and T_{90} provided in the BATSE catalog, we also use an effective duration defined as the ratio of the total energy released to the peak luminosity:

$$T_\mathrm{eff}(T) = \frac{\mathcal{F}}{f_\mathrm{P}\langle h\nu\rangle} = \frac{\mathcal{F}}{\bar{f}_\mathrm{P}(1+\Delta t/T)\langle h\nu\rangle}. \qquad (2)$$

Here $\langle h\nu\rangle$ is the average photon energy in the BATSE trigger bandpass (50–300 keV).

We have performed the t_w test for stochastic independence on several combinations of these variables and for all three trigger times $\Delta t = 64$, 256, and 1024 ms.

PEAK FLUX-DURATION CORRELATIONS

A test of the correlation between the average peak flux $\bar{f}_\mathrm{P}$ and duration is the most straightforward, but is complicated because of the variable threshold $\bar{f}_\mathrm{lim}$. Consequently a clear truncation boundary cannot be delineated in the $\bar{f}_\mathrm{P}$–T plane. For the sake of simplicity and clarity we select a subsample of data which could be described with a single constant threshold $\bar{f}_\mathrm{lim}$. The truncation boundaries in the $\bar{f}_\mathrm{P}$–T plane become parallel to the axes and the t_w test then reduces down to a simple rank order correlation test.

The first three rows of Table 1 show the values of t_w for each of the three trigger times. The first obvious feature in these numbers is that the values of t_w are significantly and consistently larger for correlations involving T_{90} than for the other measures of duration. The most likely explanation of this result is that the T_{90} values are underestimated at low values of $\bar{f}_\mathrm{P}$, giving rise to a larger positive value for t_w and an apparent correlation (15). We therefore ignore the results based on T_{90} here and in what follows.

The second feature of these t_w values is that they are larger for larger values of Δt. This result is most likely due to the short duration bias mentioned above and can be corrected by using f_P instead of the average peak flux $\bar{f}_\mathrm{P}$. The magnitude of this correction increases with the ratio of $\Delta t/T$, and is therefore largest for $\Delta t = 1024$ ms and $T \lesssim 1$ s. The transformed data in the f_P–T plane is no longer simply truncated but the truncation is defined by

$$f_\mathrm{P} > f_\mathrm{lim} = \bar{f}_\mathrm{lim}(1+\Delta t/T), \qquad (3)$$

The values of t_w obtained for these data are given in the second three rows of Table 1. These values are now lower, especially for the $\Delta t{=}1024$ ms data, apparently confirming the assertion that the differences between the data sets at different values of Δt are due to the short duration bias. Ignoring the

t_w values obtained using T_{90}, the rest of the values are consistent with the absence of correlation.

This result is in contradiction with the cosmological hypothesis if the peak luminosities are standard candles independent of T. It is possible that this null result could arise from correlations of opposite trends in different portions of the data. For example, equal and opposite relations between f_P and say short ($T < 1$ s) and long ($T > 1$ s) bursts can cancel each other out, giving low t_w values. We test this possibility by dividing the data into subsamples and find no obvious trends.

FLUENCE-DURATION CORRELATIONS

Like the peak flux f_P (or C_P) the fluence $\mathcal{F}$ is also subject to a bias. The bias now is against the detection of weak and long bursts which could have sufficient fluence but not a large enough $\bar{C}_P$ to exceed the preset threshold $\bar{C}_{\text{lim}}$. For a given pulse shape the fluence threshold can be obtained as

$$\mathcal{F}_{\text{lim}} = \langle h\nu \rangle \bar{f}_{\text{lim}} g(T, \Delta t), \tag{4}$$

where $\langle h\nu \rangle$ is the average energy per photon and the function g is determined by the pulse shape. Thus the truncation boundary in the $\mathcal{F}$–T plane depends on the pulse shape. It can be shown that for simple pulse forms $g(T, \Delta t) \simeq (\Delta t + T)$.

Because of this uncertainty we cannot test the correlation between the fluence and duration directly. However, we know the values of $\mathcal{F}$ and $\mathcal{F}_{\text{lim}}$ for all sources and can easily find the correlation between these two quantities. Since as shown above, $\mathcal{F}_{\text{lim}}$ is a function of the duration T we can find the $\mathcal{F}$–T correlation indirectly. For simple pulses and for $\Delta t \ll T$ from the above approximation we have $\mathcal{F}_{\text{lim}} \propto T$ so that any correlation here would be a direct test of correlation between $\mathcal{F}$ and T. The results of our test gives $t_w = 4.26, 3.35, 2.28$ for $\Delta t = 1024, 256, 64$ ms, respectively, indicating a strong positive correlation in contrast to the expected negative values for the cosmological hypothesis.

CONCLUSIONS

We have carried out a test of the cosmological time dilation by searching for correlations between brightness and duration. With a proper accounting of the data truncation we extend this test to a large subset of the bursts. We use the durations T_{50} and an effective duration which we define by the ratio of the fluence to peak flux.

Because of the short duration bias (12) the average peak flux shows a small correlation with duration in the 1024 ms triggered data. After correction for this we find that the peak flux and durations are consistent with stochastic independence, with no obvious cosmological signature. Further work is required to set quantitative limits on the amount of time dilation.

	T_{90}	T_{50}	$T_{\text{eff}}(T_{90})$	$T_{\text{eff}}(T_{50})$
64 ms	0.827	-0.355	-0.191	-0.192
256 ms	2.00	1.03	1.13	1.10
1024 ms	4.17	1.58	1.32	1.28
64 ms	0.761	-0.409	-0.376	-0.421
256 ms	1.81	0.625	0.584	0.429
1024 ms	2.03	0.59	0.244	0.040

TABLE 1. t_w values for $\bar{f}_P$ versus various measures of duration (top three rows) and for f_P versus various measures of duration (bottom three rows).

We also study the correlation between the burst fluence and duration. This test cannot be done directly. But we find a strong correlation between fluence and the fluence limit, which is a function of duration (nearly proportional to it for long durations) indicating a positive correlation in contrast to the anticorrelation expected from the cosmological time dilation. It is unlikely that such a strong signature could be a result of the sum of the biases mentioned above. We must conclude that either the bursts are not at cosmological distances or the total energy released is a strong function of the redshift.

REFERENCES

1. C. Meegan et al., Nature **355**, 143 (1992).
2. W. Wickramasinghe et al., ApJ **411**, L55 (1993).
3. E. Fenimore et al., ApJ **448**, L101 (1995).
4. B. Paczyński, Nature **255**, 521 (1992).
5. T. Piran, ApJ **389**, L45 (1992).
6. J. Norris et al., ApJ **424**, 540 (1994).
7. J. Norris et al., ApJ **439**, 542 (1995).
8. I. Mitrofanov, these proceedings (1996).
9. E. Fenimore, these proceedings (1996).
10. B. Efron and V. Petrosian, ApJ **399**, 345 (1992).
11. T. Lee and V. Petrosian, these proceedings (1996).
12. V. Petrosian et al., in AIP Conference Proceedings **307**: Gamma Ray Bursts, eds. G. J. Fishman, J. J. Brainerd, & K. Hurley, New York: AIP Press, p. 93 (1994).
13. T. Lee, V. Petrosian, and J. McTiernan, ApJ **412**, 401 (1993).
14. V. Petrosian and T. Lee, ApJ, in press (1996).
15. J. Norris, these proceedings (1996).

Intrinsic Dependence of Gamma-Ray Burst Durations on Energy

Georgia Richardson[*], Thomas Koshut[*], William Paciesas[*], Chryssa Kouveliotou[†]

[*] *University of Alabama in Huntsville, Huntsville, AL 35899*
[†] *Universities Space Research Association, NASA/MSFC, Huntsville, AL 35812*

> We have measured T_{90} and T_{50} as a function of energy for a set of bright BATSE gamma-ray bursts, selected on the basis of their peak photon flux on the 64 ms time scale. These events lie mainly on the $-3/2$ portion of the log $N(> P)$–log P curve; thus, in the cosmological scenario their measured characteristics may be relatively free from redshift and time dilation effects, so that the measured dependence of duration on energy may reflect an intrinsic dependence. We examine the compatibility of our results with recent work (1), which showed that the functional dependence of duration on energy is well-represented by a power law.

INTRODUCTION

The deficit of weak bursts in the burst intensity distribution observed with BATSE indicates that there is either a fall off in the source density at a certain distance, or that the bursts are being affected by cosmological effects. Any intrinsic burst property, such as the dependence of burst duration on energy, must be accounted for when using temporal data to search for cosmological effects. In this paper, we attempt to measure the dependence of duration on energy. We do this by measuring $T_{90}(T_{50})$ (2) in the individual energy channels for a set of bursts chosen such that they do not exhibit strong cosmological effects. The measured values of $T_{90}(T_{50})$ in each of the energy channels are fit to a power-law. The results are analyzed, and compared with previous work (1).

SELECTION OF DATASET

$T_{90}(T_{50})$ was measured in four broad energy channels for 72 intense gamma ray bursts observed with BATSE. The energy ranges of the four channels are 25–50 keV, 50–100 keV, 100–300 keV, and >300 keV. The bursts used for this study were selected by their peak photon flux on the 64ms time scale (P_{64}) such that they fell on the $-3/2$ portion of the log $N(> P_{64})$ – log P_{64} curve. This was done to ensure that the bursts chosen are relatively free from cosmological effects. Only bursts with $P_{64} \geq 8.0$ photons cm^{-2}s^{-1} were

© 1996 American Institute of Physics

chosen. No selection was made on burst duration. Of the 82 bursts that satisfied these criteria, $T_{90}(T_{50})$ could not be measured for 10. Nine of these bursts were eliminated because of data gaps during the bursts. $T_{90}(T_{50})$ could not be measured for the other burst because of strong magnetospheric activity in the background data. The resulting dataset consists of 72 of the brightest bursts in the BATSE 3B catalog (3).

FUNCTIONAL DEPENDENCE OF $T_{90}(T_{50})$ ON ENERGY

For each burst, we fit the four values of the $T_{90}(T_{50})$ in each energy channel to a power law function given by

$$T_{90} = A_0 E^{\alpha_{90}} \quad (1)$$

and,

$$T_{50} = B_0 E^{\alpha_{50}} \quad (2)$$

using a χ^2 minimization technique. We used the lower energy thresholds to represent each energy channel in the fit. The distributions of the best fit indices, α_{90} and α_{50}, are shown in figures 1a and 1b.

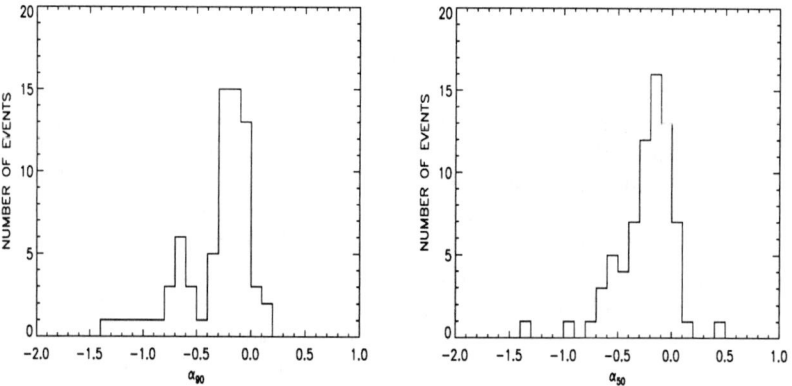

FIG. 1. Distributions of the best fit indices, α_{90} (1a) and α_{50} (1b), of power law fits to the burst durations in four energy channels.

We find that for these distributions $\langle\alpha\rangle_{90} = -0.31 \pm 0.04(0.32)$ and $\langle\alpha\rangle_{50} = -0.24 \pm 0.04(0.27)$ where the first error is the error in the mean, and the error given in parenthesis is the standard deviation of the distribution.

When we calculated the $P(\chi^2 > \chi^2_{obs})$ probabilities, we find that approximately 75% of the fits for T_{90} were acceptable, while approximately 93% of the fits for T_{50} were acceptable. There are a number of fits where $P(\chi^2 > \chi^2_{obs}) \approx 1$. This is probably due to the small number of degrees of freedom in the fit, or to a breakdown during the fitting procedure of some assumptions concerning the distribution of errors.

TABLE 1. Average values for $\langle T_{90} \rangle$ and $\langle T_{50} \rangle$

Energy Channel	$\langle T_{90} \rangle$ (sec)	$\sigma_{T_{90}}$ (sec)	$\langle T_{50} \rangle$ (sec)	$\sigma_{T_{50}}$ (sec)
1	35.8	5.9	9.8	1.7
2	31.6	5.8	8.4	1.5
3	26.8	5.3	7.5	1.7
4	20.4	4.0	7.1	1.8

Using a slighty different approach, we first averaged the values of $T_{90}(T_{50})$ for each energy channel. These values are given in Table 1.

We then fit a power law to these data as we did in the previous case. Figures 2 and 3 show the resulting best-fit for $\langle T_{90} \rangle$ and $\langle T_{50} \rangle$, respectively.

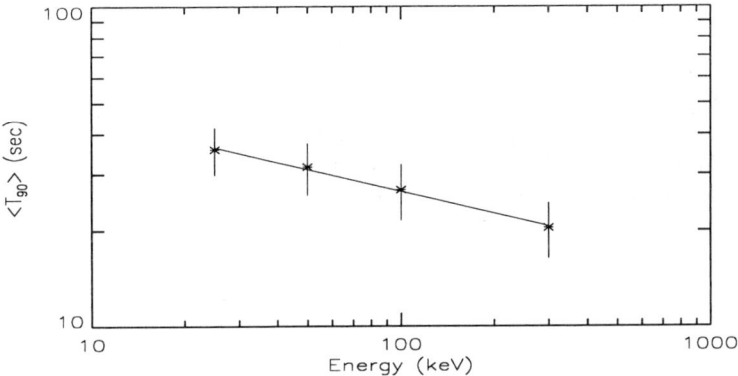

FIG. 2. Best fit of the $\langle T_{90} \rangle$ for four energy channels.

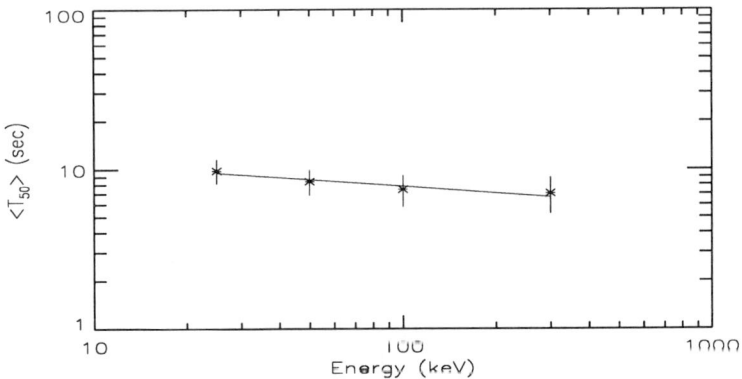

FIG. 3. Best fit of the $\langle T_{50} \rangle$ for four energy channels.

The resulting best fit indicies for the averaged data are given by $\alpha_{90} =$

-0.15 ± 0.10 and $\alpha_{50} = -0.15\pm0.12$ with $P(\chi^2 > \chi^2_{obs})_{90} = 0.98$ and $P(\chi^2 > \chi^2_{obs})_{50} = 0.97$.

DISCUSSION

In recent work (1,4) it was shown that the functional dependence of duration on energy is well represented by a power law with an index $\alpha \approx -0.4$. Their work utilized the autocorrelation function and the average pulse width. The values of α_{90} and α_{50} found using our two different methods were within 1.5 sigma of each other. The weighted means of the indices using the results of the two methods are $\alpha_{90} = -0.29\pm0.04$, and $\alpha_{50} = -0.23\pm0.04$. A quantitative comparison of our results with the previous work is difficult without knowing more about the errors and distributions of the resulting α's found in their studies.

REFERENCES

1. E.E. Fenimore, et al., ApJ **448**, L101 (1995).
2. C. Kouveliotou, et al., ApJ **413**, L101 (1993).
3. C.A. Meegan, et al., ApJ Suppl., in press (1996).
4. J.P. Norris, et al., ApJ **424**, 540 (1994).

Profiles of γ-Ray Bursts and their Component Pulses

J. D. Scargle[1], J. P. Norris[2], R.J. Nemiroff[3] and J.T. Bonnell[4]

[1] *NASA/Ames, Moffett Field, CA, 94035,* [2] *NASA/Goddard,* [3] *GMU,* [4] *USRA/GSFC*

> One physically informative regularity of the otherwise heterogeneous γ-ray bursts, is that many consist of well defined pulses. To objectively quantify burst temporal structures, we developed an automatic modeling procedure for time-tagged event (TTE) and time-to-spill (TTS) data that separates overlapping pulses and determines their energy-dependence. The crucially important initial guess solution is determined by wavelet-denoising a time series of delta-function photon arrival times, and then locating, modeling, and removing successive pulses (analogously to the *clean* deconvolution procedure of radio astronomy). The final model is then determined by minimizing a Bayesian likelihood function Pr(model|data) based on the Poisson probability distribution. This effort will yield statistical information on pulse rise-time, decay-time, peakedness, and amplitudes, plus their energy dependences – both within a single burst and for a large ensemble of bursts.

INTRODUCTION: THE PULSE PARADIGM

Understanding the γ-ray burst (GRB) *profiles* – *i.e.* the variation with time of the γ-ray intensity as the burst unfolds – will no doubt elucidate the physics of these mysterious events. For the many bursts consisting of two or more pulses (the "Pulse Paradigm"), explicit separation of overlapping pulses should provide more unambiguous information on the time-scale quantities reviewed by Norris (6,7). Toward these goals, we construct burst models that specify the background level, the number of pulses, and the parameters of each pulse – rise and decay times, plus the peak's roundness, amplitude and location in time. Such *deconvolution* is a notoriously difficult problem, mainly due to the difficulty of distinguishing overlapping pulses from a single broad one.

Our aim is to overcome the difficulties and extract all of the information inherent in the BATSE data, by marshaling these procedures:

1. Use the full time-resolution of the TTE (or TTS) data, by avoiding binning and the resulting dependence on bin size and location

2. Fit models using a maximum likelihood criterion that employs the exact

© 1996 American Institute of Physics

Poisson nature of the photon arrival-time statistics [1]

3. Use the information on the energy-dependence of the pulses

4. Determine and subtract the background in each channel

5. Carefully determine the starting solution for the optimization

6. Use this simple, proven, and flexible pulse model (5) :

$$I_n = \alpha e^{-[\beta,\gamma(n-\tau)]^\nu}, \tag{1}$$

where the notation means that β is used for times $n < \tau$ and γ for $n > \tau$.

7. Use an automatic fitting procedure, to avoid the limitations and biases of manual fitting [*e.g.* Ref. (5)].

Explanations of a few of these points are in order:

Point 1: The time-tagged event (TTE) data – in which the arrival time for each photon is recorded with 2μ sec resolution – contain the best information about variability at short time scales, and are the main object of this study. But – as they stop when 32,768 photons have been accumulated – this choice limits us to fainter, narrower bursts. The time-to-spill (TTS) data – in which the arrival time for every N_{spill}–th photon is recorded ($N_{spill} = 64$ for all observations to date) – sacrifice some short time-scale information, but are important when the TTE data do not cover the entire event.

Point 2: Maximum likelihood fitting criteria for point-processes have been known for some time in particle physics and astronomy [*e.g.*, (2)]. We adopt a modification of this approach designed for fitting models to Poisson-distributed photon arrivals.

Point 5: Our cost function (defined in Part 3 below) is not a quadratic form in the data-model residuals. Therefore we must use a non-linear optimization procedure – iteration of an operation that converges to a solution by stepwise improving the fit. As is always the case with such methods, a poor initial guess may yield convergence to an irrelevant local optimum of the cost function – rather than the desired global optimum. Hence a very important part of the procedure is the identification of an initial set of pulses comprising a good representation of the burst (Part 2 below).

PART 1: OBTAIN PROFILES FROM RAW DATA

For unbinned data (*e.g.* TTE and TTS) it is necessary to construct an estimate of the profile, evaluated at evenly spaced times. A simple way to do

---

[1] Fitting procedures applying least-squares criteria to data with non-normal errors typically yield biased parameter estimates. For example, χ^2 methods underestimate the area under a fitted curve in the presence of Poisson errors [(1), Section 9.2].

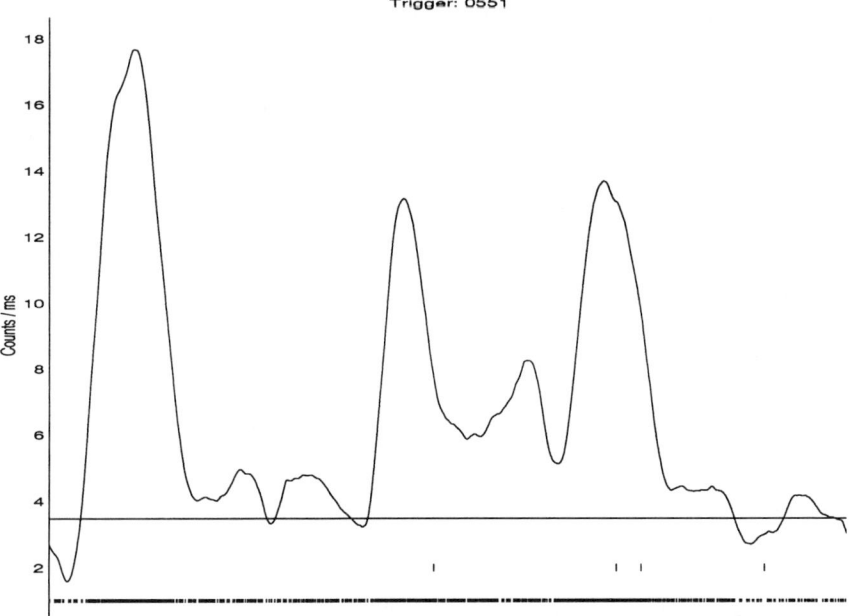

FIG. 1. The raw TTE data are indicated by vertical lines over those 2μ-sec intervals in which a photon was detected. For those few cases where two photons were recorded in an interval (presumably from different detectors) the ordinate is plotted as 2 instead of 1 (photons per interval). The profile derived from denoising this sequence of delta functions is shown as a solid line, with ordinate read in photons per millisecond.

this without binning is to difference an estimate of the *cumulative distribution function* (CDF; *i.e.*, the number of photons that have arrived by time t, as a function of t). Some kind of smoothing must be applied either to the CDF or its difference, or the resulting profile is too noisy to be useful. Even better, wavelet denoising techniques developed at Stanford (3) applied directly to the raw TTE data (without the need for the CDF) yield smooth profiles with structure consistent from channel to channel – a property important for the success of the model fitting.

Figure 1 demonstrates this denoising of sample TTE data, placed in 2μ sec bins. Since this is the resolution at which the raw photon arrivals are recorded, no temporal information is compromised and we do not regard this as binning.

The theory behind wavelet smoothing (3) assumes that the noise is additive and normally distributed. More work needs to be done to put the denoising of highly nonnormal count data on a firm footing. We have adapted the determination of the wavelet coefficient thresholds, as developed by Donoho and coworkers, to the Poisson case using numerical simulations.

PART 2: IDENTIFY PULSES FOR THE INITIAL SOLUTION

The procedure for identifying and separating pulses from the denoised profiles consists of these steps (somewhat like deconvolution using the CLEAN technique of radio astronomy), applied to the background-subtracted profiles determined as indicated above:

1. Locate the maximum of the denoised profile
2. Find the intervals from this peak to the first minima on either side
3. Fit single-sided models to the data within these two regions
4. Combine these into a two-sided model (with peakedness parameter equal to the average to the two found in the previous step)
5. Subtract this model from the data profile and replace any negative values with zero
6. Repeat the above steps, until all significant pulse structure has been removed

The model used in step (3) is Eq. (1) for $n < \tau$ or $n > \tau$, as appropriate. Because the values of α and τ are fixed in step 1, the determination of values of ν and the appropriate exponential coefficient (β or γ) is simple: fit a straight line in the $log(-log \frac{I}{\alpha})$ vs. $log(n - \tau)$ plane. A pulse is included only if it occurs – at times sufficiently close – in both channels 2 and 3, which typically have the largest signal-to-noise. [2]

The energy dependences of the pulse parameters are represented in the following way. The five parameters $\alpha^3, \beta^3, \gamma^3, \tau^3$, and ν^3 represent the pulse in channel 3, taken as the reference channel. The same parameters for the other channels are taken to be ($k = 1, 2, 3, 4$ is the channel index):

$$\alpha^k = \alpha^3 + (k-3)D_\alpha \tag{2}$$

$$\beta^k = \beta^3 e^{(k-3)D_\beta} \tag{3}$$

$$\gamma^k = \gamma^3 e^{(k-3)D_\gamma} \tag{4}$$

$$\tau^k = \tau^3 + (k-3)D_\tau \tag{5}$$

$$\nu^k = \nu^3 + (k-3)D_\nu \tag{6}$$

The energy-dependence coefficients $D_\alpha, D_\beta, D_\gamma, D_\tau$, and D_ν are determined from the differences between parameter values for channels 2 and 3, in an obvious way. This ends the determination of the initial guess for the full model.

---

[2] The BATSE TTE data is a list of arrival times of photons separated into four energy channels, channel 1: 25 – 60 keV; channel 2: 60 – 110 keV; channel 3: 110 – 325 keV; channel 4: > 325 keV.

PART 3: FIND MAXIMUM LIKELIHOOD SOLUTION

The final solution is determined by maximizing the likelihood of the fit, based on the Poisson distribution $P_X = \frac{\lambda^X e^{-\lambda}}{X!}$, giving the probability of detecting X photons if the model intensity is λ. Then, considering all times n, we adopt as the cost function minus the log-likelihood function:

$$I = -log(\prod_n P_n) = \sum_n (\lambda_n - X_n log \lambda_n) \qquad (7)$$

For infinitesimal bin width, each bin contains either zero or one photon (the former contributing cost λ and the latter $\lambda - log\lambda$), so that in this limit (2)

$$I = \sum_n (\lambda_n) - \sum_i log \lambda(t_i) \qquad (8)$$

Note well that the first sum is over all bins, while the latter is over all detected photons. The first term is typically held constant by imposing a normalization condition (which can in fact be derived rigorously). Good convergence results from optimizing the resulting simple cost function – the sum of the log of the model evaluated at the photon arrival times.

FUTURE WORK

The methods described here, and others, are being applied to the variable resolution time-to-spill (TTS) data in collaboration with Lee and Bloom (4). Ultimately we will produce a comprehensive set of pulse parameters for essentially all those bursts for which the pulse model makes sense.

REFERENCES

1. P. R. Bevington, and D. K. Robinson, "Data Reduction and Error Analysis for the Physical Sciences," 2nd. Ed., McGraw-Hill Inc., New York (1992).
2. W. Cash, Ap. J. **228**, 939 (1979).
3. Professors David Donoho, Iain Johnstone, et al., have published a number of theoretical papers on wavelet denoising and related topics. These are located on the World Wide Web at **http://playfair.stanford.edu/reports/donoho/** and **http://playfair.stanford.edu/reports/johnstone/**; and a free wavelet software system is at **http://playfair.Stanford.EDU:80/ wavelab/**.
4. A. Lee, E. Bloom, and J. D. Scargle, these proceedings.
5. J.P. Norris, et al., ApJ **459**, in press (1996).
6. J.P. Norris, these proceedings.
7. J.P. Norris, J. T. Bonnell, R. J. Nemiroff, and J. D. Scargle, these proceedings.

How Exponential are FREDs?

Bradley E. Schaefer and Samuel E. Dyson

Yale University, Department of Physics, New Haven CT 06520-8121

A common Gamma-Ray Burst light curve shape is the "FRED" or "fast-rise exponential- decay". But how exponential is the tail? Are they merely decaying with some smoothly decreasing decline rate, or is the functional form an exponential to within the uncertainties? If the shape really is an exponential, then it would be reasonable to assign some physically significant time scale to the burst. That is, there would have to be some specific mechanism that produces the characteristic decay profile. So if an exponential is found, then we will know that the decay light curve profile is governed by one mechanism (at least for simple FREDs) instead of by complex/multiple mechanisms. As such, a specific number amenable to theory can be derived for each FRED.

We report on the fitting of exponentials (and two other shapes) to the tails of ten bright BATSE bursts. The BATSE trigger numbers are 105, 257, 451, 907, 1406, 1578, 1883, 1885, 1989, and 2193. Our technique was to perform a least square fit to the tail from some time after peak until the light curve approaches background. We find that most FREDs are not exponentials, although a few come close. But since the other candidate shapes come close just as often, we conclude that the FREDs are misnamed.

INTRODUCTION

The BATSE detectors have provided the Gamma Ray Burst community with a large number of light curves of excellent time resolution and statistics. Most bursts have complicated light curves, often with structure much shorter than their duration. In contrast, the BATSE data have shown that a fraction of the bursts appear to have a simple and smooth light curve that consists of a fast-rise followed by an exponential decay. These fast-rise exponential-decay bursts have been called by their acronym as "FRED" bursts or simply as "FREDs".

Suppose that FREDs were shown to have a decay that was closely an exponential function. Then we would have a unique timescale. More importantly, we would have confidence that some single and simple physical process created that timescale. This might then suggest the physical mechanism responsible. For example, perhaps black hole formation in a NS-NS collision may never produce an exponential decay while the optical thinning phases of energetic fireballs always produce exponentials. Exponentiality might be used as a test for any proposed model. Thus, if FREDs are truly exponential, then we would know the decay is a single simple process and we would have critical tests for

© 1996 American Institute of Physics

specific proposed mechanisms.

The fitting of exponentials to the decay portion of the light curve also has utility for defining the energy dependence of the light curve. This question is of critical importance for the interpretation of time dilation in bursts (1). That is, cosmological bursts of high z will not only appear dilated but will also appear redshifted so that the observed light curve is from a shifted energy. If there is a dependence of duration with energy, then the two effects (dilation and redshift) will be confused and offset. Fenimore et al. (1) has used an autocorrelation technique to evaluate this energy dependence and finds that the time scale is proportional to $E^{-0.4}$. Yet this technique is dominated by the fast rising portion of the light curve. So the fits to the decays provide a different measure of the energy dependence. It is unclear which measure is more relevant for time dilation studies.

EXPONENTIALITY TESTS

How can we test whether a particular light curve has an exponential shape? We have selected ten simple and bright FRED bursts, and then we have fitted exponential functions to the decay portion of the light curve. These fits were made with the traditional chi-square optimization of fit parameters. These fits yield best fit parameters and chi-square values which can be used for tests of exponentiality. We have used five such tests:

(I) How low is the reduced chi-square for an exponential fit? This obvious test is to simply fit an exponential to the decay portion of the light curve, then examine the reduced chi-square value to see if the fit is satisfactory. For example, if the value is near unity for a fit over the entire post-peak time interval, then an exponential shape would be reasonable. We will adopt a threshold of 1.5 for the reduced chi-square for acceptable fits.

(II) How close to the peak can a fit be pushed? Most FREDs have a rounded top for which any exponential would not yet be relevant. So it should be acceptable to start the fit at some time well after peak. But if the fit time interval is started too late, then the light curve will cover a small dynamic range and an exponential will always give a good fit. (Remember, these light curves were chosen for smooth decays.) But if a good fit can be pushed close to the peak time, then the exponential shape becomes more compelling. Bursts for which the start time of an acceptable fit is within one e-folding time scale of the peak will have passed this test.

(III) How many e-folding times are covered? Suppose that a fit covered a small time range (in units of the exponential time scale), then the light curve would have small curvature and many functions would fit well. But if the fit covered many e-folding times, then the characteristic shape of the exponential would be unique. In practice, we will require that an acceptable exponential fit must cover more than 3 e-folding time scales.

(IV) Does the time scale change throughout the decay? If the light curve decay is really exponential, then the same time scale should be found for the last half of the decay as for the first half of the decay. This test avoids the problem that the good statistics in the first half will dominate any fit of the entire decay. This test is implemented by comparing the fitted time scales for the first and second halves of a time interval starting just after the peak and ending when the light curve goes to the background level.

(V) Do other functional forms fit as well? Our analyzed bursts were selected for having a decay shape that falls off smoothly with a decelerating rate. But might the real functional form be something other than an exponential, with the exponential fit being fortuitous? To test this, we have fit alternative shapes to the decay. If the other shapes yield comparable quality fits, then the exponential fits are likely accidental. This test is applied to all fits together, by comparing the fractions of acceptable fits produced by each functional form. Our first alternative functional form is a power law, where the zero time is allowed to vary. If this zero time is set long before the peak, then a power law function will approximate an exponential. Our second alternative functional form is an exponential of time raised to the one-third power. This specific form is identified as fitting the average peak aligned light curves of 460 BATSE bursts (2).

These five tests were applied to 10 bright FRED bursts. The light curves were provided by the Compton Observatory Science Support Center, with Robert J. Nemiroff taking time to provide the data in a convenient format. We used Large Area Detector data with a time resolution of 1.024 seconds. Nemiroff provided a background fit based on the Goddard group's extensive experience in such matters. We used data from all four BATSE discriminator channels and treated each separately. The ten bursts we examined have BATSE trigger numbers of 105, 257, 451, 907, 1406, 1578, 1883, 1885, 1989, and 2193. Burst 907 consisted of a well-separated FRED followed by a jumble of FREDs, while burst 1989 consisted of three well-separated FREDs. Thus, a dozen FREDs have been examined with four energy channels.

RESULTS

Of the 48 light curves, only 5 passed the first test, 21 passed the second test, 12 passed the third test, and 7 passed the fourth test. Only two burst/channel combinations (for channels 2 and 3 of trigger 1885) passed all four tests. For channel 2 of trigger 1885, the reduced chi-square just after peak is 1.2, an acceptable fit is obtained right up to the time of the peak, 5.0 e-folding times are encompassed in the fit, and the first and second half time scales are 26 and 19 seconds. For channel 3 of trigger 1885, the reduced chi-square after the peak is 0.7, an acceptable fit is obtained right up to the time of peak, 3.8 e-folding times are encompassed in the fit, and the first and second half time scales are 12.0 and 12.0 seconds. Thus, only 2 out of 48 light curves produced a true exponential shape.

The same four tests were applied to the power law and Stern functions. For the first test alone, the two alternative functions had 6 and 4 passed fits, as compared to 5 passed fits for the exponential function. For the first four tests combined, the two alternative functions had 3 and 1 passed fits, as

Trigger	Chan. 1	Chan. 2	Chan. 3	Chan. 4
#105	1.1	1.0	1.0	1.5
#257	21.3	28.6	27.6	20.5
#451	5.0	6.4	7.9	22.8
#907	13.8	8.0	4.3	14.5
#1406	12.6	10.9	9.9	17.1
#1578	13.9	11.9	10.1	19.9
#1883	7.3	7.4	4.9	1.4
#1885	67.4	24.8	12.4	...
#1989a	9.4	7.4	4.6	...
#1989b	10.7	9.8	8.0	...
#1989c	23.0	20.9	17.6	...
#2193	68.5	39.9	18.1	66.0

compared to 2 passed fit for the exponential function. (Note that two of the passed fits for the power law function were for trigger 1885 where the best fits had a very early zero time so as to effectively produce an exponential fit.) With these statistics, we conclude that these ten bursts fail the fifth test for exponentiality. That is, alternative functional forms produce as many good fits as does the exponential form, so there is no reason to think that there is anything special about the rare light curve that does appear to fit an exponential shape.

Our fits provide a direct measure of the decay time scale for a dozen FREDs as a function of energy. Fits have been made to identical time ranges for each BATSE channel. (Some bursts had no sigificant flux in channel 4 at any time.) The chosen time ranges started just after the peak and ended when the burst flux became insignificant in any channel. The energy range for the four BATSE discriminator channels are roughly 25-50 keV, 50-100 keV, 100-300 keV, and >300 keV for channels 1 to 4 respectively. The best fit e-folding time scales for each FRED are given in the table above.

We see that, over channels 1 through 3, the decay time scale decreases with rising energy. This is similar to the result found through an autocorrelation analysis (1). Their results are based on lag times of less than 2.5 seconds and thus relate specifically to the energy dependance of the rise time of bursts. Our results are for the decay times of bursts and thus are complimentary to the rise time energy dependance.

We were surprised that 6 out of 8 FREDs had much longer time scales in channel 4 than in channel 3. This is strongly different than was found for the rise times (1). The significance of this result is unclear.

The relative energy dependance of the light curves can be quantified by averaging the bursts together. Specifically, we have scaled all individual burst time scales by the time scale for channel 2 and then logarithmically averaged the values for all available bursts. For the rise times, we can use the power law energy dependance found by (1). These are compared in the table at the top of this page.

	Chan. 1	Chan. 2	Chan. 3	Chan. 4
Decay Time	1.15	1.00	0.83	1.61
Rise Time	1.28	1.00	0.69	0.42

CONCLUSIONS

We find that the decays of FRED bursts are not well fit by exponentials. Only two channels for one burst had an acceptable exponential fit, a fraction which is comparable to that found for other functional forms. Thus the acceptable exponential fits appear to be rare and fortuitous. This conclusion rejects the possibility that the decay time scale is a useful characteristic as well as an indication of simplicity. A short summary is that FREDs are misnamed.

The decay time scale decreases from channels 1 to 3 (similar to the decrease seen in the rise time scales), although the decay time appears to jump up for channel 4. It is unclear what effect this will have for time dilation analyses.

REFERENCES

1. E.E. Fenimore, et al., ApJ Lett. **448**, L101 (1995).
2. B.E. Stern, ApJ, submitted (1995).

Hardness Ratio versus Duration for PVO compared to BATSE and PHEBUS

I. A. Smith[1], A. Crider[1], E. P. Liang[1], B. C. Dunne[1],
E. E. Fenimore[2], H. Li[2]

[1] *Rice University,* [2] *LANL*

Previous studies have suggested that scatterplots of a measure of the burst hardness versus a measure of the burst duration might be used to separate the bursts into two classes in a cleaner way than using the duration alone. This effect has been seen by both BATSE and PHEBUS: the shorter bursts being harder than the longer bursts. However, we do not find a significant trend of this nature in the PVO bursts. We also show that the quantity used to calculate the hardness ratio can be important, because it affects the size of the error bars and consequently the confidence to which an individual burst can be assigned to a class.

INTRODUCTION

The 3B BATSE catalog appears to confirm the bimodality of the burst durations seen by previous experiments (5,6,9,7,8). The challenge has been to find other characteristic features of the bursts in the two classes, or to find a cleaner way to separate the bursts (e.g. see the papers in these proceedings by Kouveliotou et al., Belli, Pizzichini, Dezalay et al., and references therein).

One promising result from a study of the BATSE and PHEBUS catalogs was to plot a measure of the burst hardness against a measure of the burst duration. For the BATSE bursts, Kouveliotou et al. (7) determined the hardness ratios HR_{32} of the total counts in the 100 to 300 keV and 50 to 100 keV energy ranges; these counts were determined for the duration T_{90}. They plotted the HR_{32} against T_{90} for the bursts in the first BATSE catalog; their Figure 4 is reproduced in Figure 1. From this they concluded that there is a significant correlation between the hardness ratio and the duration.

A similar hardness ratio versus duration behavior was found in the PHEBUS bursts (2,3). They used higher energy channels than BATSE (320 to 7000 keV and 120 to 320 keV) and used two different measures for the hardness ratio, the mean hardness ratio (the ratio of the count rates for the spectrum integrated over the whole burst) and the peak hardness ratio (which uses the spectrum at the time of the peak). A significant hardness ratio versus duration correlation was found for both hardness measures.

© 1996 American Institute of Physics

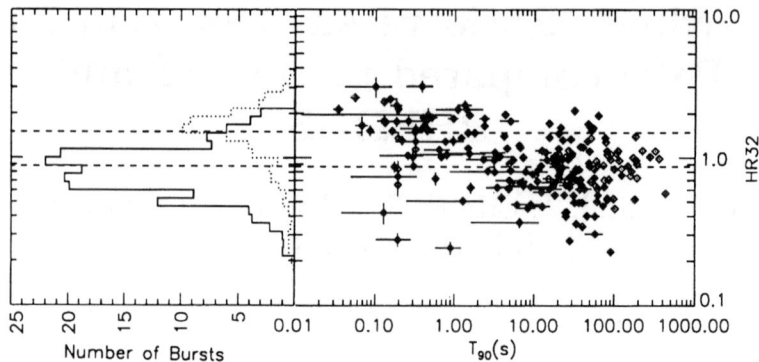

FIG. 1. Figure 4 of Kouveliotou et al. (7). The hardness ratio uses the total counts during the duration T_{90}.

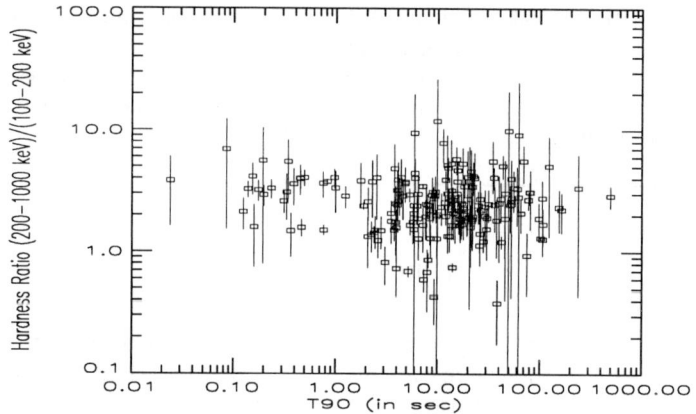

FIG. 2. PVO Catalog. The peak hardness ratio is used.

PVO RESULTS

We used several measures for the hardness ratios of the bursts listed in the PVO catalog (4); these included using fluences and peak counts, and using different energy channels. We determined T_{90} durations using an algorithm similar to the one used by BATSE; see (4) for details of the duration algorithm and errors. We then plotted the hardness ratios against T_{90}. As found previously (6) we also found that there is a bimodal distribution in the durations of the PVO bursts, with an approximate division at $T_{90} = 1.5$ s.

In our analysis so far, we have not seen a significant correlation between the hardness ratio and the duration using any of the hardness measures. As an example, Figure 2 shows the peak count hardness ratio using the 200 to 1000 keV and 100 to 200 keV channels for 247 PVO bursts (approximately one fourth of the PVO bursts were not used because of problems such as data gaps, or the burst did not return to background before the memory filled up). The sample average and sample standard deviation for the short bursts ($T_{90} < 1.5$ s) is 3.32 ± 0.93, compared to 2.75 ± 1.64 for the long bursts.

Currently, we do not have a good explanation why the PVO results disagree with the other experiments. Since we do not see a significant correlation using any of the hardness measures or energy channels, it seems unlikely that instrumental selection effects are the cause; however, this cannot be ruled out. Because our energy channels lie between those of BATSE and PHEBUS, this is unlikely to be important.

It might be expected that the correlation would be weaker when peak counts are used, because bursts tend to show a hard-to-soft evolution that would naturally make longer bursts softer in an integrated measure. Indeed, the PHEBUS result was less significant using the peak hardness ratios (3), and we find a similar result using the BATSE peak counts; however, a significant correlation is still seen for both BATSE and PHEBUS using the peak hardness ratios.

PVO primarily saw nearby bright bursts, and it will take 30 years for BATSE to detect a similar number of nearby bursts. It is therefore an interesting possibility that the bursts detected with PVO are indeed different from the ones seen with BATSE and PHEBUS, and that the difference in the hardness ratio results comes from evolution or distance effects. As a preliminary test of the bright BATSE bursts, we calculated the peak hardness ratios using the peak counts in the 100 to 300 keV and 50 to 100 keV energy channels on the 256 ms timescale for the 3B catalog. The brightest 100 bursts still show a significant hardness ratio versus duration correlation, though the effect is slightly weaker.

BATSE RESULTS

The total counts used by Kouveliotou et al. (7) are not included in the public BATSE data. Instead, the energy fluences in the same energy channels are given. We used these to determine the hardness ratio F_{32} of the energy fluence in channel 3 (100 to 300 keV) to the energy fluence in channel 2 (50 to 100 keV), and plot this versus T_{90}. Figure 3 shows the results of using the original 2B catalog (406 bursts have measured values for all the relevant quantities). The result is similar to Figure 1. The panel on the left shows that there appears to be a separation into two clusters, which might provide a cleaner way to separate the bursts into two populations compared to only using their duration.

The panel on the right in Figure 3 includes the errors in the measurements. The line roughly marks the boundary $F_{32} = 2.2 T_{90}^{0.436}$ that appears to separate the two clusters in the left hand panel (see also (1)). It can be seen that including the errors dilutes the apparent split into two clusters, although there still appears to be a hole in the scatterplot. Note that the errors are relatively larger than the ones in Figure 1. This suggests that using total counts may be

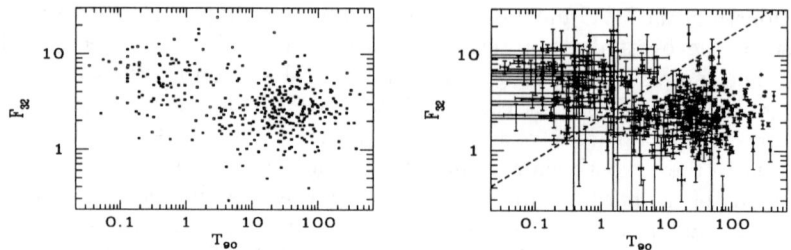

FIG. 3. 2B Catalog using the energy fluences for the hardness measure.

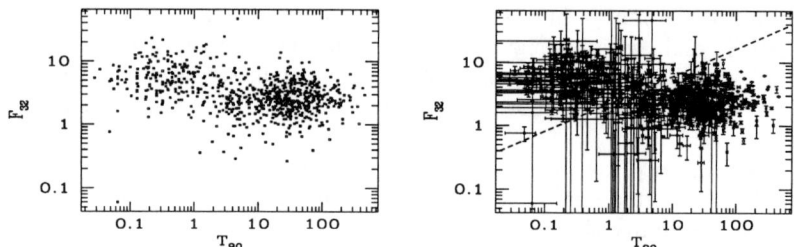

FIG. 4. 3B Catalog using the energy fluences for the hardness measure.

more reliable than using energy fluences for studies of this nature using the BATSE catalog. This becomes particularly important when the properties of sub-classes of the bursts are compared based on cuts made in the hardness ratio versus duration plots (e.g. see the papers in these proceedings by Kouveliotou et al., Belli, Pizzichini, Dezalay et al., and references therein). The confidence that a burst can be assigned to a particular class depends on the size of these errors.

Figure 4 shows the same results as Figure 3, but using the 3B catalog with the revised durations (797 bursts are plotted); see also (8). The split into two clusters is slightly less evident, especially when the error bars are included. Finally, Figure 5 shows the 385 bursts in (3B-2B), with the revised durations. It can be seen that these latter bursts appear to fill in the gap, making the separation into two clusters less clear.

We have tried using other measures for the hardness of the bursts in the 3B catalog. For example, the average photon energy of the burst can be found by dividing the total energy fluence in all four channels by the total photon fluence in all four channels. The hardness duration correlation is still present, though the errors in the values of the hardnesses are usually larger.

Acknowledgements We would like to thank the BATSE team for providing us with data that is not in the public archive. This work was supported by NASA grant NAG 5-1515 at Rice University.

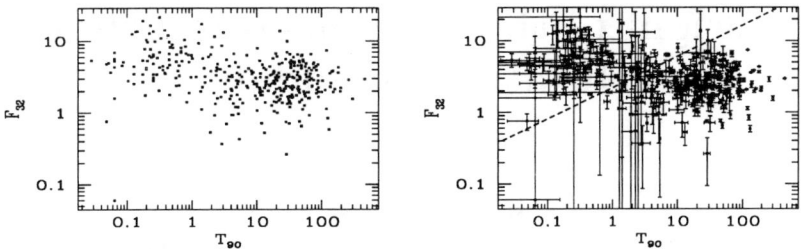

FIG. 5. (3B-2B) Catalog using the energy fluences for the hardness measure.

REFERENCES

1. B.M. Belli, Ap&SS **231**, 43 (1995).
2. J-P. Dezalay, et al., in Gamma-Ray Bursts, ed. W. S. Paciesas & G. J. Fishman (New York: AIP **265**), 304 (1991).
3. J-P. Dezalay, et al., Ap&SS **231**, 115 (1995).
4. E.E. Fenimore, et al., in preparation (1996).
5. K. Hurley, in Gamma-Ray Bursts, ed. W. S. Paciesas & G. J. Fishman (New York: AIP **265**), 3 (1991).
6. R.W. Klebesadel, in Gamma-Ray Bursts: Observations, Analyses and Theories, ed. C. Ho, R. I. Epstein, & E. E. Fenimore (Cambridge U. Press), 161 (1992).
7. C. Kouveliotou, et al., ApJ **413**, L101 (1993).
8. C. Kouveliotou, et al., in preparation (1996).
9. C.A. Meegan, et al., ApJ submitted (1996).

Origin of the Gamma-Ray Burst Duration Bimodality

V. C. Wang

Joint Science Department
The Claremont Colleges, Claremont, California 91711

The bimodality in the distribution of gamma-ray burst duration, T_{90}, observed with BATSE suggests there are two distinct subclasses of bursts, placing serious constraints on the origin of gamma-ray bursts. However, the T_{90} samples are a mixture of individual peak durations and temporal separations between adjacent peaks. The studies of individual peak durations and the peak separations in the BATSE 1B catalog show these are two distinct distributions. Each distribution can be fit by a single smooth function, but with different time scales. I propose that a bimodality can be produced by the superposition of these two distinct time scales in bursts. A preliminary Monte Carlo simulation shows that mixing these two distributions randomly in a burst can generate apparent bimodality.

INTRODUCTION

Gamma-ray bursts are among the most perplexing phenomena in the Universe. The diversity in time histories and energy spectra and the lack of observed quiescent counterpart in any other wavelength complicate the study of the underlying sources. Our understanding of gamma-ray bursts has to rely heavily on the study of their global properties.

The distribution of gamma-ray burst durations is one of the important global properties which have been studied extensively in the past with various instruments (e.g. (1–3)). The peak of the duration distribution varies from 10s to 40s, depending on the thresholds and data accumulation times of the instruments, and on the duration definition. BATSE has adopted a new definition of burst duration, the nominal T_{90}. It is defined to be the time during which the accumulated excess ("burst") counts above background increase from 5% to 95%, thus encompassing 90% of the total "burst" counts. The T_{90} distribution of 222 bursts in the first BATSE catalog ((4), hereafter 1B catalog) exhibits significant bimodality (5). There is a deficit in bursts with T_{90} around 1 s to 10 s, which morphologically separates gamma-ray bursts into two classes (Fig. 1).

If the bimodality is an intrinsic property of burst duration, it implies there may be two distinct subclasses of classical gamma-ray bursts. A study on the correlation between the duration and the spectral hardness ratio (5) suggests short bursts are predominantly harder, and longer ones tend to be softer.

© 1996 American Institute of Physics

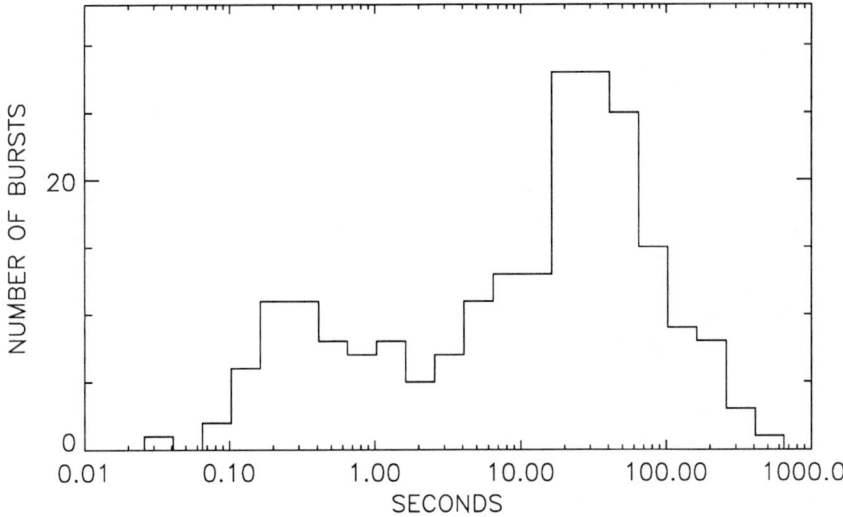

FIG. 1. Distribution of duration as measured by T_{90} in the first BATSE catalog.

However, the sky maps of both subclasses are consistent with isotropy. Moreover, the intensity distribution between these two subclasses show no significant differences. Clearly, the bimodality will place serious constraints on the sources of gamma-ray bursts. There are, however, inconsistencies in the burst duration definition (6).

PROPOSED ORIGIN OF BIMODALITY

In an attempt to understand the origin of the duration bimodality, I re-examine the time histories of 260 bursts in the 1B catalog and find an anomalous effect from multi-peaked bursts in the T_{90}. Since the majority of gamma-ray bursts in the 1B catalog consist of multiple peaks, this effect can account for the observed bimodality in T_{90}.

The adjacent peaks in a multi-peaked burst are often separated by background on the time scale of a few seconds to several hundred seconds. If such a burst has relatively strong secondary peaks, its T_{90} includes the time span across separated peaks. One example is GRB911005, listed as Trigger 869 in 1B. There are six well separated peaks in this burst with durations ranging from 2 to 17 seconds and peak separations ranging from several to 70 seconds. Its T_{90} is measured as 110 seconds, including all six peaks. Clearly in this case, the time scale of T_{90} is dominated by peak separations. On the other hand, for a single-peaked burst or a multi-peaked burst with relatively dim secondary peaks, T_{90} may only measure the brightest peak. For example, the brightest peak of GRB910522, or Trigger 219, occurs 105 seconds after trigger.

Its T_{90} is only 29 seconds, which does not include the first peak and shows the time scale of a single peak duration.

Besides the relative brightest of individual peaks, the duration measurement is also biased by the local background and trigger threshold. A multi-peaked burst can significantly lose parts of its signals to the background, depending on the distance and the local background level. Therefore a burst at greater distance may appear very different in duration from a similar burst closer-by. For a multi-peaked burst at greater distances, T_{90} often fails to include all peaks, e.g. the precursors, and only measures the width of the brightest peak; whereas a similar burst at closer distance is likely to have more of its peaks included in T_{90} and shows a different time scale. The same effect happens when the background varies and the duration is changed significantly.

Therefore, I propose an alternative origin for the observed bimodality. It is artificially produced by mixing two distinct time scales in the T_{90} measure: the individual peak duration and the temporal separation between adjacent peaks. The duration samples are a combination of bursts with only one peak included in T_{90} and bursts with multiple peaks included in T_{90}. The division between these two groups is purely instrumental as discussed above. However, these two groups are characterized by different time scales underlying the peak durations and peak separations, respectively.

This model does not require bimodality in either distribution to produce the observed bimodality in T_{90}. It can also explain the apparent correlation between duration and spectral hardness ratios. As mentioned above, a previous study (5) showed that durations are anticorrelated with their spectral hardness ratios. Short gamma-ray bursts are predominantly harder and longer ones tend to be softer. This hardness ratio was calculated by integrating the counts throughout the entire burst in each of the four discriminator channels. However, such correlation could arise from the spectral evolution. Ford et al. (7) investigated spectral evolution in 37 bright, long gamma-ray bursts observed with the BATSE spectroscopy detectors. They have found evidence that for bursts with multiple peaks, later peaks tend to be softer than earlier ones. Therefore, the averaged hardness ratios for multi-peaked bursts tend to be smaller.

PRELIMINARY TEST AND RESULT

I have studied the individual burst peak durations and temporal separations between adjacent peaks in the BATSE 1B catalog as two distinct distributions. The duration of a peak is defined to be the period of time during which burst flux continuously exceeds the local background by some threshold. More than 1000 peaks are found and analyzed in the 1B catalog, based on this definition.

The distribution of the individual peak durations in 1B bursts is presented in the upper panel of Fig. 2, which can be fit with a single smooth function. To suppress the noise fluctuation and to test the instrumental effects on the distribution, the peak durations are measured with respect to various threshold levels. The result shows the distribution is rather sensitive to the

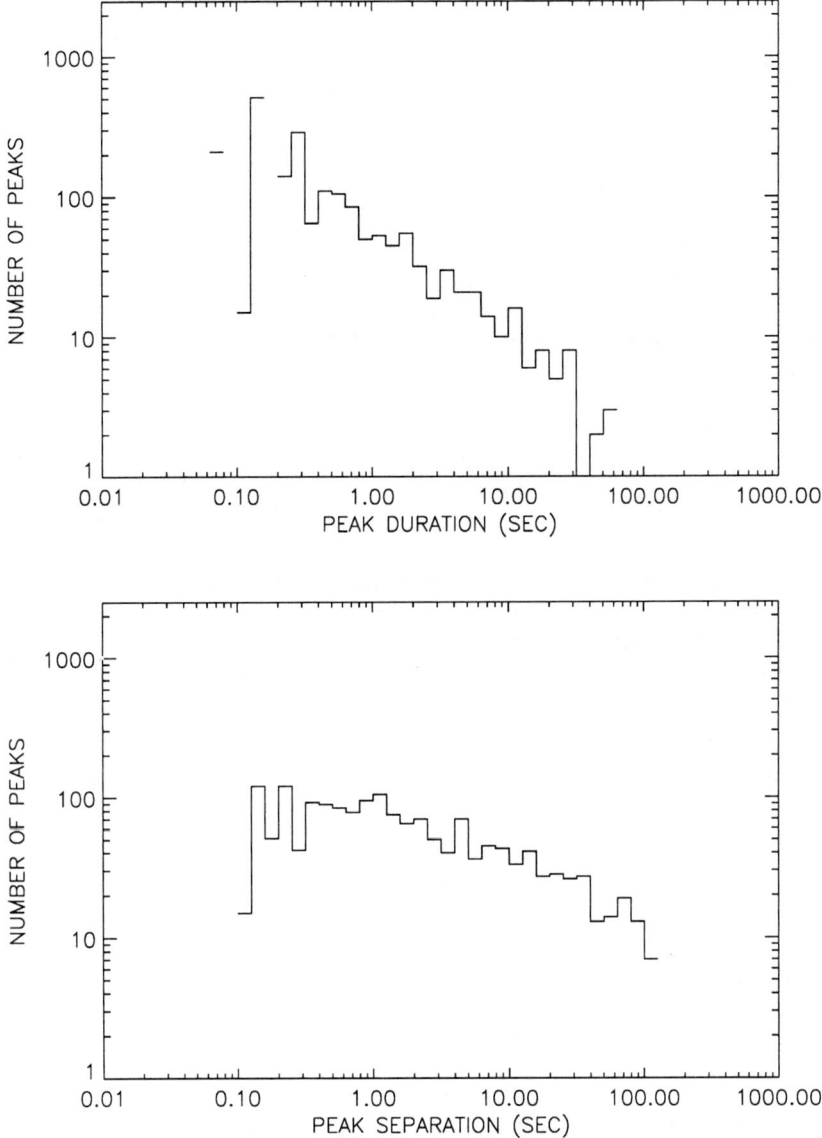

FIG. 2. Upper panel: Distribution of peak durations of 222 bursts in the 1B catalog. Lower panel: Distribution of peak separations for the same bursts.

background level. Nevertheless, the single smooth distribution is fairly robust (8). There is no bimodality observed in the peak duration distribution.

A similar study is done on the peak separation. It is defined to be the period of time when the detected flux falls below the same threshold level between two adjacent peaks. The distribution of all peak separations in the 1B bursts is presented in the lower panel of Fig. 2, which is also consistent with a single smooth function. Moreover, it has a much flatter slope than the distribution of peak durations, indicating the temporal separations between adjacent peaks and the durations of the peaks have very different time scales. Typically, peak separations are longer than peak durations. The combination of these two time scales can explain the observed bimodality. For the T_{90} samples are a mixture of individual peak durations and the time spans across multiple peaks, which include both durations and separations. Bursts with multiple peaks included in T_{90} are generally longer than bursts with only one peak included in T_{90}.

Simulations are currently being made to understand the relation among the distributions of peak duration, peak separation and the morphological shape of T_{90} distribution (8). A preliminary result supports the hypothesis that bimodality can be produced by mixing two distinct time scales. In this study, the time profiles of multi-peaked bursts are simulated with peak durations and separations randomly drawn from two distinct power-law functions. Their T_{90} distribution exhibits significant apparent bimodality, very similar to that observed. I am also exploring any possible correlation between peak duration and peak separation that may contribute to bimodality. So far no such correlation has been found.

Acknowledgements: I thank NASA for support under a CGRO Cycle 5 grant.

REFERENCES

1. T. Cline, & U. Desai, in 9th ESLAB Symp., (Noordwijk:ESRO), 37 (1974).
2. E. Mazets, et al. Astron. & Astrophys. **80**, 3 (1981).
3. K. Hurley, in Gamma Ray Bursts, AIP Conf. Proc. **265**, eds. W. Paciesas & G. Fishman, 3 (AIP, New York, 1992).
4. G. Fishman, et al., ApJ Supp. **92**, 229 (1994).
5. C. Kouveliotou, et al., ApJ Lett. **413**, L101 (1993).
6. R. Lingenfelter, V. Wang, & J. Higdon, in Gamma Ray Bursts, AIP Conf. Proc. **307**, eds. G. Fishman, J.J. Brainerd, K. Hurley, 222 (AIP, New York, 1994).
7. L. Ford, et al., ApJ **439**, 307 (1995).
8. V. Wang, in preparation (1996).

A Multifractal Study of the Scaling Property of Gamma-Ray Burst Time Profiles

Yuan Yan[1], J-P Lestrade[1], J-P. Dezalay[1]
Mitzi Adams[2], and Beverly Stark[2]

[1] *Department of Physics and Astronomy*
Mississippi State University
Mississippi State, Mississippi 39762, USA
[2] *Marshall Space Flight Center*
Huntsville, Alabama 35812, USA

> Multifractal analyses have been proven successful in exposing scaling properties of experimental data. These methods of analysis have been applied to diverse areas of research including solar magnetic fields and dissipative fluid dynamics. Moreover, they are especially useful as algorithms for the detection of fully developed turbulence. In this preliminary paper we use the observed multiplier distribution, $P(M)$, to directly calculate the $f(\alpha)$ function of gamma-ray burst time profiles. The results show the time-scaling properties of these profiles. At the very least, multifractal analyses should provide a better way to categorize GRB time profiles than currently exists. In the most-optimistic scenario, however, these algorithms could reveal fundamental physical properties of the underlying system(s) that produce the bursts.

INTRODUCTION

With the notable exception of the 8-sec periodicity in the March 5th, 1979 event, past studies of gamma-ray burst time profiles have failed to show periodicities or deterministic trends in profile structures (1,2). One approach that has not yet been extensively tried is that of fractal algorithms. A good reason for this is that most fractal algorithms, in order to give dependable results, require dynamic systems of long duration that have achieved stationarity. With gamma-ray bursts, we are confronted with transient phenomena with usually short data sets (i.e., only hundreds of data points).

Still, it could be interesting to see what results a multifractal analysis would give. The first question that we may be able to answer, for at least the longest, more complicated profiles, is whether or not they are self-similar. That is, are the features that are seen on the large scale replicated on intermediate or smaller scales? Second, we would like to know if the structures seen in GRB profiles are totally random in nature or if there is some determinism in them – a determinism that perhaps can be revealed through a multifractal analysis.

© 1996 American Institute of Physics

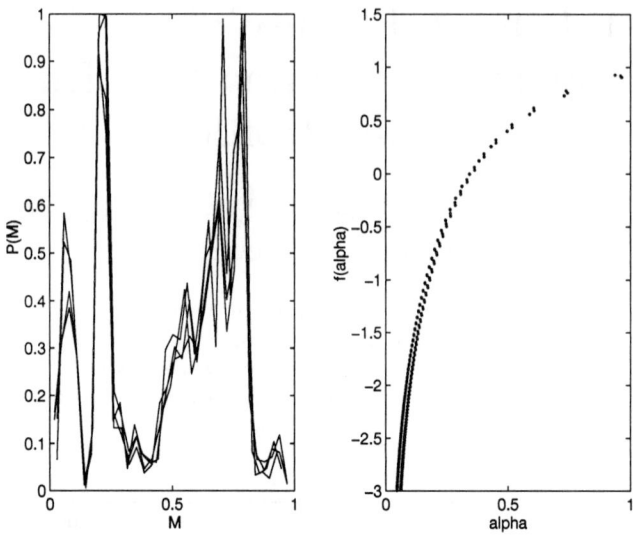

FIG. 1. Multiplier distribution $P(M)$ and multifractal spectrum from a known $M(0.2, 0.8)$ data set on four different scales.

MULTIPLICATIVE CASCADE RANDOM SAMPLING

In order to investigate the scaling nature of our data, we first begin by constructing the measure multiplier, M. This multiplier is the ratio of the number of counts in the left half of a time range to the number of counts in the complete range:

$$M = \frac{\sum_{i=1}^{L/2} X_i}{\sum_{i=1}^{L} X_i}$$

The ranges are sampled randomly on the data set. A histogram of these multipliers is the multiplier probability distribution $P(M)$. As the length of the time range is changed, this gives us a "multiplicative cascade multiplier distribution". Lawrence (3) shows that $P(M)$ is a very useful tool. From it we are able not only to determine the scaling nature of our set, but also to calculate the multifractal spectrum, $f(\alpha)$. Indeed, he claims that $P(M)$ is a more fundamental property than $f(\alpha)$. Furthermore, invariance of $P(M)$ with changing time scale is one of the criteria for the time scale invariant nature of the time series.

To test our code, we have constructed a data set which was built using a cascade multiplier process. That is, a number of counts are divided between two halves of the total time range according to a fixed ratio – in this case 1/5 – 4/5. Then, those two halves are further divided and the counts are again redistributed with the 1/5 and 4/5 ratios. Thus, we now have 4 time bins with measures of 1/25, 4/25, 4/25, and 16/25. This process continues until the dataset is of the desired length. We then apply our multiplicative cascade algorithm to these data to search for the presence of both scale invariance and the 1/5–4/5 signature.

Figure 1a shows the result. We purposely chose a short data set since our grb profiles are typically of this length. As this figure demonstrates, the lack of fine detail due to a short data set results in poor statistics.

The scale invariance, or self-similarity, of the dataset on different time scales shows up as an overlap of the 4 $P(M)$ distributions. Also evident is the signature of the 1/5 – 4/5 ratio in the count distribution

CALCULATING THE MULTIFRACTAL SPECTRUM FROM THE MULTIPLIER DISTRIBUTION

There are two benefits in constructing the $P(M)$ histograms. First, it is then easy to observe scaling features in the data (3). Second, from the moments M^q of $P(M)$, we can calculate the singularity index α and the spectrum $f(\alpha)$ through (3–5)

$$\alpha(q) = \frac{- <M^q log(M)>}{<M^q> log(2)}$$

$$f(\alpha) = q\alpha(q) + \frac{log(<M^q>)}{log(2)} + D_0.$$

In these calculations, q is an integer (≥ -1) and D_0 is a constant. All of the averages are over the multiplier distributions. Figure 1b presents the multifractal spectrum for the known multiplier artificial data set. Note that on four different time scales (64, 32, 16, and 8 data points), the $f(\alpha)$ curves overlap each other on the left slope, corresponding to the moments of $q \geq 0$. This implies that the $P(M)$ of the datasets are governed by the same power law. This is due to the fact that the artificially-generated datasets are self-similar on these scales.

MULTIPLICATIVE CASCADE MULTIPLIER DISTRIBUTIONS AND MULTIFRACTALS OF GRB TIME PROFILES

The data we are using are the BATSE DISCSC 64-msec intensity-time profiles. To date, we have applied the algorithm to only a few GRB's. Before we can run in production mode, there are many questions that have to be answered about the treatment of problems, such as noise, in the data. Therefore, our results in this paper are preliminary.

The ordinate in GRB profiles is the total count rate in all four energy channels. We normalize our profiles using the peak amplitude in the 64-sec

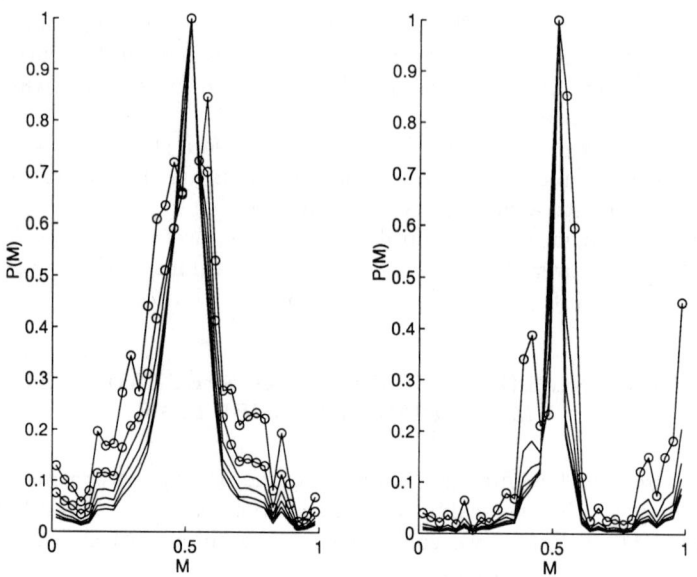

FIG. 2. Multiplier distribution $P(M)$ for a "Spikey" (left) and a "smooth" (right) GRB profile.

time interval $x_M(t)$ and the minimum in the background, $x_m(t)$.

$$X(t) = \frac{x(t) - x_m(t)}{x_M(t) - x_m(t)}$$

The measure, i.e., the normalized count rate, is thus always positive.

Figure 2a presents the multiplier probability distribution $P(M)$ on six time scales (8, 4, ..., 0.25 seconds) for a highly-structured ("Spikey") grb profile (BATSE trigger number 1288). The $P(M)$ curves overlap each other on time scale of 2, 1, 0.5, and 0.25 seconds. However, there are variations for time scales of 4 and 8 seconds (curves with "o"). Symmetry about $M = 0.5$ is expected due to the fact that we split our intervals in half to calculate the multipliers. The dominant peak in $P(M)$ centered on $M = 0.5$ is probably due to noise. Similar features in the $P(M)$ for different time scales hint at the possibility of scale invariance in this profile. However, we hesitate to conclude this until we have examined more profiles.

Shown in Figure 2b is the histogram for a smooth profile (BATSE trigger number 2389). Notably, without the large spikes present in the highly-structured profile, the multiplier distributions tend to lie closer to 0.5 and so the histograms are narrower in this case.

Figure 3 presents the multifractal spectra for the two GRB time profiles on time scales of 8, 4, 2, 1, and 0.5 seconds. The most notable difference between the two is that the overlap of distributions is more complete in the case of the highly-structured burst. We have not yet tested the significance of this observation.

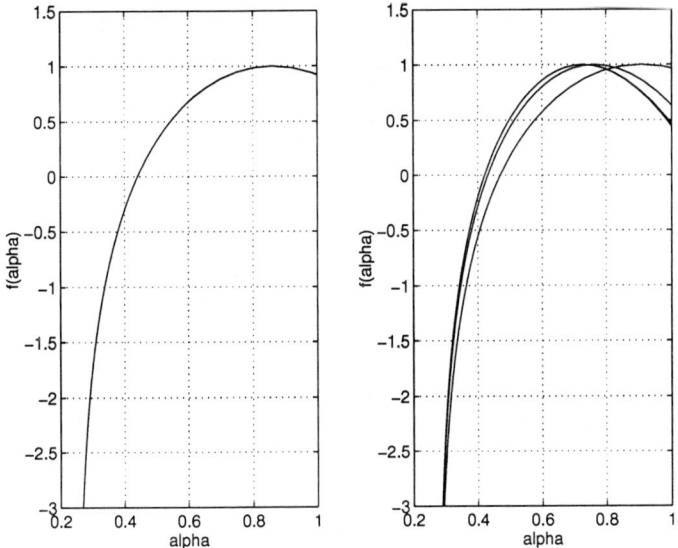

FIG. 3. Multifractal spectrum from a "Spikey" (left) and a "smooth" (right) GRB profile.

PRELIMINARY CONCLUSION

Our study is in an exploratory stage since we have not yet completed the development of our algorithm. We will be applying our code to many bursts to see if the profiles are scale invariant, (i.e., do the micro features mimic the macro features?) or if other deterministic trends are hidden in the apparently random distribution of peaks.

Fluid systems which have reached the state of fully developed turbulence (FDT) are usually scale invariant. This implies dynamic independence for each eddy in the turbulence and short correlation times. If it can be shown that gamma-ray burst time profiles possess such properties, it would have implications for models of burster sources.

REFERENCES

1. J. P. Norris, in Gamma-Ray Bursts, eds. W.S. Paciesas and G.J. Fishman, AIP Conf. Proc. **265**, 294 (AIP, New York, 1993).
2. C. K. Kouveliotou et al., in Gamma-Ray Bursts, eds. W.S. Paciesas and G.J. Fishman, AIP Conf. Proc. **265**, 265 (AIP, New York, 1993).
3. J. K. Lawrence, et al., Phys. Rev. E. **51**, 316 (1995).
4. C. J. G. Evertsz and B. B. Mandelbrot, in Chaos and Fractals: New Frontiers of Science, ed. H.-O Peitgen et al. (Springer, 1992).
5. A. B. Chhabra and K. R. Sreenivasan, Phys. Rev. A. **43**, 1114 (1991).

A Compact Representation of GRB Time Series Using Multiscale Edge Detection

C.A.Young, D.C.Meredith and J.M.Ryan

University of New Hampshire
Center for the Study of Earth, Oceans, and Space
Durham New Hampshire 03824

> Multiscale edge detection using wavelet transform maxima provides a robust method to compress information in a transient signal. We apply this method to GRB time series data from the BATSE LADs. With this technique real temporal features manifest themselves as "edges" over a wide range of time scales, whereas Poisson noise seldom extends beyond the shortest time scales. This provides a method to quantify the variability, identify structures (e.g. FREDs), significantly suppress noise and compress the volume of data by as much as a factor of 50. There is a potential for further parameterizing a GRB time series by identifying and quantifying "edge" families that extend over a range of time scales.

INTRODUCTION

We currently have three main sources of information on gamma-ray bursts (GRBs). The photon spectrum and the spatial distribution of GRBs reveal only one part of the GRB mystery. The GRB lightcurves contain considerable information, but are more difficult to interpret. These time series of individual GRBs vary from simple FREDs (fast rise exponential decay) to complex bursts with many sharp peaks. It appears that these time series may offer information that lead to an understanding of the burst problem. In order to understand what physics these time series might be telling us we need a robust method for reducing the data volume and suppressing noise. (One of the major difficulties with GRB data sets is their high noise level.)

The transient nature of GRBs makes traditional Fourier methods problematic. Wavelets are a promising tool for analyzing transient signals. In the process of studying the use of wavelets (1), we found one form of analysis which lends itself naturally to GRB lightcurves, multiscale edge detection(MSED). MSED is a process of detecting rapid variations or large gradients in a signal. These edges and their locations contain most of the interesting information in transient signals. The scale in MSED denotes the neighborhood over which these variations are detected. MSED provides a method for reducing the noise in these signals, reducing the volume of data needed to represent these signals,

© 1996 American Institute of Physics

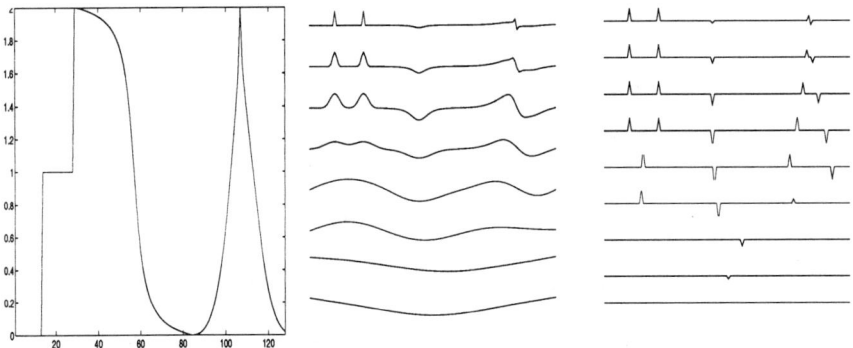

FIG. 1. (a) Signal of length 128; (b) discrete wavelet transform computed over 8 scales; (c) local maxima of the wavelet transform.

and a natural setting for discussing features, and thus the variability, of the signals.

MULTISCALE EDGE DETECTION

An edge detector is used to detect rapid variations within a signal (2,3). One form of edge detector consists of a smoothing operator and a derivative operator. An MSED detector repeats this procedure over many different scales or resolutions. The edge detector we use, $\psi(x)$, is the first derivative of a smoothing function, $\theta(x)$. The smoothing function we use is a cubic spline and so its derivative is a quadratic spline (2). In order to detect edges we perform a discrete convolution of the edge detector with our signal, $f(x)$. This convolution is performed at different scales, s, by a dilation of $\psi(x)$, i.e.

$$\psi_s(x) = \frac{1}{s}\psi(\frac{x}{s}). \tag{1}$$

This algorithm can be applied efficiently if the time series has a length $N = 2^J$, where J is an integer. The range of scales is then $s = 1$ to 2^J. This is called a dyadic scale. After computing this set of convolutions the local extrema of each set are calculated. The extrema represent the sharpest rises or falls in the signal. These final sets of extrema represent the edges in our original signal for a range of scales, s. Figure 1 shows a test signal, its wavelet transform and its wavelet extrema.

The evolution of the extrema over different scales is related to the Lipschitz coefficient that can be used to characterize singularities (4). The Lipschitz condition states that $f(x)$ is Lipschitz α on the interval $[x_1, x_2]$ if,

$$|f(x_2) - f(x_1)| \leq L|x_2 - x_1|^\alpha, \tag{2}$$

where L is a positive number. This means that the "difference quotient",

$$\frac{|f(x_2) - f(x_1)|}{|x_2 - x_1|^\alpha}, \tag{3}$$

never exceeds a fixed finite value L in absolute value. This provides a measure of a function's differentiability or a measure of its degree of singularity. If a function is Lipschitz 0 in an interval it is discontinuous but bounded (e.g. a step) but if it is Lipschitz -1 then it is singular (e.g. a delta function). Lipschitz regularity is used in general for the discussion of functions with infinite resolution, but we can still use this concept to speak of "regularity at a resolution." If our signal is sampled at unity resolution, we can address variations on a scale greater than 1.

In order to relate the extrema to a Lipschitz coefficient we apply a wavelet formalism (2,5). (For more detailed discussions of wavelets, see (6), (7) or (8).) We can show that $\psi(x)$ is a wavelet. This allows us to use the result from wavelet analysis that states that the relationship between the wavelet transform of $f(x)$ and the Lipschitz coefficient in an interval is simply,

$$|W_{2^j}f(x)| \leq K(2^j)^\alpha, \tag{4}$$

where K is a positive number and the wavelet transform of $f(x)$ is just our discrete convolution,

$$W_{2^j}f(x) = f * \psi_{2^j}(x) \tag{5}$$

Using the amplitudes of the wavelet extrema, A, we can calculate the Lipschitz coefficients, α and the constant K by minimizing the following equation,

$$\log_2(A_j) - \log_2(K) - \alpha j = 0. \tag{6}$$

DENOISING

Although no systematic study has been performed to establish that the normal orbital background is Poisson distributed, preliminary indications support this assumption (9). The Poisson noise in GRBs can be approximated as Gaussian. Gaussian noise has the properties that its wavelet transform is also Gaussian, its amplitude decreases by 2 with increasing scale and it has $\alpha > -\frac{1}{2}$ (5). In general (but not always) our real signals have positive Lipschitz regularity. These facts enable wavelet maxima due to a signal to be distinguished from those due to noise by looking at the evolution of the extrema across scales. We see empirically that noise dominates at the finest scales but the signal dominates at scales greater that 2^4. We can then reduce the extrema set by requiring extrema propagate to level 2 or 3 and have Lipschitz coefficients < -2 or we set them to zero. This technique assumes that we know the Lipschitz regularity of the "true" signal so to improve our results we also employ the wavelet thresholding techniques of Donoho (10). The new, smaller extrema set is reconstructed, producing a reduced version of the original signal.

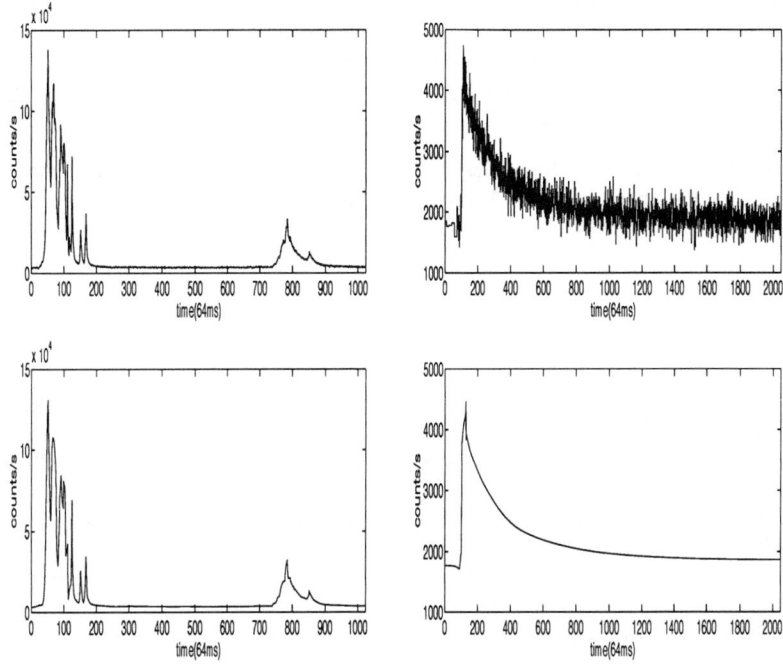

FIG. 2. The top 2 time series correspond to the original burst GRB910503(left) and the original burst GRB910602(right). The bottom 2 correspond to the reconstructed burst GRB910503(left) and the reconstructed burst GRB910602(right)

We can see in Figure 2 the results of this method applied to a FRED burst, GRB910602, and a very intense, spiky burst, GRB910503. The original burst GRB910602 is represented by 1024 data points. After applying the denoising method our GRB is reconstructed with 19 extrema. The noise in this GRB is reduced significantly, allowing the simple structure of this FRED to stand out. GRB910503 is fully represented by 1024 data points. After the denoising the GRB is reconstructed from 123 extrema. Figure 2 shows little difference between the original and reconstructed GRB, due to the higher signal-to-noise ratio. So we have reduced the amount of data while producing a very good reconstruction. These two examples show the power of the MSED.

CONCLUSIONS AND FUTURE WORK

We have applied a method for suppressing noise, compressing data volume and extracting structures in GRB lightcurves. We can now begin to classify the large variety of GRB lightcurves. The bursts can be classified by two different means. First we can classify bursts by the amount with which we are able to compress the data, e.g. simpler bursts are compressed by very large amounts while very complex bursts are not compress much at all. Second, we can classify bursts by quantifying the individual structures in the burst using their Lipschitz coefficients. This new way to represent GRB lightcurves will enable us to begin extracting the information that points directly to the underlying physics producing these phenomena.

Acknowledgments. This work was supported through NASA contract NAS5-26645 and through NASA Compton GRO guest investigation under grant NAG5-2731.

REFERENCES

1. D.C. Meredith, J.M. Ryan, C.A. Young and J.P. Lestrade, in Gamma-Ray Bursts, eds. G.J. Fishman, J.J. Brainerd and K. Hurley, AIP Conf. Proc. **307**, 701 (AIP, New York, 1993).
2. S.G. Mallat and S. Zhong, IEEE Transactions on Pattern Analysis and Machine Intelligence, **7** (1992).
3. J. Canny, IEEE Transactions on Pattern Analysis and Machine Intelligence, **PAMI-8**, 679 (1986).
4. R. Courant and F. John, Introduction to Calculus and Analysis: Volume 1, Springer-Verlag: New York, (1989).
5. S.G. Mallat and W.L. Hwang, IEEE Transactions on Information Theory, **38** (1992).
6. Y. Meyer, Wavelets: Algorithms and Applications, SIAM: Philadelphia (1993).
7. I. Daubechies, Ten Lectures on Wavelets, SIAM: Philadelphia, (1992).
8. G. Kaiser, A Friendly Guide to Wavelets, Birkhauser: Boston, (1994).
9. C.A. Meegan, Private Communication, (1995).
10. D.L. Donoho, IEEE Transactions on Information Theory, **41**, 3 (1995).

SPECTRAL STUDIES

Gamma-Ray Burst Continua: A Review

David L. Band

CASS, UC San Diego, La Jolla, CA 92093

I review our current knowledge of the burst continuum and its evolution during a burst based on the observations of *CGRO* and other missions. The burst continuum can be described adequately by a simple four parameter model which is curved at energies less than a few hundred keV and is a power law $N(E) \sim E^{-2}$ (or softer) at higher energies. However, there are indications of deviations from this simple form below ~15 keV. X-ray emission has been reported before and after the gamma ray emission, and GeV photons were detected for an hour and a half after one burst. Bursts usually (but not always) show hard-to-soft evolution over the burst as a whole and during individual intensity spikes; hardness and count rate variations generally track on a time scale of a second. Further advances will result from detectors with greater spectral resolution and effective area, and from extending the spectrum, particularly to lower energies.

I. INTRODUCTION

Since the continua of classical gamma ray bursts are a manifestation of the basic radiation physics within the burst source, the observed spectral phenomena will constrain and guide the modeling of the emission region once the origin of these events is identified. Observations by BATSE and the other instruments on *CGRO* provide a coherent description of burst spectroscopy which will be a foundation for theoretical future research.

Here I review our current understanding of the continua of classical bursts (i.e., all bursts other than soft gamma repeaters, which are covered elsewhere in these proceedings). Similarly, I do not cover the issue of narrow spectral features (e.g., absorption lines) which are reviewed separately (1). Because the origin of bursts is currently unknown, the theories of continuum formation are highly speculative, and likely to be modified or abandoned when the sources are identified. Consequently, I focus here on the observations which should be relevant whatever the solution of the burst mystery.

While for the past five years I have been a member of the BATSE instrument team specializing in spectroscopy, the views presented here are my own, and do not necessarily represent the team as a whole. Further information can be found in recent conference proceedings (2–5), elsewhere in these proceedings, and in the references cited here.

© 1996 American Institute of Physics

II. HISTORICAL OVERVIEW

The description of the burst continuum has gradually been refined since gamma ray bursts were first discovered. Spectra have been accumulated by detectors with progressively better energy resolution, greater effective area and broader energy range. In addition, the analysis techniques used to deconvolve the observed count spectra have increased in sophistication. The early detectors typically studied only a few strong bursts.

IMP 6 (6) and *IMP 7* (7) observed the spectra of 6 and 9 bursts, respectively, with a 2.25" diameter by 1.5" thick CsI crystal which covered the energy range 0.06–1 MeV with 14 channels. Both detectors found that the spectrum could be described as $N_E(E) \propto \exp(-E/150 \text{ keV})$. In addition, *IMP 7* found that the spectrum above 400 keV was a power law with an index of -2.5. Similarly the KONUS detectors on *Veneras 11* and *12* (8) covered the 0.03–2 MeV energy band with 16 channels with 8 cm diameter by 3 cm thick NaI crystals. The spectra of a sample of 143 bursts could be characterized by $N_E(E) \propto E^{-1} \exp(-E/E_0)$ where E_0 ranged between ~ 20 keV and ~ 2 MeV, peaking at 250 keV.

Observations by the GRS (seven 3" by 3" NaI crystals) on *SMM* (9) showed definitively that the exponential rolloff seen below a few hundred keV could not continue indefinitely. More than 60% of a sample of 72 bursts had significant emission above 1 MeV in excess of extrapolations from low energy.

The *Ginga* burst detector consisted of a proportional counter (covering 1.5–28 keV with 16 channels) and a scintillation counter (covering 14–375 keV with 32 channels), each with an effective area of 60 cm^2. Spectra could typically be fit by either thermal bremsstrahlung or thermal cyclotron models (10). Since *Ginga* had no burst localization capability, the angle between the detectors and a burst was unknown unless another burst instrument also observed the event. Since the detectors' response was angle-dependent, this introduced an uncertainty into the spectral deconvolution.

CGRO has brought a formidable array of spectral capabilities to bear on burst spectra. First, spectra can be observed between ~ 10 keV and a few MeV by BATSE, and when bursts fall in their fields-of-view, between 1 and 10 MeV by COMPTEL and between 35 MeV and 30 GeV by EGRET. Second, the eight BATSE Spectroscopy Detectors (SDs) are built around larger NaI crystals (cylinders of 5" diameter and 3" thickness) than previous burst detectors, and the spectra are accumulated into 256 channels which overresolve the nearly-optimal spectral resolution achievable with NaI (e.g., 19% at 60 keV). Thus the SDs provide better temporal and spectral resolution than previous instruments, which justifies more sophisticated data analysis. The SDs are well-calibrated (11) and the detector response includes scattering off the spacecraft and the Earth (12); spectra are fit using standard χ^2-minimization parameter-fitting techniques. BATSE's Large Area Detectors (LADs) have an enormous collecting area (2025 cm^2); although the spectral resolution is inferior to the SDs, it suffices for characterizing the continuum. Thus the LADs permit spectral studies with greater temporal resolution of fainter bursts.

Third, the long period over which the burst database has been accumulated (since April 1991) by detectors with greater collecting area permits studies of spectra in large burst samples: typically 50–100 bursts for studies requiring high time and spectral resolution, and many hundred events for studies with less stringent requirements.

III. BASIC CONTINUUM SHAPE

Spectra observed by BATSE and all other current detectors are described adequately between $\sim$10 keV and $\sim$100 MeV by the 4 parameter "GRB" function (13)

$$N(E) = \begin{cases} AE^\alpha \exp[-E/E_0], & \text{for } E \leq E_b, \\ A'E^\beta, & \text{for } E \geq E_b, \end{cases} \quad (1)$$

where $E_b = (\alpha - \beta)E_0$ and A' is chosen to make the function continuously differentiable. This function can be fitted over the energy range $\sim$15–1500 keV to spectra accumulated over all or part of a burst by BATSE's SDs or LADs. The parameters differ between and within bursts, with typical values $\alpha = -1$, $\beta = -2$, and $E_0 = 150$ keV. BATSE has refined the earlier description of the continuum by showing that the spectral indices of the low and high energy power laws can vary.

Note that this is a very flexible spectral form which, with the appropriate parameters, can become many of the standard spectral models such as a power law, an exponential and thermal bremsstrahlung. Although the exponential suggests a thermal origin, the GRB function is merely a parameterized spectral shape without any physical motivation. Its shape is similar to the synchrotron spectrum from a power law electron distribution with a low energy cutoff (14,15). In the synchrotron spectrum the electron power law results in a high energy photon power law, while the synchrotron kernel produces an asymptotic low energy power law with a $-2/3$ photon spectral index.

The TGRS germanium detector on the *WIND* spacecraft has an energy resolution superior to BATSE. To date the spectra accumulated by TGRS are consistent with BATSE spectra (16,17).

Since the spectral indices vary, we find that the spectral hardness is best described by E_p, the energy of the peak of $\nu F_\nu \propto E^2 N(E)$, not E_0. For the spectra of entire bursts we find a broad E_p distribution peaked at $E_p \sim$ 150 keV, with only a few bursts above 500 keV (see Figure 1). Thus there are no characteristic energies evident in the spectra (13). As I will show below, the spectrum varies during a burst, demonstrating that variations in E_p are intrinsic to the emission process, and are not a consequence of the redshifting of a characteristic energy. Whether the width of the E_p distribution is narrow enough to constrain theoretical models severely is a matter of interpretation. Selection effects may favor a particular range of E_p values (18). If the burst is too soft then the spectrum will cutoff below a detector's trigger energy (usually 50–300 keV for BATSE). If a burst's peak energy luminosity is the standard

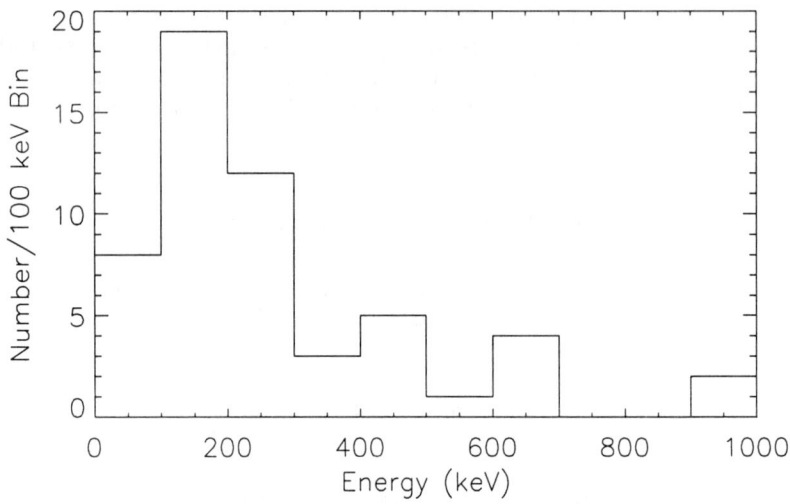

FIG. 1. Distribution of E_p, the peak of $\nu F_\nu \propto E^2 N(E)$, for a sample of 54 intense bursts.

candle, then hard bursts will have fewer counts in the trigger band than softer bursts. That a detector's trigger energy affects the observed E_p distribution is demonstrated by the high energy turnovers in the bursts accumulated by the shield on SIGMA (19), which triggers in the 0.25-2 MeV band and provides spectra covering 0.2-15 MeV.

Spectra accumulated by COMPTEL and EGRET, when intense bursts fall within their fields-of-view, are consistent with a continuation of the low energy spectra observed by BATSE. Thus far, COMPTEL has observed 18 bursts (20) and EGRET five (21). In particular, the spectra can be described as power laws with spectral index $\beta \sim -2$. When curvature is seen at the low energy end of COMPTEL spectra (e.g., in GB 910814 and GB 940217 (20)), BATSE finds a high value of E_p.

Joint fits to spectra accumulated by different instruments can all be described by the 4 parameter GRB spectral form. Specifically, spectra from Ulysses and COMPTEL have been fit for GB 910503, GB 911118, GB 920622 (22), from Ulysses, COMPTEL and the BATSE LADs for GB 920622 (23), and from COMPTEL and the BATSE SDs for GB 940217 (24). Similarly, composite spectra have been created by independently fitting the spectra from different missions and then plotting the resulting photon spectra together. These composite spectra also have a shape which is consistent with the GRB functional form: curvature at the low energy end, a power law at the high end. Such spectra have been produced for GB 910503 (25,26) and GB 940217 (27,28) incorporating BATSE SD, COMPTEL, and EGRET data.

The empirical four parameter GRB functional form in eq. [1] suffices for the

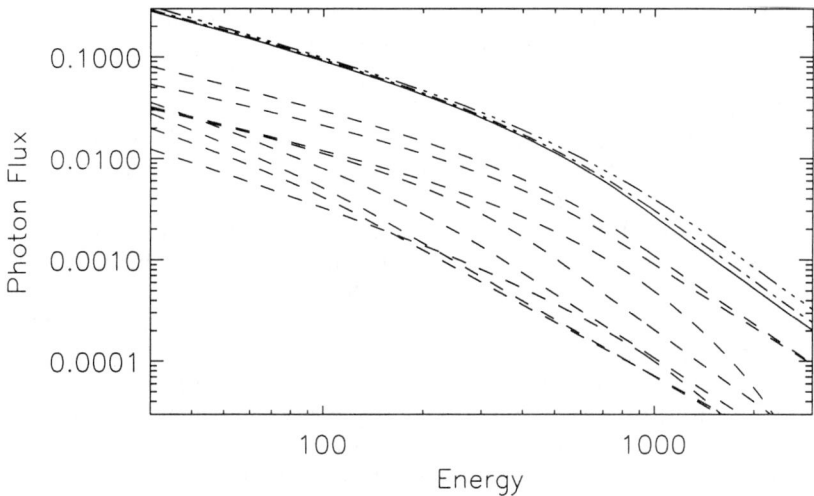

FIG. 2. The solid curve is the fit over the first 12 s of GB 910503. The dashed curves are the (scaled) fits over 8 time segments during these 12 s, and the dash-dot curve is the sum of these 8 fits. The dash-3 dot curve is the sum of fits over 37 time segments. Part of the discrepancies between the curves for the entire 12 s period results from the deadtime correction.

moderate spectral resolution of NaI detectors and proportional counters, and the signal-to-noise ratio typical of detectors with areas of a few hundred cm^2. As I will show below, the spectrum varies significantly during a burst, and the spectrum of the entire burst must be a composite of spectra with different shapes. However, the GRB function is currently sufficient to describe the spectrum of the entire burst and of short time segments during the burst. Figure 2 shows how the sum of broad band spectra with different shapes is a smooth, featureless spectrum. Undoubtedly more parameters will be needed for spectra from detectors with higher resolution and greater effective area.

IV. POSSIBLE NEW LOW ENERGY COMPONENT

In some bursts the X-ray emission below 10 keV appears to continue long after the higher energy radiation, as will be discussed in greater depth below. Is this evidence of an additional spectral component, or is this an extreme case of the softening of the primary component described by the GRB model? X-ray observations are only now beginning to address this issue.

The *Ginga* proportional counter extended the spectrum down to ~ 2 keV. The low energy data were used to study both the bursts' time evolution (10) and their low energy spectrum (29). Because the burst angle was unknown for the *Ginga* bursts and the detector efficiency had a very large angular dependence at low energies as a result of the composition of the detector case, the true shape of the low energy continuum is uncertain (this was also a problem for the analysis of the *Apollo* burst (31)). However, there was no indication of a deviation from the four parameter form in eq. [1] (29). *HEAO 1* accumulated the spectra above 2 keV of the particularly soft burst GB 780506; the spectrum of the first 10.24 s has a significant downturn below 4 keV (30).

In addition to the regular spectra produced by pulse height analyzers (PHAs), the BATSE SDs also report the counts in four integral discriminator channels. The second lowest discriminator also triggers the PHA, although the usable PHA spectrum begins a few keV above the discriminator energy because of an instrumental artifact (11). The lowest discriminator is at approximately half the energy of the second lowest discriminator. Thus the difference between these two discriminator channels results in an additional spectral channel between 5–8 keV and 10–16 keV. The analysis of this new spectral channel is fraught with instrumental issues (e.g., the detectors' low energy response and the discriminators' energy calibration). Nonetheless, in 12 of the 86 bursts analyzed thus far the number of counts in this additional spectral channel is significantly greater than predicted by a GRB model fit to the regular PHA spectrum and this new channel (32). Whether this additional spectral component below ~ 15 keV is present simultaneously with the higher energy radiation, or results from an X-ray tail late in the burst that is averaged with higher energy emission earlier in the burst, remains to be determined. If confirmed, this new low energy component is a significant new addition to our understanding of the burst continuum.

V. NEW HIGH ENERGY COMPONENT

At GeV energies a new component has been observed in some bursts; in at least one case this high energy emission lasted significantly after the lower energy radiation. Evidence for this component consists of less than one hundred counts from the EGRET spark chamber which is sensitive to energies between 35 MeV and 30 GeV. Consequently, it is difficult to characterize the spectrum and temporal characteristics of this component.

During 5 bursts 53 spark chamber counts were observed while the low energy event was in progress (21). A spectrum formed from this small number of counts appears to be harder, $\beta \sim -1.95$, than lower energy emission. However, the count spectrum can also be (over)interpreted as a steeper power law below 1 GeV, and an excess above 1 GeV. In addition, the high energy counts frequently occur late in the burst. That this high energy component lasted after the lower energy radiation is undeniable for GB 940217, although this component may also have lingered in GB 910601 (21).

GB 940217 was a truly incredible burst (27). BATSE and EGRET observed a multi-spike burst lasting 160 s at energies from 15 keV to 2 GeV. The BATSE burst showed hard-to-soft spectral evolution, and the low energy spectrum had a break energy of $E_\mathrm{p} \sim 1$ MeV. During these 160 seconds EGRET detected 10 spark chamber events, two with energies above 1 GeV. During the next 15 minutes, before the Earth occulted the source, an additional 8 photons were observed. When the source reappeared from behind the Earth a further 10 photons were detected, a full 90 minutes after the burst began. One photon of this last set had an energy of 18 GeV! The probability that this photon is a chance background event is $P = 5 \times 10^{-6}$.

VI. SPECTRAL EVOLUTION

Burst spectra evolve over the course of a burst. Two trends characterize this variability (33), although the universality of these trends is not yet clear. First, the spectral hardness and the count rate vary together on time scales of seconds (34): the spectrum hardens when the count rate increases. Second, on top of this correlated variation of hardness and count rate is a hard-to-soft trend within and between intensity spikes (35). Bursts generally begin with hard spectra, and successive spikes are usually softer, even when the count rate is greater. Finally, spikes often start hard, with E_p peaking before the count rate, although the temporal resolution of our detectors is often insufficient to observe this effect. In the absence of an identified burst source and a scenario for the emission of the observed radiation, a physical explanation for this spectral evolution is highly speculative. Nonetheless, the hard-to-soft trend, particularly over the course of the burst, suggests that a single emission region retains a memory of preceding emission events, or that multiple regions communicate.

Liang et al. (36) report a potentially physically significant relationship between the hardness and the energy emitted since the beginning of the burst. Specifically, when the logarithm of E_p is plotted against the fluence from the beginning of the burst, the decrease of E_p during a spike and the minima between spikes follow straight lines (exponentials in linear E_p). These trends may be related to the filling and draining of the burst's energy reservoir. This phenomenon and its implications will be studied further.

Hard-to-soft evolution is seen in the LAD spectra of bursts with simple time structures which rise rapidly and fall more slowly (FREDs—Fast Rise, Exponential Decay) (37). This spectral evolution has also been found by other instruments for specific bursts, e.g., both BATSE SD and COMPTEL observations of GB 940217 (24). On the other hand, counterexamples exist: both BATSE (33) and SIGNE (38) find a preponderance of bursts exhibit hard-to-soft evolution, but there are examples of the opposite.

The spectra accumulated over 0.1 seconds or longer are broad-band: photons are observed over a broad energy range. However, early fireball models predicted continuum spectra which were somewhat broader than black bodies but much narrower than the observed spectra (39,40). Thus the question is whether some other emission mechanism operates to produce broad band spectra, or do the spectra we observe result from many short-lived thermal events which sum rapidly to the observed broad band shape. Unfortunately our detectors do not have the necessary temporal resolution to accumulate spectra on very short time scales. However, the BATSE SDs do provide lists of photon arrival times and energies. If there is short time scale narrow band emission then the photon energies should be more tightly correlated on short time scales than if the emission is inherently broad band. Such a study is currently underway (41); it will most likely result in constraints on the fraction of the narrow band emission on millisecond time scales.

At low energies the emission often lasts long after the gamma ray photons. Both P78 (42) and *Ginga* (10) observed X-ray tails which lingered more than 10 seconds after the harder emission. In addition, the P78 data shows X-ray precursors. *HEAO 1* observed 2–10 keV emission for approximately an hour after GB 780506 (30). Similarly, $\sim 10\%$ of the 70 bursts observed by the WATCH detector on *Granat* found X-ray precursors or tails (43).

VII. POPULATION TRENDS

Bright bursts are harder than dim bursts, as has been shown by averaging the hardness ratios (44,45) and comparing the hardness distributions (46) for different intensity classes. This correlation between intensity and hardness is consistent with a cosmological origin, since the spectra of more distant (presumably dimmer) bursts should be redshifted (therefore softer). Cosmological models can be fit to the data; one study (46) derives a redshift factor between dim bursts (peak flux of 1 photon cm^{-2} sec^{-1} assumed to be at z_1) and bright bursts (100 photons cm^{-2} sec^{-1} at z_{100}) of $(1 + z_1)/(1 + z_{100}) = 1.86^{+0.36}_{-0.24}$.

Finally, short bursts tend to be harder than long duration bursts (47).

VIII. SUMMARY AND ISSUES FOR THE FUTURE

1. Continuum—The continuum is fitted satisfactorily between ~ 10 keV and ~ 100 MeV by a simple four parameter model consisting of a low energy power law with an exponential cutoff followed by a steeper high energy power law. The actual shape (i.e., spectral indices, break energy) varies between and within bursts. Parameterized by the energy of the peak of $E^2 N(E) \propto \nu f_\nu$, the hardness is characterized by a broad distribution both during a single burst and between bursts, indicating the absence of characteristic energies that are evident in the spectrum.

An issue for further study is how the observed hardness is affected by selection effects. Another issue which should be determined in the future is whether the spectrum is truly broad band, or whether it is composed of short duration narrow band events that sum rapidly to a broad band spectrum.

2. Low Energy Component—Missions before BATSE found that the spectrum continued smoothly down to ~ 2 keV with no evidence of additional spectral components. However, the deconvolution of these spectra was affected by the uncertainty in the angle to the burst. The BATSE SDs have a discriminator channel between ~ 5 and ~ 10 keV which can be used for spectroscopy. In approximately 10% of the bursts this additional channel shows an excess above an extrapolation of the higher energy spectra. Ultimately the presence of this component should be verified by a dedicated X-ray detector.

3. New High Energy Component—The EGRET spark chamber detects high energy photons (∼a GeV) in some bursts after most of the emission observed by BATSE. In one burst the high energy emission lingered for 90 minutes, with the detection of an 18 GeV photon!

4. Spectral Evolution—Bursts are often characterized by hard-to-soft spectral evolution both within and between intensity spikes, although the intensity and hardness track on time scales of ∼1 s. Sufficient temporal resolution is necessary to detect these trends, and additional studies from detectors and data types with the necessary temporal resolution are needed. A key question is the universality of these evolutionary trends. In addition, a better description or parameterization of the spectral evolution is required.

5. Population Characteristics—Intense bursts are harder, which is consistent with a cosmological origin. Also short bursts tend to be harder.

Acknowledgements. I thank my colleagues on the BATSE instrument team and at UCSD for their assistance over the past five years. Burst spectroscopy research at UCSD is supported by NASA contract NAS8-36081.

REFERENCES

1. M. S. Briggs, et al., these proceedings (1996).
2. W. S. Paciesas and G. J. Fishman, eds., Gamma-Ray Bursts, AIP Conf. Proc. 265, AIP: New York (1992).
3. M. Friedlander, M. Gehrels and D. J. Macomb, eds., Compton Gamma-Ray Observatory, AIP Conf. Proc. 280, AIP: New York (1993).
4. G. J. Fishman, J. J. Brainerd, and K. Hurley, eds., Gamma-Ray Bursts, Second Workshop, AIP Conf. Proc. 307, AIP: New York (1994).
5. K. Bennett and C. Winkler, *Towards the Source of Gamma-Ray Bursts, Proc. of the 29th ESLAB Symposium.*, Astr. Sp. Sci. **231** (1995).
6. T. L. Cline, et al., Ap. J. Lett. **185**, L1 (1973).
7. T. L. Cline and U. D. Desai, Ap. J. Lett. **196**, L43 (1975).
8. E. P. Mazets, et al., Ap. Sp. Sci. **82**, 261 (1982).
9. S. M. Matz, et al., Ap. J. Lett. **288**, L37 (1985).
10. A. Yoshida, et al., P.A.S.P. **41**, 509 (1989).
11. D. L. Band, et al., Exp. Astron. **2**, 307 (1992).
12. G. Pendleton, et al., NIMS, (1995).
13. D. L. Band, et al., Ap.J. **413**, 281 (1993).
14. M. Tavani, Ap. J., in press (1996). Also M. Tavani, Phys. Rev. Lett. **76**, 3478, (1996). Also M. Tavani, in these proceedings (1996).
15. J. Katz, in these proceedings (1996).
16. B. Teegarden, et al., in these proceedings (1996).
17. D. Palmer, et al., in these proceedings (1996).
18. T. Piran and R. Narayan, in these proceedings (1996).
19. F. Pelaez, et al., Ap. J. Suppl. **92**, 651 (1994).
20. R. M. Kippen, PhD thesis, University of New Hampshire (1995). Also R. M. Kippen, et al., A. & A. **293**, L5 (1995).

21. B. Dingus, *et al.*, in (5) 187 (1995).
22. T. Aigner, *et al.*, in (5) 165 (1995).
23. J. Greiner, *et al.*, A. & A. **302**, 121 (1995).
24. L. O. Hanlon, *et al.*, in (5) 157 (1995).
25. B. E. Schaefer, *et al.*, in (4) 280 (1994).
26. G. H. Share, *et al.*, in (4) 283 (1994).
27. K. Hurley, *et al.*, Nature **372**, 652 (1994).
28. C. Winkler, *et al.*, A. & A. , (1995). Also C. Winkler, *et al.*, in (5), 153 (1995).
29. T. E. Strohmayer, *et al.*, Ap.J., submitted (1996).
30. A. Connors and G. J. Hueter, in preparation (1996).
31. D. Gilman, *et al.*, Ap. J. **236**, 951 (1980).
32. R. Preece, *et al.*, Ap. J., in press (1996). Also R. Preece, *et al.*, in (5) 207 (1995).
33. L. A. Ford, *et al.*, Ap.J. **439**, 307 (1995).
34. S. V. Golenetskii, *et al.*, Nature **306**, 451 (1983).
35. J. P. Norris, *et al.*, Ap.J. **301**, 213 (1986).
36. E. Liang, *et al.*, in these proceedings (1996).
37. N. P. Bhat, *et al.*, Ap. J. **426**, 604 (1994).
38. V. E. Kargatis, *et al.*, Ap. J. **422**, 260 (1994).
39. B. Paczynski, Ap. J. Lett. **308**, L43 (1986).
40. J. Goodman, Ap. J. Lett. **308**, L47 (1986).
41. L. A. Ford, *et al.*, in these proceedings (1996).
42. J. G. Laros, *et al.*, Ap. J. **286**, 681 (1984).
43. A. J. Castro-Tirado, *et al.*, in (4) 17 (1994).
44. R. Nemiroff, *et al.*, Ap. J. Lett. **435**, L133 (1994).
45. I. G. Mitrofanov, *et al.*, Ap. J. **459**, 570 (1996).
46. R. S. Mallozzi, *et al.*, Ap. J. Lett. **454**, 597 (1995).
47. C. Kouveliotou, *et al.*, Ap. J. Lett. **413**, L101 (1993).

Low-Energy Spectral Features in GRBs

Michael S. Briggs

Department of Physics, University of Alabama in Huntsville, Huntsville, AL 35899

> I discuss low-energy lines in gamma-ray bursts. The process of deconvolving gamma-ray spectral data and the steps needed to demonstrate the existence of a line are explained. Previous observations and the current status of the analysis of the BATSE data are described.

Spectral lines are highly informative of the conditions of the region in which they are created. Unfortunately, the observational and theoretical understanding of lines in gamma-ray bursts is confused at best. Observations made with *Ginga* provided strong evidence for absorption lines, yet BATSE has not confirmed their existence. Confirmation of the existence of lines and a better characterization of their properties might provide the essential clue to the mystery of gamma-ray bursts.

In this paper I will not tell you whether lines exist, instead I will explain what analysis needs to be done to demonstrate the existence of a line in a gamma-ray spectrum to assist you in judging the evidence for yourself. I will briefly review selected observations and the status of the analysis of BATSE data. I will concentrate on observational issues regarding spectral features below 100 keV. The views expressed are my own.

DETECTORS AND GAMMA-RAY INTERACTIONS

The detection of gamma-ray lines is much more difficult than the detection of optical lines, primarily because there is no one-to-one relationship between the energy of an incident gamma-ray photon and the energy measured by the detector, which is called the "energy loss". Because of the difference between the energy of the incident photon and the energy loss, detected events are referred to as "counts". Secondary problems are the poor signal-to-noise and signal-to-background ratios prevalent in gamma-ray astronomy.

The examples in this paper are from data collected with the Spectroscopy Detectors (SDs) of BATSE, but the concepts are true for all gamma-ray detectors—regardless of their energy resolution, the photon interaction physics is similar. The SDs are 12.7 cm diameter by 7.6 cm thick crystals of NaI(Tl) scintillator, each viewed by a photomultiplier tube (PMT) of the same diameter. When a gamma-ray interacts in the NaI crystal, a fraction of the resulting ionization energy is converted into scintillation light and measured by the PMT. In the energy range of interest, 10 keV to a few MeV, gamma-rays interact with detector matter primarily by three processes (Fig. 1): photo-

© 1996 American Institute of Physics

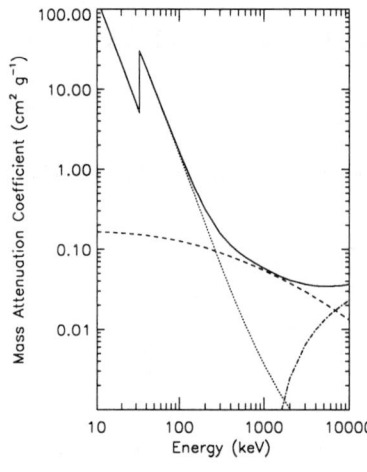

FIG. 1. The NaI gamma-ray mass attenuation coefficient μ: solid line: total; dotted: photoelectric; dashed: Compton; dot-dash: pair production. The fractional transmission is $\exp(-\mu x)$, where x is the quantity of NaI traversed in g cm^{-2}. The data are from ref. (17).

electric absorption, Compton scattering, and pair production (14,17,20).

In NaI, the photoelectric process dominates below ≈ 200 keV. In this process the gamma-ray is completely absorbed by an atomic electron. The interaction is most likely to occur with the inner-most electron that the gamma-ray has sufficient energy to ionize. Usually, an atomic cascade yields fluorescent X-rays, which will be photoelectrically absorbed unless they escape the crystal. Sometimes an Auger electron is ejected instead. The left curve of Fig. 2 illustrates a simple case, the detected energy losses expected from 20 keV photons. The fluorescent X-rays have a higher cross-section because of their lower energy (Fig. 1), so that they are unlikely to escape and thus the incident energy will be totally absorbed, leading to a simple observed spectrum. Broadening of the spectral resolution occurs, primarily due to Poisson fluctuations in the number of photoelectrons produced in the PMT (20).

A 50 keV photon is also most likely to interact via the photoelectric process, but the resulting count spectrum is more complicated (Fig. 2, bold curve). The inner-most shell of an iodine atom accessible to the 20 keV photon of the previous example is the L-shell. The K-shell becomes accessible to photons with energies above the K-shell binding energy of 33.17 keV, resulting in a cross-section increase (Fig. 1). A 50 keV photon will typically ionize a K-shell electron, resulting in a L, M or N to K shell transition and an X-ray with an energy between 28.3 and 33.0 keV (24). Sometimes this photon will interact by the photoelectric process and the total energy of the initial photon will be collected, resulting in the full-energy absorption peak at the right in the bold curve of Fig. 2. However, the fluorescence X-ray is below the K-edge and has a lower cross-section (Fig. 1) and thus a high probability of escaping the crystal, resulting in incomplete absorption of the energy of the incident gamma-ray (the left peak of the bold curve of Fig. 2).

Comparing incident photon energies just above and below the 33.17 keV K-edge of iodine, the higher energy photon has a higher interaction probability, but more importantly, a lower probability of complete energy absorption. For

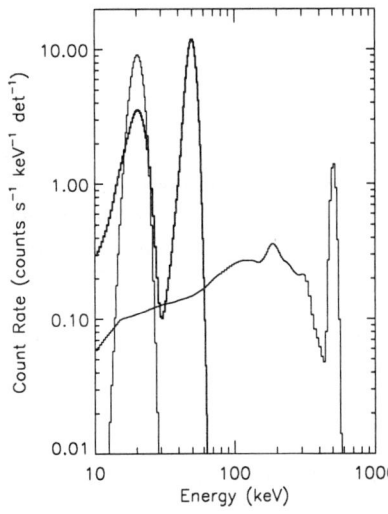

FIG. 2. Simulated instrumental energy loss spectra, commonly known as count spectra, for monoenergetic photons incident on a Spectroscopy Detector, the detector's surroundings on CGRO and the Earth's atmosphere. The simulation matches the conditions with which SD 5 observed GRB 920311 (BATSE trigger 1473), except for clarity the simulated data extend to energies slightly below where the current calibration is reliable. (Data from SD 5 and GRB 920311 are shown in Fig. 3.) In each case the input flux is 1 photon s^{-1} cm^{-2}. Left curve: photon energy 20 keV; middle, bold curve: 50 keV; right curve: 500 keV.

a hard incident spectrum, such as a GRB spectrum, this K-edge effect results in a deficit from 33 to ≈ 50 keV in the count spectrum, which appears as a peak at ≈ 30 keV and is predicted by the detector response model (see Fig. 3). This feature has been used as a verification of the calibration of the SDs (32) and should not be mistaken for an astrophysical line.

At 500 keV, photoelectric absorption occurs only ≈ 20% of time, making a contribution to the full-energy absorption peak (see Fig. 2, right curve). The most likely interaction is Compton scattering, which transfers only a portion of the incident photon's energy to an electron. Complete energy absorption will occur only if additional interactions occur, either photoelectric or Compton. In Compton scattering the maximum energy transfer to the electron, 331 keV, occurs when the incident photon is scattered 180°. A range of scattering angles approaching 180° creates the peak in the curve at 310 keV and the lack of larger energy transfers in single scattering events causes the valley above 331 keV (see Fig. 2). Correspondingly, the minimum energy of a scattered photon is 169 keV, so a range of angles approaching 180° for incident photons that *scatter into* the detector from the spacecraft or the Earth's atmosphere creates the peak at 190 keV.

At yet higher energies pair production becomes important.

SPECTRAL DECONVOLUTION AND LINE DETECTION

As described in the previous section, an observed 20 keV count could be due to a 20, a 50, or even a 500 keV photon! When a single count is observed, it is impossible to deduce the energy of the incident photon. If many counts are observed, the incident spectrum can be deduced within statistical limits in the process known as deconvolution.

We approximate the continuous (as a function of energy) detection process with the following discrete equation:

$$\vec{c} = \mathbf{D}\vec{p}. \qquad (1)$$

Here $\vec{c}$ is the vector with the counts versus energy bins. Normally these data are only available in binned form in order to reduce telemetry requirements— as long as the energy bin widths are small compared to the detector's energy resolution essentially no information is lost. Similarly, $\vec{p}$ is the vector representing the incident photon spectrum. The detector is represented by $\mathbf{D}$, the detector response matrix, which is obtained via Monte Carlo simulations of gamma-ray interactions in a computer model of the detector (e.g., (35)).

The obvious solution is

$$\vec{p} = \mathbf{D}^{-1}\vec{c}. \qquad (2)$$

This simple approach does not work if one wishes to deduce information at a resolution at or better than the intrinsic resolution of the detector, which is the goal of analyzing data for the presence of lines. The problem is that if the energy widths of the rows and columns of $\mathbf{D}$ are comparable to the detector resolution, then neighboring columns will be very similar and $\mathbf{D}$ will be nearly singular. Inverting $\mathbf{D}$ will be numerically unstable and the solution $\vec{p}$ will be unreliable, especially in the presence of statistical fluctuations in the observed count spectrum $\vec{c}_{\text{obs}}$.

Instead, the standard approach in astrophysics has become that of forward-folding (38,11,25). A parameterized spectral model is *assumed* and used to calculate $\vec{p}_{\text{model}}$, from which eq. 1 yields a model count spectrum $\vec{c}_{\text{model}}$. The model count spectrum $\vec{c}_{\text{model}}$ is compared to the observed count spectrum $\vec{c}_{\text{obs}}$ using some statistical measure such as χ^2 or likelihood and the parameters of the photon model are optimized so as to minimize the discrepancy between $\vec{c}_{\text{model}}$ and $\vec{c}_{\text{obs}}$ according to the chosen statistical measure.

It is very important to realize that a solution obtained by forward-folding is not unique but is rather photon model dependent (11,25). Even if a model fit results in a good χ^2 value, another, possibly unknown, model might result in an equal or better χ^2 value. This is shown in Fig. 3, where solutions 'a' and 'b', based upon different photon models, have essentially identical χ^2 values. What a forward-folding solution provides is the parameter values of the *assumed* photon model (25).

In gamma-ray astrophysics, deconvolved photon points are frequently obtained by scaling the model photon spectrum by the ratio of the observed over modeled counts (25)—Fig. 3 (right) is an example. This practice is potentially misleading because the deconvolved points are model dependent yet there is an almost irresistible temptation to regard them as incident spectrum "data" points. In the examples in Fig. 3, based upon alternative continuum models, the differences in the deconvolved points are subtle and the practice not too pernicious. When a line with width comparable to or smaller than the detector's intrinsic resolution is considered, the deconvolved spectra can exaggerate the significance of a line—see Fig. 4. Papers analyzing spectra in

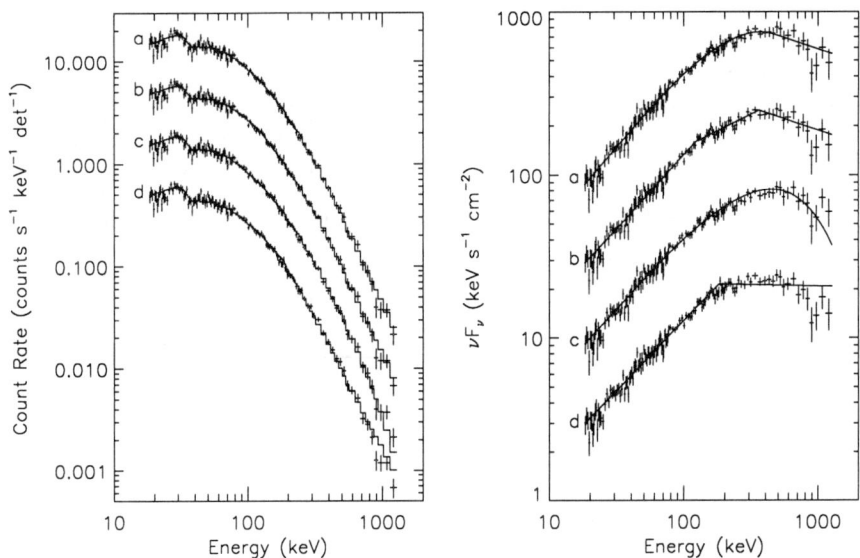

FIG. 3. To illustrate the photon model dependent nature of the forward-folding deconvolution method, identical data is analyzed using four different photon models. Model 'a' assumes Band's spectral form (2), obtaining $\chi^2=223$ for 200 degrees-of-freedom (dof). Model 'b' is a power-law with two breaks; $\chi^2=227$ for 198 dof. Model 'c' is a power law times an exponential cutoff; $\chi^2=244$ for 201 dof. Model 'd' is a power-law with one break; $\chi^2=267$ for 200 dof. The data are for the interval 1.088 to 26.624 s of GRB 920311 as observed by SD 5; above 200 keV, the data have been rebinned into wider channels for display purposes. Each group after 'a' is shifted downwards by $\times\sqrt{10}$. Left: count data (points) and models (histograms). Right: Deconvolved points and models. The νF_ν spectrum is $E^2 \times$ the photon flux spectrum. The photon models obviously differ—there are also differences in the values of the deconvolved points, e.g., the values of right-most point are 483 ± 101, 483 ± 101, 593 ± 113 and 449 ± 98 keV s^{-1} cm^{-2} for models 'a', 'b', 'c' and 'd', respectively.

the X-ray band generally show a graph of the count data and model and a graph of the residuals (e.g., (22)). Only rarely is a graph of a deconvolved spectrum presented. The gamma-ray community is advised to emulate this practice.

Because of the limitations of gamma-ray spectral data, the following steps are necessary to demonstrate that a line feature exists:

- *Deconvolve spectra using the forward-folding technique.* This is the standard approach in X-ray and gamma-ray spectroscopy.
- *Show the data and the model in counts rather than deconvolved photons.* The fit is performed by comparing observed and model counts and the data and model should be displayed in this space.
- *Show that the line is significant versus all reasonable continuum models.* Suppose that $\Delta\chi^2$ indicates that a line is a statistically significant im-

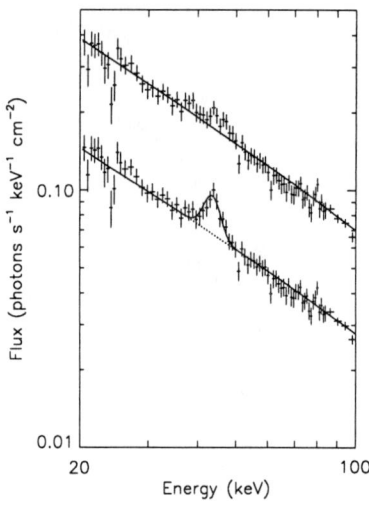

FIG. 4. To illustrate the dependence of the deconvolved photon points on the model for a model containing a line, two deconvolutions of the same data are shown. For the top curve, a continuum model is assumed, while for the bottom curve (shifted downward by ×2.5), an additive Gaussian line is also included in the model. The data are that of the line candidate in GRB 940703—for count data and models see ref. (9).

provement to continuum model A—one might be tempted to say that the line is "proven". However, if the line insignificantly improves continuum model B *and* model B has comparable χ^2 to that of A+line, then we cannot say the line must exist because continuum model B may be the correct explanation. Since we do not know the correct continuum model of GRBs, we must try all reasonable ones (11,12); since all models cannot be tried, judgement is needed. In practice, this means that a sufficiently flexible continuum model should be used (12). With current data, an appropriate choice for all bursts is Band's function (2,5), which is usually the best-fitting model for bright bursts (e.g., Fig. 3). A power law with multiple breaks is also sufficiently flexible, but has the disadvantage of not being smooth. An evolving or two-component continuum spectrum might mimic a line when a single continuum model is used for the analysis (16,36).

- *Provide a quantitative evaluation of the statistical significance of the feature.* Without such an evaluation one does not know the strength of the evidence that a line is real rather than a fluctuation. It has been traditional to use the F-test (e.g., (12,8,33)), but the BATSE team has recently realized that $\Delta\chi^2$ is more appropriate for this case, where one knows the uncertainties from Poisson statistics (23,10,7). In evaluating the significance of a feature, the number of trials should be considered.
- *Consider possible systematic errors.* Does the line centroid match that of a background line? If so, is the background subtraction adequate? Is the detector model and calibration adequate?
- *If possible, evaluate the feature for confirmation/consistency.* If more than one instrument, or more than one detector of an instrument, observed the burst with appropriate energy coverage and resolution, then the two observations should be compared. Ideally, both detectors would

show the feature to be statistically significant, thereby confirming the detection. Depending on the sensitivity of the detectors and the strength of the burst, this may not be possible. In any case, consistency between the data of the two detectors is required.

LINE OBSERVATIONS

Table 1 summarizes the analyses methods used in previous observations of gamma-ray burst spectral features below 100 keV and Table 2 presents the results of those analyses. Many of the previous analyses were done before the gamma-ray community realized the importance of all of the steps listed in the previous section and before computer improvements made the forward-folding approach practical. Probably due to improvements in the sensitivity of detectors, there has been historical progression towards continuum models with more parameters (5). Many of the earlier line analyses were done using continuum models that are now known to be too simple, raising the possibility that the line detections are an artifact of the chosen continuum model. Unfortunately, these problems mean that some of the previous analyses must now be judged as inconclusive with regard to the existence of gamma-ray burst lines. Considering the importance of this question, reanalyses are merited.

The Konus observations pioneered GRB line studies. The analysis technique was cleverly designed to minimize the computations required: the count spectra were deconvolved using a standard model and the resulting photon spectra were iteratively improved (28). The continuum model used was optically thin thermal bremsstrahlung, $\propto E^{-1} \exp -E/E_0$. The explanation of the deconvolution procedure shows one count spectrum as an example (28). It has been frequently suggested that the Konus lines might be harmonics of fundamentals that are below the typical detector threshold of 30 keV.

The most significant line in the HEAO A-4 data is in the $2nd$ peak of GRB 780325 (19). The line significance was evaluated using a simple continuum model, an exponential, fit to only the data below 200 keV (19). Using a better model and all of the data might raise or lower the significance. The deconvolution was done using photons-to-counts efficiencies derived for an E^{-2} spectrum, a procedure equivalent to approximating **D** with a diagonal matrix. This procedure was justified by the suppression of the Compton scattering re-

TABLE 1. Low-Energy Spectral Features: Analysis Methods

Instrument	References	Forward-Folding used?	Count Data & Model shown?	Flexible Continuum Model used?	Statistical Significance evaluated?
Konus	(26–28)	No	No	No	No
HEAO A-4	(18,19)	No	No	No	Yes
Ginga	(12,15,31,39,40)	Yes	Yes	Yes	Yes
Lilas	(8)	Yes	No	Yes	Yes

TABLE 2. Low-Energy Spectral Features: Results

Instrument	GRB	References	$\Delta\chi^2$	Centroids (keV)	equiv. width (keV)
Konus	many[a]	(26–28)		30–70	5–15
HEAO A-4	780325[b]	(19)	14.0	76±5	24
	780325[c]	(19)	16.6	49±3	13
	780608	(19)		66±7	14
Ginga	880205	(31,12)	48.5	19.7±0.7, 38.0±1.6	3.7, 9.1
	870303[d]	(15)	30.6	21.1±1.1	10.5
	870303[e]	(31,15)	26.2	21.4±0.7, 2×21.4	4.8, 8.4
	890929	(40)	25.6	26.3±1.5, 46.6±1.7	4, 8
Lilas	890306	(8)	89.0	11.2±0.5, 34.6±1.4	

[a] Unique amongst the results listed in this table, one GRB had an emission feature.
[b] First peak of the burst.
[c] Second peak of the burst.
[d] Interval 'S1'.
[e] Interval 'S2'.

sponse by the active shielding of the instrument and has the advantage of conservatively assuming that lines do not exist. The feature is stated to be visible in two detectors, but only summed data are shown (19).

The analysis of the *Ginga* data is excellent. (However, some papers present only deconvolved spectra.) The chance probability of obtaining a reduction in χ^2 of 48.5 by adding two lines, assuming that the lines do not exist, is 10^{-8}! The harmonic relation between the line centroids is powerful evidence for the cyclotron resonant scattering interpretation. A possible concern is that the lines were in the same channels in all 3 bursts (13)—while the centroid energies are different in GRB 890929, for this event the gains of the detectors were lower than normal (40).

The recent results of Lilas on GRB 890306 are even more statistically significant—lines at 11 and 35 keV are reported in a 68 s interval with an F-test probability of 2×10^{-13} that this is a chance result (8). The χ^2 value for the model containing the lines is somewhat high: 43.2 for 26 degrees-of-freedom. This χ^2 value, along with the location of one line near the detector threshold and the other near the iodine K-edge, are areas for concern. It would be desirable to see the count model for these data and also for a comparable spectrum not containing lines. The spectral evolution during the 68 s interval, which is essentially the entire burst, should also be investigated.

BATSE AND LINES

The Spectroscopy Detectors were added to BATSE because of the Konus line observations. The pre-launch expectations of the BATSE team were that lines would be readily found. Simulations (3) and performance tests (32) verify the ability of BATSE to detect *Ginga*-like lines, with generally poorer sensitivity than *Ginga* for 20 keV lines and better for 40 keV lines. However, no lines have been detected with BATSE (33,4). While disappointing, this lack of line detections by BATSE is not yet in strong contradiction with the observations of *Ginga* (33,4,6).

BATSE has a number of advantages for line studies:

- State of the art scintillation detectors: excellent energy resolution and advanced electronics incorporating active PMT bleeder strings and baseline restoration to handle large pulses and high count rates,
- Excellent temporal resolution so that lines can be found on whatever time scale they may exist,
- GRB locations from the LADs to aid analysis of the SD data,
- Good sensitivity,
- Extensive performance verification, on the ground and in-orbit (32),
- Multiple detectors and detector types, which enable consistency/confirmation studies,

The BATSE team is therefore confident of the instrument's ability to detect lines. Because of the lack of detections by the initial approach, visually examining burst spectra for lines (33), the team has implemented the most thorough line search ever conducted. This comprehensive computer search examines essentially all time scales and energy centroids below 100 keV. The new approach has already identified 8 candidate features with $\Delta\chi^2 > 20$ (9). Currently analysis is in progress to determine the consistency of the multi-detector data and hopefully to confirm some of the features with multiple detectors. Until that work is completed the BATSE team considers these features to be candidates rather than detections.

THE FUTURE

There is much more work to be done. On the interpretational side, bursts from galactic halo or cosmological distances have intrinsic luminosities 10^4 or more greater than deduced in the days of the galactic disk paradigm. Consequently, the physics of line creation will be different than previously envisioned. This has been examined in the context of sources in the galactic halo (e.g., (21)), but little work has been done in the context of cosmological models. Several of the past line observations merit reanalysis using techniques developed since they were published. The analysis of the BATSE data continues in order to determine the reality of the candidates identified by the comprehensive search. With eight candidates already identified, and typically several high-gain SDs viewing a burst, the BATSE team will be able to determine the consistency of the data and confirm many of the candidate features or to demonstrate that there is some problem. Data are also currently being collected and analyzed with the scintillation detectors of the Konus-W instrument (1,29) and the high energy-resolution detectors of TGRS (34). The near future will bring the launch of HETE (37) and PGS on Mars-96 (30).

REFERENCES

1. R. L. Aptekar, D. D. Frederiks, et al., in AIP Conf. Proc. **366**, 158, (1996).
2. D. Band, J. Matteson, L. Ford, et al., ApJ **413**, 281 (1993).
3. D. L. Band, L. A. Ford, J. L. Matteson, et al., ApJ **447**, 289 (1995).
4. D. L. Band, S. Ryder, L. A. Ford, et al., ApJ **458**, 746 (1996).
5. D. L. Band, these proceedings (1996).
6. D. L. Band, L. Ford, et al., these proceedings (1996).
7. D. L. Band, et al., "BATSE Gamma-Ray Burst Line Search: V. Probability of Detecting a Line in a Burst", ApJ, submitted (1996).
8. C. Barat, A&A Suppl. Ser. **97**, 43 (1993).
9. M. S. Briggs, R. D. Preece, G. N. Pendleton, et al., these proceedings (1996).
10. W. T. Eadie, D. Drijard, F. E. James, M. Roos & B. Sadoulet, Statistical Methods in Experimental Physics, New York: North-Holland (1971).
11. E. E. Fenimore, R. Klebesadel & J. Laros, Adv. Space. Res. **3**, #4, 207 (1983).
12. E. E. Fenimore, J. P. Conner, R. I. Epstein, et al., ApJ **335**, L71 (1988).
13. E. E. Fenimore, private communication.
14. E. Fermi, Nuclear Physics, rev. ed., Univ. of Chicago Press (1950).
15. C. Graziani, E. E. Fenimore, T. Murakami, et al., in Gamma-Ray Bursts, ed. C. Ho, R. I. Epstein & E. E. Fenimore, Cambridge Univ. Press, 407 (1992).
16. A. K. Harding, et al., in AIP Conf. Proc. **141**, 98 (1986).
17. J. H. Hubbell, Photon Cross Sections, Attenuation Coefficients, and Energy Absorption Coefficients from 10 keV to 100 GeV, Nat. Bureau of Stand. (1969).
18. G. J. Hueter, in AIP Conf. Proc. **115**, 373 (1984).
19. G. J. Hueter, Ph.D. Thesis, University of California, San Diego (1987).
20. G. F. Knoll, Radiation Detection and Measurement, 2nd ed., Wiley (1989).
21. M. Isenberg, D. Q. Lamb & J. C. L. Wang, these proceedings (1996).
22. K. Koyama, R. Petre, E. V. Gotthelf, et al., Nature **378**, 255 (1995).
23. D. Q. Lamb, private communication (1995).
24. C. M. Lederer, et al., Table of Isotopes, 7th ed., New York: Wiley (1978).
25. T. J. Loredo & R. I. Epstein, ApJ **336**, 896 (1989).
26. E. P. Mazets, S. V. Golenetskii, et al., Nature **290**, 378 (1981).
27. E. P. Mazets, S. V. Golenetskii, V. N. Ilyinskii, et al., Ap&SS **82**, 261 (1982).
28. E. P. Mazets, S. V. Golenetskii, et al., in AIP Conf. Proc. **101**, 36 (1983).
29. E. P. Mazets, R. L. Aptekar, D. D. Frederiks, et al., these proceedings (1996).
30. I. G. Mitrofanov, et al., Adv. Space Res. **17**, #12, 51 (1996).
31. T. Murakami, M. Fujii, K. Hayashida, et al., Nature **335**, 234 (1988).
32. W. S. Paciesas, M. S. Briggs, R. B. Wilson, et al., these proceedings (1996).
33. D. M. Palmer, B. J. Teegarden, B. E. Schaefer, et al. ApJ **433**, L77 (1994).
34. D. M. Palmer, H. Seifert, B. J. Teegarden, et al., these proceedings (1996).
35. G. N. Pendleton, et al., Nucl. Instr. & Meth. in Phys. Res. **A 364**, 567 (1995).
36. G. N. Pendleton, M. S. Briggs, R. D. Preece, et al., these proceedings (1996).
37. G. R. Ricker, J. P. Doty, S. A. Rappaport, et al., in Gamma-Ray Bursts, ed. C. Ho, R. I. Epstein & E. E. Fenimore, Cambridge Univ. Press, 288 (1992).
38. J. I. Trombka, Nature **226**, 827 (1970).
39. J. C. L. Wang, D. Q. Lamb, et al., Phys. Rev. Lett **63**, 1550 (1989).
40. A. Yoshida, et al., in Gamma-Ray Bursts, ed. C. Ho, R. I. Epstein & E. E. Fenimore, Cambridge Univ. Press, 399 (1992).

The Statistics of the Search for Absorption Lines in Burst Spectra

D. Band[1], L. Ford[1], J. Matteson[1], M. S. Briggs[2], W. Paciesas[2],
G. Pendleton[2], R. Preece[2], D. Palmer[3], and B. Teegarden[3]

[1] *CASS, University of California, San Diego, La Jolla, CA 92093*
[2] *University of Alabama, Huntsville, AL 35899*
[3] *NASA/Goddard Space Flight Center, Greenbelt, MD 20771*

> Because spectral features have not been detected in the BATSE burst data, we study the statistical implications of what could have been detected in the spectra accumulated by BATSE. Specifically, we analyze the detectability of spectral lines in burst spectra to: 1) identify the bursts in which lines would be most detectable; 2) evaluate the consistency between the absence of a line detection by BATSE and the detections reported by previous missions; and 3) constrain the rate at which lines occur based on the observations.

INTRODUCTION

No spectral lines have been detected yet in the BATSE spectra accumulated by the Spectroscopy Detectors (1–3), whereas lines have been reported by KONUS (4), *HEAO 1* (5) and *Ginga* (6). Although the visual inspection of BATSE spectra identified line candidates, none are significant enough to be considered detections (3); lines may yet be detected in new bursts observed by BATSE or by a computerized search currently under way (1). Until and unless lines are detected definitively, we must work under the assumption that lines are not present in the current BATSE database. This has led us to analyze the number and types of spectral lines which could have been detected in the observed spectra. The consistency between BATSE's observations and those of previous missions has been the primary purpose of this analysis, but the resulting methodology also constrains the frequency with which lines might be present in burst spectra; this estimate of the line frequency is relevant regardless of whether lines are ultimately detected in the BATSE spectra. Finally, these studies can identify the bursts in which lines may be most detectable. New search techniques will of course be applied first to those bursts in which they are likely to be most successful.

This work makes assumptions about the status of the search for absorption lines in the BATSE spectra. First, we assume that the visual search of BATSE spectra was complete, and the database contains no detectable lines. However, the computerized search is finding weak line candidates which were not identified by the visual search (1). Second, we base our evaluation on

© 1996 American Institute of Physics

the calculated instrument response. While we are investigating some minor unresolved instrumental issues, we are confident that the detectors function correctly (7).

The BATSE data for each burst consist of series of temporally-consecutive spectra from the detectors in four different BATSE modules. To be considered a detection, a feature must be significant in a spectrum from one detector, and the spectra from all the detectors which viewed the burst must be consistent. A feature is significant if the probability that it is a statistical fluctuation is sufficiently small as determined by a likelihood ratio test (8). Consistency between detectors means that it is plausible that a real spectral feature could have produced each detector's observed spectrum. We do not require significance in all detectors since fluctuations or unfavorable conditions (e.g., a large burst angle—the angle between the burst and the detector normal) may reduce the significance of a feature in some detectors.

We search for spectral features in individual spectra, yet we evaluate the statistics of line occurrence using the entire burst database. Therefore, we have constructed a hierarchy of probabilities. Our three goals—consistency between missions, the line frequency, and the identification of bursts worth searching—require the probability of detecting a line in each burst; this probability is constructed from the probability of detecting a line in each spectrum. The detectability of a line depends on the parameters which characterize it, such as the equivalent width, intrinsic width and energy centroid. Although we present here the methodology in terms of the analysis of a single line, it should be applied to a distribution of line types. Because of the small number of lines detected, we do not know the underlying distribution of line parameters; consequently we use the few observed lines as archetypes. In particular, we use the lines reported as detections by *Ginga*.

As a matter of notation, we define l to be the statement that a line exists, and L that a line is detected; the negation of proposition b is $\bar{b}$. Since we use conditional probabilities, we define $p(a \,|\, b)$ to be "the probability of a given b." To remind ourselves that our calculations rest on our understanding of the detector response, we include among the 'givens' the proposition I representing our knowledge of the detectors.

METHODOLOGY

$p(L_i \,|\, I)$—**the probability of detecting a line in the ith spectrum.**
A "detected" feature (proposition L_i) may be real (l_i) or spurious ($\bar{l}_i$):

$$p(L_i \,|\, I) = p(L_i \,|\, l_i I) p(l_i \,|\, I) + p(L_i \,|\, \bar{l}_i I) p(\bar{l}_i \,|\, I) \tag{1}$$

We currently assume the probability $p(L_i \,|\, \bar{l}_i I)$ of considering a fluctuation to be a detection is sufficiently small that we can set it equal to 0. We discuss below the probability $p(l_i \,|\, I)$ that a line is present in the ith spectrum.

We determined the probability of considering a real line feature a detection, $p(L_i \,|\, l_i I)$, using simulations for various line types (energy centroid, intrinsic width and equivalent width) (9). We created ~ 200 model count spectra (with noise) which were fitted as if they were real spectra; the desired probability is the fraction of the spectra which satisfied our detection criterion.

Since we used the line detections reported by *Ginga* as archetypical line types, our simulations demonstrated that BATSE would have detected the lines *Ginga* found in GB 880205 at almost all burst angles, but would have detected the line in the S1 segment of GB 870303 in only a quarter of the spectra accumulated by detectors with small burst angles.

$p(L_\alpha \mid I)$—**the probability of detecting a line in the αth burst.** A line could be detected in any of the $N(N+1)/2$ possible spectra constructed from N consecutive spectra spanning a burst accumulated by a given detector. The probability that the line is detected in a burst is found by combining the probabilities that it is detected in the individual spectra. The probability that a line is actually present in a given spectrum, $p(l_i \mid I)$ (see eqn. [1]), depends on the physics of line occurrence which is currently unknown and consequently must be modeled. In our modeling (10) we assume that this probability is a function of the time lines persist; we have constructed models where the functional dependence of the probability on the persistence time is constant, an exponential or a power law.

Burst Database. The foundation of all these probabilities is the probability of detecting a line within a given spectrum, $p(L_i \mid l_i I)$. We have found that this probability is primarily a function of the continuum's signal-to-noise ratio (SNR) and the burst angle, and therefore we need these two quantities for every spectrum accumulated by every detector which viewed each burst. The SNR can be calculated for any spectrum from the background rate and the counts accumulated for each of the constituent spectra provided by the telemetry. Thus for any burst-detector set of spectra, N count values and one background rate are required to find the $N(N+1)/2$ SNR values for all possible spectra formed from N accumulated spectra spanning a burst. Evaluated for the 25-35 keV band, the SNR is normalized to compensate for the variable energy width of an integral number of channels. Our database (10) currently includes 232 strong bursts observed by BATSE over more than 4 years.

Identifying Bursts to be Searched. Those bursts with high values of the probability of detecting a line, $p(L \mid lI)$, should have the highest priority for being searched. However, this probability is highly model dependent. For simplicity we have rank-ordered the bursts by the largest SNR for any spectrum during the burst (10).

The Line Frequency. Based on the observed bursts we estimate the likely line frequency f, the probability that a line is present in a burst (11); here we assume that f is the same for all bursts. Using a Bayesian formalism with no preconceptions about f (i.e., with a uniform "prior"), the probability distribution for f is proportional to the likelihood for f, the probability of obtaining the observations for a given value of f, $p(f \mid I)$. The probability $p(f \mid I)$ has a maximum which gives the most likely value of f, but the distribution is more informative in showing the range of reasonable values of f.

Consistency Between BATSE and Other Missions. The absence of a line detection in the BATSE spectra appears to be discrepant with the detections reported by KONUS, *HEAO 1*, and *Ginga*, but is this apparent inconsistency compelling? To answer this crucial question we constructed a number of consistency measures which require the probabilities of detecting a line in a burst (11). However, to use the methods discussed above detailed information is required about a mission's detectors and every spectrum the mission observed; such data are currently available only for BATSE and *Ginga*, and thus the comparison is between these two missions.

We derived two measures of the type used in standard "frequentist" (as opposed to Bayesian) statistics (11). Both measures require the number of bursts in which a given line would have been detectable, a number we evaluate by summing the line detection probabilities over all bursts. The first measure is the probability that if there is a single line detection, it occurs in a burst observed by *Ginga* (2): $p(n_G = 1, n_B = 0 \,|\, n_G + n_B = 1, N_G N_B I)$ where n_G and n_B are the number of *Ginga* and BATSE bursts, respectively, in which lines of the type under consideration were detected, and N_G and N_B are the number of *Ginga* and BATSE bursts in which lines could have been detected. The second measure is the probability of a result that would appear as discrepant—one or more *Ginga* detections and no BATSE detections—which is dependent on the line frequency f, $p(n_G \geq 1, n_B = 0 \,|\, N_G N_B f I)$. We use the value of this probability maximized with respect to f.

We also developed a Bayesian framework evaluating the consistency of the BATSE and *Ginga* observations (11). In this methodology we compare the probabilities in favor of different hypothesis (Bayesian statistics permits us to assign to a proposition a probability of being correct). We contrast the null hypothesis—the detectors function as understood and the *Ginga* detections are real—with contrary hypotheses: 1) BATSE is unable to detect lines even if present; 2) lines do not exist and therefore the *Ginga* detections are spurious; and 3) the bursts observed by BATSE and *Ginga* are characterized by different line frequencies. The Bayesian hypothesis comparison consists of a (usually qualitative) factor comparing our expectations (i.e., priors) and a quantitative ratio of the hypothesis likelihoods (i.e., the probabilities of obtaining the observations if the hypothesis is true). In our case the priors favor the null hypothesis while, because of the apparent discrepancy between BATSE and *Ginga*, the likelihoods generally favor the alternative hypotheses; the issue is which dominates.

RESULTS

Because the true distribution of line parameters is unknown, we use the few observed bursts as archetypes. Specifically, we use the line parameters—line centroid, equivalent width and intrinsic width—(but not the persistence times of ~ 5 s) of the two line sets reported by *Ginga*: a single line at 20.6 keV in the S1 segment of GB 870303 (12), and lines at 19.4 and 38.8 keV in GB 880205 (6). However, the SD's sensitivity drops substantially below 20 keV, and therefore the line in GB 870303 would rarely be detectable. Therefore here we consider only the lines observed in GB 880205. We approximate $N_G = 10$ for the GB 880205 lines based on a preliminary analysis of the lines' detectability by *Ginga* (13). Using the largest SNR in the spectra of each BATSE burst, we should detect lines similar to those of GB 880205 in $N_B \sim 50$ bursts.

Using these values of N_B and N_G for GB 880205, we find that the discrepancy between the absence of a BATSE detection and the *Ginga* detections is not yet compelling. The probability that detection would be in the *Ginga* observations is 17%, while the probability that the apparent discrepancy would be at least this great is 6%. For the Bayesian hypothesis comparisons, our prior expectation that BATSE and *Ginga* are consistent outweigh the quantitative likelihood ratios which favor contrary hypotheses.

The small number of line detections in the relatively large number of BATSE and *Ginga* bursts indicate that lines are a fairly rare phenomenon. If lines exist, then lines of the type *Ginga* detected (narrow lines with energy centroids between 20 and 100 keV and equivalent widths of order 5–10 keV which persist for a few seconds) occur in only a few percent of all bursts.

Acknowledgements. The work of the UCSD group is supported by NASA contract NAS8-36081.

REFERENCES

1. M. S. Briggs, et al., this conference (1995).
2. D. Palmer, et al., Ap.J.Lett. **433**, L77 (1994).
3. D. Band, et al., Ap.J., in press (1996a).
4. E. P. Mazets, et al., Nature **290**, 378 (1981).
5. G. J. Hueter, Ph.D. Thesis, UCSD (1987).
6. T. Murakami, et al., Nature **335**, 234 (1988).
7. W. Paciesas, et al., this conference (1995).
8. W. T. Eadie, et al., Statistical Methods in Experimental Physics, Amsterdam: North-Holland Publishing (1971).
9. D. Band, et al., Ap.J. **447**, 289 (1995).
10. D. Band, et al., in preparation (1996b).
11. D. Band, et al., Ap.J. **434**, 560 (1994).
12. C. Graziani, et al., in Gamma-Ray Bursts, ed. C. Ho, R. I. Epstein, and E. E. Fenimore, Cambridge: Cambridge University Press, 407 (1992).
13. E. E. Fenimore, et al., in the Compton Gamma-Ray Observatory (AIP Conf. Proc. **280**), ed. M. Friedlander, et al., New York: AIP, 917 (1993).

Testing the Compton Attenuation Theory of Gamma-Ray Burst Spectra

J. J. Brainerd, R. D. Preece, M. S. Briggs,
G. N. Pendleton, and W. S. Paciesas

University of Alabama in Huntsville

> The narrow range of energies that characterize gamma-ray burst spectra motivated Brainerd to develop the Compton attenuation theory of burst spectra. We fit model spectra to 11 bright bursts to test and confirm the theory's prediction of an x-ray excess below 10 keV.

MOTIVATION AND THEORY

All gamma-ray burst spectra have strikingly similar characteristic energies. The distribution function for the peak energies of νF_ν is itself peaked at 250 keV, with most values falling between 100 and 700 keV (5). This is unexpected for cosmological gamma-ray bursts, because such bursts must have a Lorentz factor $\gamma > 100$ (3,4), and one expects the values of γ to be broadly distributed. When one considers that other parameters such as the magnetic field strength, the electron energy distribution, and, in the case of Compton scattering, the seed photon energy will vary from burst to burst, and that they are likely to increase with γ, one realizes that the observed distribution of νF_ν peak energies is an important clue to the physical processes operating in gamma-ray bursts. Assuming that it does not have an instrumental origin, one can immediately provide two alternative physical theories to explain this effect: either there exist processes in the rest frame of the host galaxy that modify the burst spectrum, or the emission processes are tied to the Lorentz factor in such a way that the energy of peak emission is only mildly depend on the Lorentz factor. In this article, we present an observational test of a recently published theory of the first type, Brainerd's Compton attenuation theory for cosmological gamma-ray burst spectra (1).

One assumes in the Compton attenuation theory that gamma-ray burst sources produce power-law spectra, which are unaffected by a Lorentz boost, and that attenuation by intervening material in the host galaxies produces the observed spectra. Because the scattered radiation must not produce electron-positron pairs with the direct radiation—otherwise the optical depth would increase dramatically during the burst—the attenuation region must be > 0.1 pc. This makes the light travel time between scatterings much longer than the burst duration, making the scattered radiation unobservable. The two radiative processes in this theory are Compton scattering and photoelectric absorption. The spectral model produced by this theory therefore has five

© 1996 American Institute of Physics

free parameters: the Thomson optical depth τ_T, the cosmological redshift z, the metallicity, the amplitude, and the power-law index δ.

This theory places severe limits on prospective gamma-ray burst sources. The column density must be of order 10^{25} cm^{-2} to produce the observed spectrum. This and the lower limit on size suggest that the scattering bodies are molecular clouds with a density of order 10^5 cm^{-3}, which are observed only in the cores of galaxies. The burst sources would then be objects unique to the centers of galaxies. Because most of the energy emitted by the source below $\approx$ 200 keV is lost before reaching the observer, source theories requiring a highly efficient conversion of energy to observed gamma-rays can be discarded.

Observational prospects of the theory are the derivation of a cosmological redshift for each gamma-ray burst that has a well observed x-ray spectrum and the measurement of the metallicity of the interstellar medium around gamma-ray burst sources.

The theory is observationally falsifiable. First, it predicts an excess of x-rays over what is expected from a power-law continuation of the burst spectrum from the gamma-ray band—Preece et al. (6) find that such an excess exists in 15% of all bursts. Second, of the five free parameters in the spectral model, only two can vary during an outbursts: the power-law index and the amplitude. The optical depth, the metallicity, and the cosmological redshift must remain constant with time. Finally, because the scattering region is most likely a molecular cloud of high optical depth, the source is unlikely to produce an observable optical outburst, but it can produce an extended x-ray afterglow through the coherent scattering of x-rays by dust (2).

THEORY VERSUS OBSERVATION

The simplest test of the theory is to look for x-ray excesses in gamma-ray burst spectra. To do this, we define an alternative spectral form by neglecting photoelectric absorption and grafting a power-law spectrum onto the Compton attenuation spectral model at the inflection point energy of 231 keV in the host galaxy rest frame. The power-law index is chosen so that $d \log N / d \log \nu$ is continuous at the inflection point.

Both spectral models are applied to the subset of bursts that are observed by a BATSE spectroscopy detector set at the highest gain of 8x, that have a fluence above 10^{-5} ergs cm^{-2} in the energy band of 50 keV to 300 keV, and that have in this energy band a peak flux on the 256 ms timescale above 10 cm^{-2} s^{-1}. Eleven out of the first 1005 gamma-ray bursts observed by BATSE satisfy these criteria. For most bursts, we integrate spectra over the whole burst duration. In several cases the spectrum varies dramatically from peak to peak, so we integrate only over the brightest peak of the light curve. We allow each model to come to its own best fit to the data. Because only one x-ray channel is available, the metallicity is set to zero to maximize the x-ray upturn in the Compton attenuation spectral model; both models therefore have 4 degrees of freedom. The data used in these fits are the 256 channel data from the spectroscopy detectors and discriminator channel 1 of a

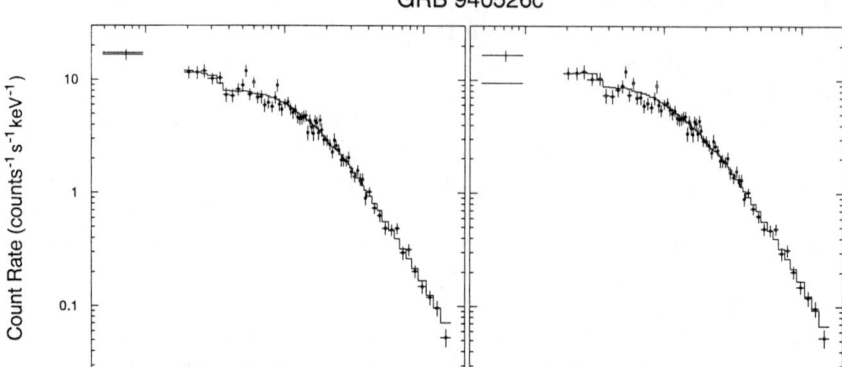

FIG. 1. Model fits to GRB 940526c (BATSE trigger 2994) count spectra. On the left is the best fit of the data to the Compton attenuation spectrum, with values of $\tau_T = 8.9 \pm 0.9$, $z = 1.5 \pm 0.2$, and $\delta = -2.6 \pm 0.2$, while on the right is the best fit to the x-ray suppressed spectrum, with values of $\tau_T = 14.1 \pm 8.2$, $z = 0 \pm 0.5$, and $\delta = -3.8 \pm 0.2$. The data are from spectroscopy detector 1, with the data point at 5–10 keV discriminator channel 1 and the remaining data points the 256 channel data binned to 8σ significance. The burst spectrum was integrated from 3.008 to 7.936 seconds after the trigger.

spectroscopy detector with an 8x gain. At this gain, the discriminator channel accumulates counts with energies between 5 and 10 keV. If the gains of two detectors observing a burst are dramatically different, 256 channel data from both are used to improve the coverage at high energy.

Figure 1 shows the best model fits for a representative burst, GRB 940526c (BATSE trigger 2994). The Compton attenuation spectral model has $\chi^2 = 193$ out of 205 degrees of freedom, and the discriminator channel is higher than the model by $0.39\,\sigma$. The x-ray suppressed spectral form has $\chi^2 = 212$ out of 205 degrees of freedom, and the discriminator channel is higher than the model by $4.13\,\sigma$.

Table 1 gives the reduced χ^2 values for the best fit models and the sigma residuals—the deviation of the count rate from the model model rate in units of counts standard deviation—for discriminator channel 1. The sigma residuals are plotted in Figure 2.

The Compton attenuation spectral model produces a good reduced χ^2 for all 11 burst spectra, and it fits discriminator channel 1 well for 10 of the 11 spectra. Sigma residuals for 9 of these 10 bursts cluster within $< 2\,\sigma$ of 0, and the tenth burst is only $3.2\,\sigma$ from 0. The only burst with a poor discriminator sigma residual is GRB 920627b, which only has an upper limit on the x-ray

TABLE 1. χ^2/ν and Discriminator Channel 1 Sigma Residuals.

Trigger	Burst	ν[c]	CASM[a] χ^2/ν	DSR[d]	XRSSF[b] χ^2/ν	DSR
1473	GRB 920311	323	2.38	3.20	12.50	55.04
1541	GRB 920406	335	1.20	1.82	1.59	11.60
1609	GRB 920517b	183	1.35	0.97	3.84	13.35
1625	GRB 920525b	372	1.09	1.72	1.29	9.02
1663	GRB 920622b	414	1.14	−1.26	1.20	2.18
1676	GRB 920627b	373	1.25	−4.26	1.25	−4.26
1711	GRB 920720a	366	1.08	0.45	1.30	8.10
2067	GRB 921123a	195	1.11	−0.07	1.13	1.52
2812	GRB 940210	163	1.08	−0.01	1.04	−0.004
2953	GRB 940429a	200	1.20	1.73	1.35	6.22
2994	GRB 940526c	205	0.94	−0.39	1.03	4.13

[a] Compton Attenuation Spectral Model.
[b] X-Ray Suppressed Spectral Form.
[c] Degrees of freedom.
[d] Sigma residual for discriminator channel 1.

flux. The 10 bursts with good discriminator sigma residuals have reasonable values for the model parameters: δ ranges between −6 and −2.5, z ranges between 1.5 and 7, and τ_T ranges between 9 and 27. As a matter of practice, the discriminator channel determines the value of z, and since the value of the metalicity affects the flux in this channel, setting the metallicity to 0 produces larger values of z than one would find with a higher metallicity. Before we can have confidence in the derived values of z, we must fit the spectral model to higher resolution x-ray data.

In contrast, the x-ray suppressed spectral form only produces good reduced χ^2 for 9 of the 11 burst spectra, and it fails to fit discriminator channel 1 in 9 of 11 spectra. For 9 of 11 spectra, the discriminator channel lies above the model, often by many standard deviations. Only one spectrum falls significantly below the model—GRB 920627b, which has no x-rays. The behavior of the discriminator sigma residuals in Figure 2 demonstrates that a spectral model must have an x-ray upturn to successfully fit the data.

SUGGESTIONS AND CONCLUSIONS

Further tests of the Compton attenuation theory require x-ray spectra of higher resolution. With better spectral resolution, the shape of the upturn and the presence of photoelectric absorption can be tested, an accurate value of z can be derived, and the constancy of z, τ, and the metallicity over the duration of a burst can be tested.

Brainerd's Compton attenuation theory produces spectral models that fit the observations well. The x-ray excess in the theory is necessary for a good χ^2 minimization to the data; when the x-ray excess is removed, the best χ^2

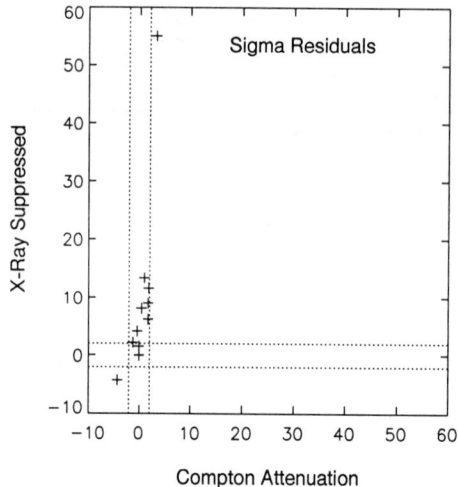

FIG. 2. Scatter plot of the deviation of the first discriminator channel from the Compton attenuation spectral model (x-axis) and the x-ray suppressed spectral form (y-axis) for eleven gamma-ray bursts. The deviations are in units of one standard deviation in photon counts, and a positive value indicates the data point is above the model. Dotted lines give $\pm 2\,\sigma$.

minimization produces model fits that consistently fall far below the data at 5–10 keV. The Compton attenuation theory therefore passes its first observational test.

REFERENCES

1. J. J. Brainerd, Astrophys. J. **428**, 21 (1994).
2. J. J. Brainerd, in Gamma-Ray Bursts, AIP Conf. Proc. **307**, ed. G. J. Fishman, J. J. Brainerd, & K. Hurley (New York: AIP), 346 (1994).
3. E. E. Fenimore, R. I. Epstein, & C. Ho, 1993, A&A Supp.**97**, 59 (1993).
4. A. K. Harding, & M. G. Baring, in Gamma-Ray Bursts, AIP Conf. Proc. **307**, ed. G. J. Fishman, J. J. Brainerd, & K. Hurley (New York: AIP), 520 (1994).
5. R. S. Mallozzi, W. S. Paciesas, & G. N. Pendleton, this volume (1996).
6. R. D. Preece, M. S. Briggs, G. N. Pendleton, W. S. Paciesas, J. L. Matteson, D. L. Band, & C. A. Meegan, this volume (1996).

A Comprehensive Search for Low-Energy Lines in BATSE GRBs

Michael S. Briggs[1], David L. Band[2], Robert D. Preece[1],
Geoffrey N. Pendleton[1], William S. Paciesas[1],
Lyle Ford[2], James L. Matteson[2]

[1] *Department of Physics*
University of Alabama in Huntsville, Huntsville, AL 35899
[2] *Center for Astrophysics and Space Science*
University of California, San Diego, La Jolla, CA 92037-0111

A computer-based technique has been developed to search bright BATSE gamma-ray bursts for spectral lines in a comprehensive manner. The first results of the search are discussed and an example line candidate shown.

MOTIVATION

Prior to the launch of CGRO, and during the early days of the mission, the BATSE team expected that GRB lines would be common and easily identified. Our expectations were based upon the observations of KONUS (15), HEAO A-4 (13) and *Ginga* (10,17). Our early efforts were based upon these expectations: we searched bright bursts visually, examining consecutive spectra and also the spectrum of the entire burst (19). Our expectations were not met: no significant lines have been found (19,5).

While the failure to find lines in BATSE GRBs does not yet imply a serious discrepancy between the *Ginga* and BATSE results (19,3), it is disappointing. Previously, lines were interpreted as strongly supporting the theory that GRBs originate nearby from a disk population. Results from BATSE on the spatial distribution of GRBs demonstrate that only a minority of the bursts can originate from a disk population (16,12,7), so if the existence of lines is confirmed, their properties might indicate the presence of a disk subclass or give clues to the physics of halo or cosmological sources.

Is the failure to detect lines in the BATSE data because lines do not exist or is the failure due to some inadequacy in the visual search approach? To answer these questions a more systematic search is needed. We have therefore implemented an automatic, computer-based comprehensive line search. The goal of the search is comprehensiveness—a brute force approach is taken so that no significant line will be missed. The purpose of the search is to identify line candidates—the search need not perfectly evaluate their significances, which will be done later under direct human control. The computer-based search has several advantages in addition to comprehensiveness: subjectivity

© 1996 American Institute of Physics

is eliminated, there is no variation in detection threshold due to human exhaustion, there is no bias towards absorption or emission features, and the search will collect statistics on all trial lines.

COMPREHENSIVE SEARCH METHOD

The addition to the BATSE instrument of the eight Spectroscopy Detectors (SDs) was largely motivated by the KONUS results. Simulations (4,11) and tests (18) of the performance of the BATSE SDs show that they should be capable of detecting KONUS or *Ginga*-like GRB lines. Each SD is a NaI(Tl) scintillator crystal, 12.7 cm diameter and 7.6 cm thick, viewed by a photomultiplier tube of the same diameter. To limit the effort to a manageable level, the initial searches are restricted to lines below 100 keV, so only detectors in high-gain mode are useful. Typically, such detectors cover the energy range of about 10 keV to about 1400 keV, although analysis begins at about 15 or 20 keV due to an electronic artifact (1). Work to model this artifact continues, so that in the future we will be able to extend the analysis to lower energies. Analysis of the data is based upon an instrument response model which accounts for the detectors, nearby spacecraft material, and scattering from the Earth's atmosphere (20). When a burst occurs, high-time resolution SHERB (Spectroscopy High Energy Resolution Burst) data are collected using a time-to-spill algorithm.

Since we do not know a priori when or how long a line will exist, we search essentially all time scales available. We search each single SHERB record, every consecutive pair, triple, and group of 4, 5, 6, 8, 10, 14, 20, ... records and the sum of all the SHERB records. Clearly, many overlapping intervals are searched, and thus the searched intervals are not all independent. Additionally, many intervals will have insufficient signal-to-noise ratio for a real line to be detectable—these intervals serve as controls. While not every possible consecutive combination of records is formed, enough combinations are searched so that no line should be missed, although its significance might not be optimized.

Similarly, we do not know a priori at what energies lines occur. Therefore, we first fit each spectrum with a continuum model, and then we perform continuum plus line fits using a closely spaced, fixed grid of trial centroids. Since the trial centroids are separated by one-third the detector resolution FWHM, no line candidate should be missed, although the exact centroid will not be found by the search.

Candidates are identified by large ($\gtrsim 20$) changes in χ^2. We have switched to using $\Delta\chi^2$ instead of the F-test because it is more appropriate when the errors are known (14,9,6). After a candidate is identified, other time intervals will be tried and the centroid will be made a free parameter in the fits.

To increase the robustness of the automatic nonlinear fits, we use a continuum model with a small number of parameters, namely the Comptonized model, which is a power-law times an exponential cutoff. This is fit to a restricted energy range of ≈ 400 keV. Our initial search is confined to lines

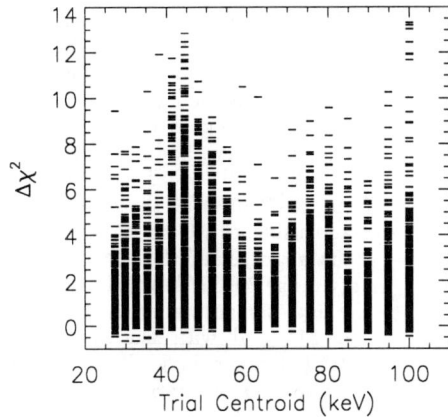

FIG. 1. Results of the comprehensive line search for GRB 920627 (trigger 1676), SD 2. The figure shows the change in χ^2 resulting from adding to the continuum model a narrow, additive Gaussian line. While SD 2 has useable data from 15–1240 keV, the search was made using the Comptonized continuum model and the restricted energy range 15–390 keV. The few points with $\Delta\chi^2 < 0$ are due to poor convergence.

which are narrow compared to the detector resolution and we use the simplest line model: an additive (or subtractive) Gaussian. So that there is continuum at energies below any line, the lowest energy trial centroid is two detector-resolution FWHM above the starting energy of the fits. Since very few of the fits begin below 15 keV, only rarely do we search for 20 keV lines.

We show as an example of a burst without a significant line the data of SD 2 for GRB 920627. The 63 SHERB records of SD 2 are summed into 693 spectra which are fit with trial lines at 20 different centroids from 27.3 to 100 keV. The largest change in χ^2 obtained by adding a line was 13.3 (Fig. 1), which has a chance probability in a single fit of a few tenths of a percent. Considering the large number of spectra searched (albeit not all independent), this is insignificant.

FIRST RESULTS OF THE COMPREHENSIVE SEARCH

The search results for GRB 940703 are quite different from those for GRB 920627. Only one high gain detector, SD 5, viewed the burst at an angle less than 100°. The data from SD 5 comprises 79 SHERB records, from which 917 spectra were formed. The largest value of $\Delta\chi^2$ identified in the search was 58.2 and many overlapping intervals had $\Delta\chi^2$ values approaching this value (see Fig. 2). When more careful fits are made, using the Band spectral form (2) and the entire available energy range, the candidate is found to have $\Delta\chi^2$ of "only" 23.4. The difference in the values of $\Delta\chi^2$ is because for this very bright burst the Comptonized model is inadequate even over a restricted energy range.

Most of the line significance appears to come from a shorter interval which excludes periods of weak emission, for which $\Delta\chi^2 = 28.1$. The spectrum of this interval is shown in Fig. 3—the candidate is seen to be an *emission* feature with an equivalent width of 2.1 keV. The line is narrow—allowing the width to vary does not significantly reduce χ^2. For 3 additional parameters,

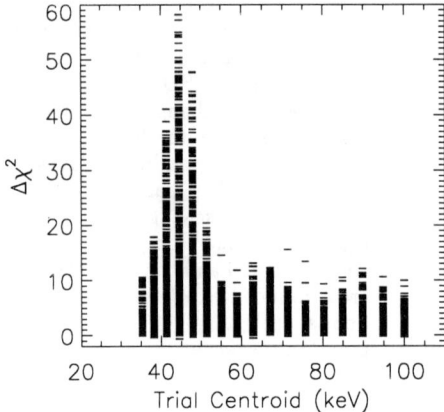

FIG. 2. Results of the comprehensive line search for GRB 940703 (trigger 3057), SD 5. The figure shows the change in χ^2 resulting from adding to the continuum model a narrow, additive Gaussian line. While SD 5 has useable data from 20–1380 keV, the search was made using the Comptonized continuum model and the restricted energy range 20–397 keV. The angle of the detector normal with the burst was 51° and with the geocenter, 79°. The few points with $\Delta\chi^2 < 0$ are due to poor convergence.

the chance probability in any particular fit of such a large $\Delta\chi^2$, if no line actually exists, is 3×10^{-6}. Additionally, the residual plots show that adding the line to the model eliminates conspicuous runs of high, then low, residuals. Even considering the large number of spectra searched, the line is highly significant—the number of effectively independent spectra is well below 917 and the line is significant in more than one spectrum.

To date, 42 of the brightest GRBs have been searched by this technique. There was an average of 2.1 detectors per burst and 53 SHERB records per detector, for a total of 4729 SHERB records. From these records 51,651 spectra were formed and 861,372 fits performed. Partial examination of the results has identified 8 candidates with $\Delta\chi^2 > 20$. There are both absorption and emission candidates and their centroids range from 40 to 70 keV.

More work remains to be done. In most cases several high-gain SDs observe a gamma-ray burst. We can therefore usually test the reality of a line candidate by examining the data of the other detectors. If the lines are real, in all cases the data should be consistent and in some cases there should be confirmation: statistically significant detections in more than one detector. The ability to perform these tests is a major advantage of BATSE, but it is also time-consuming because there are many issues. Until we have completed these tests, the BATSE team considers the features identified by the comprehensive search to be line candidates rather than detections.

REFERENCES

1. D. L. Band, et al., Experimental Astronomy **2**, 307 (1992).
2. D. L. Band, et al., ApJ **413**, 281 (1993).
3. D. L. Band, et al., ApJ **434**, 560 (1994).
4. D. L. Band, et al., ApJ **447**, 289 (1995).
5. D. L. Band, et al., ApJ **458**, 746 (1996).
6. D. L. Band, et al., ApJ, in preparation (1996).

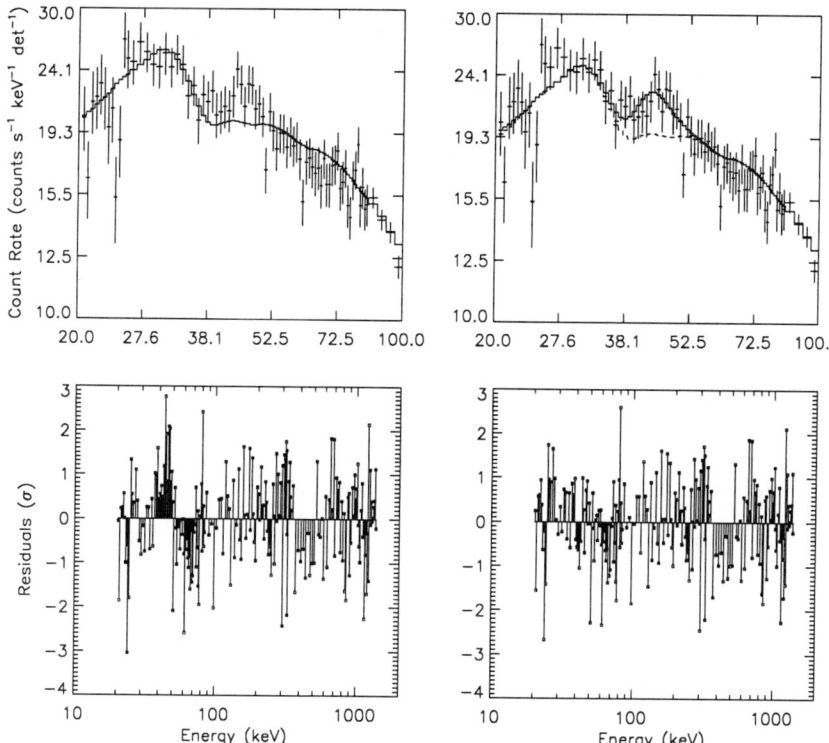

FIG. 3. Data from SD 5, GRB 940703, for the interval 42.112–64.704 s. Top panels: Count rate data and fits, shown up to 100 keV. Bottom panels: Count rate residuals (data − model) in units of σ for the entire energy range used in the fits. Left panels: A continuum-only fit, using the Band spectral form; $\chi^2 = 212.0$ with 199 degrees-of-freedom. Right panels: Added to the fits is an narrow, additive Gaussian line; $\chi^2 = 183.9$ with 197 degrees-of-freedom. The "hump" at 30 keV in the data and continuum model is expected from the detector physics (8,18). The contribution of atmospheric scattering to the counts in the line region is ≈4%.

7. M. S. Briggs, et al., ApJ **459**, 40 (1996).
8. M. S. Briggs, these proceedings (1996).
9. W. T. Eadie, et al., Statistical Methods in Exper. Physics, North-Holland (1971).
10. E. E. Fenimore, et al., ApJ **335**, L71 (1988).
11. P. E. Freeman, et al., in AIP Conf. Proc. **280**, 922 (1993).
12. J. Hakkila, et al., ApJ **1994**, 422 (1994).
13. G. J. Hueter, Ph.D. Thesis, Univ. of Calif., San Diego (1987).
14. D. Q. Lamb, private communication (1995).
15. E. P. Mazets, et al., Nature **290**, 378 (1981).
16. C. A. Meegan, et al., Nature **355**, 143 (1992).
17. T. Murakami, et al., Nature **335**, 234 (1988).
18. W. S. Paciesas, et al., these proceedings (1996).
19. D. M. Palmer, et al., ApJ **433**, L77 (1994).
20. G. N. Pendleton, et al., Nucl. Instr. & Meth. in Phys. Res. **A 364**, 567 (1995).

EGRET TASC Observations of the Bright April 25, 1995 GRB

J.R.Catelli*†, B.L.Dingus†‡, E.J.Schneid§, for the EGRET Team

*University of Maryland, College Park, MD 20742,
†NASA Goddard Space Flight Center, Greenbelt, MD 20771,
‡also Universities Space Research Association,
§Northrop Grumman Corporation, MS A01-26, Bethpage, NY 11714

> The bright burst of 1995 April 25 was observed by the EGRET instrument on the *Compton Observatory*. Although the burst was significantly off-axis (47.6°) and therefore could not be imaged at >30 MeV in the EGRET spark chamber, it was observable in the 1 to 200 MeV range by the Total Absorbtion Shower Counter (TASC) portion of the instrument. The event was highly significant in three consecutive 32.768 second spectral accumulation times, and has one of the hardest observed spectra at energies greater than 1 MeV. Spectra for shorter time intervals were unavailable for this event, as telemetry gaps caused a loss of data during the readout of the burst mode spectra.

INTRODUCTION

The EGRET instrument has measured the emission of gamma-ray bursts at high energies, with a number of interesting results. The spectra of several bright bursts extend past a GeV and the high energy emission has been observed to last for over an hour. The nature of this high energy and delayed emission is a puzzle which may hold some important clues to the nature of gamma ray bursts. Eleven bursts have been detected by EGRET (2), five of them imaged at >30 MeV in the spark chamber portion of the instrument (3).

The Total Absorption Shower Counter (TASC) on EGRET is an ~ 6000 cm² NaI(Tl) detector. It acts as a calorimeter for the spark chamber events, but it is also an independent, self-triggered detector in the energy range from 1 to 200 MeV. Its sensitivity is highly dependant on the incident angle and the amount and type of spacecraft material between it and a radiation source. The detailed GRO mass model is used with the EGS-4 Monte Carlo routine to generate response matrices for the TASC as a function of energy and incident angle. A limiting factor on TASC measurements is the 175 μs per event deadtime. The TASC has two modes of data collection, the normal mode in which 32.768 second spectra are accumulated, and the burst mode, in which 1,2,4, and 16 second spectra are taken immediately after a BATSE trigger. In addition, the TASC has rate accumulators at four different energy thresholds which are read out every 2.048 seconds. More detailed descriptions

© 1996 American Institute of Physics

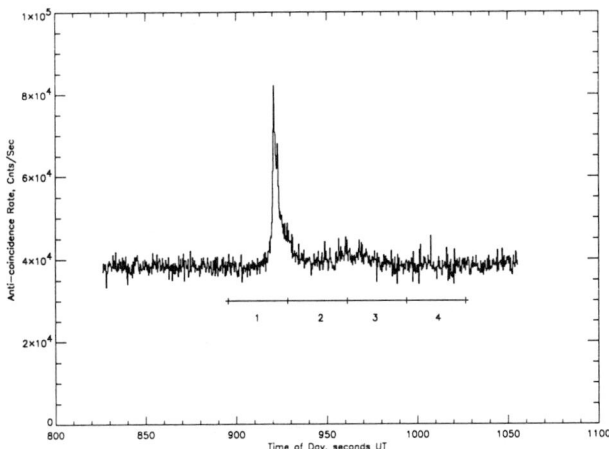

FIG. 1. The anti-coincidence dome (>50 keV) rate during the burst. The time intervals in which spectra were accumulated are indicated on the plot.

of the EGRET instrument and calibration are available in Kanbach (5) and Thompson (6).

TIME DEVELOPMENT

The rate above background in the plastic scintillator of EGRET's anti-coincidence dome (>50 keV) approximates the lightcurve of the burst, as shown in Figure 1. A large peak occurs at the BATSE trigger, about 919 sec UT. This peak appears to have a "shoulder" like feature on the falling edge and is followed by a second, lower, and more rounded peak about fifty seconds later.

This burst was clearly visible in the TASC discriminator rates, shown in Figure 2, as well as the anti-coincidence dome rates. The four energy ranges ($\sim$ >1 MeV, >2.5 MeV, >7 MeV, and >20 MeV) all show an increase in rate at the same time as the BATSE trigger, within the time resolution (2.048 seconds) of the discriminator rates.

SPECTRA

We obtained average spectra for three consecutive 32 second time intervals. Only an average livetime is available for each 32 second interval, although we expect the livetime to drop during periods of higher rate due to the 175 μsec deadtime per event. This leads to an overall normalization uncertainty, but should not affect the shape of the spectra. The flux densities are therefore slight underestimates of the actual flux.

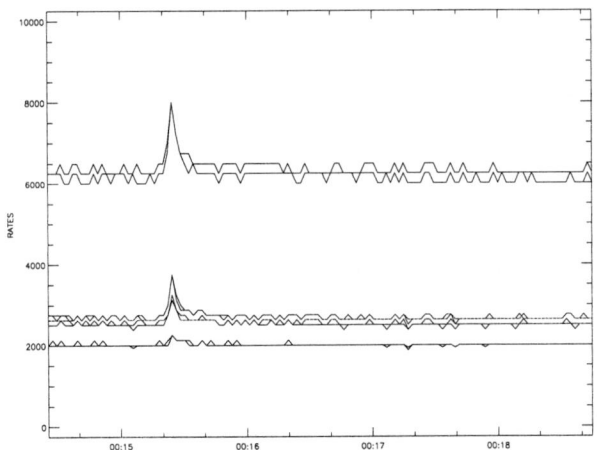

FIG. 2. The TASC rates during the burst. The four energy ranges are $\sim$>1 MeV, >2.5 MeV, >7 MeV, and >20 MeV, with >1 MeV being the highest rate.

In the first time interval, 895.625 to 928.375 sec UT, the spectrum can be fit to a power law $F(E) = kE^{-\alpha}$ with $\alpha = 1.93 \pm 0.039$, and k=2.06±0.109 photons cm^{-2} sec^{-1} MeV^{-1}. This accumulation time includes the trigger and most of the first peak. To normalize this spectrum, emission and livetime were assumed to be at background levels before the trigger and all additional deadtime and flux were assumed to come after the burst trigger. This livetime correction increased the normalization by a factor of 1.27. Also the time interval used for the burst emission began with the burst trigger at 919.21 sec UT and ended with 928.375 sec.

For the second time interval, 928.375 to 961.188 seconds UT, the spectrum is well-fit to a power law with $\alpha = 1.84 \pm 0.059$. This is one of the hardest spectra observed by EGRET for any burst, comparable to the "superbowl" burst of 930131 (7). This accumulation time includes the "shoulder" on the falling edge of the first peak and the interval between the two peaks. It is interesting to note that this is a slightly harder spectrum than that seen in the first time interval, which is against the general trend seen at lower energies (4).

The spectrum for the third interval, 961.188 to 993.938 seconds UT, which includes much of the second peak, is not well fit by a single power law. It shows evidence for a steeper slope at low energies, and a flatter high energy component. The spectrum below 10 MeV can be well-fit to a power law with $\alpha = 2.27 \pm 0.19$, but the integral flux above 10 MeV is greater than 3 sigma above what would be expected from this fit. This may be related to the delayed high energy component which has been seen previously by EGRET for 940217 (1).

The fourth interval does not have sufficient counts above background to do a spectral fit. It has an intensity at 1 MeV of 0.219±0.054 photons cm^{-2}sec^{-1}Mev^{-1}.

160

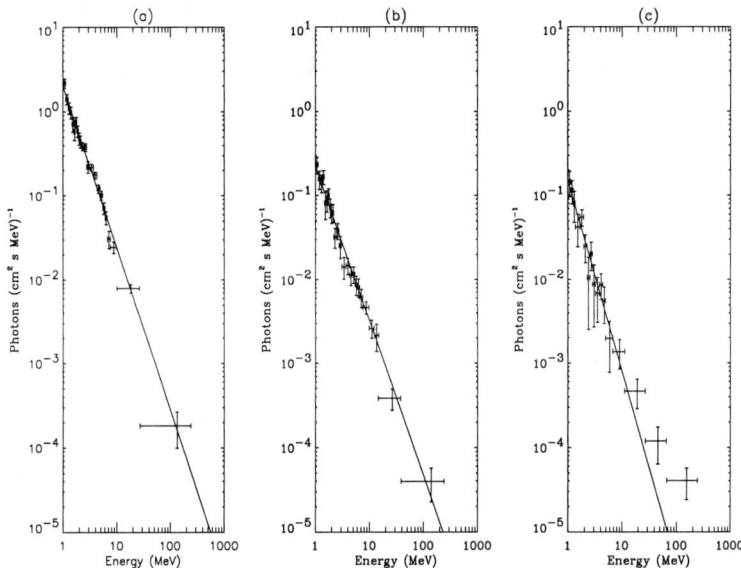

FIG. 3. a)The spectrum for 896 to 928 sec UT, with the power law fit $\alpha = 1.93 \pm 0.039$. b)The spectrum for 928 to 961 sec UT, fit to a power law of index $\alpha = 1.84 \pm 0.059$. c)The spectrum for 961 to 994 sec UT. The fit to below 10 MeV, $\alpha = 2.27 \pm 0.19$, is shown.

DISCUSSION

The April 25, 1995 burst was an interesting one for a number of reasons. It has an unusually hard spectrum at high energies, and has a slight soft-to-hard evolution during the first peak. In the later peak there is apparently an additional high-energy component such that the spectrum cannot be fit by a single power law. Unfortunately it could not be imaged above 30 MeV in the EGRET spark chambers because it was significantly off-axis (47.6°). Because of the hardness of its spectrum, this burst is visible in all of the rate intervals of the TASC. The light curve of the event appears similar in all of the available energy channels, with no evidence for a delay to the start of the higher energy components.

REFERENCES

1. K. Hurley, et al., Nature **372**, 652, (1994).
2. E. J. Schneid, et al., this workshop.
3. Dingus et al., Ap and Sp. Sci. **231**, 187, (1995).
4. Ford, et al. ApJ **439**, 307, (1995).
5. G. Kanbach, et al., Space Sci. Rev. **46**, 69, (1988).
6. D. J. Thompson, et al., ApJ Suppl. Ser **89**, 629, (1993).
7. Sommer et al., ApJ **422**, L63, (1994).

Extrapolations of Gamma-Ray Burst Spectra to the Optical-UV Band

Lyle Ford and David Band

CASS, UC San Diego, La Jolla, CA 92093-0111

Many gamma-ray burst counterpart searches are being conducted in the optical-UV band. To both predict detectability and understand the meaning of any detections or upper limits, we extrapolate gamma-ray spectra from 54 bright gamma-ray bursts to optical-UV energies. We assume optical emission is concurrent with gamma-ray emission and do not consider quiescent or fading counterparts. We find that the spectrum must be steeper (greater flux at low energy) than a simple extrapolation for more than one simultaneous optical flash to be observable per year by current searches.

INTRODUCTION

One of the curious features of gamma-ray bursts is that they have been exclusively observed in the X- and gamma-ray range. While a wealth of information about the phenomenon can be derived from these observations, they have not revealed an unambiguous distance scale nor shown much about the source environment. An excellent way to resolve both problems is to discover counterparts at other wavelengths. If bursts are consistently coincident with a particular class of objects, then the distance to those objects would set the distance scale and identify the source or source host. With this in mind, several groups (1,2) have attempted to find counterparts which flare with the burst, are in quiescence, or appear as afterglows. To date, no convincing counterparts to classical gamma-ray bursts have been found.

In this work, we consider the likelihood that a counterpart could be seen in the optical-UV energy range by extrapolating observed gamma-ray burst spectra to these energies. Although several burst models predict emission at optical wavelengths (see, e.g., (1)), the guidance they provide is slim considering that most were advanced at a time when bursts were thought to originate at distances of around 1 kpc, much closer than is now expected. Therefore, we use empirical models of counterpart emission. Although the extrapolations are not based on an explicit physical model of optical emission, they cover a broad range of possible emission scenarios so that our results should be independent of the shifting trends in burst modeling.

© 1996 American Institute of Physics

ANALYSIS

We assume that the optical spectrum from a GRB is a power law that connects to the observed gamma-ray spectrum at 1 keV. The power law below 1 keV is allowed to have many different indices and optical-UV emission is simultaneous with the gamma-rays. We do not consider fading or quiescent counterparts and ignore interstellar absorption. Essentially, we are extrapolating observed gamma-ray burst spectra to the optical regime.

Our extrapolated fluxes are based on spectral fits to 54 bright GRBs from the first 13 months of BATSE operation (3). These bursts were selected based on the peak count rate in the 50–300 keV energy band and represent the bright end of the GRB intensity distribution. Spectra averaged over the entire burst were fitted by the functional form

$$N_E(E) \left(\frac{\text{photons}}{\text{keV s cm}^2}\right) = \begin{cases} A\left(\frac{E}{100 \text{ keV}}\right)^\alpha e^{-E/E_0}, & E \leq (\alpha - \beta)E_0 \\ A'\left(\frac{E}{100 \text{ keV}}\right)^\beta, & E > (\alpha - \beta)E_0 \end{cases} \quad (1)$$

where A, α, β, and E_0 are fitted parameters and A' is chosen so that the function is continuously differentiable everywhere. The low energy end of available data in these bursts ranged from 10–30 keV so that the extrapolation to the optical is about four decades in energy. Note that this is a differential photon number spectrum and must be multiplied by energy to get the energy flux, $F_E(E) = N_E(E)E$.

Expected magnitudes for optical counterparts to GRBs were determined by extrapolating gamma-ray spectra from equation (1) down to 1 keV. Spectra were then extended to lower energies as an energy flux power law of index γ, where γ was varied from -4 to 4 in integer steps. Also, $\gamma = \alpha + 1$ (an unmodified extrapolation from gamma-rays) was used. Finally, low energy spectra were folded through U, B, and V filters and the 5–7 eV bandpass of the HETE UV Camera Array. Magnitudes were calculated as if the entire burst fluence was emitted in one second. Numerical properties of the resulting distributions for V and the 5–7 eV HETE bandpass are summarized in Table 1 (properties of the other distributions are in ref. (4)).

Although these fluence-based measures of brightness are the standard, flux-based measures are more meaningful when detector exposure times are comparable to the burst duration as is the case for many current experiments Therefore, we calculated the expected observed flux within the exposure time of several current search instruments. If a burst was shorter than the exposure time, the flux-based and fluence-based magnitudes are equal. If, however, a burst was longer, we used the average flux over the exposure time. This approach is clearly an approximation since it ignores intensity and spectral variations within a burst. The distributions of flux-based magnitudes is similar to the distribution of fluence-based magnitudes, though not equal. The $\alpha + 1$ distribution for V with a 5 s exposure is shown in Figure 1 (see ref. (4) for exact properties of the flux-based distributions).

TABLE 1. Expected Brightness for 1s Flash.

Index	V_{min}	V_{med}	min(flux)[a]	med(flux)
$\alpha + 1$	7.8	15.3	1.03×10^{-11}	1.72×10^{-14}
-4	-17.3	-12.3	3.13×10^{-3}	3.57×10^{-5}
-3	-10.7	-5.7	1.81×10^{-5}	2.06×10^{-7}
-2	-4.1	0.9	1.05×10^{-7}	1.20×10^{-9}
-1	2.5	7.5	6.22×10^{-10}	7.08×10^{-12}
0	9.1	14.1	3.70×10^{-12}	4.21×10^{-14}
1	15.7	20.7	2.22×10^{-14}	2.53×10^{-16}
2	22.3	27.3	1.34×10^{-16}	1.52×10^{-18}
3	28.9	33.9	8.21×10^{-19}	9.34×10^{-21}
4	35.5	40.4	5.06×10^{-21}	5.76×10^{-23}

[a] Values for HETE are in ergs cm^{-2} s^{-1} in the 5–7 eV band.

EXPERIMENTS

Instruments which can detect simultaneous optical emission from GRBs can be grouped into two categories. The first has large fields-of-view with long exposure times (> 30 min) and have sensitivity to transients < 6 m_V. These telescopes were designed for meteor patrols but their large fields-of-view make them well suited for GRB detection. Many of these instruments are outlined in ref. (5).

The second class of instruments we consider is characterized by short exposure times. These instruments are much more sensitive to GRBs, which seldom last more than a few minutes (the longest burst in our sample is 130 s). We consider four experiments here: the Explosive Transient Camera (ETC (6)), the Gamma-Ray Optical Counterpart Search Experiment (GROCSE (7)), the Gamma-ray To Optical Transient Experiment (GTOTE (10)), and the Ultraviolet Transient Camera Array on the HETE spacecraft (8). These instruments have exposure times of 0.4–5 s and are very sensitive to optical transients. The ground based instruments have sensitivities to transients $< 11 - 14$ m_V while HETE's limiting flux is 8×10^{-9} ergs cm^{-2} s^{-1} in the 5–7 eV band. All of these instruments monitor a portion of the sky but ETC, GROCSE, and GTOTE can slew to preliminary BATSE positions distributed by the BACODINE network (9). It can take up to five seconds to localize a GRB and another few seconds to distribute the coordinates on the network. After that, the telescope must slew to the proper position. These delays will often result in bright portions of the burst being missed by the telescope. We ignore these concerns in this work so our extrapolated brightness may be somewhat optimistic.

DISCUSSION

Although our predicted counterpart emission is a simple extrapolation over four decades, the results do provide some interesting implications for counter-

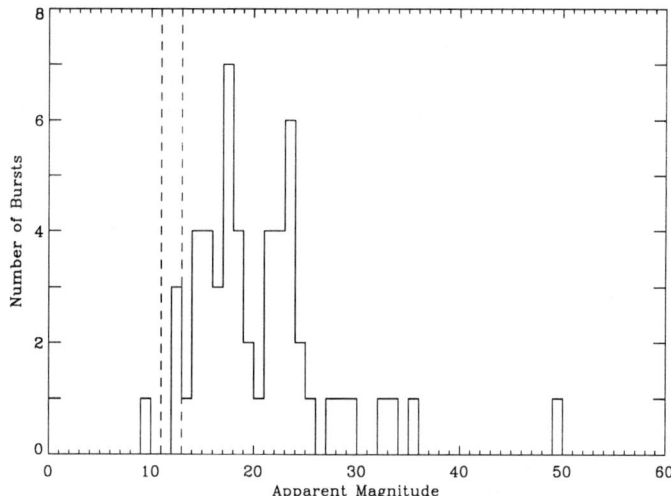

FIG. 1. The number of bursts at a given flux-based magnitude for a five second exposure assuming the flux extrapolates as a power law of index $\alpha + 1$ from the gamma-rays. The dashed lines indicate the sensitivities of the ETC ($m_V = 11$) and GTOTE and GROCSE ($m_V = 13$) experiments. Note: Only bright bursts included. If all BATSE bursts had been included in the sample, this differential distribution would continue to rise as the magnitude increased, not peak as indicated here.

part emission. One case with an immediate physical interpretation is a power law of index 2. This spectral form would result from a thermal distribution of particles with temperature greater than 1 keV, which is certainly possible in a burst environment. Table 1 shows that if this is the only component of concurrent GRB emission at lower energies, GRBs will not be observable at optical or UV wavelengths (assuming no afterglow).

The results of previous optical searches can be used to rule out some of the spectral indices considered in this simple model. In particular, Greiner (5) examined sky patrol plates from several sites which were coincident with GRBs. Three plates from the Ondřejov patrol were coincident with bursts in our sample. These plates had four hour exposures with sensitivity to events brighter than $m_V \sim 3$. The three bursts which overlap with our sample (2B911127, 2B920525, and 2B920530) would have produced optical images at least this bright if the spectrum was a power law of index less than -3 below 1 keV. This rules out more steeply falling power laws in the empirical model.

Lacking physical guidance for our expectations, the simplest hypothesis is that spectra in the optical band have the same index as at higher energies (the $\alpha + 1$ case). If this is true, then ETC, GROCSE, and GTOTE have a chance of detecting the very brightest bursts. ETC would see only the brightest burst in the sample while GROCSE and GTOTE would see about five. The number of bursts which can be slewed to is smaller though. Assuming that each site can image half the sky half the time (night only), then there is a 25% chance a BATSE GRB can be slewed to. Therefore, GROCSE and GTOTE might be able to detect one burst per year for the $\alpha + 1$ extrapolation and ETC might record a positive detection every four years. HETE would not fare as well, being unable to detect any GRBs in the UV if the $\alpha + 1$ extrapolations hold.

The instruments discussed here would have difficulty observing simultaneous optical emission from a GRB if optical photons are generated with the same distribution as gamma-rays. This implies that if concurrent emission is observed more than about once a year, the energy flux must follow a power law of index $> \alpha + 1$, indicative of another spectral component. This result would be surprising since at present there are no indications that GRB spectra deviate from their concave-down shape. Therefore, aside from the discovery of a GRB counterpart, the observation of simultaneous optical emission would indicate a multifaceted emission process, providing new and important information about the burst environment.

Acknowledgments. We thank S. Barthelmy and R. Vanderspek for sharing details of their instruments. We also thank R. Narayan for suggesting this project. BATSE research at UCSD is supported by NASA contract NAS8-36081.

REFERENCES

1. B. E. Schaefer, B. E., in AIP Conf. Proc. **307**, eds. G. J. Fishman, J. J. Brainerd, & K. Hurley, (AIP: New York), 382 (1994).
2. J. Greiner, Proc. of the 17th Texas Symp., in press (1995).
3. D. Band, et al., ApJ **413**, 281 (1993).
4. L. A. Ford & D. L. Band, ApJ, submitted (1996).
5. J. Greiner, et al., in AIP Conf. Proc. **307**, eds. G. J. Fishman, J. J. Brainerd, & K. Hurley, (AIP: New York), 408 (1994).
6. R. Vanderspek, H. A. Krimm, & G. R. Ricker, in AIP Conf. Proc. **307**, eds. G. J. Fishman, J. J. Brainerd, & K. Hurley, (AIP: New York), 438 (1994).
7. C. Akerlof, et al., in AIP Conf. Proc. **307**, eds. G. J. Fishman, J. J. Brainerd, & K. Hurley, (AIP: New York), 633 (1994).
8. G. R. Ricker, et al., Gamma-Ray Bursts, Observations, Analyses, and Theories, C. Ho, R. I. Epstein, & E. E. Fenimore, (Cambridge University: Cambridge), 288 (1992).
9. S. D. Barthelmy, in AIP Conf. Proc. **307**, eds. G. J. Fishman, J. J. Brainerd, & K. Hurley, (AIP: New York), 643 (1994).
10. S. Barthelmy, private communication, (1995).

Short Time Scale Characteristics of Gamma-Ray Burst Continua

Lyle Ford, David Band, and James Matteson

CASS, UC San Diego, La Jolla, CA 92093-0111

Using a new analysis method, we search for narrow energy band, millisecond flares in gamma-ray bursts. This type of emission may occur if burst continuum spectra are composed of multiple short-lived black bodies. We use time-tagged count data from the BATSE Spectroscopy Detectors to determine the probability that a pair of counts came from the same energy band. If burst energy spectra are composed of many short duration black bodies, the average correlation between counts should be larger when the counts are separated by small time intervals. We discuss how the correlation between counts is calculated and apply this method to three bright bursts. We find no evidence that millisecond, narrow energy band flares compose a significant fraction of the emission in these bursts.

INTRODUCTION

BATSE Spectroscopy Detector Time-Tagged Event (STTE) data provide a marvelous opportunity to study gamma-ray burst spectra with unprecedented time resolution. Unfortunately, few have worked with this type of data in its native form; most prefer to bin counts in a given time interval and study the resulting spectrum. This allows analysis using standard techniques (i.e., model fitting via maximum likelihood methods) but at the cost of time resolution since a large number of counts are required to build meaningful spectra. Irregular variability in both count rate and spectral shape has been observed for bursts on all time scales considered. Therefore, it is crucial that any analysis of burst properties be done with the finest time resolution the data will allow. We have developed a method for dealing with the STTE data which preserves its inherent 128 μs time resolution. This method is specifically designed to find short time scale, narrow-band spectral emission from bursts.

BATSE has observed that bursts are distributed isotropically on the sky but bounded in space (1). This has led to a shift in the assumed source locations for most burst models from local Galactic ($\sim$ 1 kpc) to extreme Halo or cosmological distances (> 50 kpc). One consequence of this shift is that the energy requirements for the burst are so large in the small volume bursts are believed to originate from ($\sim$neutron star dimensions) that opacity from $\gamma\gamma \rightarrow e^+e^-$ becomes large and the radiating particles rapidly thermalize. Spectra of bursts on time scales greater than $\sim$0.1 second are observed to

© 1996 American Institute of Physics

be broad band so the source must either suppress thermal emission until another process causes the energy to be radiated nonthermally or produce many thermalized regions at different temperatures which, taken together, produce a broad band spectrum. It is the latter class of models we wish to test using STTE data.

If burst spectra are composed of many black bodies, individual thermal regions may produce observable radiation at different times. At any given instant only a small number of regions may be responsible for emission, resulting in a narrow band spectrum. If the duration of individual black bodies is short, then this effect would be difficult to find using standard spectral analysis techniques since there would be very few counts to work with. Therefore, we examine individual counts to find evidence of short time scale, narrow band emission.

THE METHOD

We consider the probability that a given pair of counts was caused by a pair of photons of similar energy. Quantitatively, this can be expressed as

$$L = \int dE \int_a^b d(\Delta E) p(E|E_1', I) p(E + \Delta E|E_2', I), \qquad (1)$$

where we use standard probability notation [$p(A|B)$ is the probability of proposition A given B]. E is the energy of an incident photon, ΔE the difference in energy between a pair of incident photons, E_1' and E_2' are the energies assigned to a pair of counts (the observed energy), and a and b define the range of energy separation we are interested in. For a single pair of counts, the degree of correlation L is not particularly useful. For many pairs though, the average behavior is interesting. If counts separated by small arrival times tend to be more correlated than pairs separated by large times, then there must be a short time scale, narrow energy band process occurring.

While calculating the probabilities for equation (1) may seem a straightforward task, it is complicated by scintillation detector processes. A photon of energy E can be assigned to a number of different channels corresponding to observed energy E'. This process is quantitatively modeled by a detector response matrix and calibration algorithm (2,3). Unfortunately, a photon of energy E can be assigned to a number of different channels, making simple comparison of individual counts impossible. Therefore, we must carefully determine the probability distribution of incident photon energies which produced a given count.

The Bayesian formalism provides a clear prescription for calculating the probability that a count E' came from a photon of incident energy E. From Bayes' Theorem (4),

$$p(E|E', I) = \frac{p(E|I) p(E'|E, I)}{p(E'|I)}, \qquad (2)$$

TABLE 1. Test Characteristics

Test	Spectrum[a]	Duration (s)	Total Counts[b]	O_{21}
1	1 BB	1	3714	5×10^{129}
2	2 BB	0.128	592	6×10^9
3	PL	1	4072	3×10^{-5}
4	n BB	0.128	442	2×10^{-3}

[a] Burst composed of n simultaneous black body (BB) spectra or a power law (PL).
[b] source + background

where I represents all information we have other than the photon or count energies. The terms on the right side of this equation are easy to understand. $p(E|I)$ is the probability that a photon had incident energy E. Since low energy photons are more numerous than high energy photons in burst spectra, we can incorporate that knowledge here. Essentially, $p(E|I)$ is the broad-band spectrum observed on long time scales, which we will call $f(E)$. $p(E'|E, I)$ is the likelihood that a photon of incident energy E will end up in count bin E'. This is exactly the process described by the detector response matrix, $R(E, E')$. Finally, $p(E'|I)$ is the probability we will see anything at all in bin E', a normalizing term. Putting all this together, we find

$$p(E|E', I) = \frac{f(E|I)R(E'|E, I)}{\int f(E|I)R(E'|E, I)dE}. \tag{3}$$

TESTS

Several tests of this method have been performed, verifying that we can detect significant millisecond, narrow energy band emission in bursts. In these tests, we found the average correlation of pairs of counts as a function of the time separation of the pairs. Time interval steps of 128 μs (the time resolution of STTE data) were used and the average correlation was just the mean correlation in each time bin.

For the first test, we simulated a burst as a series of black bodies with a typical background ($\sim$1200 counts/s in the 15–1400 keV region). The temperatures of the black bodies were distributed as a power law of index -2 below 200 keV and of index -3 above this energy. The black bodies had a randomly determined duration of 1–5 ms. For this simulation, only one black body was active at a given instant and the burst lasted one second. A second test was performed for a burst of 0.128 s duration with two black bodies active at a given time. Details for all the tests are summarized in Table 1. In both cases, the average correlation as a function of time was clearly larger for small time separations, the expected result. Figure 1 shows the average correlation as a function of time separation for the second test.

To ensure that a similar trend would not appear in broad band spectra, we simulated a burst having a power law spectrum of index -2. The average correlations were constant as a function of time separation. The last test shows

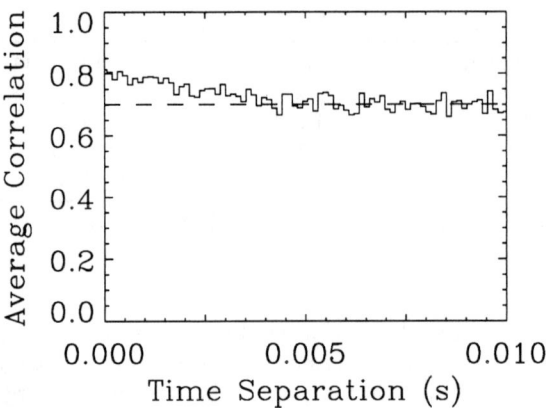

FIG. 1. Average correlation as a function of time separation for the second test. The dashed line is the median for time separations greater than 5 ms. Note the excess at small time separation.

that for this method to work, the black bodies must have short duration. For this test, we simulated a burst of black bodies with the temperature distribution of the first two tests. The black bodies turned on at 1–5 ms intervals and remained on throughout the burst. As in the third test, the average correlation was constant as a function of time separation.

SIGNIFICANCE

Although the tests reveal the sensitivity of this method, we must quantify the statistical significance of a potential signal. We use a Bayesian formalism to calculate the odds that a model of the average correlation which is constant as a function of time (the null hypothesis, no narrow band emission) is better than a model which has larger correlations at small time separations. Essentially, this means integrating the likelihood over the parameter space of each model and finding which "marginalized" likelihood is larger (4). This is expressed mathematically as

$$O_{21} = \left[\frac{p(2|I)}{p(1|I)}\right] \frac{\int d\theta_2 p(\theta_2|I_2) p(D|\theta_2 I_2)}{\int d\theta_1 p(\theta_1|I_1) p(D|\theta_1 I_1)} \qquad (4)$$

where the factor in brackets is called the prior odds, 1 and 2 refer to the models in question (null hypothesis and narrow band emission respectively), θ_i are model parameters, D is the data, and I_i is any additional information we may have. The prior odds are a way of expressing our biases quantitatively. If we have reasons to consider model 2 more likely than model 1, it is reflected in this term. The quantity O_{21} is the odds that model 2 is more appropriate

TABLE 2. Characterisicts of Searched Bursts

GRB	Time Range (s)	SD	Energy Range (keV)	O_{21}
3B920524	−1.15:13.26	5	15.4-1327	8×10^{-10}
3B931229	−0.06:0.42	0	36.4-3033	8×10^{-3}
3B940707	0.0:0.7	2	12.1-1138	10^{-2}

than model 1. If O_{21} is much greater than unity, model 2 is preferred. If O_{21} is much less than unity, model 1 should be used.

We consider a model in which the average correlation is constant (Model 1, one parameter) in time and another in which the average correlation decreases linearly in time until it settles to a constant (Model 2, three parameters). Model 1 is the case of no short time scale, narrow band emission and Model 2 implies the existence of millisecond black bodies. To favor one model over the other, we require the odds ratio to be either greater than 10 or less than 0.1. The odds for the various tests are given in Table 1. The results are consistent with our expectations in every case.

THREE BURSTS

Having shown that our method for using time-tagged counts performs as expected, we apply it to STTE data from GRBs 3B920524, 3B931229, and 3B940717. Characteristics of the data and the odds ratios are given in Table 2. From the table, it is clear that there is no evidence for millisecond, narrow energy band emission. There are several possible reasons for this. Perhaps the most obvious is that burst spectra are broad band on all time scales, but we cannot rule out the possibility that many thermally radiating regions exist. These kind of regions could persist throughout the burst (in which case, this method would fail) or may be another weaker component which occurs in addition to a strong, broad band component.

In the near future, we will work out upper limits on the fraction of emission which could have been short time scale and narrow energy band. These upper limits will constrain models which depend on multiple black body emission to produce observable radiation from gamma-ray bursts.

Acknowledgements. BATSE research at UCSD is supported by NASA contract NAS8-36081.

REFERENCES

1. G. J. Fishman, et al., ApJS **92**, 229 (1994).
2. G. N. Pendleton, et al., NIM **364**, 567 (1995).
3. D. L. Band, et al., Exp. Astron. **2**, 307 (1992).
4. T. J. Loredo, Maximum Entropy and Bayesian Methods, P. F. Fougère, (Kluwer Academic: Dordrecht), 81 (1990).

BATSE SD Observations of Hercules X-1

P. E. Freeman[1], D. Q. Lamb[1], R. B. Wilson[2], M. S. Briggs[3],
W. S. Paciesas[3], R. D. Preece[3], D. L. Band[4], and J. L. Matteson[4]

[1] *Dept. of Astronomy and Astrophysics, University of Chicago, Chicago, IL 60637*
[2] *NASA Marshall Space Flight Center, Huntsville, AL 35812*
[3] *Dept. of Physics, University of Alabama Huntsville, Huntsville, AL 35899*
[4] *CASS, University of California San Diego, La Jolla, CA 92093*

The cyclotron line in the spectrum of the accretion-powered pulsar Her X-1 offers an opportunity to assess the ability of the BATSE Spectroscopy Detectors (SDs) to detect lines like those seen in some GRBs. Preliminary analysis of an initial SD pulsar mode observation of Her X-1 indicated a cyclotron line at an energy of $\approx$ 44 keV, rather than at the expected energy of $\approx$ 36 keV. Our analysis of four SD pulsar mode observations of Her X-1 made during high-states of its 35 day cycle confirms this result. We consider a number of phenomenological models for the continuum spectrum and the cyclotron line. This ensures that we use the simplest models that adequately describe the data, and that our results are robust. We find modest evidence (significance $Q \sim 10^{-4}$–10^{-2}) for a line at $\approx$ 44 keV in the data of the first observation. Joint fits to the four observations provide stronger evidence ($Q \sim 10^{-7}$–10^{-4}) for the line. Such a shift in the cyclotron line energy of an accretion-powered pulsar is unprecedented.

INTRODUCTION

BATSE Spectroscopy Detector (SD) observations of the cyclotron line in the spectrum of the accretion-powered pulsar Hercules X-1 provides an opportunity to assess the capability of the SDs to detect such lines in the spectra of GRBs. Observations of Her X-1 have been performed by many groups (1–5). In particular, analysis of data from the *HEAO 1* A4 instrument (4), which is a NaI detector (like the SDs) and has an effective area and resolution that are similar to a single SD, yields a line center energy $E = 36$ keV and equivalent width $W_{\rm E} = 8$ keV. However, preliminary analysis (6) of an initial SD pulsar mode observation of Her X-1 indicated a line at $\approx$ 44 keV, rather than at the expected energy of 36 keV. We confirm this result by applying rigorous statistical methods developed for analysis of GRB spectra (7) to the analysis of four BATSE SD observations of Her X-1. A companion paper (8) describes studies that have been performed which confirm that the SDs are functioning as expected.

© 1996 American Institute of Physics

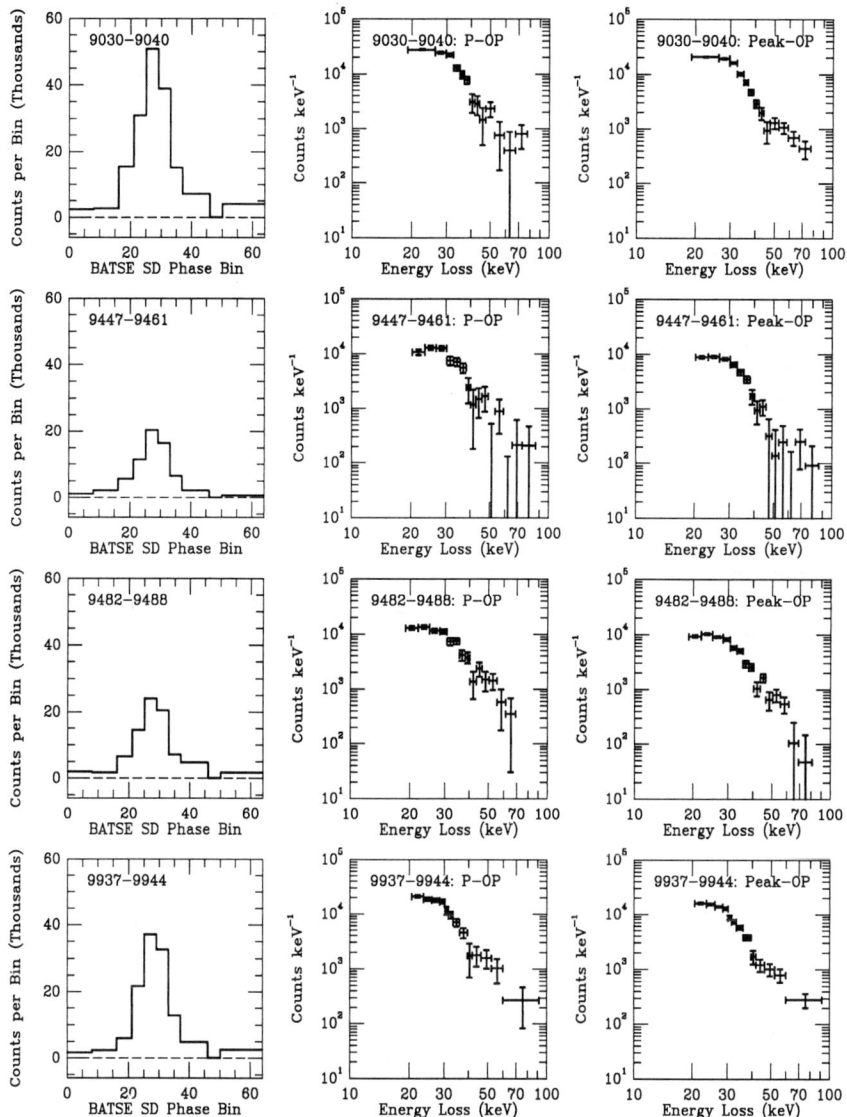

Figure 1. *Left Column:* the pulse-phase spectra for the four BATSE SD observations. We have re-binned the 64 SD phase bins into the 10 phase bins defined by Soong *et al.*, and subtracted counts in order to highlight the pulse. The seven phase bins with the largest number of counts comprise the "On-Pulse" or P phase interval, and the remaining bins comprise the "Off-Pulse" or OP interval; the three phase bins with the largest number of counts comprise the "Peak" interval. *Middle Column:* the P−OP difference spectra. *Right Column:* the Peak−OP difference spectra.

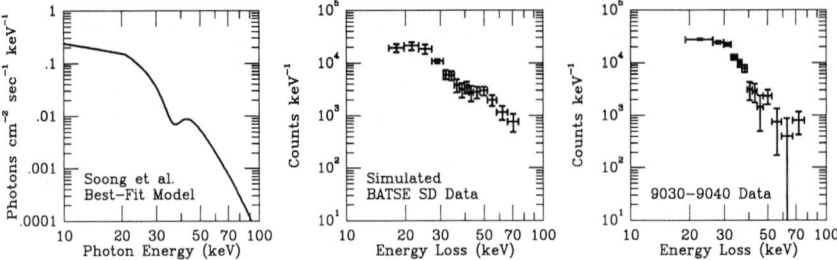

Figure 2. *Left Panel:* the best-fit model used by Soong *et al.* to fit the P−OP difference spectrum seen by *HEAO 1* A4. *Middle Panel:* the expected P−OP difference spectrum for an 80 ks observation by a BATSE SD, created by folding the Soong *et al.* best-fit spectrum through the SD response matrix. *Right Panel:* the observed P−OP difference spectrum for the 194 ks first observation, 9030−9040.

TABLE 1. Analyzed BATSE SD Observations of Her X-1

Observation (TJD[a])	SD	$\theta_{\rm inc}$	$t_{\rm obs}$ (ks)
9030–9040	0	19°	194
9447–9461	6	8°	158
9482–9488	6	23°	93
9937–9944	1	24°	161

[a] Truncated Julian Date

ANALYSIS

In Table 1 we list the four highest quality SD observations of Her X-1 made to date; these observations were undertaken during high-states of the Her X-1 35 d cycle. Figure 1 shows the SD pulse-phase data rebinned into the 10 phase bins defined by Soong *et al.*, and the "On-Pulse" minus "Off-Pulse" (P−OP) and the "Peak" minus "Off-Pulse" (Peak−OP) difference spectra. We analyze the P−OP difference spectra and, to improve S/N, the Peak−OP difference spectra. During the observations, the Low Level Discriminator (LLD) was set to ≈ 10 keV. Because of non-linearities in the energy-to-channel conversion within ≈ 10 keV of the energy of the LLD (9), we fit only to energy-loss bins above 20 keV.

Soong *et al.* fit the spectrum of Her X-1 using a continuum model which is a power law up to a break energy, and an exponential above this energy times a Gaussian line (Figure 2). The count spectrum seen by BATSE differs in three ways from that expected from folding the best-fit Soong et al. spectrum through the SD response matrix and adding simulated noise: the break energy of the observed spectrum is ≈ 10 keV higher than expected, the slope of the observed spectrum above the break is greater than expected, and the observed spectrum does not show the expected plateau around ≈ 36 keV (Figure 2).

Because the observed spectra differ from that expected, we investigate not only the Soong *et al.* continuum model, but other continuum models as well.

TABLE 2. Analysis of Single Observations: Peak Phase Interval

Obs. (TJD)	Soong			BPL		
	E (keV)	W_E (keV)	Q	E (keV)	W_E (keV)	Q
9030–9040	44.0	7.5	2.5×10^{-4}	44.3	6.1	8.7×10^{-3}
9447–9461	No Improvement in χ^2			No Improvement in χ^2		
9482–9488	40.5	1.9	0.50	No Improvement in χ^2		
9937–9944	42.8	5.6	0.013	43.6	4.4	0.068

We seek the simplest model that adequately fits the data. We use a statistical criterion based on the maximum likelihood ratio (MLR) test[1] (7,10). Use of this criterion often leads to the selection of a broken power law (BPL) model for the continuum, rather than the Soong et al. model. Joint fits indicate that this preference becomes stronger as we include the data from additional observations. The preference for the BPL over the Soong et al. model may be a consequence of the fact that we are unable to include data below 20 keV. Because the line lies on the steeply-falling part of the Her X-1 spectrum, there can be a large difference between the values of the Soong et al. and BPL continuum models at the line, and therefore in the significance of the line. We therefore give the results of continuum-plus-line fits to the data using both the Soong et al. and BPL continuum models.

We use an exponentiated Gaussian absorption model to fit the line. In this model, the line full-width at half-maximum $W_{\frac{1}{2}} = \eta\, W_E$, and $\eta \approx 1$ represents a saturated line. Using a procedure analogous to the one that we use to select continuum models, we select the simplest continuum-plus-line model that adequately fits the data. We find that a one-parameter saturated line model in which we fix the line center energy at ≈ 36 keV *never* leads to a reduction in χ^2 from the best continuum fit. We find that a two-parameter saturated line model adequately fits all of the data. We use the MLR test to determine the significance of the line.

In fits to the P−OP difference spectra, we find no evidence for a line in any of the four individual SD observations. The largest line significance ($Q = 0.18$) occurs for the first observation. Fits to simulated SD data created using the best-fit Soong et al. model for the Her X-1 spectrum, but with the line energy shifted to ≈ 43 keV, show that this result is not inconsistent with a line at ≈ 43 keV. Using joint fits, we find that the largest line significance ($Q = 0.05$) occurs when we combine data from the first and fourth observations.

In fits to the Peak−OP difference spectra, we detect a line at ≈ 44 keV with modest significance in the data from the first observation (Table 2), but not in the data from any other observation. Joint fits indicate that this line becomes more significant as we add the data from more observations

---

[1] This test assumes that the difference $\Delta\chi^2$ resulting from two model fits to data is distributed like χ^2 for N degrees of freedom, where N is the number of *additional* parameters in the more complicated of the two models. Q is the area under this distribution for $\chi^2 > \Delta\chi^2$, and small Q values indicate that it is unlikely that the improvement $\Delta\chi^2$ between the two fits would occur by chance.

TABLE 3. Analysis of Joint Observations: Peak Phase Interval

Obs. (TJD)	Soong			BPL		
	E (keV)	W_E (keV)	Q	E (keV)	W_E (keV)	Q
9030	44.0	7.5	2.5×10^{-4}	44.3	6.1	8.7×10^{-3}
9030/9937	43.6	6.7	4.7×10^{-7}	44.1	5.4	9.8×10^{-5}
All − 9482	43.8	6.7	3.2×10^{-7}	44.1	5.3	8.8×10^{-5}
All	43.1	5.3	2.3×10^{-6}	43.5	3.9	4.6×10^{-4}
9447/9937	43.7	5.6	0.010	44.9	4.8	0.054
All − 9030	41.1	2.1	0.21	No Improvement in χ^2		

(Table 3). However, the addition of data from the third observation reduces the significance of the line. This behavior may reflect the fact that each of the four observations covers a slightly different phase of the Her X-1 35 d cycle. If we exclude the first observation from the joint fits, we find that we cannot detect the line.

We estimate the statistical error for each model parameter using Monte Carlo simulations of the best-fit models. The typical 1σ uncertainties for both E and W_E are $\approx$ 1.0–1.5 keV.

CONCLUSIONS

Joint fits to four SD observations of Her X-1 show strong evidence for a cyclotron scattering line at $\approx$ 44 keV, rather than at the expected energy of 36 keV. A cyclotron line energy shift is unprecedented in observations of accretion-powered pulsars. A companion paper (8) describes studies that have been conducted which confirm that the SDs are functioning as expected.

REFERENCES

1. J. Trümper, W. Pietsch, C. Reppin, W. Voges, R. Staubert, and E. Kendizorra, Ap. J. **219**, L105 (1978).
2. D. E. Gruber, et al., Ap. J. **240**, L127 (1982).
3. W. Voges, W. Pietsch, C. Reppin, J. Trümper, E. Kendizorra, and R. Staubert, Ap. J. **263**, 803 (1982).
4. Y. Soong, et al., Ap. J. **348**, 641 (1990).
5. T. Mihara, K. Makashima, T. Ohashi, T. Sakao, and M. Tashiro, Nature **335**, 234 (1990).
6. D. M. Palmer, et al., in AIP Conf. Proc. **307**, eds. G. J. Fishman, J. J. Brainerd, and K. Hurley (New York: AIP, 1994), p. 247.
7. P. E. Freeman, et al., Ap. J., submitted.
8. W. S. Paciesas, et al., these proceedings.
9. D. Band, et al., Exp. Astr. **2**, 307 (1992).
10. W. T. Eadie, D. Drijard, F. E. James, M. Roos, and B. Sadoulet, Statistical Methods in Experimental Physics (Amsterdam: North Holland, 1971).

Fitting of combined Ulysses/COMPTEL GRB spectra

J. Greiner[1], T. Aigner[1], M. Sommer[1], O.R. Williams[2],
R.M. Kippen[3], K. Hurley[4], M. Boër[5], M. Niel[5]

[1] *Max-Planck-Institut für Extraterrestrische Physik, 85740 Garching, Germany*
[2] *Astrophysics Division, ESA-ESTEC, 2200 AG Noordwijk, The Netherlands*
[3] *Space Science Center, University of New Hampshire, Durham, NH 03824, USA*
[4] *Space Science Laboratory, University of California, Berkeley CA 94720, USA*
[5] *Centre d'Etude Spatiale des Rayonnements, 31029 Toulouse, France*

We have analysed the Ulysses and COMPTEL spectral data from several strong and hard γ-ray bursts which occurred in the COMPTEL field of view. Single instrument data with their limited energy range often can be fit with single power law models. In contrast, the composite spectra obtained by a joint deconvolution of the Ulysses and COMPTEL spectra (ranging from 20 keV up to 10 MeV) need models with continuous curvature. The applicability of standard emission processes to these composite spectra is discussed.

INTRODUCTION

Historically, γ-ray burst spectra have been fitted by various models such as power law, optically thin bremsstrahlung (OTTB) or Comptonisation. During the first two years of CGRO operation burst spectra of all 4 individual instruments have been generally fitted with single power law models, whereas only in very few cases broken power laws were necessary to describe the spectral shape. A detailed investigation of BATSE SD spectra has used a generalisation (two smoothly connected power laws), the so-called "Band GRB" model (1) which was shown to adequately fit the whole variety of GRB spectra.

While separately fitted BATSE, OSSE and COMPTEL spectra have been combined earlier into wide-band photon spectra (5,6), the first combined fit of multi-instrument data has been performed only recently for GRB 920622 (3). The continuously curved time-averaged spectrum of this burst derived from the combined Ulysses/BATSE/COMPTEL data in the 20 keV to 10 MeV range was shown to be fitted only by Band's "GRB" model, whereas either OTTB or Comptonisation models are too strongly curved. This work has been extended to more bursts, and the present status is given here.

© 1996 American Institute of Physics

DATA SELECTION

We have analysed five γ-ray bursts detected by Ulysses and COMPTEL, namely GRB 910503, GRB 911118, GRB 920622, GRB 930131 and GRB 940217. Aboard the Ulysses spacecraft two hemispherical CsI(Tl) detectors are capable of measuring photon events in 16 energy channels covering an energy interval from 20 to 150 keV. Once triggered, the integration time is set to 1 s for 14 consecutive spectra and after that to 2 s for the following 16 spectra. Finally, another 28 spectra are accumulated over 16 s each. A part of these late data is used in the background subtraction.

The COMPTEL telescope on board the Compton Gamma-Ray Observatory (CGRO) detects high energy photons above 300 keV. COMPTEL collects spectral information either in the "telescope" mode, which is the usual imaging mode of the telescope and provides spectra between 0.75 and 30 MeV, or in the "single detector" mode using two modules with overlapping energy intervals (0.3 to 1.3 MeV, 0.6 to 10 MeV). After receiving a trigger message from BATSE, COMPTEL collects 6 spectra with 1 s accumulation time followed by 133 spectra with 6 s accumulation time. Then it returns to the background mode with 100 s accumulation. The default background spectrum for the burst analysis is the last complete background spectrum, recorded before the burst trigger. It is used for all our analysed bursts. In our investigation only the burst mode, but not the telescope data are included.

The spectral coverage of Ulysses and the COMPTEL module data leaves a small gap between 150–300 keV which is considered to be small enough to achieve reasonable fits for the burst spectra. To eliminate time dependent effects, we select time intervals of each burst which are approximately equal for both instruments. Because of the fixed accumulation times an inaccuracy of ≈1 s remains for bursts lasting longer than ≈15 s. The Ulysses detector generally triggers later than BATSE according to its lower sensitivity (even if both spacecrafts were at the same location). Therefore, some of the first spectra measured by COMPTEL had to be excluded from the joint deconvolution.

Several models have been applied to the burst spectra: a single power law, OTTB, Comptonisation (7), and Band's "GRB" model (1) which smoothly connects two power law models:

$$N_E(E) = A \left[\frac{E}{(100 keV)}\right]^\alpha \exp\left[-\frac{E}{E_o}\right], \qquad (\alpha - \beta)E_o \geq E$$

$$= A \left[\frac{(\alpha - \beta)E_o}{(100 keV)}\right]^{\alpha - \beta} \exp(\beta - \alpha) \left[\frac{E}{(100 keV)}\right]^\beta \qquad (\alpha - \beta)E_o \leq E$$

GRB 910503: This burst shows a complex time structure with two different peaks: the intense first one lasts about 10 s while the second occurs ≈45 s after the BATSE trigger. Only the first bright peak is included in this investigation. The common interval selected starts 1 s after the BATSE trigger and extends for 11 s. Due to discriminator thresholds the three lowest energy channels of Ulysses were omitted. The dead time correction for Ulysses is

26%. The COMPTEL energy channels are combined to ensure $\geq 3\sigma$ detection significance above background in each bin. The spectrum is consistent with Band's GRB model fit ($\chi^2/\nu = 2.2$). The other spectral models give unacceptable fits ($\chi^2/\nu = 4.7$ for OTTB, 6.1 for Comptonisation and 6.9 for a power law).

The BATSE data of this burst had been fit earlier with Band's GRB model (1). A comparison the the results obtained here shows that the well determined high-energy part (above 1 MeV) reduces the error in the estimate of β, and the inclusion of the low-energy part shifts the break energy to higher values.

GRB 911118: With a trigger delay of 1 s, the selected time intervals are 1 to 24 s for COMPTEL and 0 to 22 s for Ulysses (both relative to the corresponding trigger times). The two lower energy channels of Ulysses were rejected. Unfortunately, from mid-1991 to mid-1992 the low range detector module of COMPTEL was switched off due to increased high-voltage noise leading to a wider energy gap between the Ulysses and the COMPTEL data, and thus to increased systematic errors. The energy channels of the COMPTEL high range detector are rebinned to ensure $\geq 1\sigma$ detection significance above background in each bin. The Ulysses deadtime correction fraction is estimated to 24%. Though all models give acceptable fits (see Tab. 1), Band's GRB model turns out to give the lowest χ^2/ν of 0.6 (due to the low number of photons).

Also for this burst, the BATSE data have been fit earlier (1). Despite the large gap between the Ulysses and COMPTEL data due to the missing low-energy COMPTEL module the results of our fit and that of the BATSE data are in surprising agreement (<10% difference in best fit values).

GRB 920622: The trigger delay of Ulysses is 4 s. The included time intervals are 4 to 24 s for COMPTEL and 0 to 20 s for Ulysses (relative to corresponding trigger times). The three lowest Ulysses energy bins were rejected. A deadtime correction of 20% has been applied to the Ulysses data. Also this spectrum is best fit by Band's GRB model (Tab. 1). The reduced χ^2 values are 1.6 (Band's GRB), 3.0 (OTTB), 4.6 (Comptonisation) and 7.3 (power law).

A combined deconvolution including the BATSE data (3) resulted in fit parameters being consistent with those derived from the Ulysses/COMPTEL data. The shift of the break energy towards lower energies by about 20% and a corresponding flatter α is due to the fact that the fit down to about 100 keV is completely determined by the BATSE data as compared to the Ulysses data with its lower statistics.

GRB 930131: This burst, known as the Superbowl Burst, had a duration of two seconds with the emission separated in two peaks. The first peak lasted only 0.3 s, while the second started about 0.7 s later. The included time intervals are 0 to 2 s for COMPTEL and for Ulysses (relative to corresponding trigger times). Due to a 0.57 s trigger delay of Ulysses there is an about 20% uncertainty in the flux calibration. The three lowest Ulysses energy bins were excluded from the fitting. A deadtime correction of 24% has been applied to the Ulysses data. Based on the reduced χ^2 values only Band's GRB model provides an acceptable fit (Tab. 1).

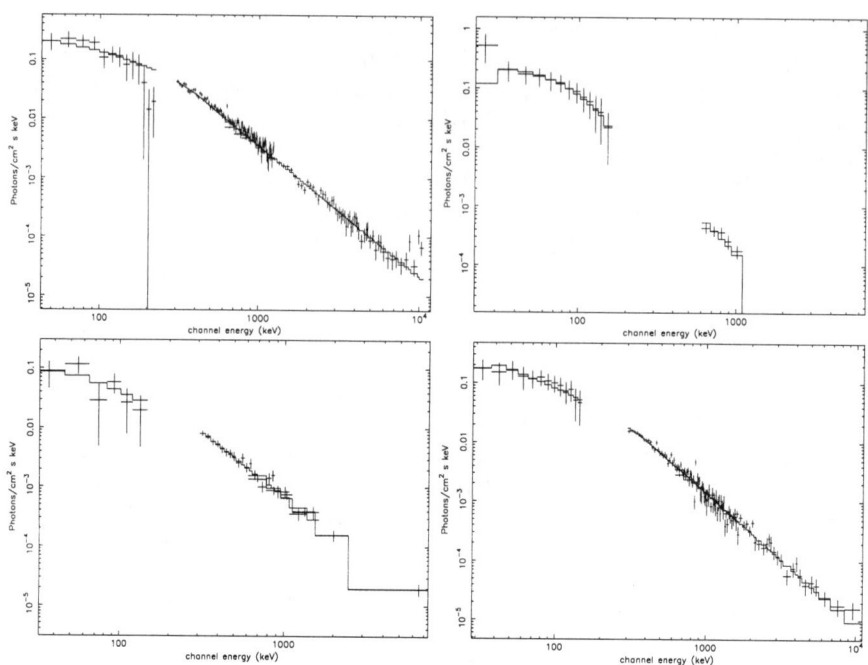

FIG. 1. Joint deconvolution of Ulysses and COMPTEL spectra with Band's GRB model: GRB 910503 (top left), 911118 (top right), 930131 (bottom left) and 940217.

GRB 940217: This burst, famous for its long-lasting high-energy emission detected with EGRET, has a rather long duration with the main peak starting only 110 s after the BATSE trigger. The Ulysses instrument triggered on the second main peak which occurred about 86 s after the BATSE trigger. Since this peak was short and the internal Ulysses trigger delay was 0.88 s, only the third main peak was used for the joint fitting. A dead time correction of 27% was applied to the Ulysses data, and the three lowest Ulysses channels were again excluded from the fit. As in the case of the other bursts, GRB 940217 (i.e., third peak) is best fit with Band's GRB model (Tab. 1).

TABLE 1. Band's GRB model fits to combined Ulysses/COMPTEL data

GRB	Norm ph/(cm² s keV)	E_p (keV)	α	β	Energy range (keV)
910503	0.21±0.02	405±25	−0.27±0.19	−2.21±0.19	42–10600
911118	1.15±1.00	105±30	+0.42±0.43	−2.76±0.18	20–10400
920622	0.11±0.03	315±25	−0.27±0.26	−2.54±0.15	29–10600
930131	0.06±0.05	370±150	−0.77±0.75	−2.20±0.13	28–10600
940217	0.11±0.02	390±50	−0.78±0.23	−2.23±0.04	28–10800

DISCUSSION

GRB spectra measured in the pre-BATSE era have often been successfully fit with thermal bremsstrahlung or comptonisation models. Observations with COMPTEL and EGRET have demonstrated that many bursts exhibit high-energy power law tails, excluding these emission processes as viable models of GRB spectra. This result is confirmed with our joint fits of Ulysses and COMPTEL data: the GRB spectra are much less curved than these models. However, the broad energy coverage achieved in our analysis also convincingly shows that GRB spectra are continuously curved, and that single or broken power laws do not adequately fit the composite Ulysses/COMPTEL spectra.

While Band's GRB model gives satisfactory fits to the combined Ulysses/-COMPTEL spectra of all five bursts investigated so far, it would be more interesting if physical models could be tested against the data. In addition, looking at the residuals of these fits suggests that the burst spectra might be continuously curved in contrast to this model of combined power laws. The blast wave model of Meszaros and Rees (4) is still too crude to be usable for detailed spectral fitting. The Compton attenuation model proposed by Brainerd (2) seems to provide the right curvature and shall be applied as the next step. Also, the proposed shock emission model (8) with only two free parameters is predictive enough to be tested.

Acknowledgements. JG is supported by the Deutsche Agentur für Raumfahrtangelegenheiten (DARA) GmbH under contract FKZ 50 OR 9201 and is grateful to the Deutsche Forschungsgemeinschaft for a travel grant. KH gratefully acknowledges assistance from JPL Contract 958056 and NASA Grant NAG5-1560. The COMPTEL project is supported by DARA No. 50 QV 90960, NASA NAS5-26645 and the Netherlands Organisation for Scientific Research (NWO). The Ulysses GRB experiment was constructed at the Centre d'Etude Spatiale des Rayonnements in France with support from the Centre National d'Etudes Spatiales and at the Max-Planck-Institute in Germany with support from Contracts 01 ON 088 ZA/WRK 275/4-7.12 and 01 ON 88014.

REFERENCES

1. D. Band, J. Matteson, L. Ford, et al., Astrophys. J. **413**, 281 (1993).
2. J. J. Brainerd Astrophys. J. **428**, 21 (1994).
3. J. Greiner, M. Sommer, N. Bade, et al., Astron. Astrophys. **302**, 121 (1995).
4. P. Meszaros, M. J. Rees, Astrophys. J. **418**, L59 (1993).
5. B. E. Schaefer, B. J. Teegarden, T. L. Cline, et al., 2nd Huntsville GRB workshop 1993, AIP Conf. Proc. **307**, p. 280 (1994).
6. G. J. Share, W. N. Johnson, J. D. Kurfess, et al., 2nd Huntsville GRB workshop 1993, AIP Conf. Proc. **307**, p. 283 (1994).
7. R. A. Sunyaev, L. G. Titarchuk, Astron. Astrophys. **86**, 121 (1980).
8. M. Tavani, PRL, in press (1995).

Effects of Compton Scattering on BATSE Gamma-Ray Burst Spectral Analysis

X.-M. Hua* and R. E. Lingenfelter[†]

*NASA Goddard Space Flight Center/National Research Council
Greenbelt, Maryland 20771
[†]Center for Astrophysics & Space Sciences
University of California, San Diego, La Jolla, California 92093

> The question of whether or not there are significant absorption lines in the BATSE spectra of gamma-ray bursts is one of the major remaining problems in the analysis of the BATSE data. Such lines, reported by KONUS, HEAO-1 and GINGA, have been interpreted as cyclotron lines in the teragauss magnetic fields of neutron stars and can set strong constraints on the origin of gamma-ray bursts. But searches of BATSE Spectroscopy Detector (SD) data show possible absorption features in the burst spectra of one detector which are not evident in that of another detector which also sees the burst. Preliminary analyses of BATSE burst data have shown (5), however, that Compton scattering of the gamma-ray burst emission by both the surrounding spacecraft and the Earth's atmosphere can greatly dilute possible absorption line features in the spectra measured by the BATSE Spectroscopy Detectors and selectively reduce their detectability, depending on the direction of the burst with respect to individual detectors. Here we present new, more systematic and complete analyses which we are currently making of azimuthal variations in the detector response resulting from Compton scattering that greatly affect not only the detection of possible absorption features but the burst spectral analysis in general.

INTRODUCTION

Absorption lines in the spectra of gamma-ray bursts, reported by KONUS, HEAO-1 and GINGA, have been interpreted as cyclotron lines in the teragauss magnetic fields of neutron stars (see, e.g., the review by Higdon & Lingenfelter (3)) and can set strong constraints on the origin of gamma-ray bursts. But so far, visual searches of the BATSE Spectroscopy Detector (SD) data show (1) possible absorption features in the burst spectra of one detector which are not evident in that of another detector which also sees the burst. Thus, there is general confusion and uncertainty about the existence of these features in the BATSE data.

We have shown (5), however, from preliminary analyses of BATSE burst

© 1996 American Institute of Physics

data that Compton scattered burst emission from both the surrounding spacecraft and the Earth's atmosphere can greatly dilute possible absorption line features in the spectra measured by the BATSE Spectroscopy Detectors and selectively reduce their detectability, depending on the direction of the burst with respect to each of the detectors.

MONTE CARLO SIMULATIONS

We are presently carrying out a more systematic and complete analysis of the effects of Compton scattering on the gamma-ray burst spectra measured by the BATSE detectors. Through this analysis we can provide an independent test of the uncertainties in the detector response matrices currently used for BATSE spectral analysis. Moreover, we can generate for the first time new detector response matrices that include not only the zenith angle dependence on burst position but also the azimuthal angle dependence, not considered in the current BATSE analysis (9). These new response matrices can be used to modify the existing detector matrices to include the azimuthal angle dependence in BATSE spectral analysis. Our preliminary analyses show that large azimuthal variations in the detector response can result from Compton scattering and attenuation that affect not only the detection of possible absorption features, but the burst spectral analysis in general.

The BATSE gamma-ray burst Spectroscopy Detector array consists of eight cylindrical NaI(Tl) scintillators, which were added on by the UCSD/GSFC groups to supplement the original MSFC Large Area Detectors after the high resolution GRSE spectrometer was removed from the GRO in 1982. These detectors were designed to provide better energy resolution and more uniform sensitivity to higher energies. To obtain broader sky coverage, the Spectroscopy Detectors were unshielded on their sides, making them nearly all-sky detectors, limited only by blockage due to the spacecraft. Because of this, however, the Spectroscopy Detectors are also exposed to Compton scattered burst emission from all of the visible portions of the Compton Observatory spacecraft, which fills the backward hemisphere and part of the forward hemisphere of each detector, and to Compton scattered burst emission from the atmosphere of the Earth, which fills roughly 33% of the detector sky.

In order to investigate the effects of such Compton scattering on the measured spectrum of gamma-ray bursts, we developed a Monte Carlo simulation program to calculate the spectra of Compton scattered photons incident upon each of the Spectroscopy Detectors, as a function of energy and both zenith and azimuthal angles with respect to both the burst source and the geocentric directions. From our Monte Carlo simulations for a number of configurations, we find that the effects of such reflected spectra depend strongly on the directions of the burst and Earth with respect to the axis of each detector, and thus the effects differ greatly from detector to detector. We also find that, in general, the reflected spectra are significantly steeper than the incident spectra of the bursts, so that the effects of reflected photons are much more important at lower energies than they are at higher energies.

These reflected photons can thus greatly reduce the detectability of spectral features at energies less than ~100 keV, making it very difficult to confirm the detection of an absorption line by detecting it in two separate detectors.

The Monte Carlo simulation program which we have developed to analyze the effects of Compton scattering on the BATSE spectra of gamma-ray bursts, calculates the spectra of Compton scattered photons incident upon each of the BATSE Spectroscopy Detectors, as a function of energy and both zenith and azimuthal angles with respect to both the burst source and the geocentric directions. This program draws on our experience developed in the study of Compton scattering in astrophysical sources (7,4,6), where we have shown the importance of Compton backscattering in producing line-like features by the reflection of annihilation line emission.

The azimuthal variations in the detector response, which are not accounted for in the present BATSE spectral analysis program, can produce serious complications in determining the incident burst spectra. The importance of including the azimuthal variations in the Compton scattering and attenuation of burst emission, particularly at burst zenith angles > 60°, which are common for the second brightest detectors, can be clearly seen in Figure 1. Here we show Monte Carlo simulations of the energy loss spectra in the BATSE Spectroscopy Detectors from a sample burst, compared with the energy loss spectrum determined with the current BATSE spectral analysis program which assumes cylindrical symmetry and thus does not include azimuthal variations. In this simulation, we assumed an incident burst spectrum of the form $(E + E_o)^{-1}$ with $E_o = 150$ keV.

As can be seen in this example, there are strong variations in the detector energy loss spectrum as a function of azimuthal angle, resulting primarily from the Compton scattered attenuation and photoelectric absorption of the burst emission by the Large Area Detector. This effect is naturally larger at larger burst zenith angles. At zenith angle of 90° the reduction in the measured energy loss spectrum can be as much as a factor of 5, and even at zenith angles with cosines between 0.25 and 0.5, where most of the bursts are observed in the second brightest detectors, the reductions still range from about 30% to 60%. More important, comparing these azimuthal variations in detector response with the single azimuthally independent response assumed in the present BATSE spectral analysis program clearly shows that significant uncertainties are introduced if the azimuthal variations are not considered.

We also plan to make systematic Monte Carlo simulations of the energy loss spectrum in the Spectroscopy Detectors resulting from monoenergetic burst photons and their accompanying Compton scattered photons from both the Earth and the spacecraft for the full range of observable burst photon energies and zenith and azimuthal angles with respect to the detector axes. Using these simulations we will generate new detector response matrices as a function of both the zenith angle and the azimuthal angle of the burst direction. We will initially generate these matrix elements directly from the Monte Carlo simulations, but we will also explore the possibility of developing analytic functions to define them, as we were able to do using Green's functions for astrophysical accretion disk models (4). These new detector response matrices can then be used to modify the matrices in the BATSE spectral analysis program to include the effects of azimuthal variations in the burst direction.

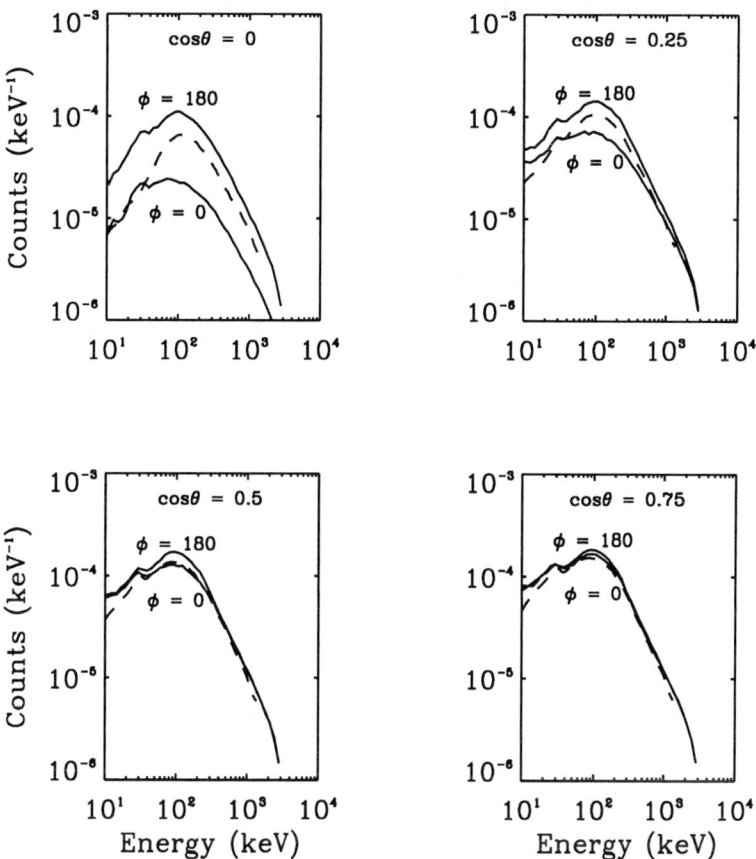

FIG. 1. The effects of spacecraft Compton scattering and absorption of gamma ray burst emission, showing Monte Carlo simulations of the energy loss spectra (solid lines) in a BATSE Spectroscopy Detector at zenith angle θ cosines of 0, 0.25, 0.5, and 0.75 and azimuthal angles ϕ of 0° and 180° measured from the Large Area Detector. These are compared with the energy loss spectrum determined with the current BATSE spectral analysis program (dashed line) which assumes cylindrical symmetry and thus does not include azimuthal variations.

CONCLUSIONS

Thus, we see that Compton scattering of the gamma-ray burst emission by the spacecraft and the Earth can dilute possible absorption line features in the measured spectra and significantly reduce their detectability, depending on the direction of the burst with respect to each of the detectors. In the detector response matrices developed for the BATSE burst spectral analysis, approximations of the effects of Compton scattering have been considered, but only for the zenith angle dependences of the burst and Earth directions and the azimuthal dependence of the burst source direction has not been included. The systematic Monte Carlo simulations which we are carrying out will allow us to make independent analyses of the burst spectra and more complete determinations of these response matrices, as functions of both zenith and azimuthal angle, so that the uncertainties in the burst spectra and the question of the existence of absorption lines can be better understood.

We also suggest that a similar failure to consider azimuthal variations in the response of the Large Area Detectors in the backward direction (9) may account for the unexpectedly large errors (2,10) in the 3B BATSE burst positions (8) determined from a new location algorithm which was modified to include emission below the plane of the detectors where the spacecraft azimuthal variations are very large.

REFERENCES

1. D. L. Band, et al. Proc. 29th ESLAB Symp., Towards the Source of Gamma-Ray Bursts, (New York: Kluwer) (in press 1995)
2. C. Graziani & D. Q. Lamb, these proceedings.
3. J. C. Higdon & R. E. Lingenfelter, Ann. Rev. Astron. Astrophys. **28,** 401 (1990).
4. X-M. Hua & R. E. Lingenfelter, Astrophys. J. **397,** 591 (1992).
5. X-M. Hua & R. E. Lingenfelter, in AIP Conf. Proc. **280,** (New York: AIP), p. 927 (1993a).
6. X-M. Hua & R. E. Lingenfelter, Astrophys. J. **416,** L17 (1993b).
7. R. E. Lingenfelter & X-M. Hua, Astrophys. J. **381,** 426 (1991).
8. C. Meegan, et al., these proceedings.
9. G. Pendleton, private communication (1995).
10. V. C. Wang & R. E. Lingenfelter, these proceedings.

Cyclotron Line Formation in a Relativistic Outflow

Michael Isenberg[1], D. Q. Lamb[1], and John C.L. Wang[2]

[1] *Department of Astronomy and Astrophysics, University of Chicago, 5640 South Ellis Avenue, Chicago, IL 60637*
[2] *Department of Astronomy, University of Maryland, College Park, MD 20742-2421*

> There is mounting evidence that, if gamma-ray bursters are Galactic in origin, they are located in a Galactic corona at distances greater than 100 kpc. This has created a need to explore new models of cyclotron line formation. In most previous calculations the line-forming region was modeled as a static slab of plasma, optically thin to continuum scattering, and threaded by a magnetic field of the order 10^{12} gauss oriented normal to the slab. Such a model is appropriate, for example, for the magnetic polar cap of a neutron star with a dipole field. However, if bursters lie at distances farther than several hundred parsecs, the burst luminosity exceeds the magnetic Eddington luminosity, and the plasma in a line-forming region at the magnetic polar cap would be ejected relativistically along the field lines. Mitrofanov and Tsygan have modeled the dynamics of such an outflow, and Miller *et al.* have calculated the properties of the cyclotron second and third harmonics, approximating them as due to cyclotron absorption. Here we describe Monte Carlo calculations of cyclotron resonant scattering at the first three harmonics in a relativistic outflow from the magnetic polar cap, and show that such scattering can produce narrow lines like those observed by Ginga.

INTRODUCTION

One of the most compelling pieces of evidence that gamma-ray bursts are Galactic in origin is the observation of absorption-like features in the spectra of some bursts and the interpretation of these features as cyclotron lines (1). In particular, the Ginga observations of three gamma-ray bursts with harmonically spaced lines (2) strongly supports this interpretation. Recent reports of several highly significant line candidates by the BATSE (3) and Konus (4) groups have further heightened interest in cyclotron line formation.

Most theoretical models of line formation in gamma-ray bursts assume physical conditions that are appropriate for burst sources in the Galactic disk. For example, in the model of Wang *et al.* (5), the line-forming region is a static slab of plasma, optically thin to continuum scattering, and threaded by a uniform magnetic field $\sim 10^{12}$ gauss oriented along the slab normal. Such a model is suitable, for example, for the magnetic polar cap of a neutron star

© 1996 American Institute of Physics

with a dipole field. Lamb, Wang, and Wasserman [LWW, (6)] pointed out that if the line-forming region is indeed at the polar cap, the static model is valid only if the bursters lie at distances less than several hundred parsecs. Otherwise, the burst luminosity is sufficient to create a relativistic plasma outflow along the field lines.

However, the BATSE burst brightness and sky distributions (7,8) suggest that if the bursters are Galactic, they lie in a Galactic corona at distances of 100–400 kpc. In light of the BATSE results, it is important to explore line formation models that are appropriate for sources at these distances. One possibility is line formation in a static slab located at the magnetic equator of a neutron star, where the plasma is magnetically confined near the surface (9,10). In the present work, however, we explore another possibility: that the lines are formed in a relativistic outflow.

In an outflow, the variation of the magnetic field and plasma velocity with altitude tends to broaden the lines. Miller et al. (11,12) calculated the properties of the second and third harmonics, approximating them as due to cyclotron absorption. They showed that narrow lines can be formed at these harmonics. Chernenko and Mitrofanov (13) calculated the properties of the first harmonic line, also approximating it as due to absorption, and found that the formation of a narrow line is possible. However, such an approximation is not valid for the first harmonic. Thus the question of whether narrow first harmonic *scattering* lines can be formed in an outflow has remained open. In the present work we use a Monte Carlo radiative transfer code to calculate the properties of the first three harmonic lines and address this question.

PHYSICS OF THE LINE FORMING REGION

In our model, photons are injected into the line forming region at a circular hot spot, with radius r_{hot}, located on the surface of a neutron star and centered on the magnetic pole. The photons are distributed uniformly over the hot spot and their directions are distributed isotropically. We choose an injected photon number spectrum that varies inversely with photon energy ($dN_\gamma/dE \propto E^{-1}$). The field strength decreases with altitude, z, as a dipole:

$$B(z) = B_o \left(1 + \frac{z}{R}\right)^{-3}, \qquad (1)$$

where B_o is the field strength at the stellar surface, R is the stellar radius, and z is the altitude above the surface. Although the lines of force in a dipole field flare outwards as the altitude increases, we assume for simplicity that the field lines remain perpendicular to the surface. This is a good approximation for $z \ll R$ and $r_{hot} \ll R$. Since our Monte Carlo simulations for $r_{hot} = 0.1R$ show that $> 90\%$ of scatters occur at $z < 0.1R$, we do not expect this assumption to have a significant effect on the emerging spectrum.

The radiation force accelerates an electron-proton plasma to a flow velocity, β_f, which varies with altitude. Mitrofanov and Tsygan (14) derived the radiation force due to resonant scattering of an electron located above the center of the hot spot. At any given altitude, there is an equilibrium velocity at which the the radiation force on the electron, averaged over the energies and directions of the photons, is equal to zero. The equilibrium velocity is:

$$\beta_e(z) \approx \frac{1}{2}\left(1 + \frac{z}{\sqrt{z^2 + r_{hot}^2}}\right). \quad (2)$$

An electron injected at the surface with an initial velocity of zero accelerates quickly. From the magnitude of the radiation force, we can estimate the distance the electron travels before reaching β_e. For a magnetic field strength $B_o = 1.7 \times 10^{12}$ gauss and an x-ray luminosity between 1 keV and 1 MeV equal to 10^{40} ergs s^{-1} (i.e., 1% of the total burst luminosity) the distance to β_e is $\sim 10^{-7}$ of the stellar radius. Once the electron reaches β_e, its velocity continues to increase according to eq.(2) until it reaches an altitude where the radiation becomes too diffuse to provide sufficient energy for acceleration to continue at the rate required by the equation. At this point, the electron's velocity starts to lag behind the equilibrium velocity. We estimate that this happens at an altitude $\sim$ a few stellar radii. Since most scatterings take place far below this point, we take $\beta_f = \beta_e$ throughout the line-forming region.

We emphasize that we calculate β_e using the *unscattered* radiation spectrum. We have not attempted in the present work to account for the effect on β_e of the reduction in photon flux at the cyclotron energy due to scattering.

We assume that in the frame of reference co-moving with the flow the distribution of electron velocities is Maxwellian. LWW showed that the heating and cooling of the electrons by scattering with the radiation balances at the Compton equilibrium temperature, T_c. Applying the single-scattering model of LWW to the angular distribution of radiation in the co-moving frame, we find that $kT_c \approx \hbar\omega_B/4$, where $\hbar\omega_B$ is the cyclotron energy.

For burster distances $\sim$ 100 kpc, the time scales for energy and momentum exchange between the electrons and the radiation field are much shorter than the time scale for establishing a Maxwellian electron velocity distribution by particle collisions. The actual electron distribution is therefore likely to be much narrower than a Maxwellian, which would narrow the cyclotron lines in the emerging spectrum. Thus our assumption of a Maxwellian electron velocity distribution in the present calculation is conservative.

Following Miller *et al.* (11,12), we calculate the density of the plasma as a function of altitude from the continuity equations for conservation of mass and magnetic flux:

$$n_e(z) = n_{e,o}\frac{B(z)}{B(0)}\frac{\beta_f(0)}{\beta_f(z)} \quad (3)$$

where $n_{e,o}$ is the plasma density at the stellar surface.

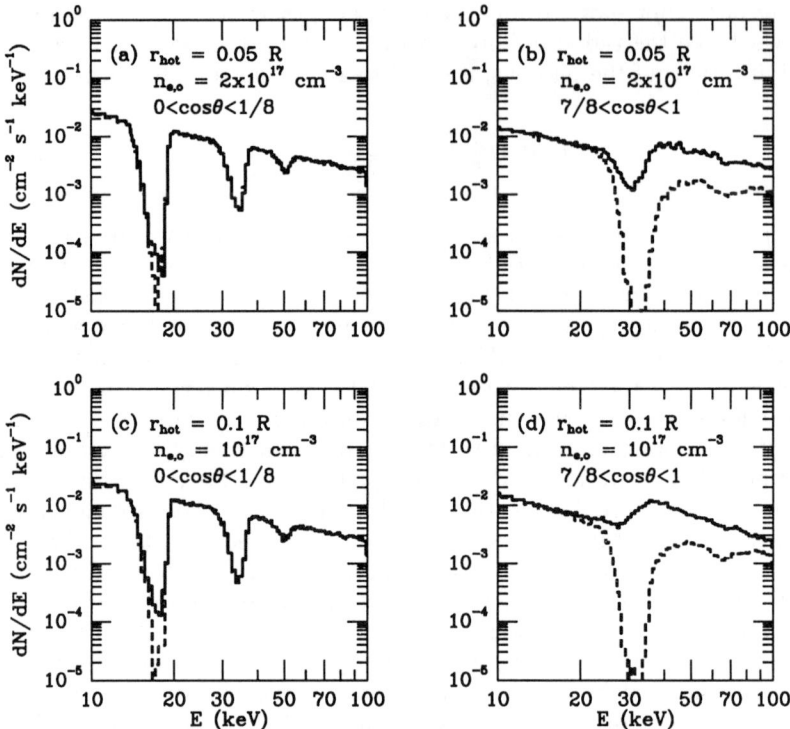

FIG. 1. Theoretical photon number spectra for $B_o = 1.7 \times 10^{12}$ gauss and electron column depth $n_{e,o} r_{hot} = 10^{22}$ cm^{-2}. Resonant scattering (solid line) and pure absorption (dashed line) spectra are shown for two viewing angles, θ with respect to the magnetic field. Top panels: hot spot radius, $r_{hot} = 0.05R$, where R is the stellar radius; bottom panels: $r_{hot} = 0.1R$. Narrow absorption-like lines occur for most viewing angles, while broad absorption-like or emission-like features occur when the viewing angle lies along the field.

RESULTS AND DISCUSSION

We calculate the emerging radiation spectrum using a Monte Carlo radiative transfer code similar to the one described by Wang et al. (5). The code is valid for line forming regions where the cyclotron first harmonic is optically thick in the line core but optically thin in the wings. The cross sections are summed over final and averaged over initial polarizations. We use exact relativistic kinematics and zero natural line width. We include scattering at the first three harmonics and photon spawning.

The emerging spectrum of radiation is shown at two viewing angles in Figure 1. $r_{hot} = 0.05R$ in the top panels and $0.1R$ in the bottom panels. In each case $B_o = 1.7 \times 10^{12}$ gauss and the electron column depth $n_{e,o} r_{hot} = 10^{22}$ cm^{-2}.

The behavior of the spectra is explained by the high velocity of the plasma, which causes scattered photons to be beamed along the field. Consequently, when the spectra are viewed perpendicular to the field (left panels) the scattered spectra are almost identical to pure absorption spectra. Although we expect this in the second and third harmonics, it is also the case for the first harmonic. In both spectra, the equivalent widths in the first and second harmonics are $W_{E1} \approx 4.7$ and $W_{E2} \approx 6.2$ keV. By comparison, in GB880205, observed by Ginga, $W_{E1} = 3.7$ and $W_{E2} = 9.1$ keV (15). The narrowness of the lines is due to the finite radius of the hot spot. Photons redward of the line are normally capable of scattering at high altitudes where the cyclotron energy is smaller, thus broadening the line. However, a photon moving at a large angle to the field escapes through the sides of the cylinder of outflowing plasma before reaching the altitude where it would scatter.

The beaming of scattered photons also accounts for the properties of the lines when viewed along the field (right panels). In Figure 1b, the first harmonic scattering line has been almost entirely filled in, compared with the first harmonic absorption line. Only a shallow line remains. When the hot spot radius is larger (Figure 1d), photons scatter at higher altitudes where the magnetic field strength is smaller. Consequently, scattered photons emerge at lower energies and fill in the absorption line entirely, forming a broad emission-like feature.

Our calculations suggest that a relativistic outflow is able to form cyclotron scattering lines with properties similar to the lines observed by Ginga, provided the hot spot is a small fraction of the stellar surface. In the future we propose to confirm this suggestion by more detailed calculations and a fit of the model spectra to the Ginga observations. If confirmed, our results would imply that the interpretation of the observed features as cyclotron lines does not rule out burst sources in a Galactic corona.

REFERENCES

1. E. P. Mazets, et al., Nature **290**, 378 (1981).
2. T. Murakami, et al., Nature **335**, 234 (1988).
3. M. S. Briggs, 1996, these proceedings.
4. E. P. Mazets, et al., 1996, these proceedings.
5. J. C. L. Wang et al., Phys. Rev. Lett. **63**, 1550 (1989).
6. D. Q. Lamb, J. C. L. Wang, and I. M. Wasserman, Ap. J. **363**, 670 (1990).
7. C. A. Meegan et al., Nature **355**, 143 (1992).
8. M. S. Briggs et al., Ap. J., **459**, 40 (1996).
9. M. Isenberg, J. C. L. Wang, and D. Q. Lamb, submitted to Ap. J. (1996).
10. P. Freeman, submitted to Ap. J. (1996).
11. G. S. Miller et al., Phys. Rev. Lett. **66**, 1395 (1991).
12. G. S. Miller et al., *Gamma-Ray Bursts: Observations, Analyses, and Theories*, Cheng Ho, R. I. Epstein, and E. E. Fenimore, eds., (Cambridge: Cambridge University Press), p. 423 (1992).
13. A. Chernenko and I. Mitrofanov, *Isolated Pulsars*, K. A. Van Riper, R. Epstein, and C. Ho, eds., (Cambridge: Cambridge University Press), p. 215 (1993).
14. I. G. Mitrofanov and A. I. Tsygan, Astrophys. Sp. Sci. **84**, 35 (1982).
15. E. E. Fenimore et al., Ap. J. Lett. **335**, L71 (1988).

The Spectral Evolution of Gamma-Ray Burst Pulses

V. E. Kargatis

Hughes STX, Code 631, Goddard Space Flight Center, Greenbelt MD 20771

E. P. Liang

*Dept. of Space Physics & Astronomy, Rice University
P.O. Box 1892, Houston TX 77251-1892*

We perform time-resolved spectral analysis on 26 multi-pulse BATSE GRBs. By fixing two parameters of a 3-parameter fit to the time-integrated values, we achieve a robust spectral hardness measure and physical fluxes. 28 pulse decay phases in 15 GRBs show significant correlation between peak power energy and instantaneous energy flux (characterized by $F_E \propto E_p^{1.7}$). We compare peak energy fluxes, peak hardnesses, and pulse durations and separations for pulse pairs in our sample. There may be some indication of a "recovery timescale" for the bursters' energy source or environment.

INTRODUCTION

Many different spectral evolution trends common to gamma-ray bursts (GRBs) have been identified. The most common tendency is for the spectra to evolve from high-energy dominated (hard) to lower-energy-dominated (soft), or hard-to-soft (e.g., (1)). Despite the frequency of this trend, a large variety of patterns is observed (2–4). Hardness–intensity correlations have been reported in several studies (e.g., (2,5,6)).

In this study, we analyzed individual GRB pulses and their relationships within multiple pulse GRBs, which required long, bright, and relatively simple bursts. 26 GRBs from the BATSE Large Area Detectors and Spectroscopy Detectors are included in this study. Time resolution for the datatypes used was ≥ 128 ms. Analysis was performed using the BATSE spectral analysis software package WINGSPAN, developed at MSFC.

A REVISED SPECTRAL HARDNESS MEASURE

We use Band's "GRB" model (GRBM) (7) due to its flexibility. This model has three varying spectral parameters: a lower energy power law, a peak power energy (related to the break energy of the smoothly varying curve), and a higher energy power law (plus a varying normalization). Due to the obliging

© 1996 American Institute of Physics

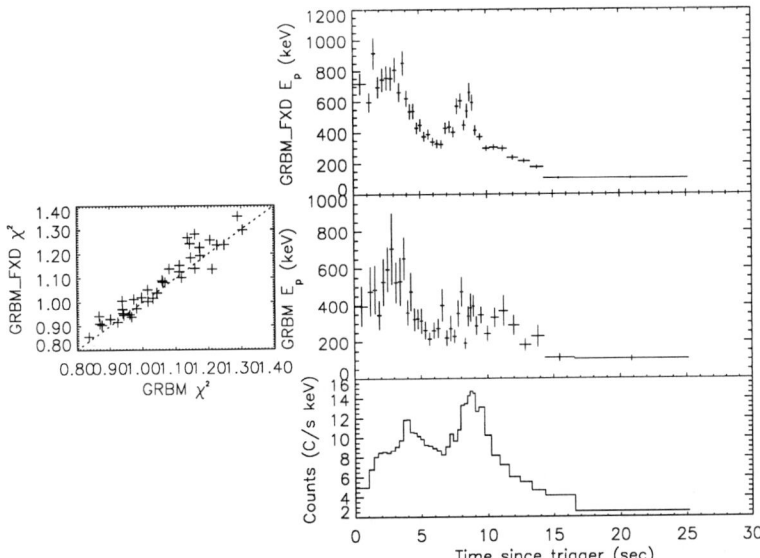

FIG. 1. χ^2 and time history comparisons for a sequence of fits to GRB 1141 using 3-parameter (GRBM) and 1-parameter (GRBM_FXD) models. For the latter, upper and lower power law indices were fixed to their time-integrated values. GRBM_FXD χ^2 values are only slightly larger, but E_p dispersion is lower, and more spectral variation is apparent.

nature of gamma-ray spectra, three varying spectral parameters often have large errors in best-fit calculations. A single varying parameter shows much less variance, usually at the cost of goodness of fit.

To obtain a single global hardness measure that largely retains the goodness-of-fit achieved by the three-spectral-parameter fits, we fit the time-integrated spectrum of the entire burst and fixed the lower and upper power law indices to their average value given by this fit (which we term GRBM_FXD). This collapses all spectral variation into the peak power energy parameter, E_p. The advantage of this method is that it tracks the overall energy content of the spectra in a robust way but still preserves a good estimate of the derived physical parameters like photon and energy flux. Goodness-of-fit is usually nearly as good as the 3+1 varying-parameter fits. Figure 1 shows spectral fits to GRB 1141 using both GRBM and GRBM_FXD, comparing E_p and χ^2 values. The rest of the results shown in this paper use the GRBM_FXD model.

HARDNESS—INTENSITY CORRELATIONS

Using the 1-varying-parameter GRBM_FXD model, we found that in many cases, a strong correlation between the "hardness" E_p and energy flux F_E appears in the decay phases of individual pulses and sometimes the burst.

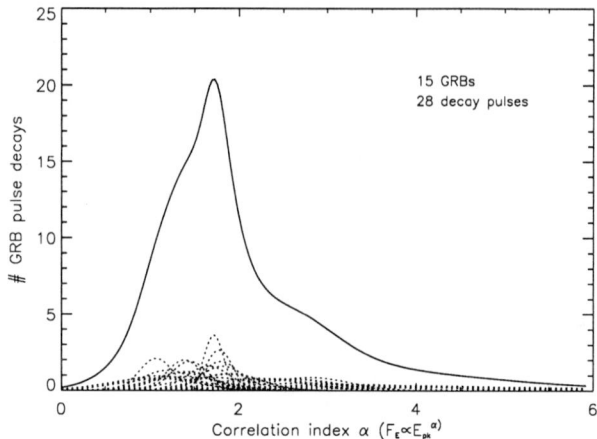

FIG. 2. Continuous histogram of correlation indices (α in $F_E \propto E_p^\alpha$). In order to view the distribution of correlation indices with errors, we assume that the errors are normally distributed, assign a unit Gaussian to each index, and add them together to get a "continuous histogram". In this way, small-error indices receive more weight than those with large error. The solid line is the summed total of the 28 pulses; the dotted lines indicate individual pulses. The distribution peaks at $\alpha \sim 1.7$.

We define a decay phase of a pulse as those intervals following the peak of the energy flux. We fit the simple functional form $F_E \propto E_p^\alpha$ to the E_p–F_E data using a linear fitting routine (FITEXY) which accounts for errors in both variables (8). Out of 26 GRBs analyzed in this study, 15 had a total of 28 pulses which showed significant correlation between hardness and intensity (where correlation is loosely defined as Spearman rank coefficient $R_s \gtrsim 0.7$ and fractional error of the correlation index < 50%). The distribution peaks at ~1.7 with a substantial spread (Fig. 2).

MULTI-PULSE BURSTS

We analyzed 26 GRBs with prominent pulse pairs in order to investigate the effects earlier pulses may have on later ones. These results may indicate whether the burster environment is impacted in a systematic way, or if the burst mechanism producing the pulses has a "reset" timescale or is random in nature. We calculate and compare several parameters—pulse durations, total "active" duration (the time over which there is significant burst activity), pulse separation, pulse peak hardnesses, and peak fluxes. In the case of overlapping pulses (10 of 26 in the sample), only total duration was calculated, since separate pulse durations are unavailable.

Fig. 3 shows durations, separations, and peak hardnesses compared between the first and second pulses in each burst. Pulse durations show significant scatter, but peak hardnesses appear associated. Most first pulse peak hardnesses are larger, as is typical. These results are consistent with those

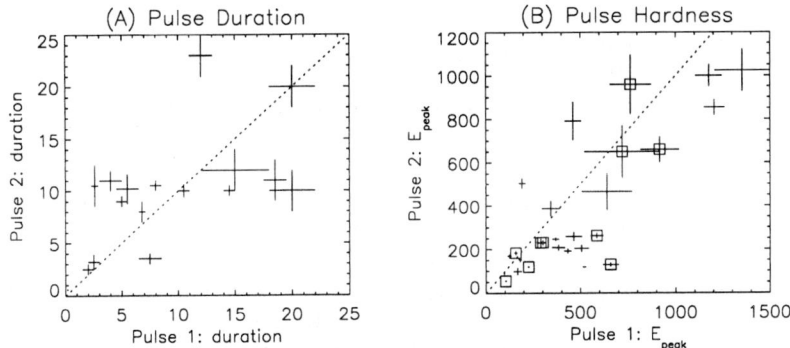

FIG. 3. Comparing pulses within each burst. (A) Durations of the first vs. second pulse. There is significant scatter, and little indication of correlation. (B) The peak hardnesses of the two pulses. There is significant correlation indicating that hard first pulses are generally followed relatively hard second pulses. The dotted line marks equal values. The squares mark overlapping pulses, the others are separated.

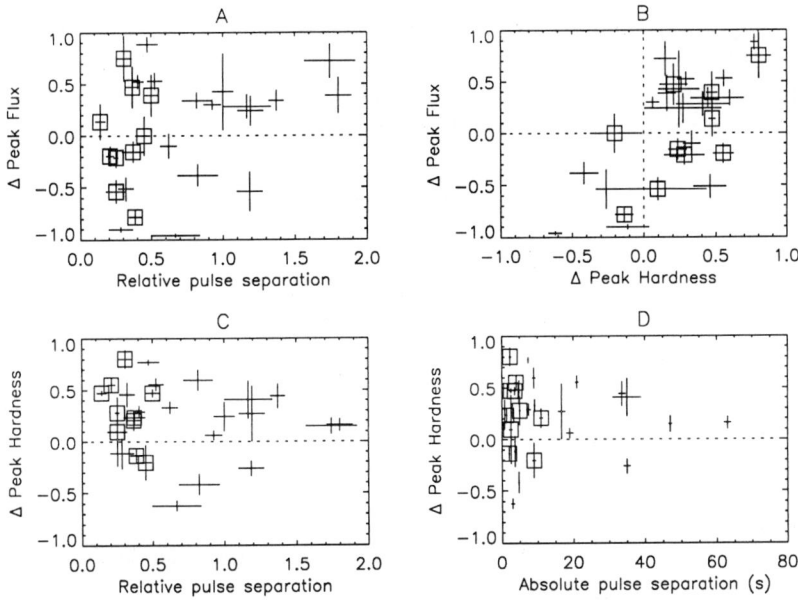

FIG. 4. Comparisons between multi-pulsed bursts with relative parameters calculated to remove absolute scale from separation, flux, and hardness variables. Squares mark overlapping pulses. (A) Relative separation vs. change in peak flux from pulse 1 to 2. Bursts above the dotted line have the first pulse brighter than second; below have the second brighter than the first. (B) Change in peak hardness vs. change in peak flux. The dotted lines mark where the peak fluxes and hardnesses of the two pulses are the same. There are no bursts in which a dimmer second pulse is harder (upper left quadrant). (C, D) Change in peak hardness against absolute and relative pulse separations.

found in a similar study of PVO multi-pulse events (9).

Fig. 4 compares scaled parameters to allow comparison between the bursts in the sample. Pulse separation scaled by the burst's active duration, and fractional changes in peak flux and hardness are parameters that remove the absolute scales of each burst and instead consider the relative effects within each burst. There appears to be no relationship between pulse separation and change in peak energy flux (Fig. 4A). The results show a typical softening across the duration of a burst. However, there may be an indication in Fig. 4C that longer separations reduce the effect a hard-to-soft trend has on the second pulse, possibly indicating that the burster may "reset" to initial, more energetic conditions after time. More long-separated bursts are required to test this hypothesis. Change in flux and change in hardness are loosely correlated in Fig. 4B. This reflects a common hardness–intensity correlation in that brighter pulses tend to be harder. This trend is countered by the trend for later pulses to be softer (brighter but softer second pulses appear in the bottom right quadrant).

Bursts with overlapping pulses show no substantive difference in flux or hardness differences when compared to separated pulses.

Acknowledgements. This work was supported by NASA grants NAG5-1515 (EPL) and Graduate Student Researchers Program fellowship NGT5-0924 (VEK). We wish to thank the BATSE team for their thoughtful assistance with software, data, and our questions.

REFERENCES

1. J. P. Norris, et al., Astrophys. J. **301**, 213 (1986).
2. V. E. Kargatis, et al., Astrophys. J. **422**, 260 (1994).
3. P. N. Bhat, N., et al., Astrophys. J. **426**, 604 (1994).
4. L. A. Ford, et al., Astrophys. J. **43**, 307 (1995).
5. S. V. Golenetskii, et al., Nature **306**, 451 (1983).
6. W. S. Paciesas, et al., in AIP Conf. Proc. **265**, ed. W. Paciesas & G. Fishman (AIP: NY), 190 (1992).
7. D. L. Band, et al., Astrophys. J. **413**, 281 (1993).
8. W. H. Press, et al., Numerical Recipes (Cambridge University Press: Cambridge), 660. (1992).
9. J. Lochner, J., in AIP Conf. Proc. **265**, ed. W. Paciesas & G. Fishman (AIP: NY), 289 (1992).

COMPTEL Measurements of MeV Gamma-Ray Burst Spectra

R.M. Kippen*, J. Ryan*, A. Connors*, M. McConnell*,
C. Winkler†, L.O. Hanlon†, V. Schönfelder‡, J. Greiner‡,
M. Varendorff‡, W. Collmar‡, W. Hermsen§ and L. Kuiper§

*Space Science Center, University of New Hampshire, Durham, NH 03824
†Astrophysics Division, ESA/ESTEC, NL-2200 AG Noordwijk, NL
‡Max-Planck-Institut für Extraterrestrische Physik, D-85748 Garching, FRG
§SRON-Utrecht, Sorbonnelaan 2, 3584 CA Utrecht, NL

We present results from the on-going spectral analysis of gamma-ray bursts measured by the COMPTEL instrument in its main Compton "Telescope" observing mode (0.75–30 MeV). Thus far, 18 bursts have been analyzed from three years (April 1991 – April 1994) of observations. The time-averaged spectra of these events above 1 MeV are all consistent with a simple power law model with spectral index in the range 1.5–3.5. Exponential, thermal bremsstrahlung and thermal synchrotron models are statistically inconsistent with the burst sample, although they can adequately describe some of the individual burst spectra. We find good agreement between burst spectra measured simultaneously by BATSE, COMPTEL and EGRET, which typically show a spectral transition or "break" in the BATSE energy range around a few hundred keV followed by simple power law emission extending to hundreds of MeV. However, the temporal relation between MeV and GeV (e.g., as measured by EGRET) burst emission is still unclear. Measurement of rapid variability at MeV energies in the stronger bursts provides evidence that either the sources are nearby (within the Galaxy) or the gamma-ray emission is relativistically beamed.

INTRODUCTION

With relatively few exceptions, most detailed measurements of gamma-ray burst (GRB) spectra have been made in the energy range around a few hundred keV. Several experiments have shown that burst spectra in this range can be well described by a combination of two variable power laws connected by a smooth (but variable) transition (1). Unfortunately, our present knowledge of the characteristics of burst emission below ~10 keV and above ~1 MeV is limited. In this study, we use the COMPTEL instrument aboard the *Compton* Gamma Ray Observatory to characterize the properties of burst emission at the high-energy end of the spectrum. Data from 18 GRBs (2) detected

© 1996 American Institute of Physics

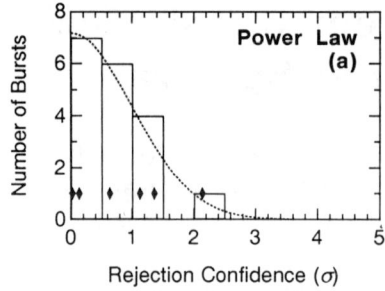
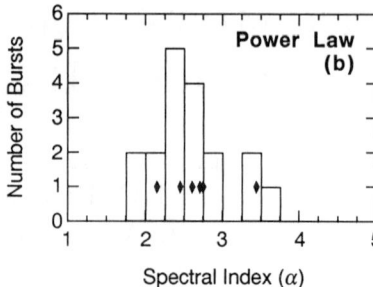

FIG. 1. The distribution of power law goodness-of-fit estimates (a) and best-fit spectral indices (b) for 18 COMPTEL bursts. Diamonds represent values for the six largest bursts and the dotted curve shows the average distribution expected from random statistical fluctuations.

TABLE 1. The acceptability of different simple spectral models based on the full sample of COMPTEL time-averaged burst measurements.

Model	Differential Spectrum	Q^a
Power Law	$N_E(E) \propto E^{-\alpha}$	9.1×10^{-1}
Thermal Synchrotron	$N_E(E) \propto e^{-(4.5E/E_c)^{1/3}}$	1.1×10^{-2}
Thermal Bremsstrahlung	$N_E(E) \propto E^{-1} e^{-E/kT}$	$< 10^{-6}$
Exponential	$N_E(E) \propto e^{-E/E_o}$	$< 10^{-6}$

[a] Probability that the distribution of 18 COMPTEL model fits can be explained simply by random statistical fluctuations of the model.

in the main COMPTEL "telescope" observing mode (0.75–30 MeV) are used to supplement earlier MeV burst measurements of *SMM* and to complement contemporaneous *Compton–* BATSE and EGRET observations.

ANALYSIS AND MODELING OF TIME-AVERAGED SPECTRA

To examine the global properties of burst emission at MeV energies, COMPTEL telescope data have been selected from each of the 18 bursts forming individual time-averaged count spectra. These raw spectra, combined with background, instrument response and livetime information are studied through fitting to model analytic functions. Small numbers of counts make the standard χ^2 model fitting method inapplicable. Hence, the spectra are fitted using a "forward-folding" maximum likelihood technique and model acceptability is evaluated through Monte Carlo simulation. In this analysis, empirical background models based on data from similar orbital conditions as during each burst are employed and detector response functions are obtained through Monte Carlo simulation of the instrument (3).

Each of the 18 time-averaged burst spectra were fit with several simple model functions (Table 1). Due to typically low count-rates in the individual burst spectra, particular models can be rejected on statistical grounds only for the strongest and hardest bursts. However, by examining the *distribution* of goodness-of-fit estimates from the full burst sample we can characterize global properties of MeV emission with enhanced sensitivity. For instance, the distribution of power law model fits is statistically consistent with random deviations, indicating that the power law model adequately describes time-averaged MeV burst emission (Figure 1a; Table 1). While the simple thermal models can adequately describe many of the individual burst spectra, they are statistically inconsistent with the full COMPTEL burst sample. These findings agree with the earlier results of *SMM* (4). The distribution of best-fit power law spectral indices (Figure 1b) is consistent with that of the *SMM* bursts and with extrapolations of BATSE spectra into the MeV energy range (1).

Most of the bursts detected by COMPTEL have been observed simultaneously by other *Compton* instruments, thus wide-band spectral comparisons are possible. COMPTEL spectra for particular strong bursts have been re-computed to match the accumulation times of these other observations. In most cases we find good agreement between BATSE, COMPTEL, OSSE and EGRET in overlapping regions of the spectra (5–8). These wide-band comparisons show that above a variable turnover (typically lying in the BATSE energy range around a few hundred keV), time-averaged burst spectra can be well described by a single power law "tail" out to $\gtrsim$100 MeV.

EXTENDED MEV EMISSION?

In at least two bright bursts detected by COMPTEL (GRB 930131 and GRB 940217), the EGRET spark chamber (SC) observed GeV photons well after the main impulsive parts of the bursts at keV energies subsided (9,10). Power law spectral indices based on fits of the EGRET data are consistent (within large uncertainties) between the impulsive and extended portions of the bursts. However, the temporal and spectral relation between GeV and keV burst emission is unclear due to the limitations of the EGRET SC (severe deadtime and few counts). The observed GeV emission is either a low-intensity *extension* of that present during the main burst (i.e., little or no change in the spectral shape) or a distinctly harder spectral component that has a *delayed* turn-on. Spectral measurements in the MeV energy range can, in principle, solve this important question.

We have examined COMPTEL data during the extended/delayed emission intervals of GRB 930131 and GRB 940217 where EGRET observed GeV photons in an attempt to identify any change in the spectrum from that of the impulsive burst. No significant MeV burst emission has been detected in these intervals. The COMPTEL upper limits are compared with the extrapolated EGRET SC spectra and TASC measurements in Figure 2. The COMPTEL sensitivity cannot rule out the possibility of *extended* MeV burst emission with the same spectral shape as measured by EGRET (i.e., the COMPTEL upper limits are not inconsistent with the EGRET SC extrapolations). Thus, we can find no evidence that there is significant spectral change at MeV energies

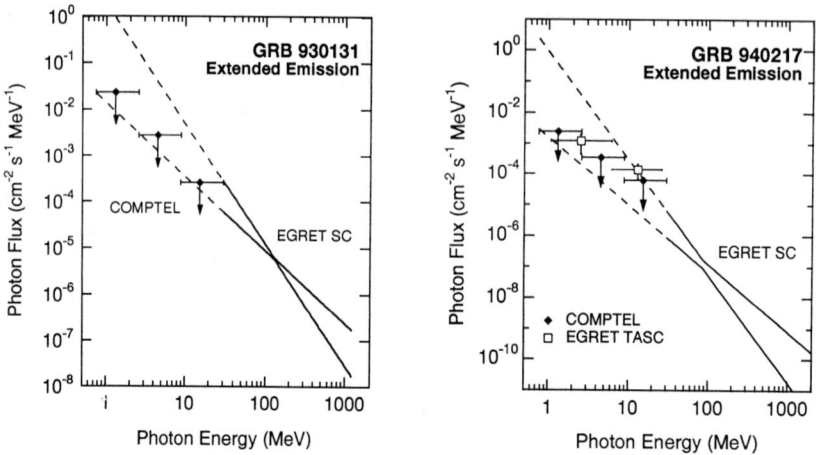

FIG. 2. Extended emission spectra of two bursts comparing COMPTEL upper limits (2σ confidence) with EGRET measurements (9,10).

in the post-impulsive intervals of these bursts.

RAPID VARIABILITY AT MEV ENERGIES

The intensity of MeV burst emission measured by COMPTEL is in many cases observed to vary on short time scales. Several of the stronger bursts contain short, high-intensity pulses of emission that are observed up to several MeV (see Figure 3). Poor statistics, electronics deadtime and telemetry gaps limit the ability of COMPTEL to measure such pulses. However, using the available measurements we are able to put conservative limits on the variability time-scale ($\delta t \lesssim$ 50–100 ms), the instantaneous peak flux ($F_p \gtrsim$ 10–20 photons cm^{-2} s^{-1} MeV^{-1}) and the maximum energy at which the pulses are observed ($E_{max} \gtrsim$ 2–5 MeV).

If the density of photons at burst emission sites is great enough, γ–γ pair production will attenuate the spectrum above ~1 MeV. The lack of a spectral cutoff in the COMPTEL data (power law spectral index $\alpha \sim$ 2.0–2.5), combined with intense and rapid variability measured simultaneously *at high energies* suggests that γ–γ pair production is an inefficient attenuation mechanism for GRBs. If the MeV burst emission is isotropic, the COMPTEL observations indicate that the sources must be well within a distance of $D_{max} \lesssim$ 1 kpc in order to avoid γ–γ attenuation (11). If the burst sources are at extragalactic distances, significant anisotropic beaming of the photons is required (e.g., due to bulk relativistic motion of the emitting plasma; (12)). For instance, bulk Lorentz factors $\Gamma_{min} \gtrsim$ 100 are required if the sources are at cosmological distances ~1 Gpc and moderate beaming ($\Gamma_{min} \gtrsim$ 3) is required even if the sources are within 100 kpc. It should be emphasized that the mere existence of high-energy gamma rays (such as measured by EGRET) does

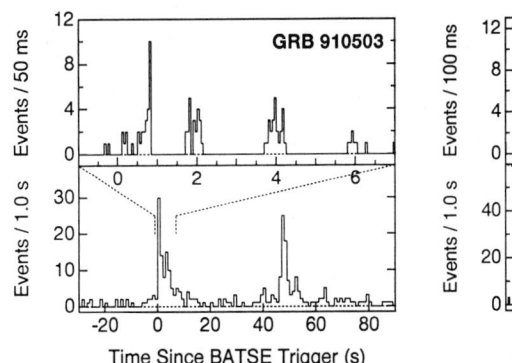 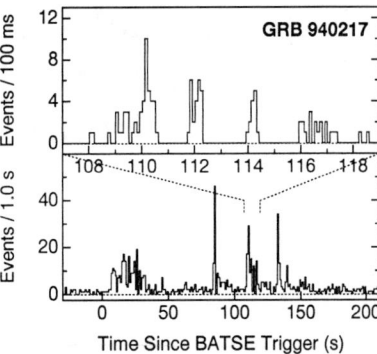

FIG. 3. COMPTEL intensity–time profiles (>1 MeV) of two bursts showing rapid intensity variations. Gaps in the profiles are due to the limited capacity of on-board telemetry buffers.

not constrain severely the source distance or the amount of beaming. Rapid variability of the high-energy flux such as shown here is required (13).

Acknowledgements. COMPTEL is supported by NASA under contract NAS5-26645, by the German government through DARA grant 50 QV 90968 and by the Netherlands Organization for Scientific Research (NWO). JG acknowledges the support of DARA under contract FKZ 50 OR 9201.

REFERENCES

1. D. Band et al., ApJ **413**, 281 (1993).
2. R.M. Kippen et al., ApSS **231**, 231 (1995).
3. R.M. Kippen et al., ApJ in preparation.
4. S.M. Matz et al., ApJ **288**, L37 (1985).
5. J. Greiner et al., A&A **302**, 121 (1995).
6. C. Winkler et al., A&A **302**, 765 (1995).
7. B.E. Schaefer et al., in AIP Conf. Proc. **307**, eds. G.J. Fishman, J.J. Brainerd & K. Hurley, 280 (AIP, New York, 1994).
8. G.H. Share et al., in AIP Conf. Proc. **307**, eds. G.J. Fishman, J.J. Brainerd & K. Hurley, 283 (AIP, New York, 1994).
9. M. Sommer et al., ApJ **422**, L63 (1994).
10. K. Hurley et al., Nature **372**, 652 (1995).
11. W.K.H. Schmidt, Nature **271**, 525 (1978).
12. M.G. Baring, ApJ **418**, 391 (1993).
13. J.M. Ryan et al., ApJ **422**, L67 (1994).

Exponential Decay of Gamma Ray Burst Spectral Break Energy with Photon Fluence

E. P. Liang[1], V. Kargatis[2]

[1] *Rice U.*, [2] *Hughes STX*

We study spectral evolution of GRBs by plotting the spectral break energy E_{peak} defined by the peak of the time-resolved νF_ν (power per decade of energy) distribution versus photon fluence integrated from burst trigger time. For pulses within bursts that are sufficiently isolated and that have clean decays, we find that E_{peak} generally decreases exponentially as a function of the photon fluence.

We model time-resolved count spectra of GRBs with the "Band" model (1), which consists of two power laws joined smoothly by an exponential. The model has 3 parameters plus an overall normalization: the two power law slopes plus the exponential constant, which is proportional to E_{peak}. To focus on the evolution of E_{peak}, we fix the two power-law slopes by the time-averaged spectrum of the whole burst. Experience tells us that such reduced models are more robust than the full model yet the reduced chi-squares are not worse than those of the full models. The time intervals are chosen to be the minimum that still give enough photons for credible spectral fitting. When we compare the time history of E_{peak} with those of the photon and energy flux (cf. (3)), we see that most pulses either evolve from "hard-to-soft" (7) or E_{peak} "tracks" the flux, but there is no universal relation between E_{peak} and flux. Such behaviors are well known among GRB pulses, e.g., (5) (2).

However, if instead of plotting E_{peak} versus time, we plot E_{peak} versus the photon fluence, we find something remarkably interesting. For most clean, isolated pulse decays in a burst, we find that E_{peak} generally decreases exponentially with photon fluence. This trend, if confirmed by future higher-time resolution results, promises to shed important light on the GRB physical mechanism. We have analyzed over 34 bright BATSE bursts with long smooth decay profiles most relevant to time-resolved spectroscopy. We find that an overwhelming majority of the pulses are consistent with such exponential decays.

The above result is open to a variety of interpretations. However, the simplest and most tantalizing one is that the GRB pulse decay is governed by simple radiative cooling. For an isolated plasma of a fixed number of emitting particles cooling only via gamma rays, the above exponential decay of break enegy with photon fluence would be expected if the average escaping

© 1996 American Institute of Physics

photon energy is proportional to the average emitting particle energy, such as in the case of thermal bremsstrahlung or multiple Compton scattering (e.g., (8). In that case the exponential decay constant is simply the total number of emitting particles modulo the source distance squared.

In many multi-pulse bursts we find that the decay constant is approximately invariant between the leading pulse and the following pulses. If this is true then it suggests that the physical parameters determining the decay constant is invariant from pulse to pulse. In the context of the above cooling interpretation this means that the number of emitting particles is invariant from pulse to pulse. This suggests that the repeating pulses are from the same burst site and the plasma is likely confined. All of these have grave consequences for the currently popular astrophysical models.

Details of the above results are located elsewhere (6) (4).

Acknowledgements. This work was partially supported by NASA grant NAG 5-1515. VK was the recipient of NASA graduate student research program fellowship NGT5-0924. We thank the BATSE team for help in this work.

REFERENCES

1. D. Band, et al. Ap. J. **413**, 281 (1993).
2. L. Ford, et al. in AIP Conf. Proc. **307**, ed. G. Fishman et al., 298 (AIP, NY) (1993).
3. V. Kargatis and E. Liang, Ast. Ap. **231**, 177 (1995).
4. V. Kargatis et al., Ap. J. Lett., in preparation (1996).
5. E. Liang in AIP Conf. Proc. **265**, ed. W. Paciesas et al., 246 (AIP, NY) (1992).
6. E. Liang and V. Kargatis, Nature **381**, 49 (1996).
7. J. Norris, et al. Ap. J. **301**, 213 (1986).
8. G. Rybicki and A. Lightman, Radiative Processes in Astrophysics (Wiley, NY) (1979).

The $\nu\mathcal{F}_\nu$ Peak Energy Distributions of Gamma–Ray Bursts Observed by BATSE

Robert S. Mallozzi[1,2], William S. Paciesas[1,2],
Geoffrey N. Pendleton[1,2], Michael S. Briggs[1,2],
Robert D. Preece[1,2], Charles A. Meegan[2], Gerald J. Fishman[2]

[1] *University of Alabama in Huntsville*
Department of Physics, Huntsville, AL, 35899

[2] *NASA/Marshall Space Flight Center*
Huntsville, AL, 35812

The majority of gamma-ray bursts exhibit a peak in their $\nu\mathcal{F}_\nu$ photon energy spectra at an energy E_p that is in the energy range ~20–2000 keV of the Large Area Detectors of the Burst and Transient Source Experiment (BATSE) on the Compton Gamma–Ray Observatory. If gamma–ray burst sources are at cosmological distances, then the spectra of dim bursts should be redshifted to lower energies relative to those of bright bursts. The magnitude of the shift is a function of the cosmological redshifts z of both the dim and bright burst sources and hence yields the range of redshift available to the bursts; this range is further constrained by considering cosmological model fits to the burst number–intensity distribution. We produced photon energy spectra for ~400 bursts using data from BATSE to investigate if this expected shift in the $\nu\mathcal{F}_\nu$ peak is observed. We find that the mean peak energies of the burst spectra are correlated with intensity: lower intensity groups of burst spectra exhibit a lower average peak energy, although the distributions of E_p are quite wide. Denoting the redshift of an event with an observed peak photon flux P (photons cm^{-2} s^{-1}) by z_P, we find that the maximum range consistent with the bursts of our sample is $(1+z_1)/(1+z_{100}) = 1.86^{+0.36}_{-0.24}$.

INTRODUCTION

Using a single count spectrum averaged over the approximate total duration interval for each event, we have produced photon energy spectra for a large ensemble of gamma–ray bursts using CONT data from the BATSE Large Area Detectors. These data consist of count rates in 16 energy channels spanning a range of approximately 20–2000 keV, with temporal resolution of 2.048 seconds. We limited our sample to bursts with a peak flux greater than ~1 photon cm^{-2} s^{-1}, where the peak flux was derived from count rate data in

© 1996 American Institute of Physics

256 ms time bins in the energy band 50–300 keV, and adopt this peak flux as our definition of burst intensity in this paper. Although the photon spectra were created from 2.048 second resolution count rate data, we categorized burst intensity using the peak fluxes on a 256 ms timescale because these flux values are routinely calculated and available for all bursts observed by BATSE for which sufficient data exist. The sample of bursts was truncated because the uncertainty in E_p increases as burst intensity decreases. Below this limiting value, we found the estimated errors to be on the order of E_p itself, so that the $\nu\mathcal{F}_\nu$ peak energy is unconstrained. Between 1991 April and 1994 May, BATSE recorded 1005 bursts. Application of the flux cutoff to those bursts for which sufficient data were available (i.e., no telemetry gaps), yielded 402 events. We have omitted an additional three events since the best-fit values of E_p for these three were outside the energy range available to the detectors, resulting in 399 events in the data set.

Although the burst sample was limited, cosmological effects, if present, should still be evident in the reduced data set, since the minimum flux level of our sample of bursts is well below the turnover in the integral number–intensity distribution.

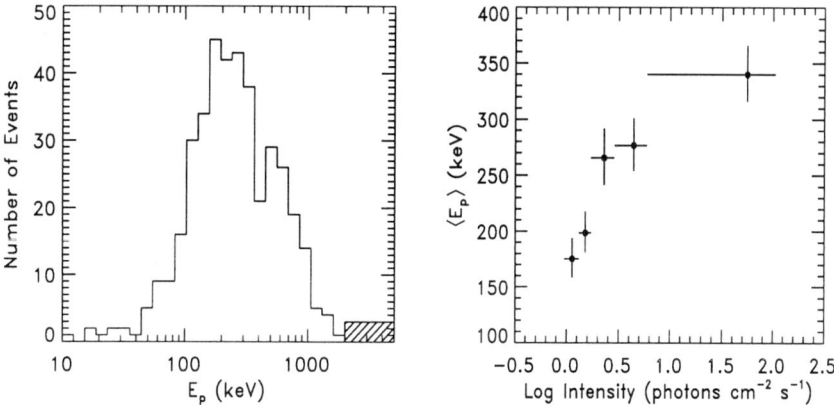

FIG. 1. a) Distribution of E_p for 399 gamma–ray bursts. b) Average E_p for five burst intensity groups.

Figure 1a shows the differential distribution of E_p for 399 bursts. The hashed portion of the histogram represents a lower limit to E_p of ∼2 MeV. We grouped the bursts into five variable width intensity bins of ∼80 events each, and ranked them such that group 1 had the lowest peak flux values and group 5 had the highest values. We found a marginally significant correlation between burst intensity and average peak energy: lower intensity groups exhibited a lower average peak energy (Spearman rank–order correlation coefficient $r_s =$ 0.9, which would occur by chance 4% of the time if no correlation existed). The logarithmic average peak energies $\langle E_p \rangle$ of each of the five distributions as a function of intensity are shown in Figure 1b.

METHOD OF ANALYSIS & RESULTS

We investigated if the $\nu\mathcal{F}_\nu$ peak energies of the different burst intensity groups are consistent with the same (redshifted) parent distribution by multiplying the E_p distributions of groups 1–4 by some factor (>1) and comparing to the brightest group via the K–S test for two data sets. Table 1 gives the flux ranges and values of the mean peak energies of the five intensity groups, and the range of shift factors for which the four groups of bursts are consistent with the brightest bursts in the sample. We define the range of acceptable shift factors as those that result in a K–S consistency probability between the shifted data set and the brightest data set of 0.25 or more (but see below). There is a clear trend for the average peak energy to be lower for groups of lower intensity, qualitatively consistent with the expected cosmological signature.

TABLE 1. Summary of Burst Group Properties

Burst Group	Peak Flux (γ cm^{-2} s^{-1})	Number of Events	$\langle E_p \rangle$ (keV)	Shift Range
1	0.95—1.30	80	$175.38^{+18.25}_{-16.53}$	$1.86^{+0.36}_{-0.24}$
2	1.30—1.70	79	$198.66^{+18.94}_{-17.29}$	$1.62^{+0.34}_{-0.24}$
3	1.70—2.90	79	$265.72^{+26.26}_{-23.82}$	$1.27^{+0.15}_{-0.10}$
4	2.90—5.90	79	$276.58^{+24.43}_{-22.44}$	$1.17^{+0.20}_{-0.11}$
5	5.90—105.0	82	$339.96^{+25.53}_{-23.75}$	

For sources at redshift z, the observed peak energy E_{obs} of the $\nu\mathcal{F}_\nu$ spectrum is related to the emitted $\nu\mathcal{F}_\nu$ peak energy E_{emit} by $E_{\text{obs}} = E_{\text{emit}}/(1+z)$. The results of Emslie & Horack (1) have shown that for sources near their limiting redshift value of $z_{0.5} \sim 1.25$, the observed burst luminosity function must be narrow. Therefore, we consider burst sources to be monoluminous; the intensity then serves as a surrogate for distance, and hence redshift. Denoting values of dim/bright events by subscripts max/min and assuming the intrinsic peak energy distribution of burst sources is not a function of redshift $[(E_{\text{dim}})_{\text{emit}} \approx (E_{\text{brt}})_{\text{emit}}]$, one obtains

$$1 + z_{\text{max}} = (1 + z_{\text{min}}) \left(\frac{E_{\text{brt}}}{E_{\text{dim}}} \right)_{\text{obs}}. \qquad (1)$$

We used the observed E_p distributions to compute the ratio of peak energies in equation 1, deriving ranges of redshifts compatible with the bursts of our sample. The broad ranges of shift factors result in a considerable overlap of allowable redshifts between the intensity groups. Figure 2 shows the ranges of z consistent with the data. The minimum redshift z_{100} corresponds approximately to the brightest bursts of the sample (group 5). The circles represent

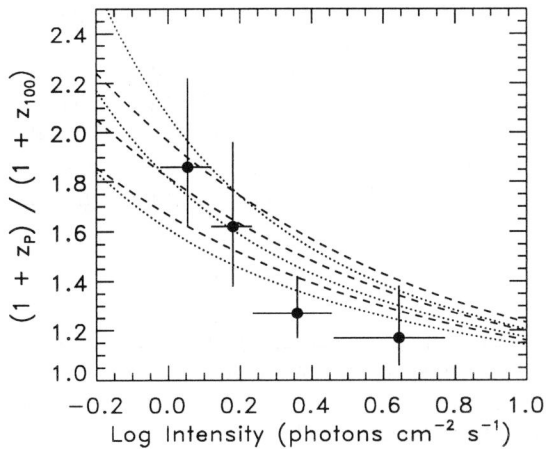

FIG. 2. Ranges of z consistent with the relative E_p shifts.

the ratio of the redshifts of the two burst groups being compared, derived from the shift where the K–S probability is a maximum. The maximum redshift of each group scales according to the specific value of z_{100}. The curves in this figure are standard cosmological models (see below). The largest redshift range consistent with the data set is $(1+z_1)/(1+z_{100}) = 1.86^{+0.36}_{-0.24}$, in agreement with the value of ∼1.75 found in an analysis of time dilation effects within burst temporal structures and durations (2). The errors simply reflect our choice for the value of the K–S consistency probability used to delineate where the two distributions being compared are different, and will increase if one chooses a value less than 0.25. Using $z_{100} \approx 0.1$ found from model fits to the burst differential number–intensity distribution (3), we infer $z_1 = 1.05^{+0.39}_{-0.27}$.

We examined the consistency of the redshift ranges with Friedmann cosmological models of non–evolving standard candles. The brightness of an event of luminosity L at a redshift z is (4)

$$P = \frac{(1+z)^{-\alpha} L}{4\pi S_0^2 r^2}. \qquad (2)$$

The burst spectral index is α, S_0 is the scale factor at the current epoch, and the radial coordinate $r = r(\sigma_0, q_0, z)$ depends on the choice of cosmology. For power law spectra, the brightness ratio of two sources at redshifts z_{min} and z_{max} is

$$\frac{P_{min}}{P_{max}} = \left(\frac{1+z_{min}}{1+z_{max}}\right)^{-\alpha} \left(\frac{r_{max}}{r_{min}}\right)^2. \qquad (3)$$

We assumed that the brightest sources with a flux of 100 photons cm^{-2} s^{-1} are at a redshift z_{100}, and inverted equation 3 to construct the curves $(1+$

$z_P)/(1+z_{100})$ as a function of the flux P for several values of σ_0 and q_0. We used three values for z_{100} (0.08, 0.10, 0.12), and a photon power law index of 2. The resulting curves are shown in Figure 2. The dashed lines are models with $\Lambda = 0$ defined by $\sigma_0 = 1$, $q_0 = 1$, and the dotted lines are models with $\Lambda > 0$ ($\sigma_0 = 1$, $q_0 = -1$). The redshift ranges derived from the shifting of the E_p distributions are consistent with those derived from brightness considerations; however, the data cannot distinguish between cosmological models.

DISCUSSION

One effect of burst sources at cosmological distances would be the redshifting of the photon energy spectra of weak events relative to strong events. We have shown that the BATSE data display a trend that is consistent with this effect. These and other recent results that have probed redshift effects support the hypothesis that GRB sources are indeed at cosmological distances, although different interpretations of the data are possible (5).

We found that the maximum range of redshifts of bursts spanning a peak flux interval of $\sim$(1–100) photons cm^{-2} s^{-1} is $(1+z_1)/(1+z_{100}) = 1.86^{+0.36}_{-0.24}$. The actual redshift of the dimmest bursts observed by BATSE is expected to be larger than this, due to the restriction of our data set to bursts with a peak flux greater than $\sim$1 photon cm^{-2} s^{-1} (the dimmest bursts observed by BATSE have peak fluxes $\sim$0.3 photons cm^{-2} s^{-1}) (6).

The results presented in this paper show that observed burst spectra are consistent with, but not proof of, sources at cosmological distances. The maximum value of redshift derived from the burst spectra is in general agreement with redshifts obtained from other analyses: model fits to the burst number-brightness distribution (7–10), moments of this distribution (11), and duration distribution time dilation effects (2). In addition, the data agree with simple cosmological models, although individual models cannot be distinguished.

REFERENCES

1. A. G. Emslie & J. M. Horack, ApJ **435**, 16 (1994).
2. J. Norris, et al., these proceedings (1996).
3. R. S. Mallozzi, G. N. Pendleton, & W. S. Paciesas, submitted, (1996).
4. S. Weinberg, Gravitation and Cosmology, New York: Wiley, (1972).
5. I. Mitrofanov, et al., these proceedings (1996).
6. G. J. Fishman, et al., ApJS **92**, 229 (1994).
7. E. E. Fenimore, et al., Nature **366**, 40 (1993).
8. E. E. Fenimore, et al., Nature **357**, 140 (1992).
9. S. Mao & B. Paczyński, ApJ **388**, L45 (1992).
10. W. A. D. T. Wickramasinghe, et al., ApJ **411**, L55 (1993).
11. J. M. Horack & A. G. Emslie, ApJ **428**, 620 (1994).

Statistical Analysis of BATSE Gamma-Ray Burst Spectra

I. Mitrofanov[1], A. Chernenko[1], A. Pozanenko[1],
M. S. Briggs[2], R. D. Preece[2], W. S. Paciesas[2], G. J. Fishman[3]

[1] *Space Research Institute (IKI), Profsoyznaya 84/32, 117810 Moscow, Russia,*
[2] *Dept. of Physics, University of Alabama in Huntsville, Huntsville, AL 35899,*
[3] *NASA Marshall Space Flight Center, Huntsville, AL 35812*

We apply peak aligned averaging to BATSE 16 channel 2.048 s accumulation time spectra from the brightest LAD. We present average peak count and photon spectra. These signatures are used to compare bright and dim GRBs.

AVERAGE SPECTRAL PROPERTIES AT THE PEAK

In this paper we deal with a subset of GRBs from the 3B catalog (2), the 260 bursts with peak flux measured on the 256 ms time scale $f_{256} > 1.3$ photons cm^{-2} s^{-1} and duration $t_{90} > 2$ s. We analyze 16 channel 2.048 s resolution continuous data recorded by the BATSE LAD detectors (CONT data type). First of all, we build a conventional average emissivity curve (ACE) in the broad energy range 50–300 keV (4,5). It is presented in top left panel of Fig. 1. The errors correspond to the variance of the mean.

To build average count spectrum we perform the following steps: consider a time history of count rate in the energy range 50–300 keV; identify the main peak, determine the peak rate $C_{\max}$; normalize the rates in all 16 energy channels by $C_{\max}$; align all bursts at the times of their main peak; average the aligned, normalized time histories separately for each energy channel. The 16 average rates at each time interval relative to the main peak determine the average count spectrum for this interval. The count spectrum for the main peak of the ACE is presented in top right panel of Fig. 1.

To obtain the average photon spectrum for the main peak of the peak aligned time history we used the following algorithm: perform model dependent deconvolution of the spectrum at the peak of each event, normalize it by the photon flux integrated over the 50–300 keV band, and average over the set of GRBs. The photon spectrum νF_ν for the main peak averaged over the bursts with $f_{256} > 1.3$ photons cm^{-2} s^{-1} is presented in bottom left panel of Fig. 1. This spectrum generally agees with the shapes typically found in individual GRBs (1). The spectrum peaks at about 250 keV.

In the course of averaging of photon spectra it is possible to analyse average properties of residual counts. Such study could hopefully reveal narrow spectral features provided they occur systematically in many GRBs at some

© 1996 American Institute of Physics

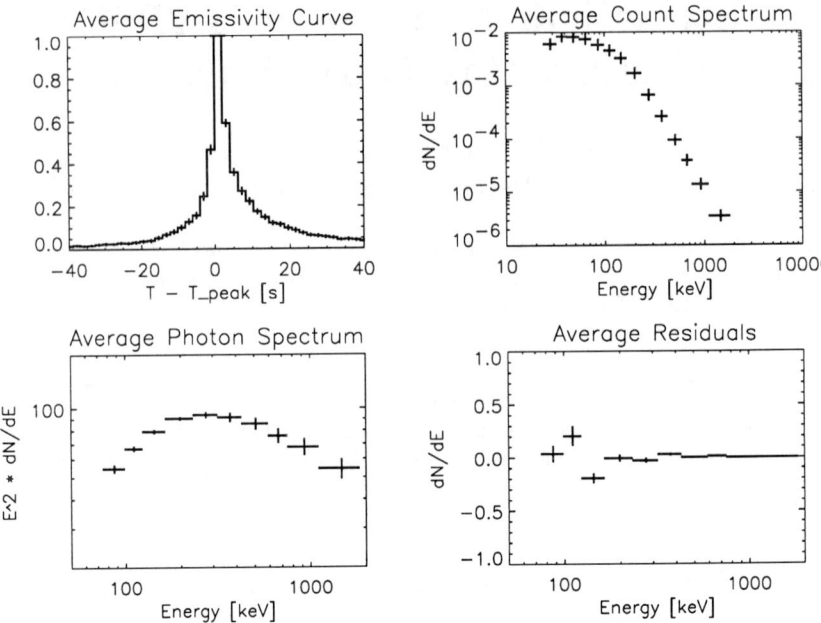

FIG. 1. Results for GRBs with $f_{256} > 1.3$ photons cm^{-2} s^{-1}. Top left panel: Average Emissivity Curve (ACE) for the energy range 50–300 keV. Top right panel: the average count spectrum for the main peak of the ACE. Bottom left panel: the average photon spectrum for the main peak of ACE. Bottom right panel: average residuals for the main peak of the ACE.

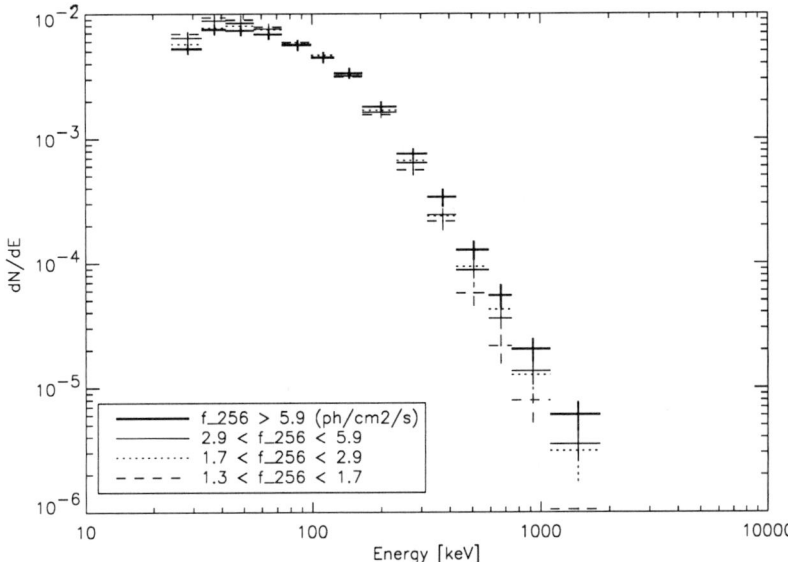

FIG. 2. The average normalized count spectra of the main peak for the 4 intensity groups. The correspondence between the line styles and intensities of the sets of GRBs is shown.

particular energy. In bottom right panel of Fig. 1 the average residuals are presented. The pattern is dominated by the known systematics of the energy-to-channel conversion (6). Due to these systematics we had to refrain from using the lowest energy channels for the study of photon spectra since the χ^2 test gave unacceptably large values.

COMPARISON OF BRIGHT AND DIM GRBS

To investigate how the average spectral properties of GRBs depend on their intensity we obtained the average count and photon spectra for the peaks of 4 of the 5 intensity groups defined in (3). In Fig. 2 we show the overlapped average peak count spectra. It is clear that brighter GRBs are harder than dimmer ones.

In Fig. 3 average photon spectra νF_ν of the peaks are presented. It is evident that the peak of the spectrum shifts to higher energy as intensity increases. This indication of hardness/intensity correlation is similar to that described in (3). However, the results presented here show the spectral properties at the peak on the 2.048 s time scale while the results obtained in (3) were obtained for integrated spectra of the GRBs. Therefore, the effect of hardness/intensity correlation is found for both cases.

It is interesting in this respect to compare the average values of E_p at the peak and those for the integrated spectra. Visually inspecting Fig. 3 one can estimate the positions of the E_p and then compare them with those obtained

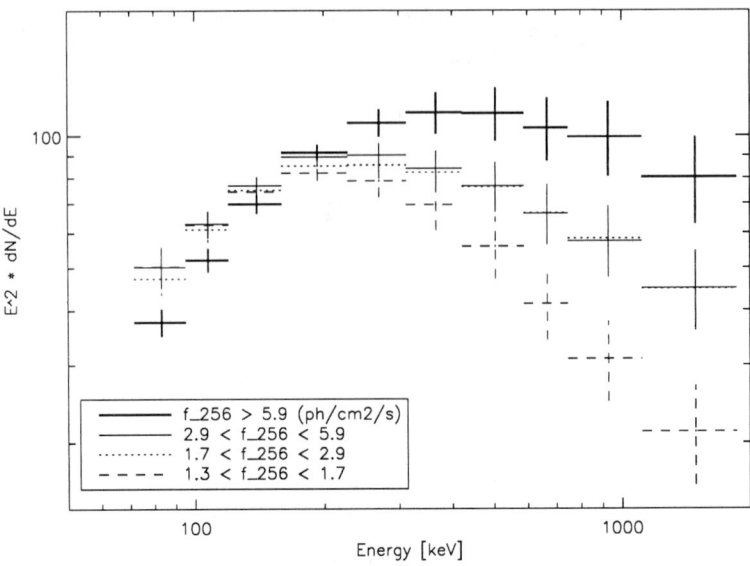

FIG. 3. The average normalized photon spectra of the main peaks for the 4 intensity groups. The correspondence of the line styles and intensity of the sets of GRBs is shown.

in (3). For all intensity groups our values of E_p are greater than or equal to those of the integrated spectra from (3): about 400 keV for the brightest group compared to 335 keV from (3), about 250 for the next 2 groups (compared to 256 and 239 keV), and about 200 keV compared to 155 keV for the fourth intensity group.

The systematic excess of E_p for the peak over the E_p of the integrated spectrum is most likely associated with well known correlation of hardness and flux within a given burst that usually makes the peak spectrum of the burst harder than the integrated spectrum.

REFERENCES

1. D. L. Band, ApJ **432**, L23 (1994).
2. C. A. Meegan, et al., ApJ Supp., in press (1996).
3. R. S. Mallozzi, et al., ApJ, in press (1995).
4. I. G. Mitrofanov, et al., in AIP Conf. Proc. **265**, eds. W.S. Paciesas & G.J.Fishman, (New York: AIP), 163 (1992).
5. I. G. Mitrofanov, et al., ApJ, **459**, 570 (1996).
6. G. N. Pendleton, et al., in AIP Conf. Proc. **304**, (New York: AIP), 749 (1994).

Performance Evaluation of the BATSE Spectroscopy Detectors

W.S. Paciesas[*], M.S. Briggs[*], R.B. Wilson[†], R.D. Preece[*], J.L. Matteson[‡] and D.L. Band[‡]

[*] *University of Alabama in Huntsville, AL 35899*
[†] *NASA/Marshall Space Flight Center, Huntsville, AL 35812*
[‡] *University of California, San Diego, La Jolla, CA 92093*

The BATSE spectroscopy detectors (SDs) were specifically designed to combine good energy resolution and sensitivity for observations of gamma-ray burst (GRB) spectra. One of the prime goals of BATSE is to search for and, if found, to characterize spectral line features in GRBs. Thus far, no convincing line features have been detected by BATSE. We summarize the in-orbit performance of the SDs in order to show that these detectors are operating as designed and have the sensitivity to detect line features if they exist.

INTRODUCTION

As originally designed, BATSE/CGRO was intended primarily to detect and locate weak gamma-ray bursts (GRBs). Spectroscopy was a secondary objective, so the original BATSE detectors were not optimized for this purpose. Furthermore, the CGRO payload initially included a wide-field instrument employing germanium detectors that would have had excellent spectroscopy capability for detailed studies of the spectral features which the KONUS and *Ginga* observations indicated as rather common in GRBs. After budgetary contraints forced the deletion of the germanium instrument, the lack of high quality GRB spectroscopy was identified as a significant shortcoming of CGRO and the spectroscopy detectors (SDs) were added to BATSE in order to remedy this situation at relatively low cost.

After nearly five years of BATSE operation, we have yet to observe any convincing line-like features in GRB spectra (1,2). If the BATSE SDs are working properly, the upper limits are consistent with *Ginga* only if the fraction of GRBs with line features is rather low (3,4). Our calculations of the SD sensitivity are based on pre-launch data which have been verified by in-flight measurements. In this paper we summarize evidence from pre-launch and in-flight data that indicates that the BATSE SDs are working as designed. Additional evidence from observations of the cyclotron absorption feature in the X-ray pulsar Her X-1 is summarized in a separate paper (5).

© 1996 American Institute of Physics

PERFORMANCE EVALUATION

The SDs were subjected to extensive pre-launch calibrations using standard lab reference sources and also calibrations at SLAC using a monochromatic beam at a series of energies from 6–60 keV. Energy resolution and uniformity were shown to be nearly optimal for NaI, as illustrated in Table 1 for several relevant energies. Absolute and relative efficiency measurements were obtained and the non-linear light output of the NaI crystal was mapped in detail (6). The detector response matrices were constructed using Monte Carlo simulations validated by the laboratory measurements (7).

The non-linearity of the pulse height analysis electronics is an additional complication. After launch, a significant electronic artifact was discovered that affected low pulse heights and was not completely addressed by pre-flight calibrations. A methodology to correct for this effect, which has been termed the SLED (for Spectroscopy Low-Energy Distortion), was developed by combining the available pre-flight calibrations with certain in-flight data and laboratory measurements from an engineering prototype detector (6). This correction is now routinely included in BATSE data analysis. However, most of the BATSE upper limits on spectral line features were obtained from SDs operated at high gain, where the SLED is confined to a few pulse height channels below ~15 keV and is therefore not a significant source of systematic error.

The following sections summarize various in-flight tests that we have performed in order to verify the SD performance.

Energy Resolution

We have verified the in-flight energy resolution of the SDs in several ways. Firstly, the widths of in-flight background lines are consistent with pre-launch calibrations. Table 1 shows typical measurements for three background lines. The expected widths from pre-launch calibrations are shown together with the maximum widths consistent with the data at the 1σ and 2σ confidence levels. Secondly, the observation of steep low-energy pulse height spectra, such as in solar flares, implies good energy resolution. Figure 1 shows a count spectrum from one such flare, on 1993 Dec 22. The low-energy spectral cutoff is due to the electronic threshold. The drop at high energies is consistent with the pre-launch resolution, which was 5.5 keV FWHM at 14.4 keV. This is an important addition to the background line measurements because the 67 keV line is the lowest energy line usable for direct resolution measurements.

Since the counting rates during bright GRBs can be rather large, one might question whether such high rates might somehow degrade the energy resolution. However, we have observed many spectra during bursts prior to background subtraction and find no indication of such an effect. When counting statistics are sufficiently good, the widths of background lines during GRBs are consistent with the same lines in the non-burst background.

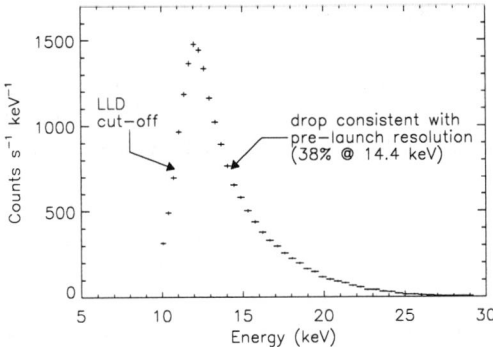

FIG. 1. Count spectrum of a solar flare (1993 Dec 22) with a soft spectrum. The cut-off at low energy is due to the electronic threshold. The drop at high energies is limited primarily by the detector resolution, which is consistent with the measured pre-launch resolution (5.5 keV FWHM at 14.4 keV).

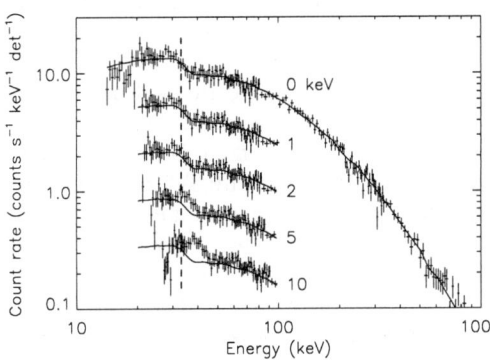

FIG. 2. Effect of simulated gain shifts on the K-edge feature. The original data (0 keV shift) from GRB 920525 are shown with the best-fitting model spectrum superimposed (solid curve). Portions of the data near the K-edge are shown for simulated gain shifts of 1, 2, 5, and 10 keV. The vertical dashed line is a guide to the eye. A trend in the residuals is evident for gain shifts as small as 2 keV.

Energy Calibration

After correction for the known non-linearities, the measured energies of in-flight background lines are as expected. Table 1 shows some typical values. The K-edge feature due to the iodine in the scintillator may also be used to provide an additional verification at a lower energy. Figure 2 illustrates this, where we show the count spectrum for a GRB for hypothetical energy calibration offsets ranging from 0 keV (i.e., no error) to 10 keV. An error of 2 keV or more shows a clear trend of the residuals in the K-edge region relative to a smooth model spectrum. Thus, an error of this magnitude would produce many apparent line features in BATSE data around 30 keV if it were present.

TABLE 1. BATSE In-Flight Background Lines

Origin	True energy (keV)	Meas. energy (keV)	Pre-launch FWHM (keV)	Max. in-flight FWHM (keV)	
				1σ	2σ
^{125}I	67.3	70.1	9	9.7	9.9
^{123}I	193	191	20	20.5	20.7
e^+-e^-	511	507	40	41.6	42.4

Detector Response

We have verified the accuracy of the SD detector response matrices by intercomparison of the data with several other detectors. An extensive program of observations of the Crab pulsar using the BATSE on-board pulsar folding memories has been performed. Data were accumulated from both the SDs and the large area detectors (LADs). Detailed evaluation of the data is in progress. However, preliminary spectral fitting of a single power-law to combined data from 5 SDs resulted in a photon index $\alpha = -1.99\pm0.02$ versus $\alpha = -2.07\pm0.01$ for a similar LAD spectrum. Given the known systematic errors these values are quite consistent. On the other hand, there is a noticeable difference in normalization between the two detector types, with the LAD spectrum being $\sim$20–30% more intense. The origin of this difference is not understood at this time; however, similar offsets have been seen in comparisons between LAD data and other instruments, suggesting that the error is in the LAD response and not the SDs. In any case, an error in absolute normalization, though relevant for some analyses, cannot explain the absence of spectral features.

Further comparison between SDs and LADs have been performed for GRBs and solar flares. The values of spectral parameters determined independently by SDs and LADs are generally in good agreement (except for the absolute normalization offset discussed above). A typical example is shown in Figure 3 for a GRB. Simultaneous count rate spectra from one LAD and one SD are shown for a GRB together with the model spectrum that best-fits the data of both detectors, with the model allowing a normalization difference between the detectors. Except for the normalization difference, the agreement is quite good. The SD data have also been compared with GRB spectra from other instruments and again we find good agreement in general [e.g., (8)].

SUMMARY

We have verified by a number of in-flight tests that the SD performance is nominal:

1. Background lines, K-edge and low-energy threshold all have expected widths and energies.

2. Steep count spectra are observed in sufficiently soft solar flares.

3. Crab pulsar spectra are consistent between the BATSE LADs and SDs.

4. Simultaneous LAD/SD solar flare and GRB spectra are in good agreement.

5. One electronic artifact not noted before launch, the SLED, has been accounted for in data analysis, but in most cases it only affects data below $\sim$15 keV, which have not been used in the search for GRB spectral line features.

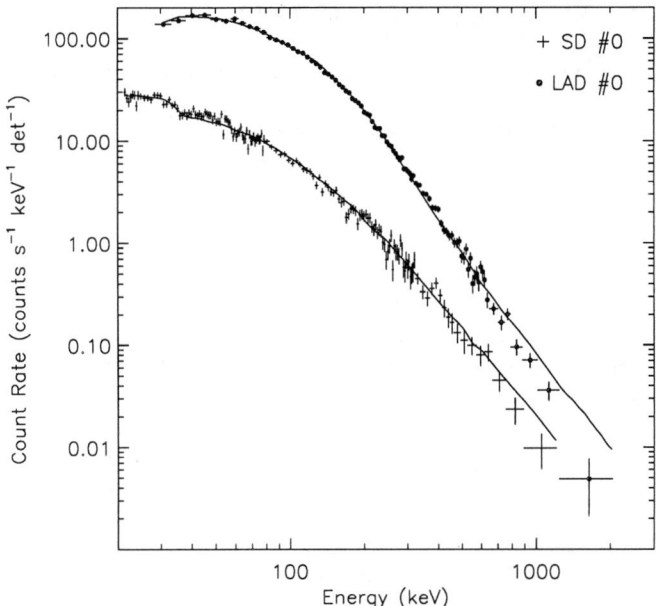

FIG. 3. Simultaneous SD & LAD count rate spectra during GRB 921207. The solid curve is the model spectrum that best fits both data sets, allowing for a normalization difference of ~20% (see text). The agreement in spectral shape between the two detectors is good.

We conclude that systematic errors do not significantly affect the BATSE observations of Her X-1 (5) nor can they account for the BATSE non-detection of line features in GRBs (1,2). Several avenues for more detailed validation may be pursued in the future, including comparison of simultanous solar flare observations with *Yohkoh*, derivation of Sco X-1 spectra using Earth occultation, analysis of pulsed spectra from 4U 0115+63 and A 0535+26, and comparison of spectra of Her X-1 with RXTE and BeppoSAX observations.

REFERENCES

1. D.M. Palmer *et al.* ApJ, **433**, L77 (1994).
2. M.S. Briggs, these proceedings.
3. D.L. Band *et al.* ApJ, **434**, 560 (1994).
4. D.L. Band *et al.* ApJ, **447**, 289 (1995).
5. P. Freeman *et al.* these proceedings.
6. D.L. Band *et al.* Exp. Astr., **2**, 307 (1992).
7. G.N. Pendleton *et al.*, Nucl. Instr. & Meth. **A 364**, 567 (1995).
8. J. Greiner *et al.* A&A, **302**, 121 (1995).

High Resolution GRB Spectra from the Transient Gamma-Ray Spectrometer (TGRS)

D.M. Palmer,[†‡] H. Seifert,[†‡] B.J. Teegarden,[†*] N. Gehrels,[†]
T.L. Cline,[†] R. Ramaty,[†] K. Hurley,[¶] R. Pehl,[§] N. Madden,[§]
and A. Owens[◊]

[†] *NASA/Goddard Space Flight Center, code 661, Greenbelt MD 20770,*
[‡] *Universities Space Research Association,* [*] *on leave to CESR,* [¶] *University of California, Berkeley,* [§] *Lawrence Berkeley Labs,* [◊] *U. of Leicester*

The Transient Gamma-Ray Spectrometer (TGRS) is a germanium spectrometer designed to produce high-resolution (2–3 keV) spectra of bright gamma-ray bursts. In its first year of operation it has triggered on 62 GRBs, of which ~33 were bright enough for spectroscopy. Almost all of these bursts were also seen by the Konus detectors on the same spacecraft, and about half were also observed by BATSE. This allows the instruments and their results to be compared and will allow line candidates seen by any of the instruments to be confirmed or refuted by the others. This paper presents some typical spectra from GRBs observed by TGRS, and includes for comparison the corresponding BATSE and Konus spectra.

INTRODUCTION

The Transient Gamma-Ray Spectrometer (TGRS) is a 215 cm^3 (6.7 cm dia. × 6.1 cm) germanium spectrometer on the WIND spacecraft, launched on 1 Nov. 1994 towards an eventual orbit at the Sun-Earth L$_1$ point (1,2). Its field of view is roughly the southern ecliptic hemisphere. During its first year its performance has been excellent, with the passive radiator maintaining its temperature below the 85 K design temperature since shortly after launch. Below 1 MeV, the energy resolution remains at 2–3 keV and the effects of radiation damage on gain and resolution, although detectable, are negligible. At higher energies the energy resolution has slightly increased, e.g., from 6.0 to 8.6 keV at 2.8 MeV. The only other significant feature that complicates the analysis is an instrumental electronic artifact at 30 keV produced by saturating cosmic rays. Since the detector bias voltage was turned on, on 16 Nov. 1994, TGRS has been in continuous operation except for a total of 65 hours in Nov. and Dec. 1994 for passages through the Earth's radiation belts. As of 16 Nov. 1995 it had triggered on 62 GRBs, of which ~33 were bright enough to make spectroscopy worthwhile. For more details on the current

© 1996 American Institute of Physics

state of the instrument, see (3), in this volume.

The primary purpose of TGRS was to study the absorption and emission lines which, at the time the instrument was developed, were thought to be common in the spectra of GRBs. The behavior and detailed profiles of the features would provide insight into the processes by which they are produced. These features have been reported, notably, in spectra from Konus and Ginga. However, the 8 BATSE Spectroscopy Detectors (SDs) on GRO did not confirm the existence of GRB spectral lines, contrary to original expectations (4).

TGRS is being used to search for line features in GRBs in an attempt to resolve this apparent discrepancy. TGRS's high energy resolution makes it more sensitive to narrow line features than a NaI scintillator of similar size. In addition, WIND carries a pair of Konus detectors, sharing much of the technological heritage of the earlier Konus detectors that detected many GRB lines, but enlarged to the size of a BATSE SD. Most strong GRBs will now be observed by multiple detectors, allowing independent confirmation of candidate spectral lines.

GRB SPECTRA

This paper presents examples of detector energy loss (count) spectra from three different GRBs, identified here by the date and TGRS trigger time in seconds. Two of these were observed by all three instruments, and the third, which is the brightest GRB seen by TGRS in its first year, was seen by both TGRS and Konus. For each GRB, a spectrum is presented for each instrument over time periods which are as close as possible to identical, subject to the constraint that each instrument bins the data in different time bins. The Konus spectra were provided by the Konus team (5). Some of these spectra are discussed by the Konus team in this volume (6). Positional information is from BATSE.

About count spectra

Only count spectra are currently available for TGRS and Konus because the software for producing photon spectra for these two instruments is still under development (3). There are several factors that complicate the comparison of count spectra observed by different instruments, or even by a single instrument observing at different angles.

Below $\sim$40 keV, attenuation from the material housing the detectors becomes noticeable at the $\sim$30% level. TGRS is enclosed in an aluminum cryostat which is roughly 0.5 mm thick in most places, while the sides of the Konus housing are 2 mm thick aluminum, with alloying elements that contribute significantly to the material's attenuation. However, the face of the Konus detector is covered by a thin beryllium window which provides virtually no attenuation if the burst direction is close to the normal axis of the detector. The BATSE SD housing is similar except that its Be window covers only half of the detector's face. Thus, the relative detector efficiencies are both energy and angle dependent.

The different sizes and materials of the detectors also change the frequency of partial-energy events caused by Compton escape. Below a few hundred keV this is less significant, as the detectors discussed here are optically thick by photoelectric absorption, which usually leads to total energy deposition.

Finally, the count spectra shown here are normalized by dividing the number of counts per second by the frontal area of the detector. However, the geometric area varies significantly as a function of angle, and so this normalization is only approximate. At 50° from the detector normal, typical of the GRB spectra presented here, TGRS presents a geometric cross-section that is 153% of its frontal area, while Konus or a BATSE SD presents 123%. For this angle at low energies, therefore, this effect will boost the TGRS count spectrum by ∼25% compared to the Konus spectrum, even though the axes of these two instruments mounted on Wind are parallel and thus present the same angle to each GRB. At higher energies this difference is reduced somewhat, as the γ-rays can cut the corner, passing through the thin edge of the geometric cross section.

These differences will be removed by the construction of detector response models which will allow us to directly compare derived photon spectra among instruments.

GRB 950325_63393

This burst (BATSE trigger #3481) occurred at an angle of 58° to the TGRS and Konus detector normals, and 7° from the BATSE SD2 detector normal. Fig. 1 compares spectra from these instruments for a subinterval of this GRB. The spectra are consistent with each other in shape and normalization (after adjusting for geometric cross section) above 35 keV, where the electronic artifact disturbs the TGRS spectrum.

GRB 950425_00924

This burst (BATSE trigger #3523) occurred at an angle of 51° to the TGRS and Konus detector normals, and 53° from the BATSE SD4 detector normal.

Fig. 2 compares spectra from the different instruments. Due to transmission errors, TGRS data for the period corresponding to the last two seconds of the BATSE and Konus spectra are not available. However, a comparison of BATSE spectra for the two different time periods are almost identical except for a ∼10% shift in normalization. A BATSE SD spectrum just over that 2 second interval shows no features of interest.

The count spectra for the three instruments are consistent with each other in both shape and normalization above 200 keV, with a ∼25% TGRS geometry enhancement from 40-150 keV. The BATSE SD spectrum is intermediate between the two.

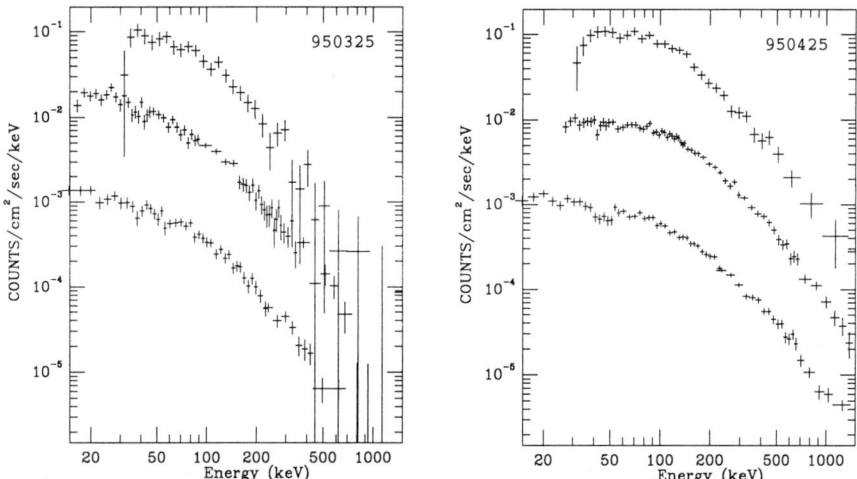

FIG. 1. GRB 950325 spectrum as seen by TGRS (top), BATSE SD2, and Konus (bottom). This spectrum was accumulated by TGRS 35.7-40.6 seconds after its trigger, and at similar times for the other two instruments. The Konus and SD2 spectra have been shifted downwards by 1 and 2 decades, respectively, for clarity.

FIG. 2. GRB 950425 spectra for TGRS, BATSE SD4, and Konus, plotted as in Fig. 1 for times relative to the TGRS trigger of 4.4-10.9 seconds for TGRS, and 4.9-12.4 seconds for BATSE SD4 and Konus.

GRB 950822_13750

This burst was the strongest seen by TGRS in its first year. BATSE did trigger on this burst (#3767), but 5 seconds later, before the strongest emission, GRO entered the South Atlantic Anomaly and the BATSE detectors automatically turned off. The early BATSE data did, however, produce a position for this burst which is 46° from the TGRS and Konus detector normals.

Fig. 3 compares TGRS and Konus spectra for a subinterval of this burst. Once again, the count spectra for the two instruments differ by up to 20% at low energies, consistent with geometry. This is more readily apparent due to the good statistics from this high-fluence GRB.

Fig. 4 shows the TGRS count spectrum over the entire GRB. The dip at 30 keV is due to the electronic artifact mentioned earlier. Otherwise, the spectrum is very smooth, showing no significant absorption or emission features with very high resolution and good sensitivity.

DISCUSSION

This comparison indicates that the different instruments produce similar count spectra, after adjusting for geometry. A more rigorous comparison will require photon spectrum deconvolution software, currently being developed.

None of the TGRS GRB spectra shown, nor any other TGRS GRB spectra

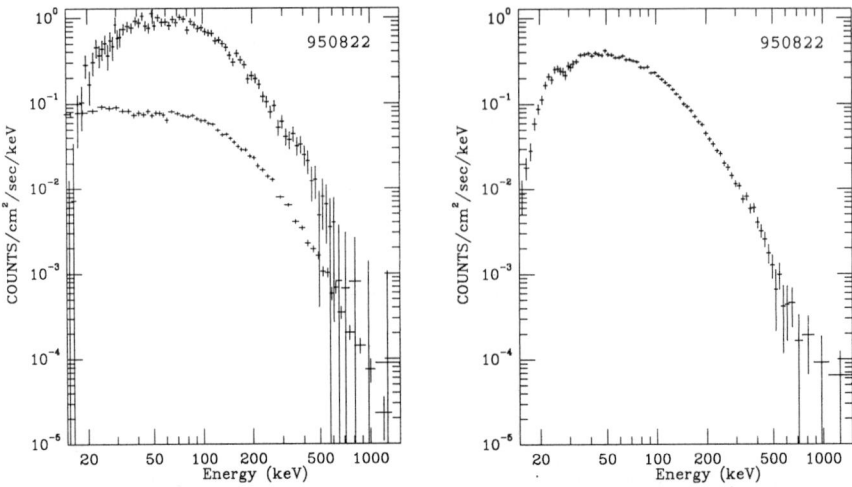

FIG. 3. GRB 950822 spectrum as seen by TGRS (top) and Konus (shifted down 1 decade) for a time corresponding to 12.4-13.8 seconds after the TGRS trigger.
FIG. 4. TGRS spectrum for the entire GRB, 0-33.8 seconds. Energy bins are $\Delta E/E \sim 5\%$ below 500 keV

that have been examined, show significant line features in either absorption or emission. TGRS does detect lines in the background spectrum with its expected resolution. An occulter, a block of lead that rotates with the spacecraft to shield the detector from each point near the ecliptic plane for a quarter of each 3 second spacecraft revolution, has allowed us to obtain very good spectra of the Galactic center 511 emission line, demonstrating our ability to detect lines in variable sources.

We therefore conclude that the TGRS instrument is working properly and that none of the GRBs that TGRS has observed in its first year contain spectral line features with a strength exceeding our predicted sensitivity. This is consistent with the rate in the combined BATSE SD and Ginga data.

If a candidate line feature is reported in a GRB spectrum by one of the instruments currently operating, it is now likely that the same GRB will be seen one or more additional instruments, allowing the candidate to be either confirmed or refuted. If TGRS sees a confirmed line, its high resolution spectra may provide insight into the physical processes of a gamma-ray burster.

REFERENCES

1. H. Seifert, et al., Astrophys. & Space Sci. **231**(1), 475 (1995).
2. D. Palmer, et al., Astrophys. & Space Sci. **231**(1), 161 (1995).
3. H. Seifert, et al., this volume (1995).
4. D. Palmer, et al., ApJ **433**, L77 (1994).
5. S. V. Golenetskii, private communication (1995).
6. E. Mazets, et al., this volume (1995).

Recent Gamma-Ray Burst Continuum Observations and Their Implications for Line Features in Gamma-Ray Bursts

Geoffrey N. Pendleton, Michael S. Briggs,
Robert D. Preece, Robert S. Mallozzi, William S. Paciesas

Department of Physics, University of Alabama in Huntsville, Huntsville, AL 35899

Gerald J. Fishman, Charles A. Meegan

NASA/Marshall Space Flight Center, Huntsville, AL 35812

Chryssa Kouveliotou

*Universities Space Research Association
at NASA/Marshall Space Flight Center, Huntsville, AL 35812*

In the pre-CGRO era, most apparent features in Gamma-Ray Burst (GRB) spectra were commonly interpreted as cyclotron absorption lines produced near strongly magnetized neutron stars distributed in the Galactic disk. Since the advent of the BATSE observations, resulting in the abandonment of the Galactic disk model, the significance and implications of spectral features in bursts are being reexamined. Recent observations of Gamma-Ray Burst spectra indicate that even narrow absorption line-like features in burst spectra can be produced as a natural consequence of continuum superposition. Since the overlapping of distinct continuua can be shown to produce apparent features in bursts, and the flux histories of bursts clearly support burst pulses overlapping, the conventional single continuum model may not be adequate for calculating the statistical significance of apparent line features in burst spectra.

INTRODUCTION

Before the discovery by BATSE (2) that bursts are isotropic and inhomogeneous, the identification of absorption line features superimposed on smooth continuum spectra in gamma-ray bursts (GRBs) had been considered definitive proof that a highly magnetized neutron star population distributed in the Galactic disk was responsible for bursts (1). In the CGRO era, however, it has become clear that the energetics of even relatively nearby Galactic halo burst

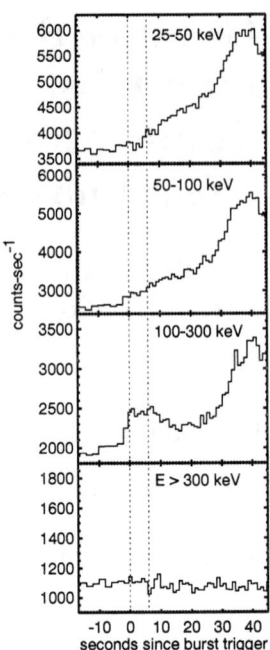

FIG. 1. Count rate time history of GRB 940330 in 4 energy channels. Between 0 and 6 s (marked by vertical dotted lines), note that large flux between 100 and 300 keV compared to the other channels.

populations require the development of new models. Since absorption lines in burst spectra are no longer a natural consequence of a generally accepted gamma-ray burst model, the interpretation of features in their spectra should be made from a more general and objective viewpoint.

In the light of recent continuum observations, we will focus on how the choice of the spectral continuum model effects the significance of features in burst spectra. Several studies have suggested that multiple independent spectral components contribute simultaneously to the continuum spectra of GRB's and that this behavior can cause spectral features (7,8,10). Analyses of BATSE burst data indicate that the superposition of the smooth continua that are observed at different times during bursts produce broad dips in some of the observed spectra (8). Recent observations of GRB spectra with BATSE have revealed the presence of a spectral continuum with a sharp cutoff in flux below roughly 50 keV. The combination of this spectral form with a softer spectrum commonly observed during many burst intervals produces a summed continuum with a significant narrow line feature at 40 keV.

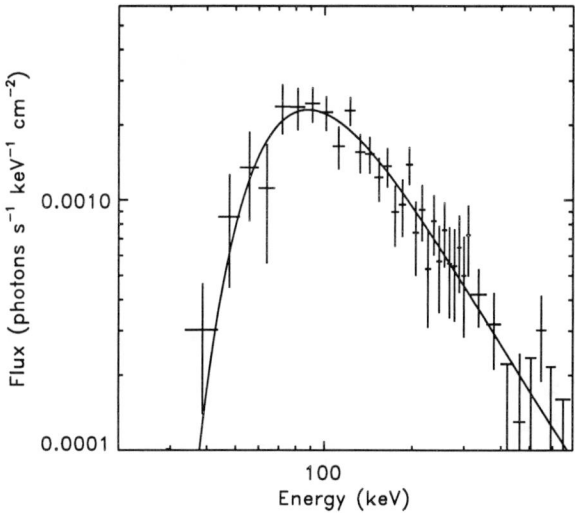

FIG. 2. Hard spectrum from GRB 940330, 0–6 s.

ANALYSIS

Figure 1 shows the count rate history for GRB 940330 in four broad energy bands from two BATSE Large Area Detectors (LADs). The burst triggered at t = 0 s due to the increase in flux in the 50–300 keV range. The observed flux is at the background rate for t < −10 s in all energy ranges. The interval from 0 s to 6 s is marked with vertical dotted lines in each energy range. It is clear that the flux in the 100–300 keV range has increased significantly in the 0–6s interval while the increase in the 25–50 keV range has been negligible.

A spectral fit to the high energy resolution data from LAD #6 for this interval is shown in figure 2. A flexible spectral form, consisting of a lognormal function with a FWHM that is a linear function of the log of the energy (9), is used that clearly reveals the low energy cutoff in this spectrum. Fits with other models, such as a doubly broken power law, produce the same general spectral shape. This continuum form is consistent with the data from four separate detectors of BATSE: two of the LADS and two of the smaller Spectroscopy Detectors.

Quite frequently, burst emission peaks evolve to softer continuum forms that persist while the peak intensity gradually diminishes. For example, a much softer spectrum is produced by GRB 940330 in the 75–77 s interval. A lognormal functional form is fit to this data (9) that produces an acceptable fit to the softer burst continuum, while allowing for the observed X-ray paucity (11).

Since the intensities of GRBs often exhibit complex patterns during emission, with multiple, rapidly changing peaks mixed together, it is reasonable to expect that their continua will overlap in some intervals. To see whether

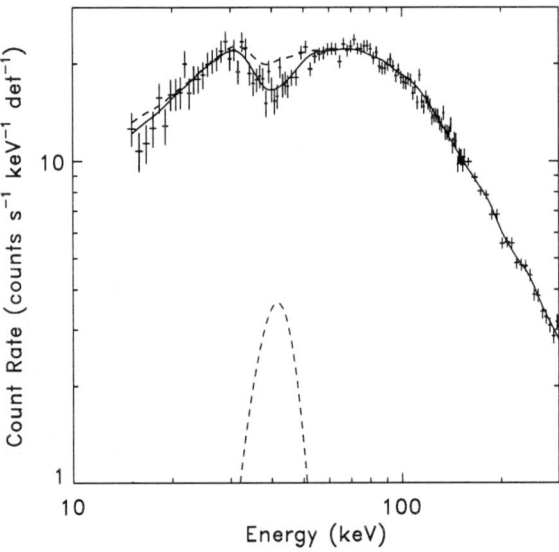

FIG. 3. Synthetic line from combined GRB 940330 continuum spectra.

two observed continuum forms could be combined to produce a spectrum with an apparent narrow line feature, simulated counts data were produced for a BATSE Spectroscopy Detector using the superposition of the two spectral forms that were fit to intervals 0–6 s and 75–77 s of GRB 940330. The two photon continua were summed and folded through a Spectroscopy Detector response matrix (13) and statistical fluctuations added. The BATSE Spectroscopy Detector is similar in performance and design to other detectors (5,11,12) that have observed absorption lines in bursts at 40 keV.

This simulated data set was then fit with a continuum model, and a continuum plus a Gaussian absorption line model, to assess the significance of an apparent line feature measured in this way. The continuum model used was a power law with two breaks like that used in some previous line evaluations (3). This continuum model alone yielded a χ^2 of 163.11 with 121 degrees of freedom, with a chance probability of 6.4×10^{-3}.

In Figure 3, a fit to the simulated data using the continuum model described above plus a gaussian absorption line is displayed as a counts spectrum in the energy range 14 to 300 keV. The line centroid is 41.3 ± 2.4 keV with a FWHM of 10.5 ± 5.5 keV, consistent with other narrow line features reported in this energy range (3). The addition of this Gaussian absorption line resulted in a χ^2 of 132.8 for 118 degrees of freedom, with a chance probability of 0.18. The F-test probability that the χ^2 improvement due to the addition of the line occurred by chance is 2.0×10^{-5}. Significant line features (i.e., at above the 3σ level) of various widths are obtained in about 50% of the simulated realizations of this spectral continua combination.

CONCLUSIONS

This analysis shows that the choice of the continuum model used in an absorption line significance analysis can dominate the apparent statistical significance of an identified line feature. In the case presented here, the choice of an inappropriate (albeit extremely flexible) continuum model transformed a 1σ fluctuation into a highly significant line detection. The impact of the continuum model choice does not have to be as dramatic as it is here to severely effect the interpretation of features found in GRB's, particularly since potentially significant line features have been reported in only a small fraction of burst spectra analyzed. It is clear that a thorough analysis of the burst spectral continuum behavior at all energies is necessary before the physical significance of features in burst spectra can be addressed.

REFERENCES

1. D. Q. Lamb, et al, Ap.J. **363**, 670 (1981)
2. C. A. Meegan, et al, Nature, **355**, no. 6356, p. 143 (1992)
3. E. E. Fenimore, et al, Ap.J.Lett., **335**, L71 (1988)
4. T. Murakami, et al., Nature, **335**, 234 (1988)
5. C. Barat, et al., Astron. Astophys. Suppl. Ser. 97, 43 (1993)
6. G. N. Pendleton, et al., NIMS A **364**, 567 (1995)
7. Mitrofanov et al., in Gamma-Ray Bursts: Observations, Analyses, and Theories, ed. C. Ho, R. Epstein, and E. Fenimore, 209 (1992)
8. G. N. Pendleton, et al, in Proceedings of Compton Gamma-Ray Observatory Symposium, ed. M. Friedlander, N. Gehrels, D. J. Macomb, 1040 (1993)
9. R. S. Mallozzi, et al., in Proceedings of the Second Huntsville Workshop on Gamma-Ray Bursts,ed. G. Fishman, J. Brainerd, K. Hurley, p. 308 (1994)
10. A. Chernenko, et al., Mon. Not. R. Astron. Soc. 274,361-368 (1995)
11. T. Murakami, in Gamma-Ray Bursts: Observations, Analyses, and Theories, ed. C. Ho, R. Epstein, and E. Fenimore, 239-248 (1992)
12. E. P. Mazets, & S. V. Golenetskii, Astrophysics and Space Science, **75**, No. 1, 47-81 (1981)
13. G. N. Pendleton, et al., NIMSa, **364**, p. 567 (1995)

Detailed Spectral Analysis Revealing Two Distinct Classes of Pulses in Gamma-Ray Bursts

Geoffrey N. Pendleton[1], William S. Paciesas[1],
Robert D. Preece[1], Michael S. Briggs[1]
Chryssa Kouveliotou[2], Charles A. Meegan[3]

[1] Department of Physics, Univ. of Alabama in Huntsville, Huntsville, AL 35899
[2] Universities Space Research Association, NASA/MSFC, Huntsville, AL 35812
[3] NASA/Marshall Space Flight Center, Huntsville, AL 35812

A model independent spectral analysis technique has been developed to study the internal spectral behavior of most of the gamma-ray bursts observed by BATSE, including the weaker ones. Spectral indices are calculated in the 20–100 keV, 50–300 keV, and E > 300 keV energy ranges with 64 ms time resolution data. A subset of bursts is identified that exhibits a marked lack of emission above 300 keV, and these bursts have luminosities about an order of magnitude lower than bursts with significant emission above 300 keV. They exhibit an effectively homogeneous intensity distribution as opposed to the rest of the burst population. In addition, evidence is presented indicating that both types of emission are common in many bursts, suggesting that a single source object is capable of generating both. These results strongly favor a source object that produces two different types of emission with varying degrees of superposition.

INTRODUCTION

We have developed a low energy resolution, model-independent spectral continuum analysis technique and applied it to BATSE Large Area Detector (LAD) data of a set of 882 gamma-ray bursts. These spectral results are derived from the LAD four channel discriminator data taken in burst trigger mode with 64 ms time resolution. The procedure converts the data from counts bin^{-1} s^{-1} to photons cm^{-2} bin^{-1} s^{-1}. These discriminator data are particularly convenient for performing continuum spectral analysis robustly on large sets of bursts, including the weakest ones observed by BATSE. The spectral analysis procedure is described in detail elsewhere (1,2).

Recently, a number of investigators have used the spectral properties of gamma-ray bursts to identify subclasses that exhibit distinct intensity distributions (3–5). Some burst subsets have been identified, using cuts on hardness ratios of the BATSE 100–300 keV fluence over the 50–100 keV fluence in ergs and counts, whose intensity distributions are nearly homogeneous (3–5). In some cases, information about the burst durations have been used to determine how the hardness ratio cut is applied (4,5). These analyses all show that subsets of bursts that are softer in the 50–300 keV range have $\langle V/V_{\max}\rangle$ values close to 0.5, the value expected for a homogeneous distribution.

Using the spectral database described above, we have isolated a subset of bursts identified by a marked lack of fluence above 300 keV which exhibits homogeneity as strongly as, or more strongly than, any burst subset identified by spectral cuts at lower energies. In addition, close examination of flux histories of gamma-ray bursts reveals that the type of emission present in bursts with a lack of fluence above 300 keV (hereafter referred to as No-High-Energy or NHE bursts) is often clearly present during particular intervals of bursts that do exhibit significant higher energy fluence for much of their duration. (bursts referred to hereafter as High-Energy or HE bursts). This observation leads one to hypothesize that a single astrophysical object is responsible for both types of emission and that sometimes only the less luminous emission is observed.

DEFINITION AND COMPARISON OF HE AND NHE BURSTS

We can establish a correspondence between an effective power law index α (for the energy bins of interest) and an observed hardness ratio using the following expression,

$$\frac{\int_{300}^{6500} E^\alpha\, dE}{\int_{100}^{300} E^\alpha\, dE} = \frac{P_4}{P_3}. \tag{1}$$

Here P_4 is the Fluence in Photons cm^{-2} in the $E > 300$ keV bin and P_3 is the Fluence in the 100–300 keV bin. In this case the high energy integral energy range covered 300 keV to 6500 keV. A spectral index cut of $\alpha < -5.5$ was used to isolate the NHE bursts.

Figure 1 displays the integral intensity distributions of the HE bursts (thick histograms) and NHE bursts (thin histograms) on the 64 ms timescale. The dashed, slanted lines tangent to each distribution represent the $-3/2$ power law shape of a homogeneous intensity distribution. It is qualitatively clear that the NHE distributions appear to be homogeneous to lower intensities than the HE bursts, although they deviate from homogeneity near threshold on the longer BATSE trigger timescales. To explore the shape of these distributions, we use an expression for the first moment of the intensity distribution expressed in physical units, Photons cm^{-2} sec^{-1}, similar to one developed by Horack & Emslie (7).

$$V_\mathrm{p}/V_\mathrm{plim} = (P/P\mathrm{lim})^{-3/2} \tag{2}$$

In this case, we add the constraint that the intensity distribution should be measured only down to flux levels, plim, where the instrument sensitivity

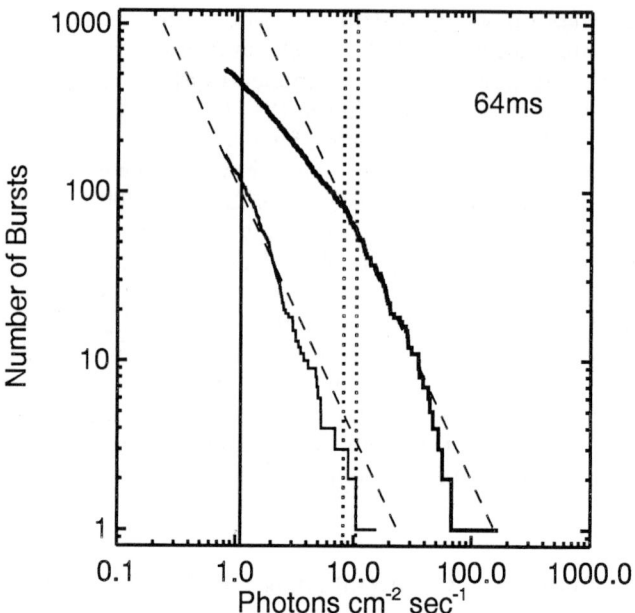

FIG. 1. Peak flux intensity distributions for the HE (thick histogram) and NHE bursts (thin histogram) on the 64 ms timescale.

starts to significantly effect the shape of the observed intensity distribution, and not below that intensity level. If the intensity distribution above plim is homogeneous, then $\langle V_p/V_{plim}\rangle = 0.5$. For this analysis, a plim value of 1.2 photons cm^{-2} s^{-1} has been chosen so that the exposure for the weakest bursts is at least 70% of the exposure for the brightest bursts. Above this threshold there were 103 NHE and 406 HE bursts. The $\langle V_p/V_{plim}\rangle$ values were 0.54 ± 0.026 and 0.34 ± 0.014 for these sets of bursts, respectively.

At this point it is useful to develop a method for defining the range over which the observed intensity distributions break from homogeneity. Given the relatively low statistics for the burst sets in the homogenous intensity range, this definition will be somewhat coarse. However, the method is adequate for out present purposes. We calculate $\langle V_p/V_{plim}\rangle$ as a function of intensity for each distribution. We pick as the upper edge of our homogeneity break range the lowest intensity where $\langle V_p/V_{plim}\rangle \geq 0.5$. The lower bound is the lowest intensity where $\langle V_p/V_{plim}\rangle + 1\sigma \geq 0.5$. For the HE burst distributions, these ranges are shown by the dotted vertical lines in figure 1. On the 64 ms timescale the NHE break never even starts, although studies of the intensity distributions on the other BATSE timescales indicates that the NHE intensity distribution breaks around the BATSE threshold (8). If we make the assumption that the homogeneous parts of the intensity distributions represent bursts that occupy the same region of space (i.e., that the distance at which deviation from homogeneity becomes apparent is the same for all bursts) then we can make first order estimates of the relative luminosities and source densities of the HE and NHE Bursts. Using the HE break range specified in figure 1, we obtain an HE/NHE luminosity ratio range of 8.8–10.0 and a source density ratio range of 1.2–1.7. These results indicate that the NHE emission is about an order of magnitude weaker than the HE emission, and the the NHE bursts may be more numerous per unit volume. Systematic effects that could influence these results are studied in detail elsewhere (8).

NHE PEAKS WITHIN HE BURSTS

Closer inspection of some of the bursts classified as HE due to their total emission reveals that they frequently contain peaks that lack high-energy emission (called NHE peaks), as well as peaks with high-energy emission (HE peaks). An analysis technique was developed to identify peak emission intervals within bursts, so that spectral analysis could be performed on the data within each interval. The technique divides the entire burst into intervals so that most of the emission within an interval is due to an emission peak that started inside that interval. It is described in detail elsewhere (8).

We compared the properties of NHE peaks within HE bursts and NHE peaks within NHE bursts and found that they were quite similar (8). Comparison of the HE and NHE peaks within HE bursts revealed more differences between these two types of peaks from the same bursts than between the NHE peaks from the distinct NHE and HE bursts. These results indicate that the NHE emission in both kinds of burst are produced by the same mechanism and hence the NHE and HE bursts originate from the same source.

We performed this peak decomposition analysis on 750 bursts. The similarity of the NHE peaks amongst the bursts leads us to examine the intensity distribution of NHE peaks within HE bursts.

In the analysis presented here we have calculated the peak flux of the HE bursts, using the flux data only from those peak intervals that contain NHE emission exclusively. In figure 2 the upper thick histogram shows the intensity distribution for the HE bursts that were analyzed using the peak decomposition technique. The upper, thin, solid histogram that matches this thick one at higher intensities, but drops below it at lower intensities, represents those bursts that could be separated into two or more peaks. As we get to lower intensities bursts are not decomposed into separate peaks as effectively.

The lower, solid, thick histogram shows the 64 ms intensity distribution for the NHE bursts from the set analyzed here. The lower, solid, thin histogram in this panel is the intensity distribution for the HE bursts calculated using only the NHE peaks within those bursts with $\langle V_\mathrm{p}/V_\mathrm{plim} \rangle = 0.48 \pm 0.041$ This result supports the argument that the NHE peaks in both HE and NHE bursts come from the same source type. In figure 2, the homogeneity break ranges for the intensity distributions are delineated in the same fashion as they are in figure 1. The HE/NHE luminosity derived from the break ranges spans

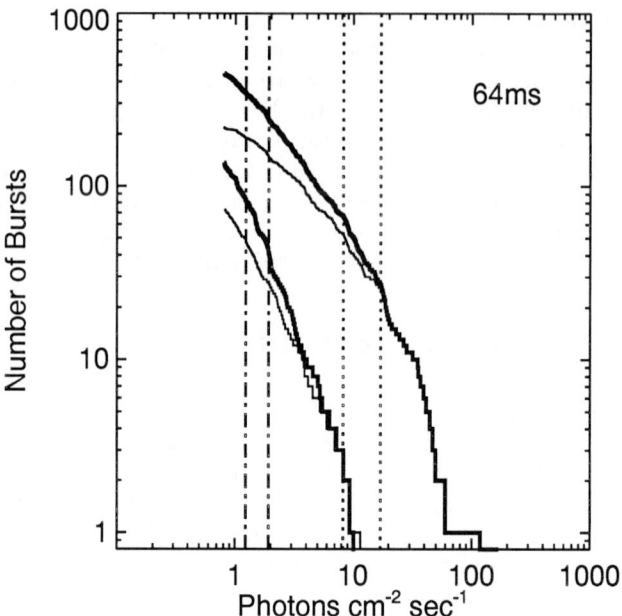

FIG. 2. Peak flux (64 ms) intensity distributions for the HE peaks (upper pair of histograms) and NHE peaks (lower pair) from within the HE bursts. Thick histograms: all bursts in the sets; thin histograms: those bursts that could be separated into two or more peaks.

the range from 4.3–14.0. If we average the intensities of the twenty brightest peaks of each type and take the HE/NHE ratio, we obtain 8.3. All the analyses support a difference between the HE and NHE luminosities of about an order of magnitude. The most likely source for these observations is one capable of producing both the brighter HE emission and the fainter NHE emission, possibly with different beaming angles for the two types of emission.

REFERENCES

1. G. N. Pendleton, et al., ApJ **431**, 416 (1994).
2. G. N. Pendleton, et al., ApJ **464**, 606 (1996).
3. G. Pizzichini, Cospar Meeting, Amburg-Germany (1994).
4. B. M. Belli, Astrophys. Sp. Sci. **231**, 43 (1995).
5. C. Kouveliotou, et al., these proceedings (1996).
6. G. Pizzichini, these proceedings (1996).
7. J. M. Horack & A. G. Emslie, ApJ **428**, 620 (1994).
8. G. N. Pendleton, et al., in preparation (1996).

Are There MeV Gamma-Ray Bursts?

Tsvi Piran[1,3] and Ramesh Narayan[2,3]

1. The Racah Institute for Physics, The Hebrew University, Jerusalem, Israel
2. Harvard-Smithsonian Center for Astrophysics, Cambridge, MA, U.S.A.
3. ITP, UCSB, Santa Barbara,, U.S.A.

It is often stated that gamma-ray bursts (GRBs) have typical energies of several hundred keV, where the typical energy may be characterized by the hardness H, the photon energy corresponding to the peak of νF_ν. Among the 54 BATSE bursts analyzed by Band et. al. more than half have 100 keV $< H <$ 400 keV. Is the narrow range of H a real feature of GRBs or is it due to an observational bias? We consider the possibility that bursts of a given bolometric luminosity occur with a distribution: $p(H) d \log H \propto H^\gamma d \log H$. We model the detection efficiency of BATSE as a function of H and calculate the expected distribution of H in the observed sample for various values of γ. The Band sample shows a paucity of soft (X-ray) bursts, which may be real. However, because the detection efficiency of BATSE falls steeply with increasing H, the paucity of hard bursts need not be real. We find that the observed sample is consistent with a distribution above $H = 100$ keV with $\gamma \approx 0$ (constant numbers of GRBs per decade of hardness) or even $\gamma = 0.5$ (increasing numbers with increasing hardness). Thus, we suggest that a large population of unobserved hard gamma-ray bursts may exist. It is important to extend the present analysis to a larger sample of BATSE bursts and to include the OSSE and COMPTEL limits. If the full sample is consistent with $\gamma \gtrsim 0$, then it would be interesting to look for MeV bursts in the future.

One striking feature that is common to all gamma-ray bursts (GRBs) is the fact that most of the observed photons correspond to low energy gamma-rays, with energies of a few tens to few hundreds of keV. While other features of the bursts, in particular the temporal structure, vary significantly from one burst to another, this feature seems to be quite invariant. One wonders, therefore, whether this is a clue to the nature of GRBs — a phenomenon that theorists should strive to explain — or if it is just the consequence of an observational bias against detection of harder or softer bursts.

Band et al. (1) have analyzed 54 strong GRBs (hereafter the Band sample), fitting the spectra using a four parameter function:

$$N(E) = \begin{cases} (\frac{E}{100 \text{ keV}})^\alpha \exp(-\frac{E}{E_0}) & \text{for } (\alpha - \beta)E < E_0 \,; \\ [\frac{(\alpha - \beta)E_0}{100 \text{ keV}}]^{(\alpha - \beta)} \exp(\beta - \alpha)(\frac{E}{100 \text{ keV}})^\beta, & \text{for } E > (\alpha - \beta)E_0. \end{cases}$$

This function, which provides a good fit to most of the observed spectra,

© 1996 American Institute of Physics

is characterized by two power laws joined smoothly at a break energy $H \equiv (\alpha - \beta)E_0$. For most observed values of α and β, $\nu F_\nu \propto E^2 N(E)$ rises below the break and decreases above it. The break energy H is thus the "typical" energy of the observed radiation, in the sense that this is where the source emits the bulk of its luminosity. H is correlated with, but not equal to, the hardness ratio which is commonly used in analyzing BATSE GRBs, namely the ratio of photons observed in channel 3 to those observed in channel 2.

Figure 1 shows the distribution of observed values of H in the Band sample. There is a clear and marked maximum in the distribution for $H \sim 200$ keV, and most of the bursts lie over the range 100 keV $< H <$ 400 keV. Should we, therefore, conclude that most GRBs have hardnesses around 100–400 keV?

To answer this question we have used a simple model of the BATSE detector to calculate the *expected* hardness distribution of GRBs detected by BATSE for various assumed *intrinsic* hardness distributions. We have then compared the expected and observed distributions to see which intrinsic distributions are consistent with the data and which are not.

We assume that the *intrinsic* hardness distribution is as follows:

- All GRBs are adequately described by a Band spectrum, characterized by the parameters, α, β, and H.

- The number of GRBs in a logarithmic interval, $d \log H$, varies as: $p(H) d \log H = p_0 H^\gamma d \log H$.

- We also allow for the possibility that the intrinsic luminosity, L, is correlated with H: $L = L_0 (H/H_0)^\xi$.

We now calculate, for given γ and ξ, the distribution of H values of bursts seen by a detector like BATSE. We make use of the fact that BATSE triggers on counts in the second and third energy channels, with photon energies between 50 keV and 300 keV. We calculate, therefore, how many bursts with a given H yield a count rate in the 50 keV to 300 keV band that is larger than a threshold, $C_{\min}$. (We ignore the fact that BATSE's counts in these channels may correspond to higher energy photons and that the intrinsic spectrum is related to the observed counts via the DRM, (2)). The count rate depends on α and β, in addition to the above mentioned dependence on ξ and γ. We use the average values given by Band et al. (1): $\bar{\alpha} = -0.73$ and $\bar{\beta} = -2.22$. For reasons that will become clear shortly we also use $\tilde{\alpha} = -0.41$, which is the average value of α for GRBs with 50 keV $< H <$ 300 keV.

The count rate in the 50–300 keV energy band from a burst at a distance D is:

$$C = \frac{L(H) A}{4\pi D^2} \mathcal{C}(50, 300, \alpha, \beta, H) = \left[\frac{(L_0/H_0^\xi) A}{4\pi D^2}\right] \mathcal{C}(50, 300, \alpha, \beta, H) H^\xi ,$$

where A is the area of the detector. The quantity $\mathcal{C}$ represents the number of counts for every erg of energy incident on the detector (counts/ergs), and is easily calculated once the shape of the spectrum (α, β, H) and the detector limits (50, 300 keV) are given. The second equality follows from the assumed correlation between L and H. For simplicity we use here and in the following a Newtonian geometry and ignore cosmological effects. This is justified since the Band sample is composed mostly of strong bursts for which cosmological redshift and spatial curvature effects are small. A detailed discussion that includes these effects will be published elsewhere (3).

The number of bursts with hardness H detected by a detector with a limiting sensitivity $C_{\min}$ is:

$$N(H) = \left[\frac{4\pi}{3} \left(\frac{(L_0/H_0^\xi)A}{4\pi C_{\min}} \right)^{3/2} p_0 \right] \mathcal{C}^{3/2}(50, 300, \alpha, \beta, H) H^{3\xi/2 + \gamma} .$$

Figure 1 depicts $N(H)$, the expected number of detected bursts in a logarithmic interval of H, as a function of $\log(H)$. We have shown the calculated curves for $\alpha = \bar{\alpha}$ and $\alpha = \tilde{\alpha}$ and for $\gamma + 3\xi/2 = 0$ and $\gamma + 3\xi/2 = 0.5$; in all cases, we set $\beta = \bar{\beta}$. The observed distribution corresponding to the Band sample, with statistical error bars, is also shown on the same plot.

A comparison of the four calculated (or expected) curves with the observed distribution reveals immediately a paucity of soft bursts. In all four cases, the number of soft bursts we would have expected to see is significantly larger than the number actually observed. Therefore, unless BATSE has an unexpectedly large selection bias against detecting soft photons (i.e. significantly poorer sensitivity than we have assumed in our model), we conclude that the lower cut-off in the observed distribution of hardnesses is a real phenomenon. That is, there really are very few soft GRBs, and $N(H)$ does have a lower cutoff.

The story is, however, very different for larger values of H. The data show very small numbers of hard bursts; for instance, only two bursts out of 54 have $H > 1$ MeV. Nevertheless, this does *not* mean that GRBs intrinsically cut-off for hardnesses above 1 MeV. For instance, our theoretical model with $\xi = 0$ and $\gamma = 0$, with equal number of bursts per logarithmic hardness interval, actually predicts that even fewer bursts should have been observed. The best fit to the data is obtained with $\gamma + 3\xi/2 = 0.5$ which corresponds to a burst population with an increasing number of bursts with hardness (for $\xi = 0$) or to an equal number of bursts per logarithmic hardness interval ($\gamma = 0$) but with harder bursts being more luminous ($\xi = 1/3$).

The interpretation of the result is quite simple. There is an observational bias against detecting bursts with $H \gtrsim 500$ keV by current detectors. Two factors operate. For bursts with a fixed luminosity, harder bursts have fewer photons. This makes the detection of harder bursts more difficult in any detector that is triggered by photon counts. The decrease in sensitivity in BATSE is even more severe since BATSE triggers on photons in the 50 keV to 300 keV range and as the bursts becomes harder most of the emitted photons are further and further away from this energy range.

The last point depends, of course, on the power law index α in the low energy range. It suggests that hard bursts that are detected by BATSE will have values of α lower than average. Fig. 2 depicts the distribution of α values in different hardness regimes, and shows the effect clearly. Among the

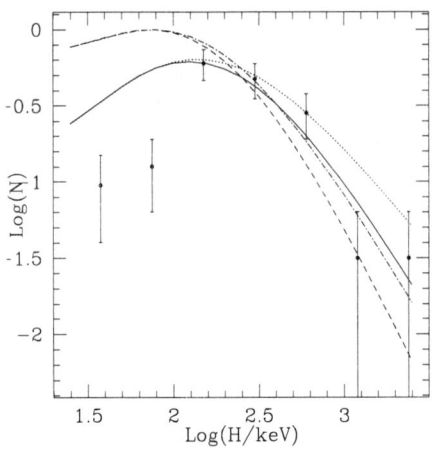

FIG. 1. Observed and expected numbers of detected bursts per interval of $\log(H)$. The the error bars represent statistical errors. The best fit (solid line) corresponds to $\alpha = \tilde{\alpha} = -0.41$, $\gamma + 3\xi/2 = 0.5$. Other curves are: $\alpha = \bar{\alpha} = -0.7$, $\gamma + 3\xi/2 = 0.5$ (dotted line), $\alpha = \tilde{\alpha} = -0.41$, $\gamma + 3\xi/2 = 0$ (dashed line), and $\alpha = \bar{\alpha} = -0.7$, $\gamma + 3\xi/2 = 0$ (dashed-dotted line).

54 bursts in the Band sample, hard bursts do have significantly more negative values of α. It is for this reason that we have introduced $\tilde{\alpha}$, the average α value over the intermediate hardness range, 50 keV $< H <$ 300 keV. If the intrinsic α distribution is independent of the hardness than $\tilde{\alpha}$ is a better estimate of the average α than $\bar{\alpha}$.

Figure 2, by itself, without any of the theoretical arguments presented earlier, suggests the existence of the selection effect we have discussed. There are two ways of interpreting the evidence in this plot. One could say that the plot reflects the true distribution of burst properties and that for some reason α happens to be correlated with H in the particular manner seen in the data. This is very ad hoc. The alternative, which we find much more attractive, is to say that BATSE has difficulty detecting hard bursts, and finds it particularly difficult to detect hard bursts with less negative, or positive, values of α; that is, such bursts do exist, but BATSE misses them. If we accept the latter explanation, then it means that BATSE is certainly missing at least some part of the hard population, namely those bursts with less negative (or positive) α. It is then but a small step to accept the entire argument presented earlier.

Our main result then is the following. When observational selection effects are taken into account, the two hard bursts with $H > 1$ MeV in the Band sample are but "the tip of the iceberg," and represent a large number of undetected hard GRBs. The observed distribution of hardness is, in fact, consistent with the possibility that *most* GRBs are harder than the bursts detected by BATSE, and that there is a large population of mostly undetected GRBs whose typical photon energy is in the MeV, or even harder, range.

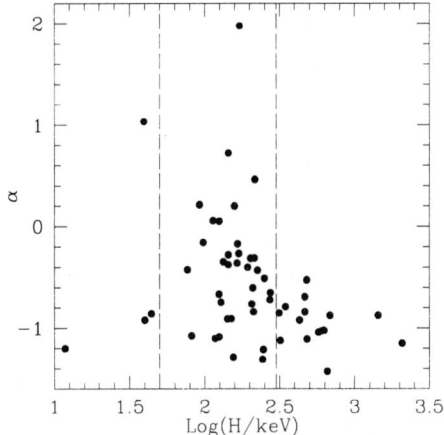

FIG. 2. The low energy spectral slope α vs. the hardness H in the Band sample. Hard bursts tend to have more negative α, which we interpret as evidence for the selection effect discussed in the paper. The vertical lines mark 50 keV and 300 keV.

So far we have compared the theoretical predictions to the hardness distribution in the Band sample. It will be interesting to perform the same analysis on a larger sub-sample (see (3)). It will also be of great interest to perform a similar analysis on data from other GRB detectors. One intriguing possibility is to search in BATSE's raw channel 4 data for untriggered events, which might be some of the missing hard bursts. This would be a complementary search to the soft GRB search reported by Kommers et. al. (4). In the meantime, before we know whether hard bursts exist or not, we should be very cautious about performing correlations between various characteristics of the observed bursts and their hardness parameters. The data are biased, as far as H is concerned, and any correlation seen (e.g., the correlation between α and H in Fig. 2) might reflect nothing more than selection effects in the data.

Acknowledgments. We thank B. Paczński for remarks that motivated this research. The research was supported by a NASA grant NAG5-1904 to Harvard University by a Basic research grant to the Hebrew University and by NSF grant PHY94-07194 to the ITP.

REFERENCES

1. D. Band, et. al., Ap. J. **413**, 281 (1993).
2. G. Pendleton, et. al., preprint (1995).
3. E. Cohen, R. Narayan, & T. Piran, in preparation (1996).
4. J. M. Kommers, W. H. G. Lewin, J. van Paradijs, C. Kouveliotou & G. J. Fishman, these proceedings (1996).

BATSE Observations of GRB Spectra at Low Energies

R. D. Preece, M. S. Briggs, G. N. Pendleton, W. S. Paciesas[*]
J. L. Matteson, D. L. Band[†] and C. A. Meegan[‡]

[*]*Dept. of Physics, University of Alabama in Huntsville, AL 35899*
[†]*CASS, Univ. of California at San Diego, La Jolla, CA 92093*
[‡]*NASA / Marshall Space Flight Center, Huntsville, AL 35812*

The BATSE spectroscopy detectors were designed to detect photons with energies as low as 5 keV. For gamma-ray burst studies, the discriminators provide a broad-band energy bin available for spectral fitting, which lies below the burst high resolution spectroscopy data. The energy range of this bin can be as low as 5–10 keV depending on the gain setting of the detector. We have surveyed roughly 100 strong bursts which satisfy the following criteria: 1) sufficient signal-to-noise, 2) a detector observing the source with a viewing angle less than 60° and 3) a detector gain sufficient to bring the low-energy data coverage down to the 5 – 20 keV range of interest. Spectra integrated over the bright portions were fit with one of several spectral forms and the agreement between the fitted model and the low-energy data is determined. We find evidence that x-ray excesses exist in $\approx$ 10% of the bursts. In most cases, the excess can be accounted for by a spectral model with upward curvature, such as has been proposed by Brainerd, or by a separate spectral component.

INTRODUCTION

The BATSE Spectroscopy Detectors (SDs) were designed to observe photons as low as 5 kev, with a low-Z thin Beryllium window in the face of each detector. In addition, the detector electronics include fast discriminators which accumulate counts below the lower-level discriminator that determines the threshold energy of the usual burst spectroscopy data (called SHERB). Depending upon detector gain, the energies covered by this single data channel (DISCSP1) can be as low as 5 – 10 keV. The actual energy thresholds of the discriminators has had to be determined after the instrument was launched. Once calibrated, the low energy datum can be included in a joint fit with the SHERB data, extending spectral coverage. We illustrate in Fig. 1 the accuracy of our calibration technique with a joint fit to solar flare data. The DISCSP1 data point lies within 2σ of the extrapolation down in energy of the thermal component of the flare's emission. Because small changes in the energy coverage translate into significant changes in the agreement of the data

© 1996 American Institute of Physics

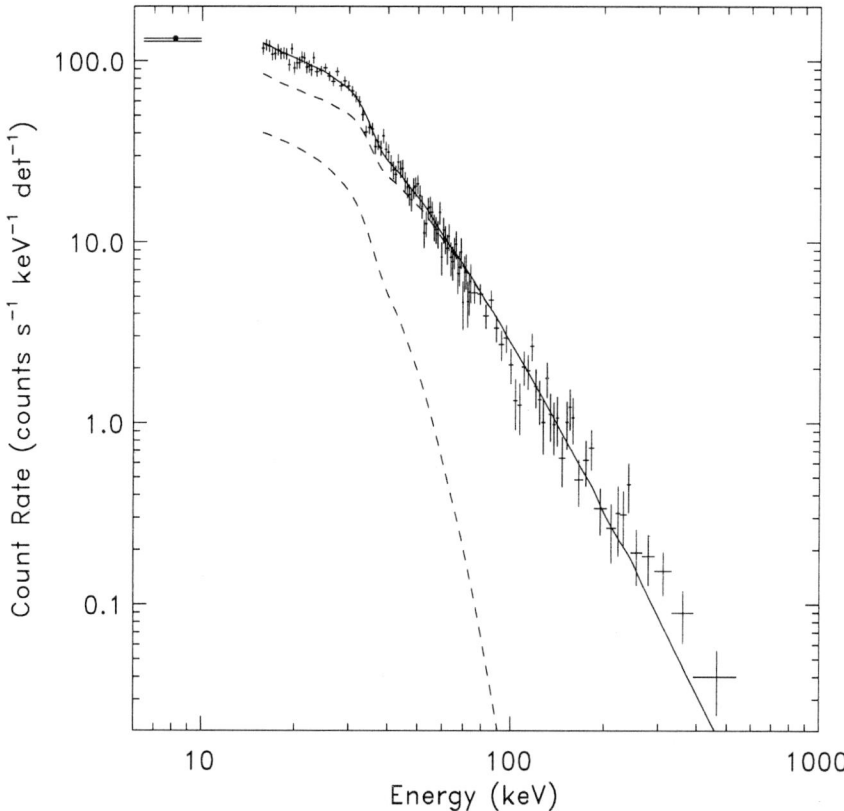

FIG. 1. Count spectrum showing a joint fit to counts accumulated in DISCSP1 and SHERB during the first 6.14 s after the trigger of the June 26, 1992 solar flare at 1715 UT. Both sets of data were accumulated by SD#0, and the DISCSP1 point is indicated by a dot. The two separate components of the fit, indicated by dashed lines, are a −1.9 spectral index power law and a 16 keV optically-thin thermal bremstrahlung (OTTB).

with the model, this demonstrates the accuracy of the energy calibration of DISCSP1.

The bursts included in our study had to satisfy a number of criteria. First, we require the fluence to be $\gtrsim 2 \times 10^{-5}$ erg cm^{-2}. The detector gain must be high enough to put the lowest energy channel into the range of interest: 5 − 20 keV. Finally, to avoid attenuation of low energy counts by the Be window, we require the source-to-detector-axis angle to be less than 60°. Roughly 100 bursts from the BATSE mission to date satisfy these criteria. For each of these bursts, we determine the time interval with the largest signal-to-noise ratio in the selected detector, over the entire usable energy range of the SHERB data ($\approx$ 20 keV − 2 MeV). We then sum up the individual spectra over this entire

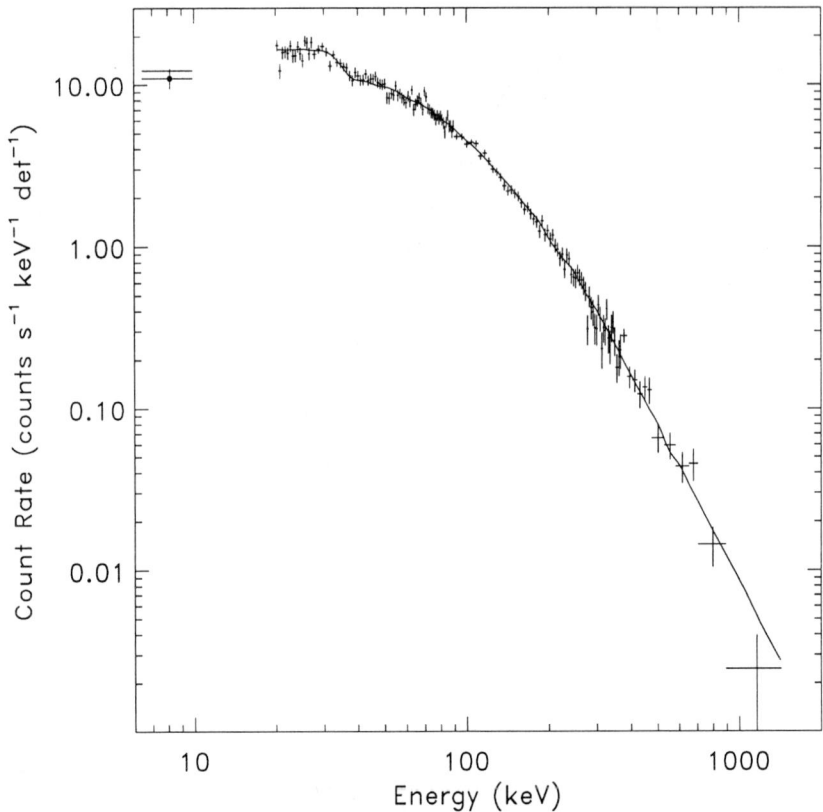

FIG. 2. Spectrum showing a joint fit to counts accumulated in DISCSP1 and SHERB during the interval 15.7 – 70.3 s of GRB941020 in SD#2. The DISCSP1 point is indicated by a dot. The model is indicated by a solid curve, except at the low-energy point, where it is a horizontal line.

time interval to obtain one spectrum in each of the SHERB and DISCSP data types. These are then jointly fit, using standard forward-folding techniques, to the canonical burst spectral form of Band (1), consisting of two smoothly-joined power-law segments. The model is concave downwards over the entire energy range. In most cases, there is good agreement between the low-energy data point and the model, jointly fitted to all the data, as seen in Fig. 2.

RESULTS

For each burst, we measure the agreement between the fitted model and the low-energy data in terms of their difference in units of sigma. The agreement for the example shown in Fig. 2 is 0.2 σ. In 10 – 15% of the bursts studied,

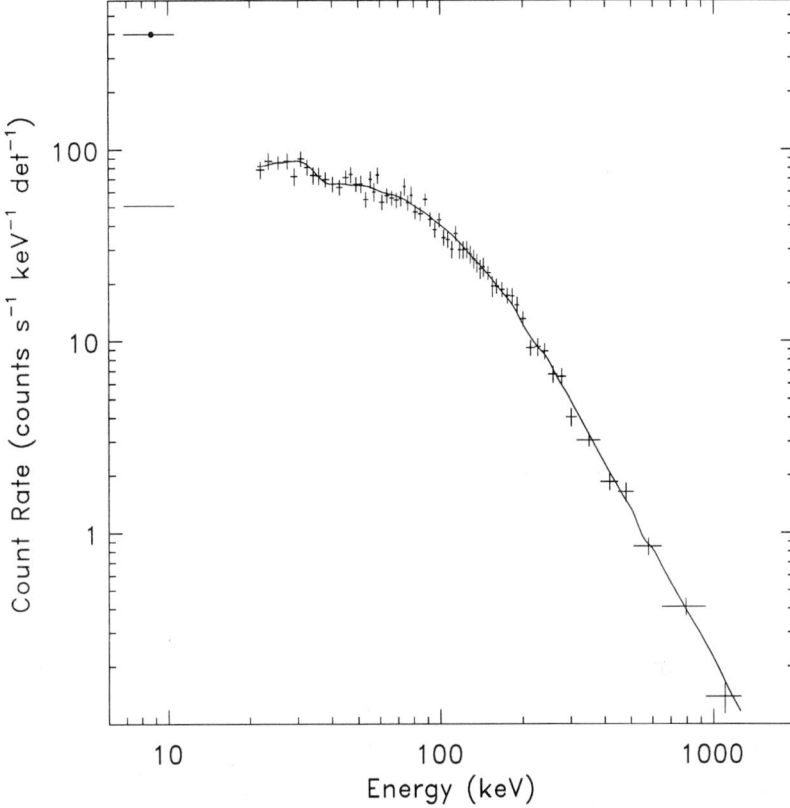

FIG. 3. Joint fit to counts accumulated in DISCSP1 and SHERB during the interval 7.6 – 8.8 s of GRB920517, in SD#7. The sum of the GRB spectral form and an OTTB model were used. The summed model is indicated by a solid curve, except at the low-energy point, where the fitted GRB model counts are indicated by a horizontal line.

the data and model disagreed at a significance greater than 5 σ. In each of these cases, there is an excess in the count rate in the low-energy data with respect to the fitted model. No bursts were observed to have deficits with respect to the fitted model more significant than $\approx$ 3 σ.

An excess at low energies is a possible indication for an additional spectral component in those bursts where it occurs. For example, GRB920517 has a 10σ excess between 6.8 and 10.5 keV, and the model fit is quite poor. With an additional OTTB spectral component, with a fitted kT = 1.0 $\pm$ 13 keV, the fit is considerably improved (see Fig. 3). Note that with only one data point, the form of the low-energy spectral component cannot be determined. Even assuming that OTTB is the appropriate model for the excess, the temperature is very poorly determined. A turn up in the spectrum at low energies is

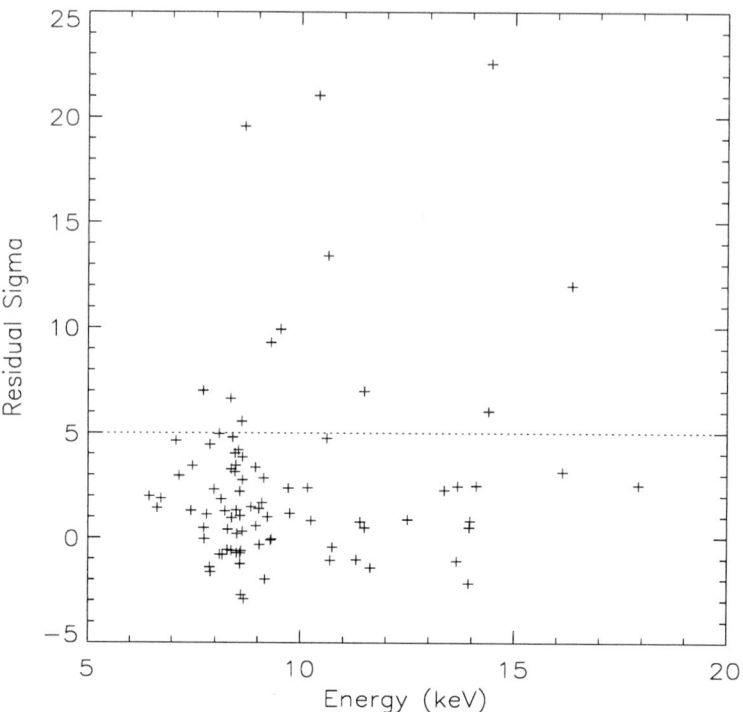

FIG. 4. Sigma Residuals (difference between observed counts and model, divided by the model error) of fitted DISCSP1 data for 86 BATSE bursts versus the DISCSP1 average energy.

predicted by at least one spectral model, which results in acceptable fits (2,3). The agreement between the observed counts and the fitted model in units of sigma is presented in Fig. 4, for a sample of 85 BATSE bursts (4). While most of the sample is consistent with the low-energy data agreeing with the fitted model, $\sim$ 10% have an excess significant at greater than the 5σ level.

REFERENCES

1. D. Band et al., ApJ **413**, 281 (1993).
2. J. Brainerd, ApJ **428**, 21 (1994).
3. J. Brainerd, these proceedings (1995).
4. R. Preece et al., ApJ **473**, in press (1996).

Time-resolved Spectroscopy with the BATSE Large Area Detectors: High-Energy Behavior

R. D. Preece, M. S. Briggs, G. N. Pendleton, W. S. Paciesas*
D. L. Band, L. A. Ford[†] and C. Kouveliotou[‡]

*Dept. of Physics, University of Alabama in Huntsville, AL 35899
[†]CASS, Univ. of California at San Diego, La Jolla, CA 92093
[‡]USRA / Marshall Space Flight Center, Huntsville, AL 35812

For burst spectral continuum studies, the BATSE Large Area Detectors (LADs) have better count statistics over their entire energy range than the Spectroscopy Detectors (SDs). The LADs also have moderate energy resolution, which means that most continuum spectral models will be well-determined over nearly 2 decades in energy. With the SD data, Ford et al. have shown that bursts exhibit a variety of evolutionary behavior in E_{peak}, the energy of the peak in $\nu \mathcal{F}_\nu$. By using LAD data for bright bursts such as those studied by Ford et al., we can track the evolution of any spectral parameter with finer time resolution and we can extend the analysis to fainter events, for a sample size of 110. Here, we concentrate upon the time evolution of the high-energy power law component (E^β) above E_{peak}, when it can be determined. The one noticeable feature from the survey is a negative correlation of high-energy power law index (steepening of spectra) with intensity during bright peaks of 21 of the bursts. In 25 of the others, the power-law evolution generally follows the same hard-to-soft pattern throughout a burst as the other spectral parameters. The absence of a hard power-law component in a group of 7 'super-soft' bursts indicates one extreme of the distribution of power-law indices across the entire sample.

INTRODUCTION

The BATSE Large Area Detectors (LADs) have 16 times the collecting area of the Spectroscopy Detectors (SDs), with moderate energy resolution (18% at 511 keV versus 8% for the SDs). The High Energy Resolution Burst (HERB) data cover the approximate energy range 20 – 2000 keV with 128 data channels in time-to-spill accumulations as short as 128 ms. Previous analyses of BATSE spectral data has focussed on SD data for bright bursts to take advantage of their good energy resolution (1,2). However, with a larger collecting area, the LADs are suited to do high-time resolution continuum studies on a larger set of bursts than is possible with the SDs. In this paper,

© 1996 American Institute of Physics

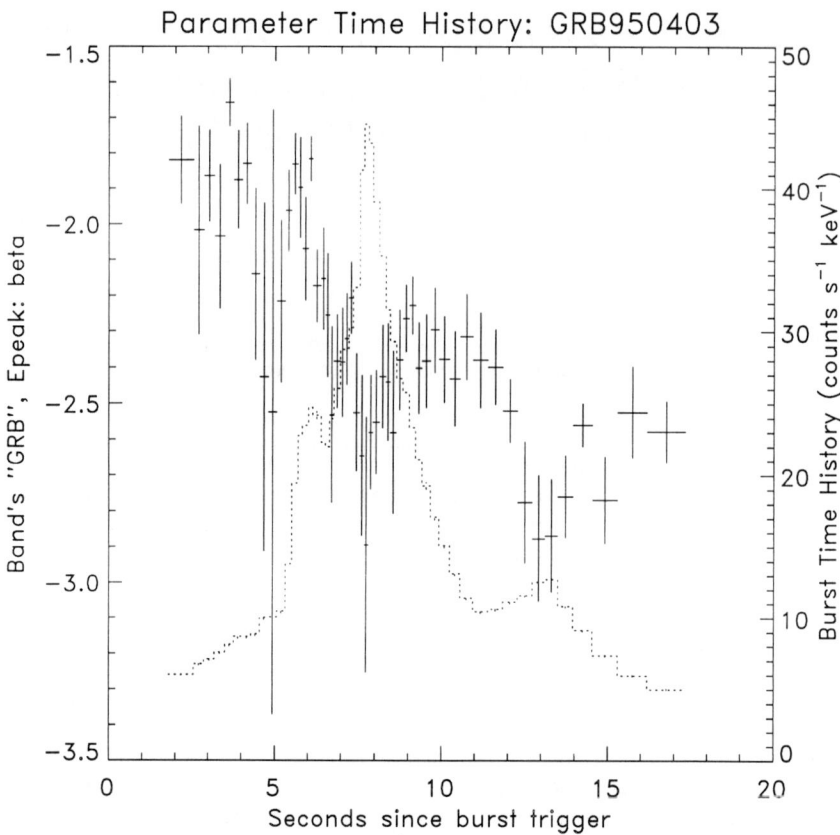

FIG. 1. Time history for GRB950403 in LAD #3 of fitted values of the high-energy power law index β plotted over the count rate history (dashed lines). The evolution of β shows an anti-correlation with intensity for peaks at 8 and 13 s after the trigger (at t = 0).

we briefly discuss the results of an analysis of LAD spectral data for 110 medium-to-bright bursts.

METHODOLOGY

Bursts are selected for inclusion in our sample by their total fluence. This ensures both that each spectrum examined will have enough signal that a meaningful fit can be made and that there will be several spectra in the burst with which to track the spectral evolution. The required fluence for these observations is greater than 2×10^5 ergs^{-1} cm^{-2}. Background-subtracted spectra from each burst are binned together to obtain a signal-to-noise ratio of 45 or better, over the entire usable energy range, which is typically 28 – 2000 keV.

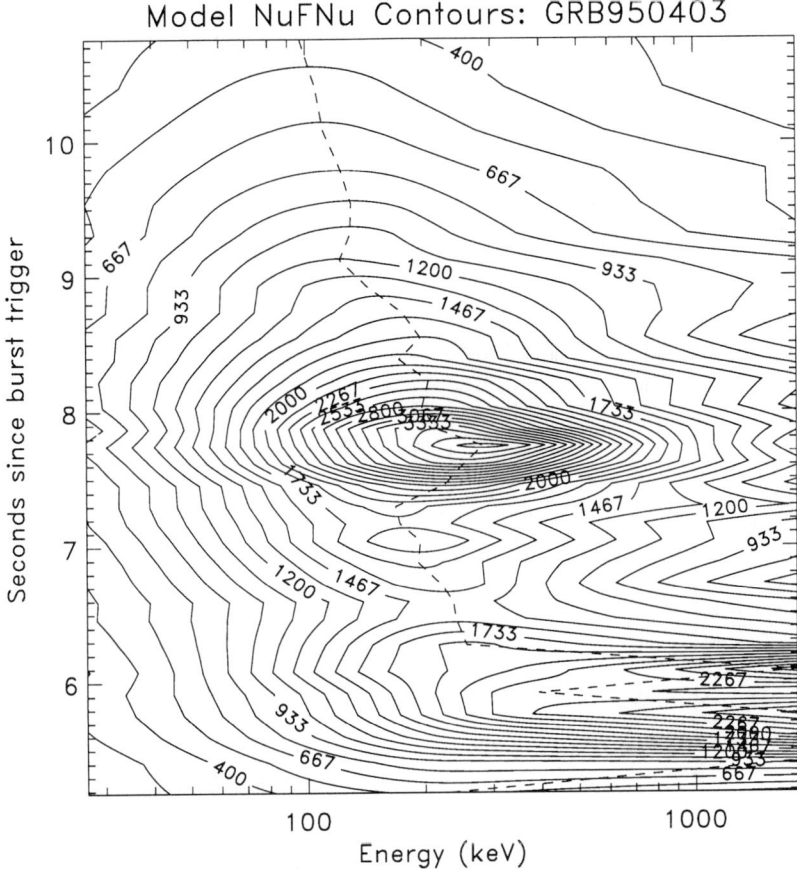

FIG. 2. Topographical map of GRB950403, consisting of the $\nu \mathcal{F}_\nu$ intensity contours of the fitted spectral model as it evolves in time. A dotted line follows the time evolution of E_{peak}. The peak at 8 s is considerably steeper on the high-energy side than the rest of the burst, due to the dip in β.

Each spectrum is fitted with one of two models: Band's GRB spectral form (1), and its restricted form in the absence of a high-energy power law ('COMP': similar to unsaturated inverse-Comptonized thermal emission). The GRB form consists of two smoothly-joined power-law segments and is parameterized in terms of the peak energy in $\nu \mathcal{F}_\nu$, E_{peak}, the low-energy power-law index, α, and the high-energy power-law index, β. The COMP model is similarly parameterized by E_{peak} and α. In the GRB model, E_{peak} is only defined where $\beta < -2$, otherwise, it is merely the energy where the high-energy power law joins to the curved portion of the model at lower energies.

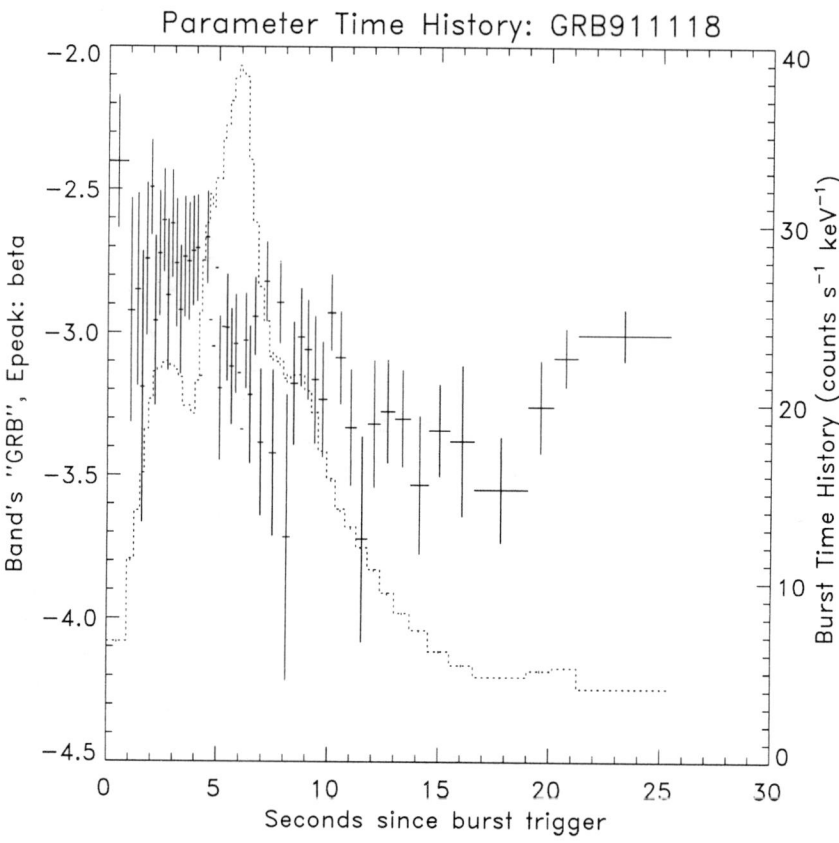

FIG. 3. Time history of fitted values of the high-energy power law index β plotted over the count rate history (dashed lines) for GRB911118. The general trend is from hard to soft.

OBSERVATIONS

Out of the 110 bursts surveyed, the majority (57) have behavior which is consistent with the fitted GRB model having a single high-energy spectral index ($\beta < -2$) over the entire burst time history, independent of intensity. The others break into 3 general classes of behavior in the high-energy power law index as a function of time. The first class, with 21 members, consists of bursts where the high-energy power-law index is inversely correlated with the intensity, as shown in Fig. 1. That is, as the burst increases in intensity, the high-energy emission becomes softer. This is seen as a steepening of the high-energy contours at the peak in the burst's topographic map (Fig. 2). This could be an indication that the peak in these events may consist of a separate, shorter and softer spectral component than the rest of the burst. Second, is a class of 25 bursts which have hard-to-soft evolution in the high-energy spectral index, similar to the hard-to-soft evolution in E_{peak} seen by

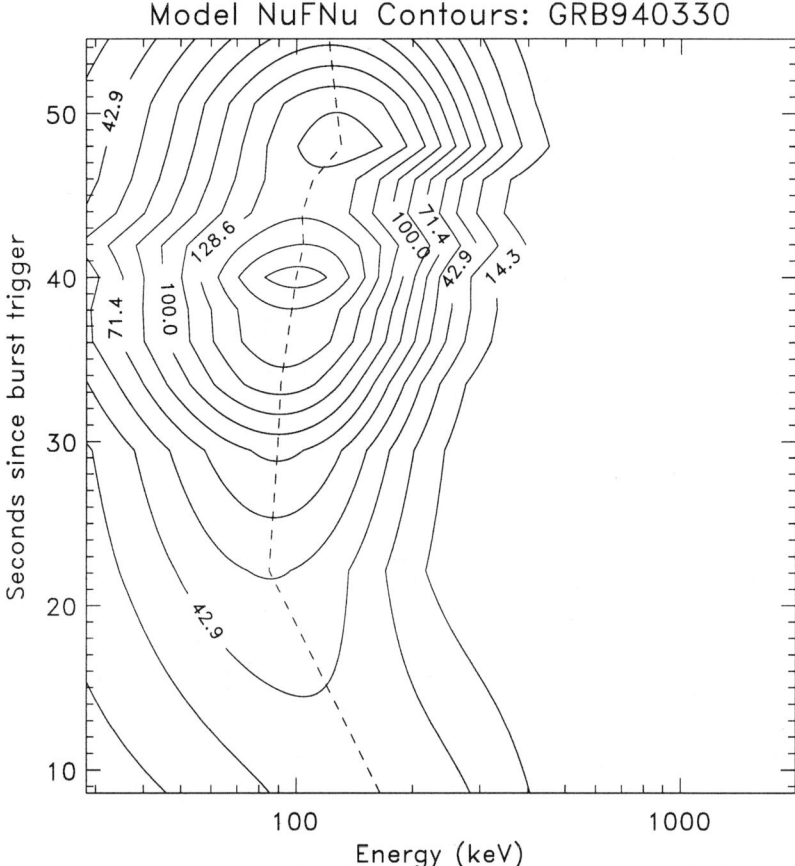

FIG. 4. Topographical map of GRB940330. Note the absence of flux at high energies (to the right in the plot) for this 'super-soft' burst.

Ford et al. (2). Fig. 3 is an example of this behavior in a burst where the hard-to-soft evolution is correlated with E_{peak}. Finally, there is a category of 7 bursts which seem to have no high-energy power-law at all. In Fig. 4, the entire burst is fitted with the COMP model, which has an exponential cut-off at high energies. In addition, E_{peak} is close to 100 keV for most of the burst. Given the lack of high-energy emission in this type of burst, this category has been called 'super-soft' (3).

REFERENCES

1. D. Band et al., ApJ **413**, 281 (1993).
2. L. Ford et al., ApJ **439**, 307 (1995).
3. G. Pendleton et al., these proceedings (1995).

Burst Spectra over a Wide Energy Range

B. E. Schaefer[1], D. Palmer[2], C. E. Fichtel[2], B. L. Dingus[2],
E. J. Schneid[3], R. M. Kippen[4], C. Winkler[5], L. Hanlon[6],
V. Schonfelder[7]

[1] *Yale University, Physics Department, New Haven CT 06520-8121,*
[2] *Goddard Space Flight Center, Code 661, Greenbelt MD 20771,*
[3] *Grumman Corporation Research Center, A01-026, Bethpage NY 11714-3580,*
[4] *University of New Hampshire, Morse Hall, Durham NH 03824,*
[5] *ESA-ESTEC, Postbus 299, NL-2200 AG Noordwijk, The Netherlands,*
[6] *University College Dublin, Belfield, Stillorgan, Dublin 4, Ireland,*
[7] *Max-Planck-Institut fur Extraterrestrische Physik, Postfach 1603, 85740 Garching, Germany*

Burst spectra over a restricted energy range look like a power law plus some curvature. Such spectra can be well fit by most models or mechanisms, and hence reveal little of the underlying physics. With the launch of the *Compton Gamma Ray Observatory*, we can construct composite spectra from tens of keV to tens of GeV. This wide spectral range yields burst spectra that can sharply confront models. We have constructed wide-range composite spectra for several bursts, two of which are presented here: GRB910601 and GRB910814. The energy ranges are 28 keV to 10 MeV and 100 keV to 200 MeV. These spectra were constructed from spectra by BATSE, EGRET, COMPTEL, and OSSE. The spectral shape of GRB910601 shows a continuous turnover between 100 keV and 1 MeV, with peak νF_ν at 500 keV. The spectral shape of GRB910814 is dominated by a sharp and strong spectral break near 2 MeV.

In addition to these two broad spectra, we have derived an 'average' spectral shape for 20 bright bursts from 40 keV to 1500 keV. This average spectra will be useful for studies of the red shifting of average spectra.

INTRODUCTION

The shape of the Gamma Ray Burst spectral continuum undoubtedly contains much information relating to the physics of the light emitting region. Historically, much effort has gone into fits of specific models or functional forms to the best available burst spectra. Until the advent of the *Compton Gamma Ray Observatory*, these spectra had a limited energy range, typically from 30 keV to 1 MeV. The best of these was obtained by SMM/GRS for GRB840805 from 10 keV to 100 MeV (1). These spectra showed a power law shape with some curvature, a shape that could be matched by virtually all models. As such, little could be learned.

However, with the launch of the *Compton Gamma Ray Observatory*, bursts can be examined from roughly 20 keV to tens of GeV. Thus, a dynamic range of energy of up to six orders of magnitude are potentially available, as an improvement on the roughly two orders of magnitude previously available. This new broad energy range is constructed from all four *Compton Gamma Ray Observatory* instruments; BATSE, OSSE, COMPTEL, and EGRET.

COMPOSITE SPECTRA

Composite spectra can be created for only a small fraction of all bursts. To get to high energy, the burst must be in the field-of-view of COMPTEL and EGRET. To get good photon statistics (especially at high energy), the burst must be among the brightest. In practice, these restrictions allow only a few bursts per year for which a composite spectrum can be created.

A composite spectrum for GRB910503 has been presented from 20 keV to 200 Mev (2). Overlays of BATSE, OSSE, COMPTEL, and EGRET spectra for GRB910601 have been presented from 40 keV to 10 MeV (3). Overlays of BATSE, COMPTEL, and EGRET spectra for GRB920622 and GRB940217 have been presented from 20 keV to 10 MeV (4,5). This paper presents composite spectra of GRB910601 and GRB910814.

The four instruments all produce burst spectra with different time bins. Nevertheless, these spectra can be combined to yield a composite spectrum. Care must be taken to ensure that the time interval of the extracted burst photons are equivalent. For GRB910601 and GRB910814, we have used time intervals which include all burst flux. This means that exact matching of the time bins is not necessary since the times with imperfect overlap contain no burst flux. Thus, the spectra are for the entire burst and can be termed 'fluence spectra'.

For each burst, we had spectra from the BATSE LADs, the BATSE SDs, the EGRET TASC, and the COMPTEL D2 detectors, while the various particular events also had spectra from the EGRET spark chamber, the COMPTEL telescope, and the OSSE detectors. The spectra used in this analysis have previously been reported by the various instrument teams (6–9,3).

The combination of each spectra was accomplished by rebinning the flux to some standard energy bins and then performing a weighted average over all contributing spectra. The results (Figures 1, 2) are composite GRB spectra over a wide range of energy. Note that our spectral plots are of νF_ν versus energy.

AVERAGE SPECTRUM

The 'average' spectrum (Figure 3) was constructed from 20 bright BATSE bursts (6). The average was formed by first rebinning all spectra into a tem-

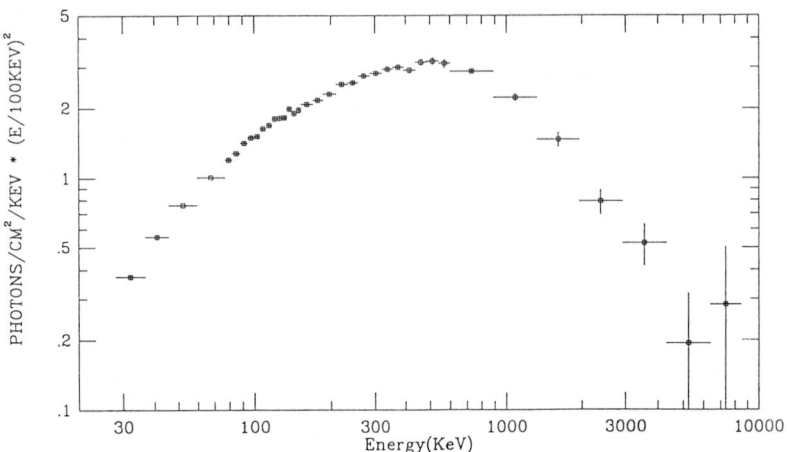

FIG. 1. GRB910601

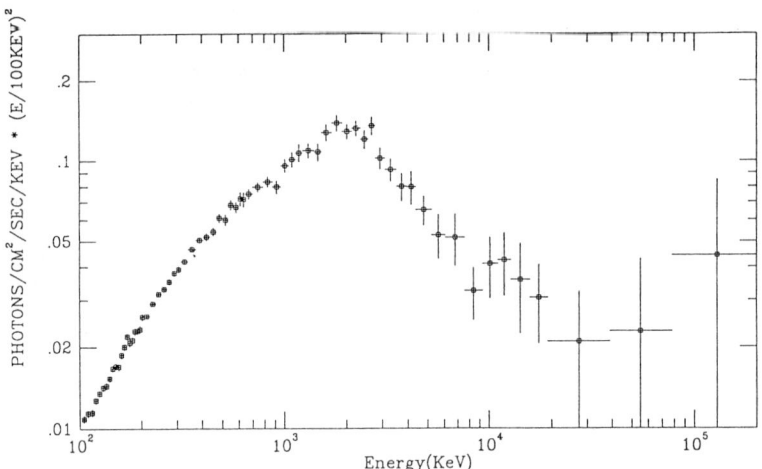

FIG. 2. GRB910814

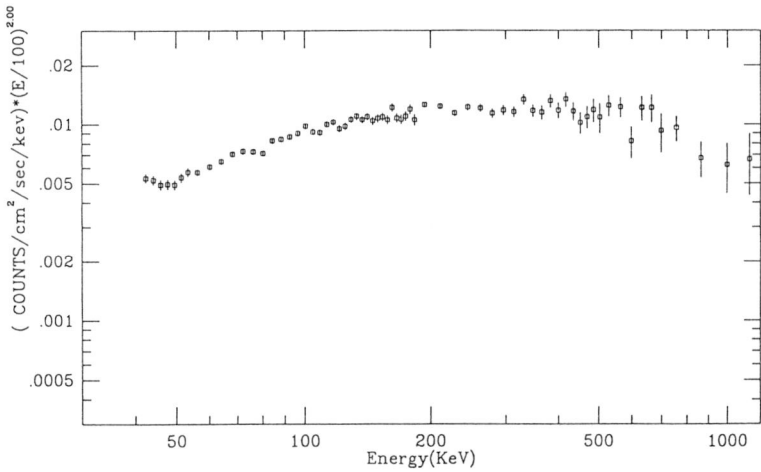

FIG. 3. Average of 20 Bursts

plate bin structure, then performing a weighted average over all spectra. This procedure will give somewhat higher weight to the brighter spectra in the sample. Nevertheless, the spectral shape is robust, since the use of various subsets (e.g., brights versus faints) produces no significant variation. Since all 20 events are among the brightest seen by BATSE, these are likely to represent an unredshifted average. This average shape can be redshifted to yield an average change in hardness or spectral slope as a function of z.

DISCUSSION

The first result from our analysis is the good agreement between all four instruments. That is, to within the stated uncertainties, there is no evidence that any instrument(s) has a significant problem with the normalization of their spectra.

We have tried fitting many of the usual functional forms to these broad band spectra. All such attempts only yield bad fits. This demonstrates that the older simple models are inadequate and must be discarded.

The most striking feature of these spectra is the very sharp and strong spectral break around 2 MeV for GRB910814. All spectra show νF_ν peaks from 0.5 to 2 MeV, with the average spectrum flatter, perhaps because it combines bursts with a range of peak energies. The individual burst spectral slopes approach $\nu F_\nu \propto \nu^{1/3}$ for energies below this peak, as predicted by expanding fireball models (10).

REFERENCES

1. G. H. Share, et al., Adv. Space Res. **6**, 15 (1986).
2. B. E. Schaefer, et al., in Gamma Ray Bursts, eds. G. J. Fishman, J. J. Brainerd and K. Hurley, AIP Conf. Proc. **307**, 280 (AIP, New York, 1994).
3. G. H. Share, et al., in Gamma Ray Bursts, eds. G. J. Fishman, J. J. Brainerd and K. Hurley, AIP Conf. Proc. **307**, 283 (AIP, New York, 1994).
4. J. Greiner, et al., Astron. Astrophys. **302**, 121 (1995).
5. C. Winkler, et al., Astron. Astrophys. **302**, 765 (1995).
6. B. E. Schaefer, et al., ApJS **92**, 285 (1994).
7. L. O. Hanlon, et al., Astron. Astrophys. **285**, 161 (1994).
8. C. Winkler, et al., in Compton Gamma Ray Observatory, eds. M. Friedlander, N. Gehrels and D. J. Macomb, AIP Conf. Proc. **280**, 845 (AIP, New York, 1993).
9. P. W. Kwok, et al., in Compton Gamma Ray Observatory, eds. M. Friedlander, N. Gehrels and D. J. Macomb, AIP Conf. Proc. **280**, 855 (AIP, New York, 1993).
10. J. Katz, ApJ, submitted (1995).

TASC Measurements/Upper Limits for Energetic Gamma-Ray Bursts Within The EGRET Field of View

E. J. Schneid[1], D. L. Bertsch[2], J. R. Catelli[6], B. L. Dingus[3],
J. A. Esposito[3], C. E. Fichtel[2], R. C. Hartman[2], S. D. Hunter[2],
G. Kanbach[7], D. A. Kniffen[5], Y. C. Lin[4],
H. A. Mayer-Hasselwander[7], P. F. Michelson[4], C. von Montigny[8],
R. Mukherjee[3], P. L. Nolan[4], P. Sreekumar[3], and
D. J. Thompson[2]

[1] *Northrop Grumman Corporation, MS A01-26, Bethpage, NY 11714*

[2] *NASA/Goddard Space Flight Center, Code 600, Greenbelt,MD 20771*

[3] *USRA, NASA/GSFC, Greenbelt, Md 20771*

[4] *W.W. Hansen Experimental Physics Lab. Stanford U., Stanford, CA 94305*

[5] *Dept. of Phys., Hampden-Sydney College, Hampden-Sydney,VA 23943*

[6] *Dept. of Astronomy, U. of MD, College Park, MD 20742*

[7] *Max-Planck-Institut für Extraterrestrische Physick, D-85748, Garching Germany*

[8] *NAS/NRC, NASA/GSFC, Greenbelt, MD 20771*

For the period from launch in April 1991 to present, EGRET had reported on the detection of seven energetic gamma-ray bursts. A systematic analysis of the Total Absorption Shower Counter (TASC) data has been made for bursts within the EGRET field of view. Spectral indexes are given for these bursts detected by EGRET and upper limits are given for bursts detected by COMPTEL and not by EGRET.

INTRODUCTION

The high-energy emission of some gamma-ray bursts is important for understanding the nature of the burst energy producing mechanisms. EGRET can measure gamma rays in the range of 1.0 to >200 MeV with the Total Absorption Shower Counter(TASC), a large NaI spectrometer, and from 30 MeV to >30 GeV with its spark chamber telescope. The EGRET spectral range therefore overlaps those for BATSE and COMPTEL in the MeV range but

© 1996 American Institute of Physics

extends these measurements for intense bursts to higher energies. EGRET with its spark chamber has measured gamma rays up to 18 GeV, with the duration of the emission extending for greater than 54000 seconds during the GRB940217 burst (3).

This paper presents a summary of the TASC data for the detected bursts and upper limits for other bursts in EGRET's field of view. A further guide for burst selection was a requirement that the burst be detected by COMPTEL (5), which has a similar field of view. Previous calculations (8) have reported that the TASC sensitivity seriously degrades when the burst is outside the field of view.

INSTRUMENT DESCRIPTION

The EGRET instrument consists of a spark chamber telescope with interleaved tantulum plates to convert gamma rays in the energy range from 30 MeV to over 30 GeV to electron-positron pairs, and to image the trajectories of the pair. The TASC measures the energy of the pair as it emerges from the bottom of the spark chamber. A large anticoincidence shield to reject charged particle events surrounds the spark chamber but does not extend to cover the TASC. More detailed description of the instrument and calibration has been given by Hughes (2),Kanbach (4), and Thompson (11).

The EGRET TASC has a special burst/flare mode for recording gamma-ray burst and solar flares. In this mode a pulse height spectrum is accumulated for all events from 1 MeV to > 200 MeV. Without any anticoincidence shield , the analysis of the TASC data must account for the large backgrounds measured by the TASC. Spectra are routinely accumulated for 32.76 seconds, but when activated by a BATSE trigger, EGRET accumulates four sequential spectra with integration times variable from 1 to 16 seconds. For energetic high energy bursts coming within the telescope field of view (±45 degrees), the spark chamber telescope can provide valuable high energy spectral, directional and temporal information.

BURST RESULTS

The sensitivity of the TASC to bursts depends on the accumulation time over the burst, the orbital backgrounds at the time of the burst, and the area x efficiency of the TASC for the direction to the burst. Figure 1 shows the area x efficiency products calculated for 1 and 10 MeV gamma rays using the CGRO Mass Model and the EGS4 code for the some of the bursts that were in the EGRET field of view. The data are plotted as a function of zenith angle only and some of the variations are due to azimuthal angle effects. The 1 MeV gamma rays are more effected by the EGRET instrument and spacecraft intervening material. The dip in area X efficiency around 30 degrees for 1 MeV results from the attenuation of the gamma rays by the presence of the spark chamber telescope walls. The curves do not extrapolate to the geometrical

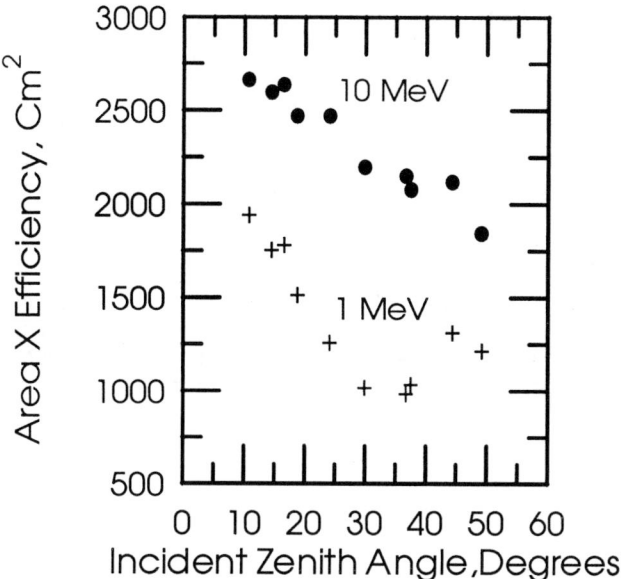

FIG. 1. Calculated TASC Area X Eff Product for 1 and 10 MeV Gamma Rays

area of the TASC at zero degrees (6606 cm^2) due to attenuation caused by the instrument material above the TASC.

EGRET detections of gamma-ray bursts are summarized in table 1. These bursts are associated with the strongest bursts above 300 keV observed by BATSE and COMPTEL. The table contains the burst identification, its BATSE trigger number, the time, the spectral index for a power law fit to the data, the flux at 1.0 MeV for the power law and the EGRET publication reference for that burst.

The upper limits at 1.25 and 10.0 MeV for nine other bursts are listed in Table 2. This table lists the burst identification, the BATSE trigger time, the direction to the burst, and the two sigma upper limits for the flux in the 1.0–1.5 and 9.5–10.5 MeV energy regions. Burst GRB930309 has a flux slightly larger than two sigma at 1.25 MeV, therefore a flux is listed rather than an upper limit.

Acknowledgments. The EGRET team gratefully acknowledges support from the following: Bundesministerium für Forschung and Technolie, Grant 50 QV9095 (MPE); NASA Grant NAG5-1741 (HSC); NASA Grant NAG5-1605 (SU) and NASA Contract NAS5-31210(NGC).

REFERENCES

1. J. R. Catelli, B. L. Dingus, and E. J. Schneid, This workshop

TABLE 1. EGRET Detected Bursts

Id.	BATSE No.	Time(UT)	Spectral Index	Flux @1.0 MeV	Refs.
GRB910503	143	07:04:13	2.24±0.03	8.71±0.47	(7)
GRB910601	249	19:22:15	3.67±0.20	0.98±0.08	(6)
GRB910709	503	11:33:22	1.76±0.14	0.53±0.15	
GRB910814	678	19:14:33	2.94±0.10	14.5±1.4	(6)
GRB911118	1085	18:57:38	No Fit	0.48±0.13	
GRB920622	1663	07:05:05	2.75±0.10	1.05±0.72	(9)
GRB930131	2151	18:57:12	1.97±0.09	1.90±0.26	(10)
GRB940217	2831	23:02:42	2.50±0.08	0.69±0.09	(3)
GRB940301	2855	20:10:37	2.48±0.10	0.26±0.05	(9)
GRB950425	3523	00:15:18	1.93±0.04	1.62±0.09	(1)

TABLE 2. EGRET Bursts Upper Limits

Id	BATSE Time(UT)	Zenith Angle	Upper @1.2 MeV	Limits[†] @10.0 MeV
GRB910425	00:37:46	44.3	<0.13	<0.10
GRB910627	04:29:18	10.8	<0.10	<0.005
GRB920830	01:45:18	16.6	<0.14	<0.007
GRB930118	17:53:47	15.7	<0.15	<0.007
GRB930309	03:07:50	29.9	0.053±0.025	<0.022
GRB930612	00:44:18	24.1	<0.13	<0.005
GRB930704	16:49:06	18.8	<0.16	<0.006
GRB931229	07:16:06	49.2	<0.05	<0.004
GRB940314	09:59:49	14.5	<0.08	<0.002

[†]Photons MeV^{-1} sec^{-1} cm^{-2}

2. E. B. Hughes, et al., IEEE Trans. Nucl. Sci NS **27**, 364 (1980).
3. K. Hurley, et al., Nature **372**, 652 (1995).
4. G. Kanbach, et al., Proc. of the Gamma Ray Observatory Workshop, ed. W.N. Johnson, (Greenbelt, Md:NASA/GSFC), 2-1 (1989).
5. R. M. Kippen, Ph.D. Thesis , U. of NH, 'Location and Spectra of Cosmic Gamma-Ray Bursts' (1995).
6. P. W. Kwok, et al., COMPTON Gamma-Ray Observatory Symp., eds. M. Friedlander, N. Gehrels and D. Macomb, AIP Conf. Proc. **280**, 222 (1993).
7. E. J. Schneid, et al., A and A **255**, L13 (1993).
8. E. J. Schneid, et al., COMPTON Gamma-Ray Observatory Symp., eds. M. Friedlander, N. Gehrels and D. Macomb, AIP Conf. Proc. **280**, 850 (1993).
9. E. J. Schneid, et al., ApJ **453**, 95, (1995).
10. M. Sommer, et al., ApJ **422**, L63 (1994).
11. D. J. Thompson, et al., ApJ Suppl. Ser **89**, 629 (1993).

The Detector Response Matrices of The Transient Gamma-Ray Spectrometer (TGRS)

H. Seifert,[†‡] T.L. Cline,[†] N. Gehrels,[†] B. Mitra,[†‡]
D.M. Palmer,[†‡] R. Ramaty,[†] B.J. Teegarden,[†*]
K. Hurley,[¶] N. Madden,[§] R. Pehl,[§] and A. Owens[◊]

[†]*NASA/Goddard Space Flight Center, Code 661, Greenbelt MD 20771,*
[‡]*Universities Space Research Association,* [*]*on leave to CESR,* [¶]*University of California, Berkeley,* [§]*Lawrence Berkeley Labs,* [◊]*U. of Leicester*

The Transient Gamma-Ray Spectrometer (TGRS) on the WIND spacecraft is a germanium detector primarily designed to measure high-resolution (~2–3 keV) spectra of bright gamma-ray bursts. The instrument response matrix, used to unfold photon spectra from the measured count spectra, has been obtained by modelling TGRS/WIND with the GEANT detector description and simulation package from CERN. The simulations were verified using laboratory measurements. We present a brief description of the TGRS instrument, as well as an outline of the measurements, simulations and software algorithms used in the generation of the detector response matrices.

INTRODUCTION

TGRS is one of eight instruments on the GGS/WIND spacecraft which was launched on November 1, 1994. After about 2 years and several deep space orbits, the spacecraft will eventually be injected into a halo orbit around the Sun-Earth L_1 point. Although TGRS is primarily intended for high resolution spectroscopy of gamma-ray bursts (GRBs) and solar flares, it also lends itself to the study of transient x-ray sources and, using an on-board passive occulter, the long-term monitoring of a few steady sources such as the Crab and the Galactic Center. As part of an interplanetary network of spacecraft, the TGRS and KONUS instruments on WIND are supporting the search for quiescent GRB counterparts in other wavelengths by contributing to the determination of accurate GRB locations from multi-spacecraft timing. Since launch, TGRS has been working as expected, and to date has proven to be very stable in its performance. In its first year of operation TGRS has triggered on 62 GRBs, ~33 of which were bright enough for spectroscopy (1).

© 1996 American Institute of Physics

TABLE 1. Principal TGRS, KONUS, and CGRO/BATSE SD detector parameters

Item	TGRS	KONUS	BATSE SD
Detectors, Material	$1 \times n$-Ge	$2 \times$ NaI(Tl)	$8 \times$ NaI(Tl)
Detector Area	35 cm^2	127 cm^2 (each)	127 cm^2 (each)
Detector Thickness	6.1 cm	7.6 cm	7.6 cm
Energy Range	$\sim$25–8000 keV	15–10000 keV	15–10000 keV
Sky Coverage	$\sim 2\pi$ sr	4π sr	$\sim 2.4\pi$ sr
Energy Resolution	1.8 at 50 keV	$\sim$6 at 50 keV	$\sim$6 at 50 keV
in keV	2.8 at 500 keV	$\sim$40 at 500 keV	$\sim$40 at 500 keV

INSTRUMENTATION

The TGRS detector is a 215 cm^3, closed-end, coaxial, high purity n-type Ge crystal in reverse electrode configuration, housed in a pressurized Al module of 20 mil side/top thickness. Since there is no annealing capability, the detector is kept at cryogenic temperature throughout the mission to minimize the effects of radiation damage. The design of the passive, two-stage radiative Be cooler is similar to the one flown on ISEE-3 (2). A 30 mil thick Be/Cu alloy passive sun shield around the sides of the cooler suppresses triggers by soft solar x-rays. The attenuation is $\sim 1/e$ at 60 keV, and $\sim 2 \times 10^{-12}$ at 20 keV. A 1 cm thick passive Mo/Pb occulter which modulates signals from within $\sim \pm 20°$ of the ecliptic plane at the spacecraft spin frequency ($\sim$20 rpm) allows TGRS to monitor sources such as the Crab and the Galactic Center, and to identify solar flares. The principal TGRS detector parameters are compared with those of KONUS (also on the WIND spacecraft) and those of the CGRO/BATSE Spectroscopy Detectors in Table 1; a complete description of the instrument can be found in Ref. (3). Over its $\sim$25–8000 keV energy range, the TGRS resolution is up to a factor of 50 better than that of scintillation detectors such as in BATSE and Ginga, which compensates for its smaller collecting area in the search for narrow spectral features.

MONTE-CARLO SIMULATIONS AND LABORATORY MEASUREMENTS

Monte-Carlo simulations of the TGRS instrument were performed using the GEANT detector description and simulation package from CERN. The package allows to simulate the transport of elementary particles through an experimental setup, and to represent the setup and the particle trajectories graphically. The model which was used for the generation of the first response matrices included only the detector and cooler assembly. A cross sectional view of the geometry is shown in Fig. 1. Simulations were performed using monoenergetic plane wave beams (Ø= 15 cm) of 1 million events each. The input energies were chosen to have a $\Delta E/E$ spacing of $\sim$10% below and 2% above 20 keV. The angle of incidence was fixed at 20°. The result of these

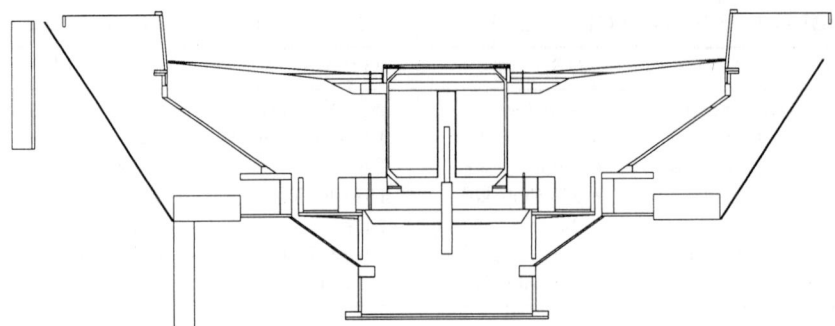

FIG. 1. TGRS Detector and Cooler model used for the first generation instrument response. A more realistic model used for the next generation matrices also includes the WIND spacecraft.

simulations is a set of grid response vectors which cover an energy range of ~10–10000 keV. The output energy resolution is 1 keV.

To study the effects of scattering from the spacecraft and to determine the angular dependence of the response, a "supergrid" of vectors at various energies and incidence angles is currently being generated. Due to the necessarily large beam sizes, and the unavailabity of suitable variance reduction techniques, the total number of events needed for these simulations is on the order of a few hundred million.

To verify and calibrate the instrument description in GEANT, comparisons between simulations and laboratory measurements using calibrated ^{241}Am, ^{137}Cs, ^{57}Co, ^{60}Co, and ^{22}Na radioactive sources were performed. For these simulations the instrument includes the detector/cooler assembly and the cryogenic test fixture. Figure 2 shows the result for a narrow beam which essentially only illuminates the immediate vicinity of the detector, and for a 360° beam with a more complete geometry description for the experimental setup. It is evident from the intensity of the low energy continuum that the contribution of scattering into the detector from the surroundings is crucial for an accurate description of the measurement data.

CONSTRUCTION OF THE DETECTOR RESPONSE MATRICES AND MODEL FITTING

Given the set of grid vectors which represent the general energy and angular dependence of the instrument response, a Detector Response Matrix (DRM) for the analysis of a particular event can be constructed. This DRM will depend both on the angle of incidence of the event and the channel/energy calibration of the instrument. The DRM will be binned according to the binning schemes chosen for both the model photon and measured count spectra. Basis for the matrix elements of the binned DRM are the full resolution vectors corresponding to each raw count spectrum energy channel. These vectors

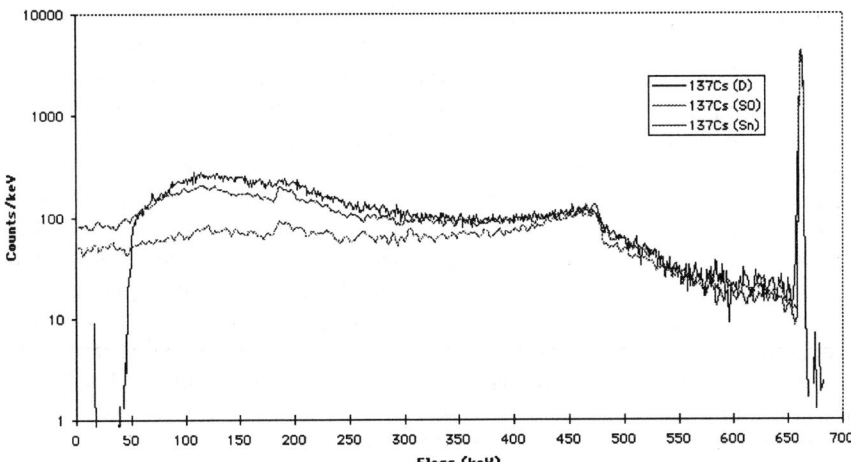

FIG. 2. Comparison of the current 662 keV instrument response (S0, lower curve) with ^{137}Cs laboratory data (D, upper curve) and a Monte-Carlo model with a larger beam where some of the surroundings are included (Sn, middle curve).

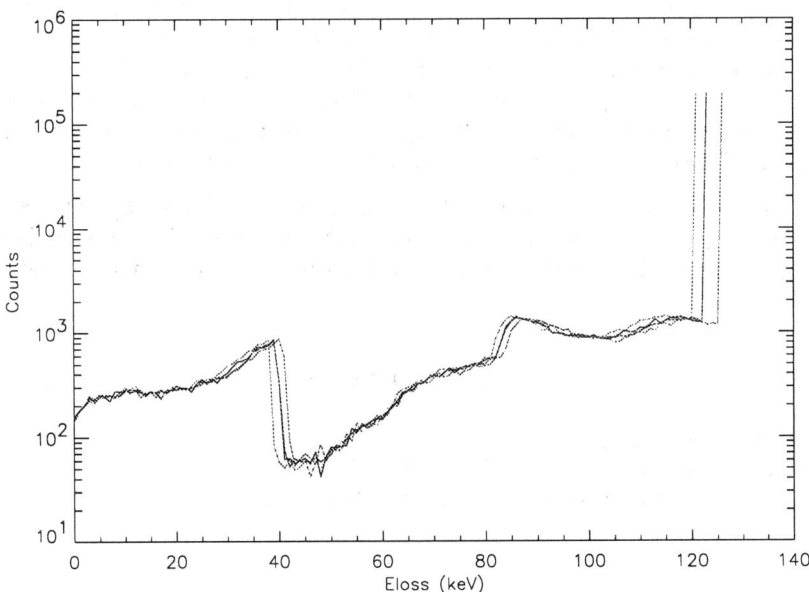

FIG. 3. Interpolation of the response vector at 123 keV (black) from two grid vectors (light grey) which are at energies $\pm 2\%$ ($\Delta E/E$) above and below 123 keV. For comparison, the simulated response at 123 keV is also shown (dark grey). The interpolation algorithm must take into account that the position of the individual spectral features in each response depend on the input energy.

D_i are interpolated from the grid vectors for each raw count spectrum channel, are corrected for angle, and then accumulated for each ("vertical") bin of the model photon spectrum. During this process each vector D_i is weighted according to $\int (D_i \times E_i^\alpha)/\int E_i^\alpha$, where E_i is the channel energy and α is an optional power-law index. Finally, each DRM row is ("horizontally") binned according to the measured count spectrum, and then converted to the correct units of counts cm^2 photons^{-1}. The algorithm used for the interpolation of a response at a particular energy takes into account the kinematics of the individual spectral features in each response. The algorithm for the angular correction uses the fact that the angular dependence of the *individual* spectral features is smooth and can be described analytically. Spectral deconvolution is performed using "forward folding", i.e., by multiplying a model source spectrum **S** with **D** and then comparing the model count spectrum **SD** with the observed count spectrum **C** in a χ^2 sense. This process and the parameters of **S** are iterated until the minimum of χ^2 has been found.

SUMMARY

A simple zeroeth-order model for the TGRS detector response matrices has been successfully developed and tested. Model fits using this version showed that the simple model does not yet adequately consider events which are scattered into the detector from the surroundings, e.g., from the spacecraft. The model underestimates the lower and overestimates the higher energies, effectively yielding spectra which are too steep. Software for the next generation matrices, which is currently being developed, will correct the instrument response for the presence of the spacecraft and incidence angle.

Acknowledgements. The authors would like to acknowledge Ms. Sandhia Bansal who has been providing invaluable programming support in the development of the matrix generation and spectral deconvolution software for TGRS.

REFERENCES

1. D. M. Palmer, et al., this volume.
2. B. J. Teegarden, T. L. Cline, ApJ **236**, L67 (1980).
3. A. Owens, et al., Space Science Reviews **71**, 273 (1995).

Cosmological Gamma-Ray Bursts and the Cosmic Ray Flux Above 10^{19} eV

Eli Waxman

Institute for Advanced Study, Princeton, NJ 08540

The objective of my talk is to present a scenario in which cosmic rays (CR's) above 10^{19} eV and cosmological γ-ray bursts (GRB's) have a common origin. This scenario is consistent with the observed CR flux provided that each burst produces similar energies in γ-rays and in protons above 10^{19} eV, with a differential power law proton spectrum $d\log N/d\log E \approx -2$. Protons may be accelerated by Fermi's mechanism to energies $\sim 10^{20}$ eV in a dissipative, ultra-relativistic wind, with luminosity and Lorentz factor high enough to produce a GRB. In this scenario, a correlation between the arrival directions of $> 10^{20}$ eV CR's and the angular positions of strong GRB's should not be detectable on a time scale $\ll 100$ yr. Alternative tests of the scenario are briefly discussed.

INTRODUCTION

Recent cosmic ray (CR) observations strongly suggest that the CR flux above $\sim 10^{19}$ eV is dominated by protons of extra-Galactic origin (1). The sources, however, remain unknown. Current models for the production of CR's are especially challenged by the recently detected events with energies $> 10^{20}$ eV (1): The high energies rule out most of the acceleration mechanisms discussed so far (2), and since the distance traveled by such particles must be smaller than 100 Mpc (3), their arrival directions are inconsistent with the position of any astrophysical objects that might produce such extremely high energy particles (4).

We show, that if GRB's are cosmological, as suggested by recent observations (5), then two remarkable coincidences arise, which suggest that GRB's and the highest energy CR's have a common origin. First, the average rate at which energy is injected into the universe as γ-rays by GRB's is comparable to the rate at which energy should be injected by a cosmological distribution of energetic proton sources in order to produce the observed CR flux above 10^{19} eV. The second coincidence concerns the energy of the most energetic CR's. Although the source of GRB's is unknown, the observational characteristics of GRB's impose strong constraints on the physical conditions in the γ-ray emitting region (6). We show, that if our understanding of these conditions is correct, then protons may be accelerated in the γ-ray emitting region up to energies $\sim 10^{20}$ eV.

© 1996 American Institute of Physics

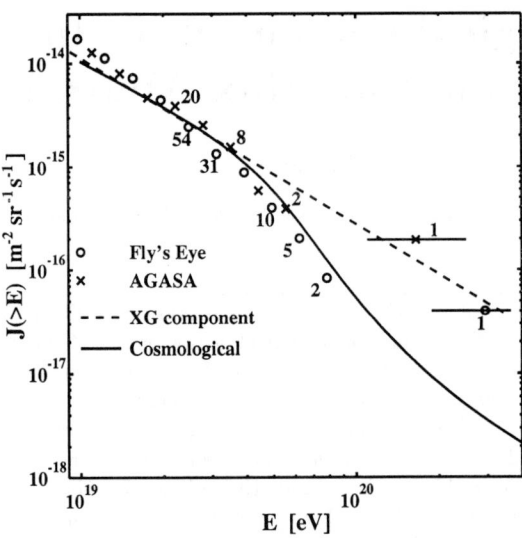

FIG. 1. The integral CR flux produced by a cosmological distribution of sources compared to the Fly's Eye and AGASA data (1). The integers shown in the figure denote the number of events observed. The flux deduced from the highest energy events is plotted with the 2σ error bars for the event energy. The dashed line denotes the fit by Bird et al. (1) for the extra-Galactic flux.

THE PRODUCTION RATE OF γ-RAYS AND HIGH ENERGY CR'S

The CR spectrum produced by a homogeneous cosmological distribution of CR sources (calculated in a method similar to (7)), is compared in Fig. 1 with current CR data. We assume that each source generates high energy protons with a power law differential spectrum $dN/dE \propto E^{-2}$ (For this calculation we have used non-evolving sources in a flat universe with zero cosmological constant and $H_0 = 75$ km s^{-1} Mpc^{-1}. The spectrum is insensitive to the cosmological parameters and to source evolution, since most of the cosmic rays come from distances < 500 Mpc). Bird et al. find that the Fly's Eye spectrum and composition data in the energy range $4 \times 10^{17} - 4 \times 10^{19}$ eV can be fitted by a sum of two power laws (1): A steep Galactic component $J \propto E^{-2.5}$ composed of iron, and a shallow extra-Galactic component $J \propto E^{-1.6}$ composed of protons. The Bird et al. fit to the extra-Galactic component is also shown in Fig. 1.

The data in the energy range 2×10^{19} eV $\leq E < 10^{20}$ eV is consistent with the cosmological model (for the binned data given in (1) the model gives $\chi^2 = 0.7$ per degree of freedom). The deficit in the number of events detected above 5×10^{19} eV, compared to a power-law extrapolation of the flux at lower energy, is consistent with that expected from a cosmological "black-body cutoff". The flux deduced from the two events observed at $> 10^{20}$ eV is higher than that predicted from the cosmological model. However, it should be noted that the statistical significance of the apparent discrepancy is not high. For the Fly's Eye exposure, for example, the model predicts an average of ~ 1.3 events above 10^{20} eV, and the probability that the first event observed at this energy range is above 2×10^{20} eV is $\sim 15\%$. With 1 event observed by each experiment, the possibility that these events are also produced by the cosmologically distributed power law sources can not be ruled out.

The present rate at which energy should be produced as $10^{19} - 10^{21}$ eV protons by the cosmological CR sources in order to produce the observed flux is $4 \pm 2 \times 10^{44}$ erg Mpc^{-3} yr^{-1}. This rate is comparable to that produced in γ-rays by cosmological GRB's: The rate of cosmological GRB events is $\nu_\gamma \sim 3 \times 10^{-8}$ Mpc^{-3} yr^{-1} (8), each producing $\approx 2 \times 10^{51}$ erg, corresponding to a γ-ray energy production rate of $\sim 10^{44}$ erg Mpc^{-3} yr^{-1}.

THE $> 10^{20}$ eV EVENTS AND TIME DELAYS

While the energy output of cosmological GRB's may be similar in gamma-rays and high energy protons, such bursts still could not account for the $> 10^{20}$ eV CR events without a significant dispersion in the proton arrival times. The high energy CR experiments observe a cone around the zenith of opening angle $\sim 45°$ over a period of < 10 yr. The rate of cosmological GRB events in a cone observed by CR experiments extending to a distance of ~ 100 Mpc, which is the maximal distance likely to be traveled by a CR of 10^{20} eV, is ~ 1 per 50 yr. The rate in a cone extending to 50 Mpc, the maximal distance traveled by the 3×10^{20} eV CR observed by the Fly's Eye, is 8 times smaller. Thus, the probability of the experiments to observe a CR pulse from a GRB event over < 10 yr period is small, unless the CR pulse is broadened in time, due to propagation through the inter-galactic medium (IGM), over a time scale ≥ 100 yr.

Time broadening is likely to occur due to the combined effects of deflection by random magnetic fields and energy dispersion of the particles. Consider a proton of energy E propagating through a magnetic field of strength $B_{\rm IG}$ and correlation length λ. The typical deflection angle for propagation over a distance d is $\theta_s \sim (d/\lambda)^{1/2} \lambda / R_L$, where $R_L = E/eB_{\rm IG}$ is the Larmor radius. This deflection results in a time delay, compared to propagation along a straight line, of order $\tau(E) \approx \theta_s^2 d/c \approx (eB_{\rm IG}/Ed)^2 \lambda / c$. The random energy loss of $> 10^{20}$ eV protons, due to interaction with MBR photons, results in a RMS energy spread comparable to the average energy over propagation distances in the range $10 - 100$ Mpc (3). Since the time delay τ is sensitive to the particle energy, this spread results in a time broadening of the CR pulse over a time $\sim \tau(E)$. The field required to produce $\tau > 100$ yr,

$$B_{\rm IG} > 2 \times 10^{-12} E_{20} d_{100}^{-1} \lambda_{10}^{-1/2} \text{ G} \tag{1}$$

(where $E = 10^{20} E_{20}$ eV, $d = 100 d_{100}$ Mpc and $\lambda = 10 \lambda_{10}$ Mpc), is consistent with our knowledge of IGM properties (9). A time broadening over $\tau > 100$ yr is therefore reasonable.

FERMI ACCELERATION IN COSMOLOGICAL GRB'S

Whatever the ultimate source of GRB's is, observations strongly suggest the following similar scenario for the production of cosmological GRB's (6). A compact source, $r_s \sim 10^7$ cm, produces a wind characterized by an average luminosity $L \sim 10^{51}$ erg s^{-1} and mass loss rate $\dot{M} = L/\eta c^2$, with $\eta \sim 300$. At small radius, the wind bulk Lorentz factor, γ, grows, until most of the wind energy is converted to kinetic energy and γ saturates at $\gamma \sim \eta$. At some large radius r_d, in the range $10^{13} - 10^{16}$ cm, the kinetic energy is re-randomized due to internal collisions or collision with the surrounding medium, and then radiated as gamma-rays by synchrotron emission and inverse-Compton scattering of relativistic electrons. The relativistic relative motions in the wind rest frame, necessary for a substantial dissipation of kinetic energy when $\gamma \gg 1$, and the existence of a magnetic field B, necessary for the synchrotron emission, are likely to give rise to Fermi acceleration of charged particles. We derive below the constraints that should be satisfied by the wind parameters (L, γ, r_d, B) in order to allow acceleration of protons to $\sim 10^{20}$ eV.

The most restrictive requirement, which rules out the possibility of accelerating particles to energies $\sim 10^{20}$ eV in most astrophysical objects, is that the particle Larmor radius R_L should be smaller than the system size (2). In our scenario we must apply a more stringent requirement. Due to the wind expansion, the wind internal energy is decreasing and therefore available for CR acceleration (as well as for γ-ray production) only over a comoving time $t_d \sim r_d/\gamma c$. The typical Fermi acceleration time, $t_a \sim R_L/c$ (2), should be smaller than t_d, leading to the requirement $R_L < r_d/\gamma$. This condition sets a lower limit for the required comoving magnetic field strength, which may be put in the form

$$\left(\frac{B}{B_{\text{e.p.}}}\right)^2 > 0.15\gamma_{300}^2 E_{20}^2 L_{51}^{-1}, \qquad (2)$$

where $\gamma = 300\gamma_{300}$, $L = 10^{51}L_{51}$ erg s^{-1}, and $B_{\text{e.p.}}$ is the equipartition field, i.e., a field with comoving energy density that produces a luminosity similar to L. The accelerated proton energy is also limited by energy loss due to synchrotron radiation and interaction with the wind photons. The energy loss time is greater than the acceleration time t_a provided (10)

$$r_d > 10^{13}\gamma_{300}^{-2}E_{20}^3 \text{ cm}. \qquad (3)$$

CONCLUSIONS

We have found that the observed flux of CR's above $> 10^{19}$ eV is consistent with a scenario in which these particles are protons produced in cosmological GRB's, provided that each burst produces similar energies in gamma-rays and CR's beyond 10^{19} eV, with a differential power law proton spectrum

$d\log N/d\log E \approx -2$. We have also demonstrated that a dissipative ultra-relativistic wind, with luminosity and bulk Lorentz factor implied by GRB observations, i.e., $L_{51} \sim 1$ and $\gamma_{300} \sim 1$, satisfies the constraints necessary to allow the acceleration of protons to energies $\sim 10^{20}$ eV by Fermi acceleration, provided that the dissipation of kinetic energy occurs at radii $r_d \geq 10^{13}$ cm (eq. 3), and that the magnetic field is close to equipartition (eq. 2).

Future investigations of the scenario should aim at obtaining characteristic signatures that may be compared with observations. Since cosmological GRB's can account for the $> 10^{20}$ eV events only if the CR pulse produced in a GRB is delayed and broadened over a time scale ≥ 100 yr, no correlation between the arrival directions of $> 10^{20}$ eV CR's and GRB's should be detected on a much shorter time scale. However, the number of bright CR sources at various energies and the angular distribution of CR events predicted by this scenario is in marked contrast to the predictions of models where the CR sources are steady (11). With the large catalogue of CR events, expected to be constructed with future CR experiments (12), this would provide a strong test for the model.

Acknowledgments. This research was partially supported by a W. M. Keck Foundation grant and NSF grant PHY 92-45317.

REFERENCES

1. D. J. Bird et al., Phys. Rev. Lett. **71**, 3401 (1993); S. Yoshida et al., Astropar. Phys. **3**, 151 (1995).
2. A. M. Hillas, Ann. Rev. Astron. Astrophys. **22**, 425 (1984).
3. F. A. Aharonian, and J. W. Cronin, Phys. Rev. D **50**, 1892 (1994).
4. J. W. Elbert, and P. Sommers, Astrophys. J. **441**, 151 (1995).
5. C. A. Meegan et al., Nature **355**, 143 (1992); B. Paczyński, Nature **355**, 521 (1992); J. P. Norris et al., Astrophys. J. **423**, 432 (1994).
6. T. Piran, in *Gamma-ray Bursts*, eds. G. Fishman et al. (AIP 307, NY 1994); P. Mészáros, to appear in Proc. 17th Texas Conf. Relativistic Astrophysics, NY Acad. Sci. 1995.
7. E. Waxman, Astrophys. J. **452**, L1 (1995).
8. E. Cohen, and T. Piran, Astrophys. J. **444**, L25 (1995).
9. E. Asseo, and H. Sol, Phys. Rep. **48**, 309 (1987).
10. E. Waxman, Phys. Rev. Lett. **75**, 386 (1995).
11. J. Miralda-Escudé, and E. Waxman, Astrophys. J. **462**, L59 (1996).
12. J. W. Cronin, Nucl. Phys. B (Proc. Suppl.) **28B**, 213 (1992).

GLOBAL STUDIES

Gamma Ray Burst Models and the angular distribution of 3B

Dieter H. Hartmann*

*Department of Physics and Astronomy
Clemson University, Clemson SC 29634-1911

Gamma-ray burst observations with the COMPTON Observatory continue to reshape our modeling of this phenomenon. The impact of the angular distribution of the data is emphasized.

INTRODUCTION

Not long ago most of us believed that a successfull GRB model involves neutron stars, and that the typical source would be located less than ~ 1 kpc from the Sun. BATSE was launched in 1991, and quickly cast doubts on the paradigm. With over 1100 bursts detected with BATSE the question of the burster distance scale is as open as ever. Nearby Galactic neutron stars are no longer deemed acceptable due to the combined observations of an isotropic angular distribution and reduced source counts of faint bursts (6,33). Some recent reviews of these and related issues are (5,13,19).

At present there is no single GRB model that provides an acceptable explanation, consistent with all major observational constraints, for the bulk of classical bursts. This lack of a credible model reflects both continued uncertainty about the distance scale and a general difficulty, both in Galactic and cosmological models, in producing either well ordered, relativistic jets or matter characterized by exceptionally high entropy per baryon. Relativistic motion seems to be required in all models with sources beyond 10 kpc. This author now favors cosmological models, but there still remain suggestions that some classical GRB's may be associated with Galactic neutron stars: the possible existence of repeaters; controversial cyclotron line features (1,5,37,38); and observations of quasi-thermal x-rays which imply $D(kpc)/R(km) \sim 1$, where R is the size of the emitting region and D the source distance. The BATSE data on isotropy and log N - Log P provide the strongest constraints on local models, but there are still loopholes. If both the kick velocity of the progenitor neutron stars and the uncertain potential of the Galactic halo can be adjusted at will, an extreme Galactic halo model consistent with BATSE data might still be constructed (7,15,41). The detection by EGRET of continuing high energy emission, for as much as 90 minutes, of photons with energies up to ~ 10 GeV (11,22,48) provides the most recent challenge to model builders, local and cosmological alike.

© 1996 American Institute of Physics

THE DISTANCE SCALE

For isotropic emission, the total energy released in a burst is

$$E_r \sim 10^{33} \left(\frac{F}{10^{-6} \text{ erg/cm}^2 \text{ s}}\right) \left(\frac{\tau}{10 \text{ s}}\right) D_{pc}^2 \text{ erg} \qquad (1)$$

For an average recurrence time τ_r during a total lifetime T_r, the total energy required is $E_{tot} \sim E_r \, T_r/\tau_r$. Consider distance scales of 10^4 AU (solar system), 100 pc (local stars), 1 kpc (Galactic disk), 100 kpc (extended Galactic halo), and 1 Gpc (cosmological). For non-repeating sources, the corresponding energies range from 10^{30} ergs to 10^{51} ergs. Simply providing this energy without consideration of efficiency for conversion to gamma-rays poses no great difficulty, even at cosmological distances. The formation of a neutron star releases gravitational binding energy of $\sim 3 \times 10^{53}$ ergs. Accretion into a black hole provides similar specific energies. However, for the requisite large accretion rates into black holes or for neutron star formation, one expects most of this energy to be carried away by neutrinos. If the source number is limited recurrence is an essential part of any GRB model. Depending on the active life of a typical source the total energy required even for non-cosmological bursts could be tremendous. We shall consider this point in detail in the subsequent discussion of Galactic halo models.

Without counterparts the distance scale is a guessing game. What are reasonable options? Very nearby options such as the Oort cloud have fallen out of favor (but see 10), mostly because it is not believed that the Oort cloud can satisfy the isotropy requirements. The local fluff of ISM around the sun may be symmetric enough, but no energy source seems to be available. The very local disk (~ 100 pc) is constrained by the symmetry braking position of the Sun 15 pc above the plane (17), and the usual Pop I disk is ruled out by global anisotropies if the sampling distance exceeds ~ 200 pc. This situation perhaps only leaves two alternatives; a very extended halo or truly cosmological distances.

Extended Halo Models

To save the Galactic hypothesis under the tight constraints of isotropy one must argue for a very extended structure, to minimize the dipole due to the solar offset from the Galactic center. The current multipole limits (6,33,49) require galacto-centric shells with radii of at least 200 kpc. On the other hand, halos that are too large may yield an excess of bursts towards M31, which is not observed (6,14). Populations coupled dynamically to the Galaxy show strong anisotropies, but tracer components confined in the potential also show strong anisotropies (6). High velocity neutron stars that left the Galaxy a long time ago are uniformly distributed between the galaxies and constitute a cosmological model (see below). Galactic halo models are thus based on continuous injection of high velocity sources into the halo from the vicinity of the disk (7,27,41).

General arguments can be made about the energy requirements for high-velocity pulsar (HVP) models (18). A minimum kick velocity ~ 800 km s^{-1} is required for guaranteed escape from the Galaxy ("streaming" models). Some halo models allow at least a fraction of the stars to be bound to the Galaxy, but the strong anisotropy of the "virialized" population requires a maximum duration of the GRB phase of $T_{max} \sim 10^9$ yrs. In either case, one demands that essentially all stars with small velocities do not produce GRB sources. Most HVP models also require a delayed turn-on of the GRB phase. To reach a distance of ~ 30 kpc requires, at speeds v $\sim 10^3$ v$_3$ km s^{-1}, a delay $\Delta t \sim 3 \, 10^7$ yrs. High velocity pulsars are rare and we may only observe their GRB events for a short time of $\sim 3 \, 10^8$ yrs before they disappear below the detection threshold.

While cosmological burst models usually invoke singular events, Galactic models require multiple outbursts, which leads to severe energy budget constraints (3,4,18,31). We presumably observe bursts to a maximum sampling distance D = 300 D_{300} kpc. The acceptable range of D_{300} may be as large as 0.5 − 2. The oldest observable source has an age $T_{max} = 3 \, 10^8 \, D_{300} \, v_3^{-1}$ yrs. Pulsars are produced at a rate of $R_b = 10^{-2} \, r_{-2} \, yr^{-1}$. Thus, during the past 10^8 yrs a total of $\sim 10^6$ pulsars were born. Of these perhaps 10% have the required velocity to become burst sources. With a burst rate $R_\gamma \sim 10^3$ R_3 yr^{-1} we expect the average source to produce GRBs at a rate

$$r \sim 3 \, 10^{-3} \, R_3 \, R_{-2}^{-1} \, D_{100}^{-1} \, f_\gamma^{-1} \, f_v^{-1} \, v_3^{1+\alpha} \, yr^{-1} \qquad (2)$$

where f_γ is the fraction of HVPs that actually become GRB sources. The fraction of HVPs with velocity v_3 is poorly known (30). We assume f = 0.1 f_v, decreasing with velocity as $v_3^{-\alpha}$. In the halo the observed fluxes and durations imply event energies of

$$E_0 \sim 10^{43} \, D_{300}^2 \, ergs \qquad (3)$$

where we used a mean effective burst duration of $T_{eff} \sim 10$ s (26). The total amount of energy required to sustain GRBs during halo passage is thus

$$E_{tot} \sim 10^{49} \, D_{300}^2 \, v_3^\alpha \, ergs \, . \qquad (4)$$

Consider the extremely favorable case that we have all Galactic neutron stars at our disposal, and we assume a total population of N = 10^9. This requires a minimum specific rate of 10^{-6} yr^{-1}, which implies a minimum energy of 10^{47} ergs. To allow such "small" values we disregard any argument about realistic efficiencies and we rely on the enhanced production of neutron stars during early disk evolution. Any selection of a subset of the pulsar population or requiring only recently (within a few Gyrs) produced neutron stars, increases the energy requirement to the scale given above.

The perhaps largest reservoir of energy is rotation. If electromagnetic dipole radiation dominates the torque on the neutron star the period derivative of a pulsar (in units of 10^{-15} s s^{-1}) is given by the standard equation

$$\dot{P}_{-15} \sim P^{-1} \, B_{12}^2 \qquad (5)$$

so that in the limit of no field decay the period follows

$$P(t)^2 \sim P(0)^2 + 6\, B_{12}^2\, t \qquad (6)$$

The spin-down luminosity is proportional to the square of the field strength and inversely proportional to the fourth power of the period. We are only interested in the total amount of energy available for bursts between the time of turn-on, T_i, and the time they leave the sampling volume, $T_f = 3\, D_{300}\, v_3^{-1}$, or, equivalently, the maximum duration of the GRB phase. From the energetics point of view we generously estimate the available energy in the limit $T_f \to \infty$, which yields

$$E_{max} \sim 3\, 10^{45}\, B_{12}^{-2}\, T_i^{-1}\; \text{ergs} \qquad (7)$$

It is clear that in order to have enough rotational energy left at the onset of the GRB phase ($\sim 3\, 10^7$ yrs), the field should be significantly less than 10^{11} Gauss. However, this is not easily accomplished because the number of sources with such low fields will be much less than the already low number assumed above. This increases the required burst rate per source, and drives up the overall energy requirement (which in turn demands an even smaller field, and so on). We have a runaway argument against the rotation powered HVP scenario. If the energy per burst is roughly constant, the rotation-driven burst rate would be proportional to t^{-2}. Steady state injection of sources leads to a density profile that approximately scales as r^{-2}, so that the actual GRB rate density would sharply peak as r^{-4} (with a hole for r less than ~ 30 kpc). Since the inner regions of the Galaxy contribute the largest dipole anisotropy, dipole moment $\propto 8.5$ kpc/r (15), this kind of model would have major difficulties to match the observed angular and brightness distributions. Proposed models circumvent this problem by requiring a smooth increase of burst activity with time. These prescriptions are artificial.

Are there alternative energy sources ? If accretion of external matter is used, approximately 10^{-4} $M_\odot$ of matter must be stored around the neutron star, to be released in $\sim 10^6$ installments of accreted blobs of mass M $\sim 10^{23}$ g, significantly larger than the most massive known comets in the solar system. The formation of a massive accretion disk around a HVP through fall-back of material during the supernova event, and the formation of planetesimals with the right properties is conceivable (28). Elastic energy stored in the neutron star crust is estimated to be less than 10^{44} ergs (3,4,31,44). Magnetic energy is only sufficient if each neutron star contains internal fields of $\sim 10^{16}$ Gauss. Even if we ignore the requirement of energy release at a steady pace over 10^{8-9} yrs, there is no compelling reason to believe that all pulsars have such internal fields. In either case, if we require ultra-high fields the fraction of pulsars endowed with those fields should be very much less than 1%. This in turn reduces the number of sources and increases the required burst rate. The required energy would be correspondingly larger and this would demand higher fields, etc. This again is a runanway argument.

Bridging the Gap

Before embracing cosmological models, consider a generic "model" that bridges the two distance scales (20,47). Imagine very old ($\sim 10^{10}$ yr) neutron stars. If they evolve without external disturbances they may encounter an internal phase transition, resulting in large structural readjustments that could yield sufficient energy. If neutron stars have small velocities, they would be bound to the galactic disk environment, where accretion may prevent the phase transition. If they are born with velocities in excess of the escape velocity they escape into the halos of galaxies and eventually into intergalactic space. Those stars (predominantly provided by dwarf galaxies) are the right candidates for undisturbed evolution. We imagine the universe filled with intergalactic HVPs, causing GRBs once near the end of their lifes. In a sense this is a model that resembles the EGH models but is clearly a cosmological model, because the distance scale is now at least 100 Mpc, including the local group of galaxies, but could be larger. If time dilation measurements turn out to be inconclusive, a more local distance scale (10-100 Mpc) could be consistent with the data. The change of slope in Log N - Log P would simply reflect the finite extend of the local distribution of galaxies. The high end of this scale is already constrained through the absence of anisotropies in super-galactic coordinates (16), leaving the local group scale (10 Mpc) to be tested against isotropy. One expects a very nearly isotropic distribution because 10^3 km s^{-1} for 10^{9-10} yrs corresponds to 1-10 Mpc. This would be the EGH equivalent of the local group. In addition the group has collapsed by a substantial fraction since it began as a density fluctuation in the early universe. The changing potential would isotropize their orbits. Inhomogeneity and isotropy could thus be accomodated by this model, although no reliable simulations exist. Burst properties could have narrow distributions, because the initial conditions of each phase transition would very similar. This may explain the observed narrow luminosity function. The model also has predictive power: no recurrences (consistent with current limits). There would be no M31 excess. There would be no global energy crisis. There would be few bright galactic counterparts because most sources originate in dwarf galaxies and the distances traveled in either case are comparable or larger than mean galaxy separations.

Cosmological Bursts

The energy required for cosmological models is about $10^{51}\delta\Omega$ ergs, with $\delta\Omega$ the beaming fraction of the burst. Even if this energy is delivered as pure radiation, two-photon pair production ($\gamma\gamma \to e^-e^+$) would create a pair plasma optically thick to gamma-rays. If no baryonic matter is present a pair fireball will be created at temperatures of $\sim$ 100 McV with a Compton optical depth exceeding 10^{10}. The fireball will thermalize regardless of the energy injection mechanism. The resulting bulk motion of these expanding pair fluids leads to Lorentz factors $\Gamma \sim 10^3$. While these factors are sufficient to boost the spectrum into the γ band, the spectrum would be approximately thermal, in conflict with the observations.

Constraints on fireball Lorentz factors follow from demanding that the opacity is below unity at all photon energies. The maximal photon energy from observed GRBs is ~ 10 GeV. We use the observed time variability, $\delta t = 10^{-3}\, \tau_{\rm ms}$ s, to estimate the size of the source region. Allowing for relativistic expansion with bulk Lorentz factor Γ the spatial scale is given by $R = c\, \delta t \Gamma \sim 3 \times 10^7 \tau_{\rm ms}\, \Gamma$ cm. The mean observed photon energy is ϵ so that, in the rest frame of the emitting source, photons have energies of order $\epsilon_0 = \epsilon \Gamma^{-1}$. If we assume that the expansion leads to beaming into a solid angle $\delta \Omega \sim \Gamma^{-2}$, the observed GRB flux, F, yields the source luminosity, L, for a given distance D, from $F = L\, D^{-2}\, \Gamma^2$. Relativistic expansion reduces the luminosity requirements (24), which implies smaller photon number densities at the source, which in turn reduces the optical depth to pair production

$$\tau_{\gamma\gamma} \sim \sigma_T\, n_\gamma\, R \sim 10^5\, D_{\rm kpc}^2\, \tau_{\rm ms}^{-2}\, \Gamma^{-4} \quad (8)$$

where we assumed a typical photon energy of $\sim 10^{-6}$ ergs (~ 1 MeV) in the rest frame and $F_{-6} \sim 1$. At cosmological distances (say 1 Gpc) the Lorentz factor must exceed $\Gamma \sim 10^4$ to significantly reduce the opacity. An additional effect of relativistic beaming is the suppression of pair creation through a reduction of the threshold energy. A test photon with energy ϵ_0 interacting with a radiation field of photons with energy ϵ_0' will create pairs only if

$$\epsilon_0\, \epsilon_0'\, (1 - \mu) \geq 2\, \left({\rm m}_e c^2\right)^2 \quad (9)$$

is satisfied, where μ is the interaction angle in the com frame. Pair creation further increases the opacity via Klein-Nishina scattering off the additional leptons. Detailed calculations (2) suggest that $\Gamma \sim 10^3$ for cosmological models. Meszaros & Rees (34) discuss how this might be reduced to $\Gamma \sim 100$ in a non-steady jet.

The burster brightness distribution suggests a maximum source redshift $z_{\rm max} \sim 1$ (9,21,35,56). Assuming $\Omega =$ and $H_0 = 75$ km s^{-1} Mpc^{-1} this implies a typical source luminosity of $L \sim 6 \times 10^{50}$ ergs s^{-1} and an event density of $\rho = 22$ bursts yr^{-1} Gpc^{-3} h_{75}^3. An upper limit on redshift can be obtained from the requirement that the photons of highest energy not be attenuated by pair production off the intergalactic IR field. EGRET recorded several bursts with GeV emission, the highest photon energy recorded to date (GB940217) is ~ 10 GeV (22), implying a maximum source redshift ~ 1. If future observations extend burst detections into the TeV range, severe constraints on the burster distance scale would result. Some of the GeV emission from GB940217 was delayed by over one hour. Such delays are hard to explain in any "promt" scenario, but models in which delayed GeV emission may be a characteristic feature have been developed (34).

Most cosmological models are based on progenitors residing in galaxies. The B-band emissivity of the universe ($L_u \sim 2 \times 10^8$ L$_{B\odot}$ Mpc^{-3}), and the B-band luminosity of our Galaxy ($L_g \sim 2 \times 10^{10} L_{B\odot}$) thus imply a typical host density of $n \sim 10^7$ Gpc^{-3}, i.e., $n_7 \sim 1$. If emission is beamed into the fraction $\delta\Omega$, the specific event rate per host is

$$\lambda = 2.5 \times 10^{-6} \; \lambda_{\gamma 3} \; n_7^{-1} \; \delta\Omega^{-1} \qquad (10)$$

where $\lambda_{\gamma 3}$ is the observed full sky burst rate in units of 10^3 yr^{-1}. If strong beaming (say, $\delta\Omega \sim 10^{-2}$) were an essential part of cosmological burst models, specific rates as high as 10^{-4} yr^{-1} would be required. Such high rates may be hard to accomplish in conventional merger scenrios.

The interpretation of GRB statistics in the framework of cosmological models usually emphasizes the faint end of their distribution, but much can be learned from rare, bright bursts (16,43). In the BATSE sample the nearest bursts may be as close as z $\sim$ 0.1 (12,56), or D $\sim$ 300 h^{-1} Mpc. This would allow high energy photons or particles from such bursts to reach the Earth without excessive degradation through interaction with the cosmic microwave background or the intergalactic IR photons. An association of ultra-high energy photons ($\sim$ 100 TeV) with GRBs has recently been claimed for the bright burst GRB910511 (40), and a possible association of a bright GRB with an ultra-high energy cosmic rays (UHECR) at E $\sim 10^{20}$ eV was also suggsted (36, but see 53-55). However, this association requires the burst distance to be less than $\sim$ 50 Mpc, which is in conflict with statistical arguments derived from Log N - Log P, which implies only a very small probability for a GRB from such a small distance.

Substantial losses along their cosmic paths limit the universe observable above 10^{19} eV to distances less than $\sim$ 200 Mpc. On this distance scale the luminous universe is not uniform, but shows a concentration towards the supergalactic (SG) plane. Any cosmological model in which GRBs trace luminous matter would predict deviations from homogeneity and isotropy on small scales. It is currently unknown above which lengths scale homogeneity is a good representation of LSS, but for D $\sim$ few 100 Mpc the local universe is know to be very inhomogeneous (39,52). In particular, the concentration of galaxies towards the supergalactic plane could provide a "smoking gun" for cosmological GRB models. A recent search for such anisotropies (16) has not been successfull, but the sample of bursts is still very small, with too few bursts from the local universe (see also 43).

One argument against a cosmological burst paradigm is the perceived absence of bright galaxies in small GRB error boxes (12,46,57), the so-called "no-host problem". While it is possible to argue that bursts may either be located well outside of their hosts or that their hosts could be underluminous galaxies, it may not be necessary to resort to any of these "solutions". The reason is simply the fact that recent studies of IPN error boxes in the near IR (25) and optical (29) produced a substantial number of potential host galaxies in the expected magnitude range. The "no-host problem" may no longer exist.

Another critical issue is burst recurrence. Most cosmological models require single events, so that an unambiguous detection of one or more repeaters in the current burst sample would provide a severe constraint on cosmological scenarios. At this meeting there was an extensive discussion of the evidence and the data quality used for repeater searches. While all workers in this field now appear to agree that there remains no significant evidence for burst recurrence, there were arguments to suggest that the BATSE localization quality is insufficient to allow definitive conclusions. While this issue remains to be resolved in detail, the results of Tegmark (50) suggest that the BATSE data now limit the fraction of repeating sources to less than a few percent. An independent analysis of COMPTEL data (23) also does not find evidence for repetition. While there are no doubts about COMPTEL's ability to accurately localize GRBs, the sample size is still too small to limit the repeater fraction to less than $\sim 20\%$. Future burst missions will provide much better source localizations, which should resolve the repeater question once and for all.

GRBS AND DINOSAURS

The demise of the dinosaurs ~ 60 million yrs ago has been attributed to many possible causes, and Thorsett (51) recently added GRBs to the list. The idea is simple; if GRBs occur in galaxies, they also appear in the Milky Way at a rate $\sim 10^{-6}$ yr^{-1}. Thorsett estimates that a planet would receive a damaging dose of γ-rays if the burst were closer than $R_l \sim 1$ kpc. From the event rate and the critical distance it follows that a Galactic GRB could have killed the dinosaurs.

Where, however, do bursts occur ? If, GRBs are caused by tidal consumption of stars by black holes in galactic centers (8) they would pose little danger to the solar neighborhood. The likelihood of harm to the Earth is larger if GRBs were due to "failed" supernovae (58). Woosley estimates that this type of collapse may occur once every 10^3 yrs in the disk. Another disk source of GRBs is a NS/NS merger, which would occur with a rate of $\sim 10^{-6}$ yr^{-1}. The effective rate is reduced by the fraction $f_0 = (R_l/R_g)^2 \sim 10^{-2}$ of events close enough to seriously affect the Earth environment, where R_g is the galactic radius containing merger events (~ 10 kpc). This yields a nominal rate of possible extinction events of once every 100 Myr, which is suggestively close to the timescale of terrestrial extinction events (although periodicity would not be expected). The progenitors of these binary systems also trace the distribution of massive stars, but unlike supernovae they do not occur near the Galactic midplane. At the time of binary formation velocities of v $\sim 10^2$ km s^{-1} are imparted on the system. Velocities of 100 km s^{-1} (or larger) remove the systems many kpc from their birthplaces, because the merger time scale is

$$\text{Log } \tau_m(\text{yrs}) \sim 10.5 + 2.7 \text{ Log P(days)} + 3.5 \text{ Log}\left(1 - e^2\right) \qquad (11)$$

Recent simulations of binary star evolution (42) provide realistic distributions of initial periods, P, and eccentricities, e, from which we derive the distribution of merger delays. The distribution of mergers can be described with an exponential scale height $H_z \sim 2\ \sigma_{100}$ kpc, much larger than the scale height of their progenitor stars. If mass extinctions could be induced by a

GRB within $R_l \sim 1$ kpc, these estimates indicate that less than a fraction f_0 $(1 - e^{-x})$ g_r of all events occur close enough to home to be capable of affecting the atmosphere, where $x \sim 0.6 \; R_l \; \sigma_{100}^{-1}$, and $g_r \sim 0.5$ accounts for the radial spreading of merger events beyond the canonical value R = 10 kpc. The effective rate is thus less than $2 \; 10^{-9}$ yr^{-1}, which argues against the hypothesis that GRBs were close enough, at the right time. Seven mass extinction peaks in the last 250 million years are non-periodically spaced 20 to 60 million years apart, indicating a timescale for catastrophic events that is much shorter than that derived for fatal GRBs in the Galaxy. Thus, while GRBs may produce the largest explosions in the universe, it does not seem likely that they are responsible for the demise of the dinosaurs or any of the other mass extinctions.

Acknowledgements: This work was supported in part by NASA under grant NAG 5-1578.

REFERENCES

1. D.L. Band, et al., ApJ **434**, 560 (1994).
2. M. Baring, & A. Harding, these proceedings (1996).
3. O. Blaes, R. Blandford, P. Goldreich, & P. Madau, ApJ **343**, 839 (1989).
4. O. Blaes, R. Blandford, P. Madau, & S. Koonin, ApJ **363**, 612 (1990).
5. M.S. Briggs, ApSS **231**, 3 (1995).
6. M.S. Briggs, et al., ApJ **459**, 40 (1996).
7. T. Bulik, & D.Q. Lamb, in High Velocity Neutron Stars and Gamma-Ray Bursts, eds. R. Rothschild & R. Lingenfelter, AIP Conf. Proc. **366**, 258 (AIP, Ney York, 1996).
8. B. Carter, ApJ Letters **391**, L67 (1992).
9. E. Cohen, & T. Piran, ApJ **444**, L25 (1995).
10. C. Dermer, these proceedings (1996).
11. B. Dingus, et al., ApSS **231**, 187 (1995).
12. E.E. Fenimore, et al., Nature **366**, 40 (1993).
13. G.J. Fishman, & C.A. Meegan, Ann. Rev. Astr. & Astrophys. **33**, 415 (1995).
14. J. Hakkila, et al., ApJ **422**, 659 (1994).
15. D.H. Hartmann, et al., ApJ Suppl. **90**, 893 (1994).
16. D.H. Hartmann, M.S. Briggs, & K. Mannheim, ApJ, submitted.
17. D.H. Hartmann, J. Greiner, & M.S. Briggs, A & A **303**, L65 (1995).
18. D.H. Hartmann, & R. Narayan, ApJ, in press (1996).
19. D.H. Hartmann, A & A Rev. **6**, 225 (1995).
20. D.H. Hartmann, A & A, in press (1996).
21. J.M. Horack, A.G. Emslie, & D.H. Hartmann, ApJ **447**, 474 (1994).
22. K. Hurley, et al., Nature **372**, 652 (1994).
23. M. Kippen, et al., these proceedings (1996).
24. J.H. Krolik, & E.A. Pier, ApJ **373**, 277 (1991).
25. S.B. Larson, I.S. McLean, & E.E. Becklin, ApJ Letters, in press (1996).
26. T. Lee, & V. Petrosian, these proceedings (1996).
27. H. Li, & C. Dermer, Nature **359**, 514 (1992).
28. D. Lin, S.E. Woosley, & P. Bodenheimer, Nature **353**, 827 (1991).

29. C. Luginbuhl, et al., these proceedings (1996).
30. A.G. Lyne, & D.R. Lorimer, Nature **369**, 127 (1994).
31. P. Madau, in Gamma-Ray Bursts, eds. C. Ho, R.I. Epstein, & E.E. Fenimore, 9 (CUP, 1992).
32. C.A. Meegan, et al., ApJ **446**, L15 (1995).
33. C.A. Meegan, et al., ApJ, in press (1996).
34. P. Mészáros, & M.J. Rees, MNRAS **269**, L41 (1994).
35. P. Mészáros, & A. Meszaros, ApJ, in press (1996).
36. M. Milgrom, & V. Usov, ApJ **449**, L37 (1995).
37. T. Murakami, et al., Nature **335**, 234 (1988).
38. D.M. Palmer, et al., ApJ **433**, L77 (1995).
39. P.J.E. Peebles, Principles of Physical Cosmology, (Princeton Univ. Press, 1993).
40. S.P. Plunkett, et al., ApSS **231**, 271 (1995).
41. P. Podsiadlowski, M.J. Rees. & M. Ruderman, MNRAS **273**, 755 (1994).
42. S.F. Portegies Zwart, & J.N. Spreeuw, Astron. & Astrophys., in press (1996).
43. J.M. Quashnok, these proceedings (1996).
44. M. Ruderman, ApJ **382**, 587 (1991).
45. B. Schaefer, Nature **294**, 722 (1981).
46. B. Schaefer, in Gamma Ray Bursts Observations, Analyses, and Theories, eds. C. Ho, R. I. Epstein, and E. E. Fenimore, 107 (CUP, 1994).
47. B. Schaefer, these proceedings (1996).
48. M. Sommer, et al., ApJ Letters **422**, L63 (1994).
49. M. Tegmark, et al., ApJ, submitted (1996a).
50. M. Tegmark, et al., ApJ, submitted (1996b).
51. S.E. Thorsett, ApJ **444**, L53 (1995).
52. R.B. Tully, ApJ **388**, 9 (1992).
53. M. Vietri, ApJ **453**, 883 (1995).
54. E. Waxman, Phys. Rev. Lett. **75**, 386 (1995a).
55. E. Waxman, ApJ **452**, L1 (1995b).
56. W.A.D.T. Wickramashinghe, et al., ApJ Letters **411**, L55 (1993).
57. E. Woods, & A. Loeb, ApJ, in press (1996).
58. S.E. Woosley, ApJ **405**, 273 (1993).

Implications of the Observed Angular Distribution of Gamma-Ray Bursts for Galactic and Cosmological Models

D. Q. Lamb

Dept. of Astronomy and Astrophysics, University of Chicago, Chicago, IL 60637

> I consider the implications of the observed angular distribution of γ-ray bursts for Galactic and cosmological models of the bursts. The positions of individual bursts provide information about counterparts at other wavelengths. The angular distribution on small scales provides information about burst repetition. The angular distribution on intermediate scales constrains the distance scale to γ-ray bursts, if they are cosmological in origin. The angular distribution on large scales constrains Galactic corona models of the bursts. I discuss each of these topics in light of the BATSE 3B catalog.

INTRODUCTION

Gamma-ray bursts continue to confound astrophysicists nearly a quarter of a century after their discovery (1). Despite intense study by observers and theorists alike, no one knows for sure what they are or where they come from. Confirmation by the Burst and Transient Source Experiment (BATSE) on the *Compton Observatory* of a roll-over in the brightness distribution of the bursts and the discovery that the sky distribution of even faint bursts is consistent with isotropy (2) has intensified debate about whether the bursts come from sources in our own Galaxy or are cosmological in origin.

The angular distribution of γ-ray bursts provides important information about their nature and distance. Studies of individual burst positions provide information about radio, infrared, optical, ultraviolet, and X-ray counterparts. Studies of the angular distribution on small scales can place constraints on burst repetition. Studies of clustering on intermediate angular scales can place a lower limit on the distance scale to the bursts, if they are cosmological in origin and trace the large scale structure of luminous matter, while studies of the angular distribution on large scales can constrain Galactic corona models of the bursts.

In this review I summarize various methods currently used to determine burst positions. I then discuss each of the topics mentioned above in light of the BATSE 3B catalog (3).

© 1996 American Institute of Physics

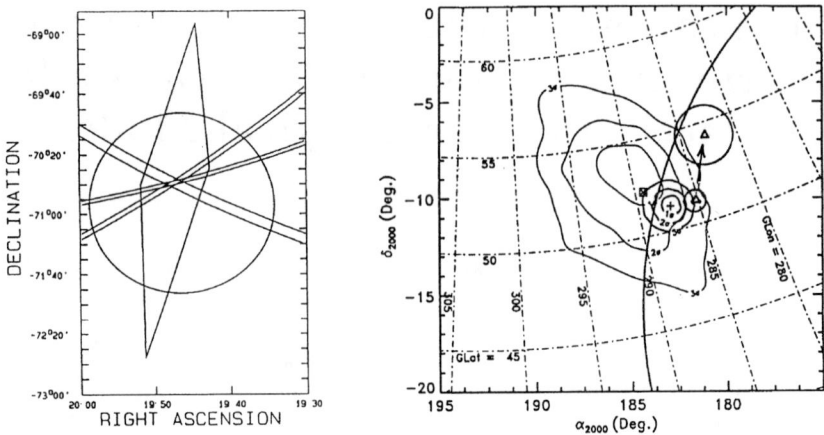

Figure 1. Localization of GB910122. The large circle is the WATCH location. The diamond-shaped error box gives the approximate boundary of the SIGMA error box. The IPN annuli are from WATCH/Ulysses, SIGMA/Ulysses, and PVO/Ulysses, and are 2' - 6' wide (4).

Figure 2. Localization of GB930131. The thin-lined contour map shows the COMPTEL 1, 2, and 3 σ confidence regions surrounding the maximum likelihood best-fit COMPTEL position (box-enclosed-cross). The thick-lined contour map shows the EGRET 1, 2, and 3 σ confidence regions surrounding the maximum likelihood best-fit EGRET position (plus sign). The smaller circle shows the BATSE 2B 1 σ error circle surrounding the best-fit BATSE 2B position (triangle), while the larger circle shows the BATSE 3B 1 σ error circle surrounding the best-fit BATSE 3B position (triangle); the arrow shows the angular displacement between the 2B and 3B positions. The IPN annulus is from BATSE/Ulysses and is approximately 1' wide. [After Ryan et al. (5)].

METHODS OF DETERMINING GAMMA-RAY BURST POSITIONS

At least four different methods are currently being used to determine the positions of γ-ray bursts. Time-of-flight measurements between instruments on the satellites in near-Earth orbit and the interplanetary spacecraft that constitute the 3rd IPN have yielded > 18 small (several arcmin^2) error boxes, and > 200 single arcs ($\lesssim$ 1 arcmin wide) (6). The imaging capabilities of the EGRET and COMPTEL instruments on the *Compton Observatory* have given us > 5 error circles with radii $\lesssim 0.5°$ (7) and > 15 error circles with radii $\lesssim 1°$ (8), respectively. Moderately accurate positions have also been achieved through the use of coded aperture masks, as in the SIGMA instrument, or rotation-modulation collimators, as in the WATCH instrument. The latter has provided > 30 error circles with radii $\lesssim 3°$ (9). The BATSE on the

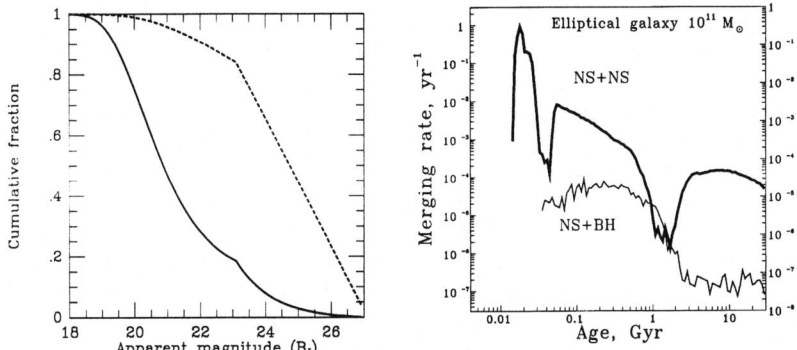

Figure 3. Local ($z = 0$) Schechter function for the number distribution (dotted line) and luminosity distribution (solid line) of field galaxies.

Figure 4. Temporal evolution of neutron star + neutron star (thick line) and neutron star + black hole (thin line) binary coalescence rates normalized to a model elliptical galaxy with baryonic mass $10^{11} M_\odot$ (20).

Compton Observatory uses detector geometry to determine the positions of the bursts it detects; this method has yielded > 1100 error circles with radii < $2° - 30°$ (3). Figures 1 and 2 illustrate the results obtained using these different techniques.

COUNTERPARTS

We expect the quiescent optical counterparts in the Galactic corona model in which the bursts come from high-velocity neutron stars to be extremely faint. This is because the neutron stars have a small radiating area ($A \approx 10^{13}$ cm^2) and lie at great distances ($d \gtrsim 100$ kpc), resulting in a small optical flux at Earth even if the neutron star is relatively hot. In contrast, bright optical counterparts are naturally expected in cosmological models, such as active galactic nuclei, failed supernova, and coalescing compact binaries.

Optical counterpart searches require accurate positions. For example, burst positions accurate to $\lesssim$ a few arcmin are needed in order to carry out counterpart searches down to $m_V \approx 22$, and to $\lesssim$ a few arcsec in order to carry out counterpart searches down to $m_V \approx 28$. Only a handful of burst positions accurate to $\lesssim$ a few arcmin currently exist; these come from the 1st and 3rd IPNs. However, missions have been proposed that could achieve positional accuracies of ≈ 30 arcsec [EXIST (10)], ≈ 10 arcsec [BASIS (11)], and ≈ 3 arcsec [ETA (12)] for hundreds of bursts.

The Schechter function for the luminosity distribution of local ($z = 0$) field galaxies, while still somewhat uncertain, implies a mean b_J of ≈ 20.5 (see Figure 3). However, observations of eight small 1st IPN error boxes find no optical counterparts down to $m_B = 21 - 25$ (13). Consequently, the γ-ray burst rate is not proportional to the blue light from galaxies, and a number of studies (13–15) conclude that the bursts cannot come from active galactic nuclei or bright blue galaxies.

While γ-ray bursts apparently cannot come from bright blue galaxies, the surface density on the sky of faint blue galaxies is very large, and there are many in even the smallest current γ-ray burst error boxes. A number of recent surveys have shed new light on the nature of these faint blue galaxies. Earlier, it was thought that the vast majority of these galaxies were nearby ($z \approx 0.1$–0.3) low luminosity ($L \approx 0.01$–$0.1\,L_*$ in the B-band) galaxies (16). However, recent surveys (17,18) suggest instead that most faint blue galaxies lie at greater distances than heretofore thought ($z \approx 0.5$–1.5) and are intrinsically bright ($L \approx L_*$ in the B-band). If so, faint blue galaxies are unlikely to be the host galaxies of the bright γ-ray bursts for which small IPN error boxes exist, since these bursts are expected to come from nearby galaxies.

Larson et al. (19) report the existence of bright ($m_K \lesssim 15.5$) infrared galaxies in or near a half-dozen small 3rd IPN error boxes. These galaxies are also bright in the optical ($m_V \approx 18$–19), and are easily visible on POSS plates. The difference in the optical content of the 1st and 3rd IPN error boxes most likely reflects the larger size of many of the 3rd IPN error boxes, which increases the probability of finding a bright galaxy inside the error box; otherwise there is a real discrepancy between the optical content of the two sets of error boxes. A comparative study of the optical contents of the 1st and 3rd IPN error boxes would help to resolve this issue.

Strong source space density and luminosity evolution is expected in the Galactic corona model of γ-ray bursts. It is sometimes said that source evolution is not expected if the bursts are cosmological. But while strong luminosity evolution may or may not occur (e.g., if the bursts come from failed supernovae or coalescing compact binaries), strong space density evolution is certainly expected. The progenitors of supernovae and compact binaries are massive stars, whose birthrates can vary strongly over the lifetime of a galaxy (cf. starburst galaxies). Consequently, the rate of failed supernovae and the birth rate of compact binaries, as well as the distribution of their initial separations, surely vary in time. Lipunov et al. (20) have carried out simulations of compact binary coalescence which illustrate this (see Figure 4).

In cosmological models, the burst rate may or may not be proportional to the blue light from galaxies (which reflects the number of massive stars in them). If γ-ray bursts come from failed supernovae, it can be, since the lifetimes of the progenitor stars are short. On the other hand, if the bursts come from coalescing compact binaries, it may not be, since the rate of γ-ray bursts produced by such binaries may lag the birth of the compact binaries, and therefore the epoch at which the progenitor massive stars of these binaries produce blue light, by 300–500 Myr.

There is increasing evidence for strong source luminosity and/or space density evolution of γ-ray bursts. For example, Atteia et al. (21) report evidence that the correlation between burst brightness (intensity) and spectral hardness seen in the BATSE sample of γ-ray bursts extends even to the brightest bursts, whose cumulative brightness distribution follows a $-3/2$ slope.

If confirmed, this correlation has several important implications. First, if γ-ray bursts are cosmological in origin, the brightest bursts supposedly come from nearby sources, where space is Euclidean, and should therefore show no cosmological redshift. The fact that these bursts nevertheless exhibit a brightness-spectral hardness correlation says that this correlation cannot be due to cosmological redshift, but is intrinsic to the bursts themselves.

Second, it says that, whether the bursts are Galactic or cosmological in origin, the $-3/2$ slope of the cumulative brightness distribution of the brightest bursts is accidental and does not imply that bursts are homogeneous or that nearby burst sources are uniformly distributed in space.

Third, it says that intrinsic source luminosity and/or space density evolution dominates the brightness distribution of γ-ray bursts (as is the case for many astronomical objects). Hence, the assumption made in many studies that the bursts are standard candles and that burst sources are uniformly distributed in space just won't do. It may also mean that the burst brightness distribution tells us very little about the distance scale to γ-ray bursts. If so and if the bursts are cosmological in origin, detection of the angular clustering expected if the bursts trace the large-scale structure of luminous matter or the identification of optical counterpart galaxies may be the only way to establish the distance scale to the bursts.

REPEATING

Repeating is expected in Galactic corona models of γ-ray bursts involving high-velocity neutron stars. In contrast, repeating would doom most cosmological models of the bursts (such as failed supernovae and coalescing compact binaries).

Spatial (22,23) and spatial-temporal (24) analyses of the BATSE 1B catalog found evidence for repeating. The repeating fraction f estimated from these studies was $\approx 12\%$ (22) and $5.5\% < f < 32.5\%$ (23). Subsequent analyses of the BATSE 2B catalog found no evidence of repeating, but the limits on f ranged from $\lesssim 20\%$ from spatial studies (25) to $\lesssim 100\%$ from spatial-temporal studies (26), due to limitations resulting from the failure of the tape recorders on-board the *Compton Observatory*.

There is only weak evidence for repeating in the BATSE 3B catalog [the excess in $w(\theta)$ at $\theta = 0°$ is $\approx 2.5\sigma$ (3)]; this is true even for the 1B sample of bursts, whose positions in the 2B and 3B catalogs differ greatly (see Figures 5 and 6).

What happened? One possibility is that the evidence for repeating in the 1B catalog is the result of a (highly unlikely) statistical fluctuation. Another is that the systematic errors σ_{sys} in the burst positions in the 3B catalog are much larger than the stated value of $1.°6$, and that consequently, the repeating signal is lost.

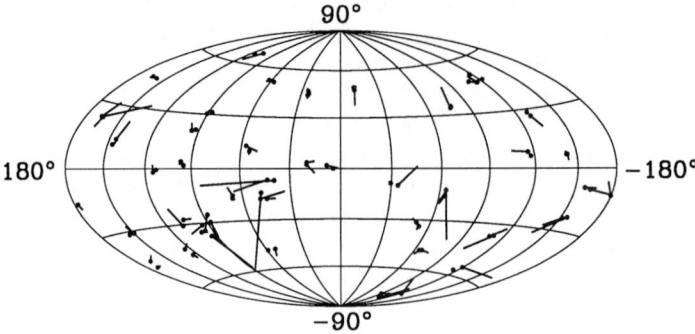

Figure 5. Angular displacements from 1B to 3B positions of bursts with nearest neighbors within 5° in the 1B catalog. These are the bursts which provided the evidence for repeating in the nearest neighbor analysis (22) of the BATSE 1B catalog.

The data used to determine burst positions in the 2B and 3B catalogs are basically the same (no new observations are possible, of course, since the bursts are a transient phenomenon). Thus the differences in the burst positions in the 2B and 3B catalogs arise, e.g., from different choices of background intervals, different choices of spectral energy-loss channels, omission of data from time intervals when the bursts are so bright that the detectors begin to saturate, and a different, more sophisticated treatment of Earth scattering (27). All such procedural changes affect σ_{sys}. Making the (conservative) assumption that the σ_{sys} for the 2B and 3B catalogs are uncorrelated, one expects the characteristic size of the angular differences θ_{2B_3B} between the burst positions in the 2B and 3B catalogs to be $\sigma = (4°^2 + 1°6^2)^{1/2}$, the root mean square of the stated σ_{sys} for the 2B and 3B catalogs. Figure 6 shows that the θ_{2B_3B} are much larger, which suggests that the σ_{sys} for the 2B and/or 3B catalogs are larger than the stated values.

We have used the BATSE data in conjunction with 196 single IPN arcs to analyze the σ_{sys} for the BATSE 2B and 3B catalogs (28). We find that the burst positions in the 1B catalog for which IPN arcs are available (and which have, by-and-large, the best-determined positions) are consistent with $\sigma_{sys} = 4°0$, the stated value for the 1B catalog (3). Fitting a constant σ_{sys} to the positions of such bursts in the 3B catalog, we find $\sigma_{sys} = 3°7$, far larger than the stated value of 1°6. But the data strongly request a more complicated model in which σ_{sys} is strongly correlated with σ_{stat}, so that $\sigma_{sys}(\sigma_{stat} = 0.1°) = 2°$ while $\sigma_{sys}(\sigma_{stat} = 10°) = 8°$ (see Figure 7).

No IPN arcs exist for the $\approx 80\%$ of bursts with larger σ_{stat}, so we are unable to characterize σ_{sys} for these bursts. However, even the σ_{stat}-dependent model is insufficient to account for the large differences between the burst positions in the 2B and 3B catalogs.

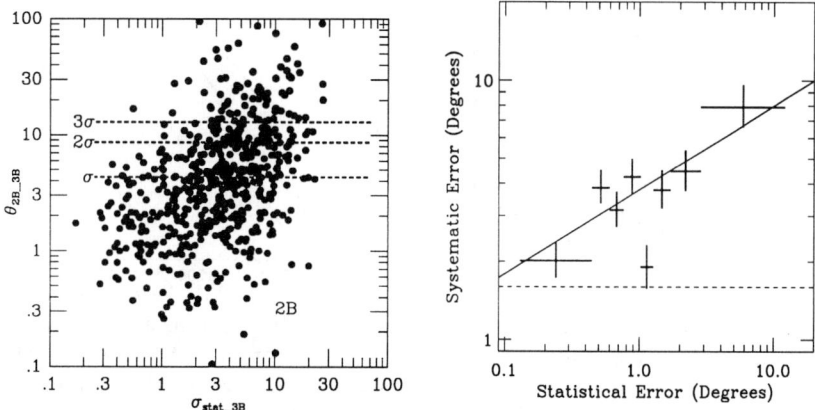

Figure 6. Scatter plot of the angular difference θ_{2B_3B} between the burst positions in the 2B and 3B catalogs versus the stated 3B statistical error σ_{stat_3B} for the 483 bursts in the 2B catalog for which positions exist. The horizontal dashed lines correspond to $\theta_{2B\_3B} = 1, 2$, and 3σ, where $\sigma = (4°^2 + 1.6°^2)^{1/2}$ is the root mean square of the stated σ_{sys} for the 2B and 3B catalogs. If the θ_{2B_3B} are due primarily to systematic errors in the 2B and 3B catalog positions and making the (conservative) assumption that the systematic errors in the 2B and 3B catalog positions are uncorrelated, one expects 187 bursts with $\theta_{2B\_3B} > 1\sigma$, whereas there are 270; 27 with $\theta_{2B\_3B} > 2\sigma$, whereas there are 132, and 2 with $\theta_{2B\_3B} > 3\sigma$ whereas there are 81.

Figure 7. Maximum likelihood best-fit σ_{stat}-dependent model of σ_{sys} for the bursts in the 3B catalog. The crosses are not data points; rather, they are the maximum likelihood best-fit constant σ_{sys} model for the given interval in σ_{stat}. The horizontal dashed line shows the stated value of σ_{sys} for the BATSE 3B catalog (28).

Adopting a total positional error of $6°.6$, which is the median total positional error for bursts in the BATSE 3B catalog σ_{stat}-dependent model for σ_{sys} described above, only a repeating fraction $f \gtrsim 15\%$ is detectable using the two-point angular correlation function $w(\theta)$. This is consistent with the limit $f \lesssim 12\%$ implied by the results of Tegmark et al. (29) taking $\sigma_{tot} = 6°.6$ in their spherical harmonic analysis, and the limit $f \lesssim 25\%$ that Bennett & Rhie (30) find using a pair-matching statistic. These limits are consistent with the estimates of f derived from analyses of the 1B catalog.

Thus we are frustrated in our attempt to answer the question of whether or not γ-ray bursts repeat by the current large uncertainties in burst positions. The hundreds of burst positions accurate to $\lesssim 100$ arcsec that would be provided by missions like EXIST (10), BASIS (11) and ETA (12) would provide a definitive answer.

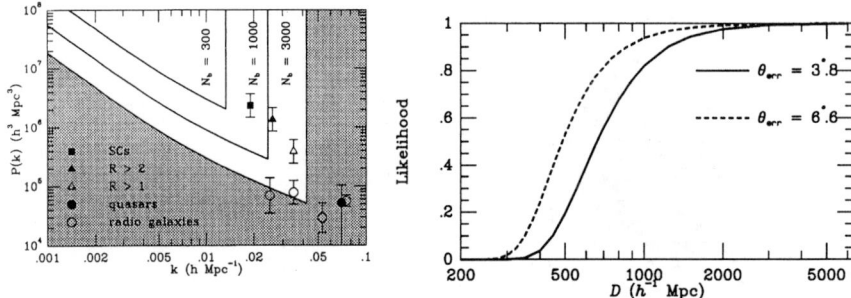

Figure 8. The power spectrum $P(k)$ of density fluctuations as a function of wavenumber k corresponding to distance scales of order $10h^{-1}$ Mpc to $3h^{-1}$ Gpc. The data points shown are from analysis of surveys of faint radio galaxies, quasars, clusters of galaxies, and from pencil beam surveys and studies of superclusters. Also shown are the regions of the $(k, P[k])$-plane that can be probed with 300, 1000, and 3000 γ-ray bursts, taking $D = 1h^{-1}$ Gpc ($z \approx 0.3$) as the effective sampling distance for illustrative purposes (32).

Figure 9. Likelihood of the BATSE 3B catalog data as a function of the effective comoving distance D to γ-ray bursts, shown for total positional errors $\sigma_{\rm tot} = 3°.8$ (solid line) and $6°.6$ (dashed line) (33). The former value of $\sigma_{\rm tot}$ is the median value for the bursts in the 3B catalog, taking the stated value for $\sigma_{\rm sys}$; the latter is the median value implied by the results (28).

CLUSTERING

If γ-ray bursts are cosmological, the burst sources are expected to trace the large-scale structure of luminous matter in the universe. One can therefore constrain the distance scale to cosmological bursts by comparing the expected clustering of bursts on the sky with that actually observed (see Figure 8) (31–33). Because the angular distribution of the bursts in the BATSE 3B catalog is consistent with isotropy (34), only a lower limit to the distance scale can be obtained, since a sufficiently large distance always leads to a sufficiently isotropic distribution on the sky.

Figure 9 shows that small values of the effective comoving distance D to γ-ray bursts are unlikely (33). Adopting the 3B median positional error of $\sigma_{\rm tot} = 3°.8$, D must be greater than $630\ h^{-1}$ Mpc, corresponding to a redshift $z > 0.25$, at the 95% confidence level. If the positional errors are better characterized by $\theta_{\rm err} = 6°.6$, the distance constraint is only slightly weakened: D must be greater than $500\ h^{-1}$ Mpc, corresponding to a redshift $z > 0.19$, at the 95% confidence level. At large values of D (the maximum value allowed is $D = R_{\rm H} = 6000\ h^{-1}$ Mpc, the size of the horizon in a closed universe), the likelihood goes to unity, because by projection a sufficiently large distance will always lead to an isotropic distribution on the sky. Note that the maximum likelihood value for D is $R_{\rm H}$, showing that the sky distribution of the bursts in the 3B catalog is consistent with isotropy.

If the lifetime of the *Compton Observatory* is extended another 6 years, yielding the locations of $\gtrsim 3000$ γ-ray bursts, it may be possible to detect the structure of luminous matter on the largest scales known (see Figure 8) (32).

LARGE-SCALE ANISOTROPIES

Large-scale anisotropies are not expected if γ-ray bursts are cosmological in origin. In contrast, such anisotropies must be present at some level, if the bursts come from high-velocity neutron stars in a Galactic corona, because the Sun is offset from the Galactic center by ≈ 8 kpc; the neutron stars are born in the Galactic plane; and the neutron stars have a circular velocity $v_{\text{circ}} \approx 200$ km s^{-1} around the Galactic center, in addition to the large kick velocity they receive at birth.

The observed peak flux distribution of the bursts and the observed small values of the Galactic dipole and quadrupole moments (34) constrain isotropic emission models to have neutron star kick velocities $v_{\text{kick}} \gtrsim 800$ km s^{-1} and the burst rate to increase smoothly over a period $\delta t \gtrsim 30$ Myr (35), while the allowed range of BATSE sampling distances is 130 kpc $\lesssim d_{\text{BATSE}} \lesssim 350$ kpc (35,36). The observed peak flux distribution of the bursts and the observed small values of the Galactic dipole and quadrupole moments constrain beamed emission models to have neutron star kick velocities $v_{\text{kick}} \gtrsim 800$ km s^{-1}, and a beaming angle $\theta_b \approx 20°$ (35,37). The allowed range of BATSE sampling distances is smaller than that allowed for isotropic emission models (35).

It has been recognized for some time that if bursts come from HVNSs in an extended corona around the Milky Way, it should also be possible to detect them from a similar corona around the nearby bright galaxy Andromeda (38). Detection of an excess of bursts from Andromeda would constitute definitive evidence that the bursts are Galactic in origin. Conversely, the observation of no excess toward Andromeda would strongly suggest that the bursts are cosmological in origin, provided that the observation is made by an instrument of sufficient sensitivity.

Recently, studies of the instrumental sensitivity and observing time required to detect such an excess have been carried out by Harrison and Thorsett (39), Li et al. (37), and ourselves (40). We find that if the bursts radiate isotropically, an experiment with a sampling distance $d_{\text{max}} \gtrsim 600$ kpc is required in order to detect the expected excess of bursts in the direction of Andromeda in ≈ 1 year of observation. If the bursts are beamed forward and backward along the direction of the neutron star's kick velocity, an experiment with $d_{\text{max}} \gtrsim 900$ kpc is required in order to detect the expected excess in a similar amount of time. Thus an instrument at least 50 times more sensitive than BATSE is needed in order to detect (or rule out) the expected excess of bursts in the direction of Andromeda, given current constraints on the BATSE sampling distance. Ways to achieve this increased sensitivity include collimation of the detectors, larger effective area, lower particle background, and longer trigger time scales (39,40).

REFERENCES

1. R. W. Klebesadel et al., Ap. J. **182**, L85 (1973).
2. C. A. Meegan, et al., Nature **355**, 143 (1992).
3. C. A. Meegan, et al., electronic catalog (grossc.gsfc.nasa.gov).
4. K. Hurley, et al., in *Gamma-Ray Bursts*, AIP Conf. Proc. **307**, ed. G. J. Fishman, J. J. Brainerd, and K. Hurley (New York: AIP), p. 364 (1994).
5. J. Ryan, et al., Ap. J. **422**, L67 (1994).
6. K. Hurley, et al., electronic catalog.
7. B. L. Dingus, et al., in *Gamma-Ray Bursts*, AIP Conf. Proc. **307**, ed. G. J. Fishman, J. J. Brainerd, and K. Hurley (New York: AIP), p. 22 (1994).
8. R. M. Kippen, Ph.D. thesis, U. New Hampshire (1995).
9. A. J. Castro-Tirade, Ph.D. thesis, U. Copenhagen (1994).
10. F. Harrison, et al., these proceedings.
11. N. Gehrels, et al., these proceedings.
12. G. R. Ricker, et al., these proceedings.
13. B. E. Schaefer, in *Gamma-Ray Bursts*, ed. C. Ho, R. I. Epstein, and E. E. Fenimore (Cambridge: Cambridge U. Press), p. 107 (1992).
14. E. E. Fenimore, et al., Nature **366**, 40 (1993).
15. E. Woods & A. Loeb, Ap. J., 453, 583 (1995).
16. K. Glazebrook, et al., MNRAS **273**, 157 (1995).
17. S. Lilly, et al., Ap. J. **455**, 108 (1995).
18. L. L. Cowie, et al., AJ **110**, 1576 (1995).
19. S. B. Larson, et al., these proceedings.
20. V. M. Lipunov, et al., Ap. J. **454**, 593 (1995).
21. J.-L. Atteia, et al., these proceedings.
22. J. M. Quashnock & D. Q. Lamb, MNRAS **265**, L59 (1993).
23. T. E. Strohmayer, E. E. Fenimore & J. A. Mirales, Ap. J. **432**, 665 (1994).
24. V. C. Wang & R. E. Lingenfelter, Ap. J. **416**, L13 (1993).
25. C. A. Meegan, et al. 1995, Ap. J. **446**, L15 (1995).
26. J. J. Brainerd, et al. Ap. J. **441**, L39 (1995).
27. G. N. Pendleton, talk at the 3rd Huntsville Symposium on Gamma-Ray Bursts (1995).
28. C. Graziani and D. Q. Lamb, these proceedings.
29. M. Tegmark, et al., Ap. J., in press (1996).
30. D. P. Bennett & S. H. Rie, Ap. J. **458**, 293 (1996).
31. D. H. Hartmann, et al., Ap. J. **367**, 186 (1991).
32. D. Q. Lamb and J. M. Quashnock, Ap. J. **415**, L1 (1993).
33. J. M. Quashnock, these proceedings.
34. M. S. Briggs, et al., Ap. J. **459**, 40 (1996).
35. T. Bulik and D. Q. Lamb, these proceedings.
36. Ph. Podsiadlowski, M. J. Rees, and M. Ruderman, MNRAS **273**, 755 (1995).
37. H. Li, et al., these proceedings.
38. J.-L. Atteia & K. Hurley, Adv. Space Sci. **6**, 39 (1986).
39. F. A. Harrison & S. E. Thorsett, Ap. J. **460**, L99 (1996).
40. T. Bulik, P. S. Coppi, and D. Q. Lamb, these proceedings.

The BATSE 3B Catalog

Charles A. Meegan[*], Geoffrey N. Pendleton[†], Michael S. Briggs[†],
Chryssa Kouveliotou[‡], Thomas M. Koshut[†], John P. Lestrade[¶],
William S. Paciesas[†], Michael L. McCollough[**], Jerome J.
Brainerd[†], John M. Horack[*], Jon Hakkila[§], William Henze[††],
Robert D. Preece[†], Robert S. Mallozzi[†], Gerald J. Fishman[*]

[*]*NASA Marshall Space Flight Center, Huntsville, AL 35812*
[†]*University of Alabama in Huntsville, Huntsville, AL 35899*
[‡]*Universities Space Research Association*
[¶]*Mississippi State University, State University, MS 93762*
[**]*Hughes STX, Huntsville AL*
[§]*Mankato State University, Mankato, MN 56002*
[††]*Teledyne Brown Engineering, Huntsville, AL*

The Third BATSE catalog of gamma-ray bursts has recently been released. It comprises 1122 GRBs spanning more than three years of operation. All of the locations (including those of the 2B catalog) have been recomputed using an improved algorithm. A few percent of the bursts in the 2B catalog had large errors, which are corrected with the new algorithm. Systematic errors in locations are $\sim 1.6°$. Several lines of evidence indicate that the size and distribution of location errors are reasonably well known. The angular distribution remains consistent with isotropy. The intensity distribution is inconsistent with homogeneity at intensities below 20 times the BATSE threshold, and consistent with homogeneity above 20 times threshold.

INTRODUCTION

The third BATSE catalog of gamma-ray bursts contains summary data on 1122 GRBs observed between April 19, 1991 and September 19, 1994. The first catalog (1B) contained 260 bursts (1) and the second (2B) contained 585 bursts (2). Bursts from the previous catalogs are included in the 3B catalog. Available data include the time of the burst, location in equatorial and galactic coordinates, C_{max}/C_{min}, peak flux, fluence, and duration. The data can be obtained electronically from the Compton Gamma Ray Observatory Science Support Center. The telnet node is grossc.gsfc.nasa.gov; the DECnet node is GROSSC or 15765. The username is GRONEWS and no password is required. The data may also be accessed from the WWW site http://cossc.gsfc.nasa.gov/cossc/batse/burstcatalog/3b_intro.html. The catalog will also appear in Astrophysical Journal Supplements (3).

A BATSE burst trigger occurs when the count rate in two or more detectors exceeds 5.5σ above background on any of three timescales: 64 ms, 256 ms,

© 1996 American Institute of Physics

and 1024 ms. The background is recomputed every 17.408 s.

INTENSITY DISTRIBUTION

The intensity distribution for the 772 bursts that were above the 1024 ms trigger threshold is shown in Figure 1. Plotted is the integral number of bursts as a function of peak flux, in units of photons cm^{-2} s^{-1} integrated over 1024 ms and from 50 to 300 keV. The solid curve represents the total number of bursts observed and the dot-dashed curve represents the number corrected for trigger efficiency. The dashed line represents the $-3/2$ power law expected for a homogeneous distribution of sources. The deviation from the $-3/2$ power law is evident, and demonstrates that the burst sources, whatever their luminosity distribution, are not distributed homogeneously in Euclidean space.

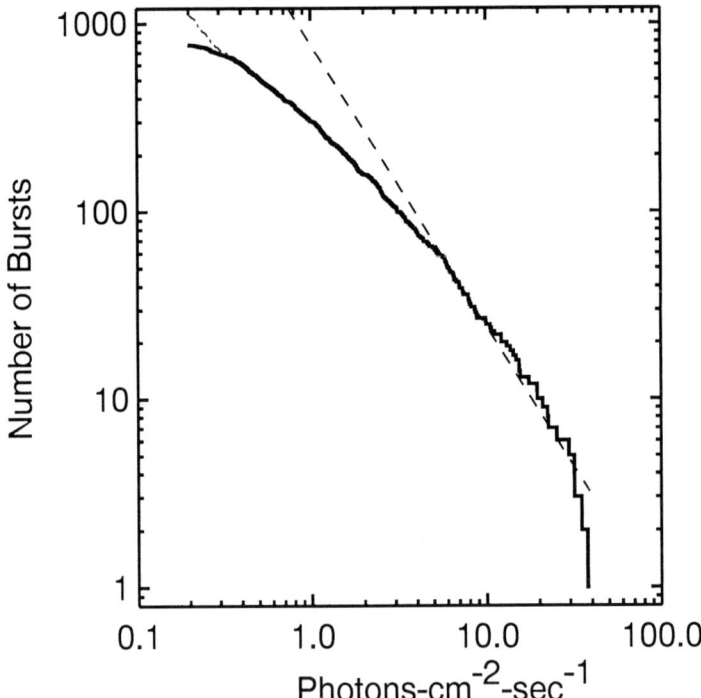

FIG. 1. The intensity distribution for 772 GRBs that exceed the 1024 ms trigger threshold.

The peak flux is computed from the count rate using the detector response matrices (DRMs) (4). This measure of intensity is defined over the same energy range and time interval as the on-board burst trigger. This is important for reducing biases in the resulting distribution. Since the time interval is the same, no correction is required to account for the wide variation in burst temporal structure. Since the energy range is the same, spectral corrections are minimized. They are not zero, however, because the conversion from counts to photons includes the effect that photons may Compton scatter in the detector, depositing only a fraction of their energy. This effect is incorporated into the DRMs. The intensity distribution in Figure 1 does contain two uncorrected biases. First, scattering by the Earth's atmosphere, which increases the detection efficiency very near the threshold, is not included in the correction for trigger efficiency. Second, the effect of statistical fluctuations, which also increases the efficiency near the threshold, is not included. The latter effect has been investigated by in 't Zand & Fenimore (5).

The standard test for inhomogeneity is $V/V_{\max}$ (6). For the 657 bursts of the 3B catalog for which this quantity can be measured, the average is $\langle V/V_{max} \rangle = 0.33 \pm 0.01$. For a galactic model employing standard candle sources in an $\alpha = 2$ halo, the BATSE sampling distance must be 4.5 times the core radius to obtain the required inhomogeneity.

LOCATIONS

BATSE's capability for locating bursts to an accuracy of a few degrees is a crucial aspect of the experiment. Only crude locations are required to determine the lower moments of the GRB distribution, which provide the most stringent constraints on galactic models. The sensitivity to repetition of burst sources, however, depends strongly on the location accuracy. The BATSE team has devoted considerable effort to refining the location techniques and providing the most accurate locations possible.

Location algorithm

Burst locations are determined by comparing intensities in different detectors, which present varying angles to the burst direction. The detector response is approximately $\cos\theta$, where θ is the angle between the burst direction and the detector normal, but the computation of location uses the more precise DRMs obtained by Monte-Carlo simulations and ground calibrations. Errors in location consist of statistical errors arising from the finite number of counts, and systematic errors presumably due to inaccuracies in the DRMs and corrections for atmospheric and spacecraft scattering. The statistical errors are determined from the standard χ^2 minimization of the burst location, and range from $< 1°$ for intense bursts to $\sim 12°$ for bursts at the BATSE threshold. Systematic errors are determined by comparing BATSE locations for the most intense events to locations more accurately determined by other instruments.

A number of improvements have been recently incorporated into the location algorithm. The most significant improvement is the use of six detectors, rather than four, in the χ^2 minimization of bursts whose locations are such that four detectors are illuminated approximately edge-on. All of the bursts in the 3B catalog have been recomputed using the new algorithm. Consequently, bursts in the 2B catalog will have different locations in the 3B catalog. As described below, the new algorithm reduces the systematic error for intense bursts from $\sim 4°$ to $\sim 1.6°$. Changes to the algorithm were not parameter optimizations based on known burst locations. They were corrections to weaknesses in the algorithm, and were completed before testing on known burst locations. Therefore, these locations do provide a fair test of accuracy.

The distribution of the changes in location of previously cataloged bursts is shown in Figure 2. The mean change is $7.2°$. The recomputed locations do not necessarily use the same source and background time intervals, so the changes also include a varying amount of statistical error.

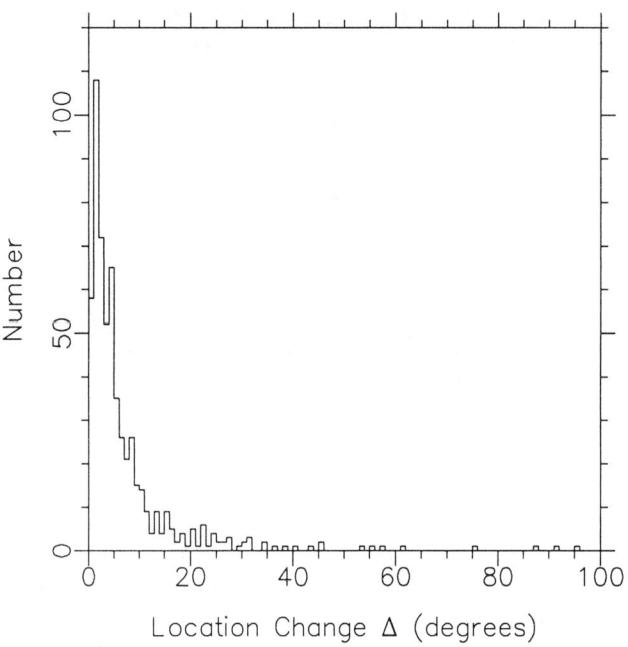

FIG. 2. The angular distance between the 2B and 3B locations.

Tests of Accuracy

The primary test of the BATSE location accuracy is a comparison to the locations of bursts with independent and accurate locations. Figure 3 shows the difference between the BATSE locations and locations of 50 bursts as determined by the IPN (7) or by WATCH (8,9). The known locations have errors of at most 0.5°. The solid line represents the distribution of total error, *i.e.*, the difference between the two locations. The dashed line represents the distribution of BATSE statistical errors. The BATSE systematic error is obtained by subtracting, in quadrature, the RMS statistical error from the RMS total error. This results in a formal systematic error of 1.6°, but values up to $\sim 2°$ are allowed.

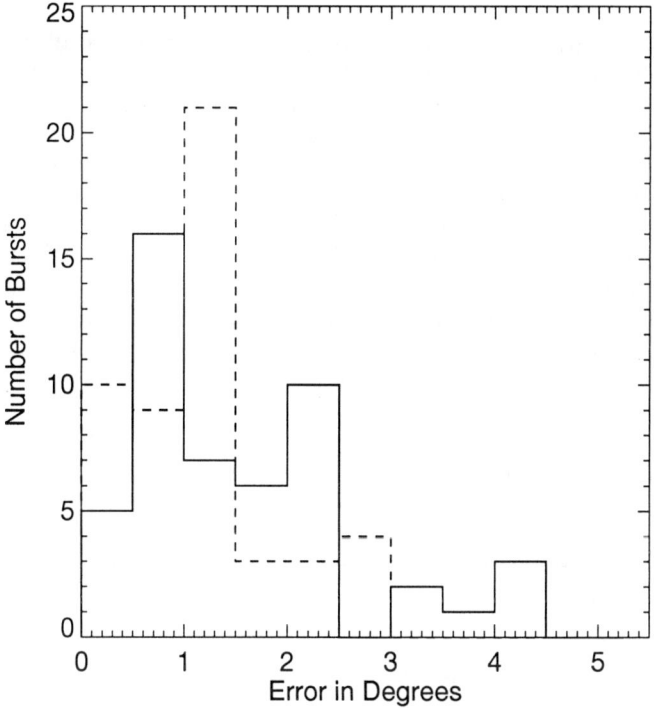

FIG. 3. Distributions of total error (solid line) and statistical error (dashed line) for bursts with independently measured locations.

There are difficulties in determining the systematic error for weaker bursts, which cannot be localized very well by other instruments. One technique consists of comparing the BATSE location to the IPN arc determined by time-of-arrival difference between BATSE and Ulysses (10). This provides a good check for bursts intense enough to generate an on-board trigger in Ulysses. However, the IPN arcs for events that do not trigger Ulysses may contain a systematic bias that underestimates the 2B errors and overestimates the 3B errors. This arises because the IPN arcs depend on the BATSE quoted location. When BATSE detects a burst that does not trigger Ulysses, the BATSE location is used to determine an approximate time that the burst would appear in the Ulysses detector. Ulysses data are then examined for excess flux in an interval around this time. Thus, a burst candidate found in the Ulysses data necessarily agrees reasonably well with the BATSE location. If the BATSE location has a large error, either no event or an unrelated rate increase will be found. Consequently, positions that are significantly improved with the 3B location will either not appear in the sample, or will be misinterpreted as errors in the 3B catalog. We conclude that IPN arcs from untriggered Ulysses bursts, while useful for narrowing the likely field of view for searches at other wavelengths, may not be reliable indicators of the BATSE systematic errors. (for further discussion of this issue, see Pendleton (11).

A test of the location accuracy for events near threshold is afforded by triggers from fluctuations from Cygnus X-1. These triggers are usually very close to threshold, of short duration, and superimposed on a fluctuating background. They are therefore among the most difficult events to locate accurately. A previous examination of 33 such events (12) found that the total error was typically 13°, and that bursts and Cygnus X-1 fluctuations could be reliably distinguished. With the new locations, the positions of 39 events have a computed RMS statistical error of 12.4° and an actual RMS error of 10.1°. The total error is consistent with the nominal systematic error of 1.6°, but inconsistent with a systematic error greater than $\sim 7°$. Regardless of the exact value of the systematic error, the Cygnus X-1 events demonstrate that the weakest events have a total error of typically 10°, and that this error is dominated by the statistical error.

The number and distribution of bursts with locations below the Earth's horizon provides another indication of the accuracy of locations (13). In the 3B catalog, 26 of the 1122 bursts had locations below the Earth's horizon. Based on the distribution of statistical errors and assuming a constant 1.6° systematic error, we would expect 21 such events, indicating consistency with the quoted error bars.

Measures of Isotropy

Figure 4 presents an Aitoff-Hammer projection in Galactic coordinates of the 3B burst locations. The Galactic dipole and quadrupole moments are within 1σ of the values expected for an isotropic distribution. The dipole moment, corrected for anisotropic sky coverage, is $\langle \cos \theta \rangle = 0.011 \pm 0.17$. The quadrupole moment, corrected for anisotropic sky coverage, is $\langle \sin^2 b - 1/3 \rangle = 0.002 \pm 0.009$. The consistency of a number of galactic models with these moments was investigated by Briggs et al. (14). An investigation of a slightly smaller sample (15) found no evidence for anisotropy in any of several subsets tested. This investigation did not use the most recent locations of the 3B dataset, but the dipole and quadrupole moments are not sensitive to small improvements in systematic error. Hartmann et al. (16) and Hakkila et al. (17) present upper limits to the repetition signal in the 3B catalog.

The low dipole moment imposes a model-dependent lower limit to the size

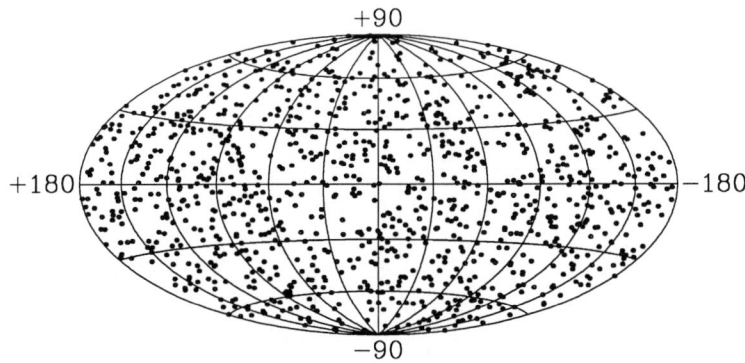

FIG. 4. The locations of 1122 bursts in galactic coordinates.

of the extended halo in Galactic models. The dipole moment of a thin shell of sources at a galactocentric radius R is given by $\langle \cos\theta \rangle = 2/3(R_{sun}/R)$, where R_{sun} is the distance of the sun from the Galactic Center (18). Using this relation, the 1σ upper limit to the BATSE dipole moment results in a shell radius of ~200 kpc. This distance may be thought of as the distance to a typical burst in galactic halo models. Bulik and Lamb (19) present a Galactic model consistent with the BATSE data that employs high velocity neutron stars.

Clustering and Repetition

The possibility that burst sources might repeat on timescales of years or less has recently become a topic of debate. Evidence for such repetition in the 1B catalog has been reported by Quashnock and Lamb (20) and by Wang and Lingenfelter (21). Repetition would appear in the BATSE data as a clustering of events on angular scales comparable to the instrumental resolution. The 3B catalog does not show a statistically significant clustering of events that would indicate repetition. Figure 5 shows the two point autocorrelation function for bursts in the 3B catalog. The dotted, dashed, and dot-dashed lines represent 1, 2, and 3σ deviations from isotropy. Repetition would appear as an excess in the last three bins.

DURATIONS

Our measures of burst durations are T_{50} and T_{90}, which are the time intervals over which 50% or 90% of the burst counts are obtained. These distributions were found to be bimodal in the 1B data (22). The effect is seen with still better statistical significance in the 3B data. Figure 6 shows a histogram of the number of bursts as a function of T_{90}, for events that exceeded the 64 ms trigger threshold. Peaks at ~0.5 s and ~30 s are apparent.

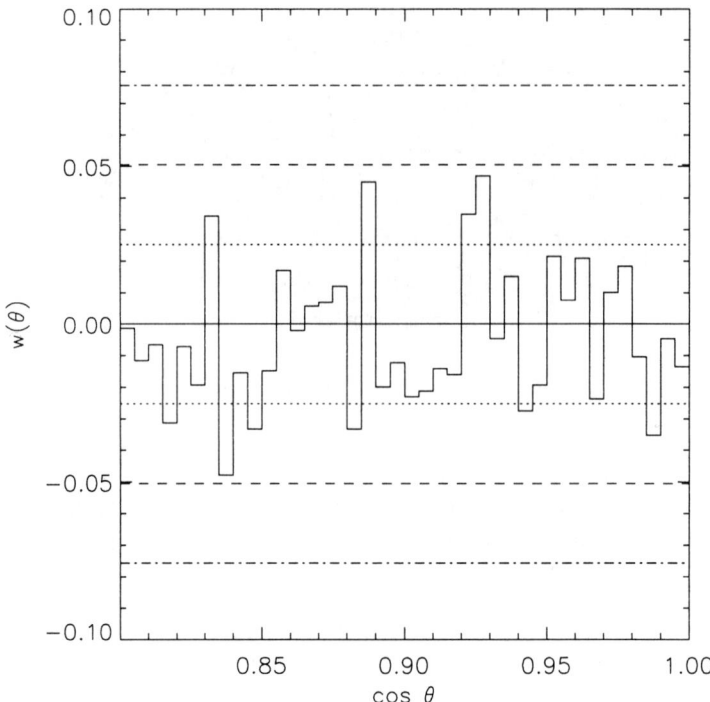

FIG. 5. The two-point autocorrelation function for bursts in the 3B catalog. Repetition would appear as an excess near $\cos\theta = 1$.

The intensity-dependent systematic effects associated with the BATSE duration measurements have been investigated (23). By reducing the signal-to-noise of a variety of observed bursts, it was found that these effects depend on the temporal structure of the burst. The largest source of error arises from the difficulty of separating background from extended weak source emission, particularly when the background and source emission vary on similar time scales. When these systematic effects are present in the BATSE data, they tend to reduce the signature of cosmological time dilation. T_{50} is less sensitive than T_{90} to intensity-dependent bias.

The durations play a role in the continuing controversy over time dilation. Norris et al. (24) (25) have reported that the BATSE bursts show the time dilation expected for sources at cosmological distances. Mitrofanov et al. (26) find no time dilation. These issues are discussed in several other papers in these proceedings (26,25).

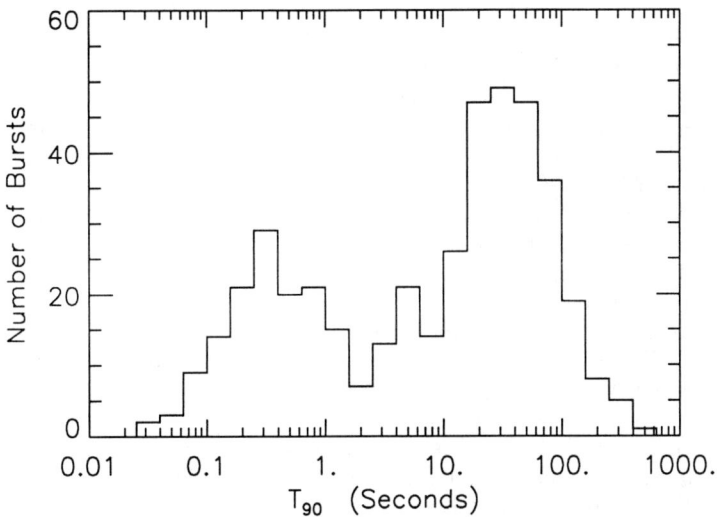

FIG. 6. The T_{90} duration distribution for bursts above the 64 ms trigger threshold.

SUMMARY

The BATSE 3B catalog, comprising 1122 GRBs, is the most extensive database of gamma-ray bursts available. The location uncertainty has been reduced to $\sim 2°$, probably the minimum feasible level. The souce spatial distribution remains isotropic to within the statistical limits, with a pronounced deficit in the rate of weak bursts. The duration distribution is bimodal with peaks at ~ 0.5 s and ~ 30 s, and ranges from <0.1 s to >100 s.

For about a year after the end of the 3B catalog we have explored different energy ranges for the burst trigger. We will produce catalogs for soft bursts (20 keV to 100 keV trigger), and for hard bursts (>100 keV).

REFERENCES

1. G. Fishman et al., ApJ Suppl. **92**, 229 (1994).
2. C. Meegan et al., "The Second BATSE Gamma-Ray Burst Catalog", available electronically from GROSSC (1994).
3. C. A. Meegan et al., ApJ Suppl., in press (1996).
4. G. N. Pendleton et al., Nucl. Inst. Meth. A **364**, 567 (1995).
5. J. J. M. in 't Zand and E. E. Fenimore, in Gamma-Ray Bursts, eds. G. J. Fishman, J. J. Brainerd, and K. Hurley, AIP Conf. Proc. **307**, 692 (AIP, New York, 1994).
6. M. Schmidt, J. Higdon, and G. Heuter, ApJ **329**, L85 (1988).
7. K. Hurley, private communication.
8. A. J. Castro-Tirado, PhD. dissertation, University of Copenhagen (1994).

9. S. Brandt, N. Lund, and A. J. Castro-Tirado, Astro.& Astrophys., in press (1996).
10. C. Grazziani, these proceedings.
11. G. Pendleton, these proceedings.
12. C. A. Meegan, G. Fishman, R. B. Wilson, and W. Paciesas, in Compton Gamma-Ray Observatory, eds. M. Friedlander, N. Gehrels, and D. J. Macomb, AIP Conf. Proc. **280**, 1117 (AIP, New York, 1993).
13. M. McCollough, C. Meegan, and G. Pendleton, these proceedings.
14. M. Briggs et al., these proceedings.
15. M. Briggs et al., ApJ **459**, 40 (1996).
16. D. Hartmann, et al., these proceedings.
17. J. Hakkila, et al., these proceedings.
18. D. H. Hartmann, E. V, Linder, and L. The, in Compton Gamma-Ray Observatory, eds. M. Friedlander, N. Gehrels, and D. J. Macomb, AIP Conf. Proc. **280**, 1003 (AIP, New York, 1993)
19. T. Bulik and D. Lamb, these proceedings.
20. J. Quashnock and D. Lamb, MNRAS **265**, L59 (1993).
21. V. Wang and R. Lingenfelter, ApJ **441**, 747 (1995).
22. C. Kouveliotou et al., ApJ **413**, L101 (1993).
23. T. Koshut et al., ApJ, **463**, 570 (1996).
24. J. Norris et al., ApJ **424**, 540 (1994).
25. J. Norris, these proceedings.
26. I. Mitrofanov, these proceedings.

Gross Spectral Differences Between Bright and Very Bright Gamma-Ray Bursts

J-L. Atteia*, C. Barat*, M. Boër*, J-P. Dezalay*, M. Niel*,
R. Talon*, G. Vedrenne*, K. Hurley†, M. Sommer‡, R. Sunyaev+,
A. Kuznetsov+, and O. Terekhov+

*CESR (CNRS/UPS), BP 4346, 31029 Toulouse Cedex, France
†SSL, University of California Berkeley, Berkeley CA 94720, USA
‡MPE, Postfach 1603, D-85740 Garching, Germany
+IKI, Russian Acad. of Sciences, Profsoyuznaya 84/32, 117810 Moscow, Russia

We consider a sample of 77 GRBs recorded by Phebus on GRANAT and the GRB detector on Ulysses in four years of operation. The comparison of the peak counts measured by these two instruments clearly shows the existence of a correlation between the intensities and the spectral hardnesses of these bursts. As sources homogeneously distributed in an Euclidean space should not exhibit any Hardness-Intensity correlation, we conclude that our sample is made of bursters which are not distributed homogeneously in space. This conclusion applies to bright (nearby ?) bursters with peak flux above 8 ph cm^{-2} s^{-1}.

We also note that, while non-homogeneous, our GRB sample exhibits the canonical value $\langle V/V_{\max}\rangle= 0.5$, confirming (if needed) that the test $V/V_{\max}$ is efficient to reject homogeneity but not to prove it.

I. THE GRB HR-INTENSITY CORRELATION

The great diversity of GRBs makes their global properties difficult to assess.
Yet, one of the well established feature of these events as a class is the tendency for bright bursts to have, on average, harder spectra than faint events. This trend has been pointed out in GRBs observed by Konus (1), Apex and Lilas (2,3), Phebus and Ulysses (4), and by BATSE (5–7). In the rest of this paper, this property of GRBs is called HIC (for Hardness-Intensity Correlation). A consequence of HIC is that it can be used to reject the spatial homogeneity of a burst sample. It is well known that the instrinsic properties (e.g., spectral or temporal characteristics) of sources homogeneously distributed in space are independent of the observed intensity. Therefore the existence of a correlation between the intensity and the spectral hardness in a GRB sample is an evidence that this sample is not spatially homogeneous.

While it has been clearly established with the $V/V_{\max}$ test that faint and moderately intense GRBs are not homogeneously distributed in space (e.g., (8)

© 1996 American Institute of Physics

and ref. therein), $V/V_{\max}$ is compatible with homogeneity for bright bursters (e.g., (9)). It is thus interesting to use complementary information like the existence of HIC to check the spatial homogeneity of these sources. In Sect. 2. we study a sample of 77 bright bursts detected by the GRB detectors on Phebus and Ulysses and we demonstrate that they exhibit a definite Hardness-Intensity correlation. We conclude that the bursters which emitted these bursts are not spatially homogeneous. The consequences of this observation on our general understanding of GRB sources are discussed in Sect. 3.

II. A SAMPLE OF BRIGHT BURSTS

Phebus and Ulysses are mid-size instruments detecting moderate and bright GRBs with peak fluxes above a few photons $cm^{-2}\ s^{-1}$. Phebus is made of 6 BGO detectors sensitive above $\approx$100 keV (10). The Ulysses GRB detector consists of 2 thin CsI crystals, sensitive in the range 25 to 150 keV (11). Both instruments have a fairly uniform sky coverage (see (4) for the details). The original aim of our study was to compare the burst properties in the two instruments, but the hardness intensity correlation soon appeared as a prominent feature, so we decided to concentrate on it.

In all the following we define GRB intensities in Phebus (I_P) and Ulysses (I_U) by their peak counts in 1 sec. The corresponding energy ranges are 150-1000 keV for Phebus (after gain correction) and 25-150 keV for Ulysses (without gain correction). the spectral hardness (HR) is measured by the ratio of these two intensities ($\frac{I_P}{I_U}$). The definition of the GRB sample used in this study is discussed below.

The Sample

From November 1990 to September 1994, Phebus detected 132 GRBs of cosmic origin. In order to avoid selection effects connected to the detection of faint events, we restricted our analysis to GRBs having a peak count rate greater than 160 in the energy range 150 keV to 1 MeV. This cut reduced the number of events to 98 (75% of the original sample). Since short events seem to have distinct spectral characteristics, we also removed from the sample 13 GRBs which had total durations shorter than 2 seconds, leaving 85 GRBs. Finally, Ulysses data were not available for 8 of these bursts (mainly at the beginning of the Ulysses mission).

The 77 remaining events were searched in Ulysses Real Time data, and *all of them* were identified with excesses larger than 4 standard deviations (on a time scale which varied from 1 to 20 sec)[1].

We were then able to measure I_P and I_U for each of the 77 GRBs.

---

[1] When a localization was available, usually from BATSE or IPN, we checked the compatibility of the time of arrival on Ulysses with the burster position on the sky.

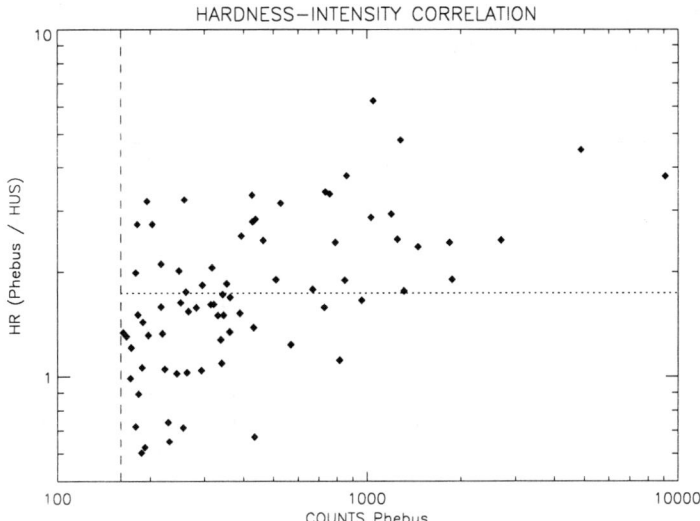

FIG. 1. Observed HR-Intensity correlation for 77 GRBs. The vertical line indicates the threshold in Phebus. The horizontal line shows the average Hardness Ratio.

The significance of the Hardness-Intensity Correlation

We checked the data against HIC by computing the coefficient of correlation between the logarithm of the counts in Phebus (log I_P), and the logarithm of the HR (log HR). This coefficient of correlation is given in Table 1 for the full sample, as well as for smaller samples made of bright events (the number of GRBs in each subsample is indicated in the first column of Table 1).

The same quantity was also computed for simulated samples having no HIC (see (12) for details on the generation of simulated samples). The results are reported in column 3 of Table 1. Being the ratio of I_P over I_U, the HR is obviously not independent of these numbers. This is clear in column 3 of Table 1, which shows that, even in the absence of HIC, the HR is positively correlated with I_P. In this study, we consider the difference between the observed correlation and the correlation expected in the absence of HIC. The significance of HIC in each sample is derived from the probability (column 4) to get a coefficient of correlation larger than the observed value, in the absence of HIC.

It is clear that significant Hardness-Intensity correlation is observed in the full sample (proba $< 2\ 10^{-6}$), and in subsamples restricted to the brightest bursts. GRBs brighter than 8 ph cm^{-2} s^{-1}, for instance[2], are hardly compatible with a population having no HIC (proba $\approx 2\ 10^{-4}$).

---

[2]The peak fluxes given here are estimated from $C_{max}^{Ulysses} = 34 * P_{BATSE}$, where P_{BATSE} is the peak flux in 1 sec in units of ph cm^{-2} s^{-1}.

TABLE 1. Measure of HIC and $\langle V/V_{\max}\rangle$ for various samples of GRBs detected by Phebus and Ulysses (see text for the details).

Number of GRBs	Corr. (obs)	Corr.[a] (simu)	Proba.[b]	$\langle V/V_{\max}\rangle$	$C_{\lim}$ (Ulysses)	Peak Flux[c] ph cm^{-2} s^{-1}
77	.56	.00	<2 e-6	—	60	1.8
74	.63	.06	<1 e-5	—	80	2.4
72	.66	.10	<1 e-5	—	100	2.9
71	.67	.11	<1 e-5	.42 ± 0.03	120	3.5
66	.69	.19	1 e-5	.43 ± 0.04	131	3.9
63	.69	.24	1 e-5	.46 ± 0.04	144	4.2
55	.74	.34	2 e-4	.45 ± 0.04	157	4.6
51	.77	.39	1 e-4	.48 ± 0.04	172	5.1
47	.78	.43	5 e-4	.50 ± 0.04	188	5.5
41	.81	.47	8 e-4	.50 ± 0.05	206	6.1
37	.83	.49	6 e-4	.52 ± 0.05	225	6.6
33	.86	.50	2 e-4	.53 ± 0.05	246	7.2
27	.90	.51	1 e-4	.50 ± 0.06	269	7.9
25	.90	.51	2 e-4	.53 ± 0.06	294	8.6
20	.85	.51	6 e-3	.49 ± 0.06	322	9.5
19	.87	.52	6 e-3	.54 ± 0.07	352	10.

[a] Assuming no HIC.
[b] Probability to measure the observed value, assuming no HIC
[c] see footnote 2

III. DISCUSSION

Our observations imply the existence of HIC, except maybe for the brightest 20 GRBs (with peak fluxes greater than 10 ph cm^{-2} s^{-1}). In fact we cannot reject the hypothesis that *all GRBs*, including the most intense, show HIC, since the low significance in favour of HIC for these bursts is mostly due to the small size of our sample.

One important consequence of these observations is that these rather intense bursters are not spatially homogeneous (or they have properties which vary with the distance to the Earth). Interestingly, HIC allows us to reject homogeneity for GRBs which would have been considered homogeneous from their $\langle V/V_{\max}\rangle$ only (col. 5 of Table 1). We believe that this is the first time that a test other than $V/V_{\max}$ is used to disprove the spatial homogeneity of a burster sample.

A second consequence of HIC is that the GRB intensity distribution is not independent of the energy range in which it is observed. This should be kept in mind when using this distribution to infer the GRB distance scale.

In view of the implications of this observation it is important to search a confirmation in the data recorded by other instruments. Having accumulated more than 1000 GRBs to date, BATSE should easily be able to confirm (or refute) the existence of HIC for bright GRBs.

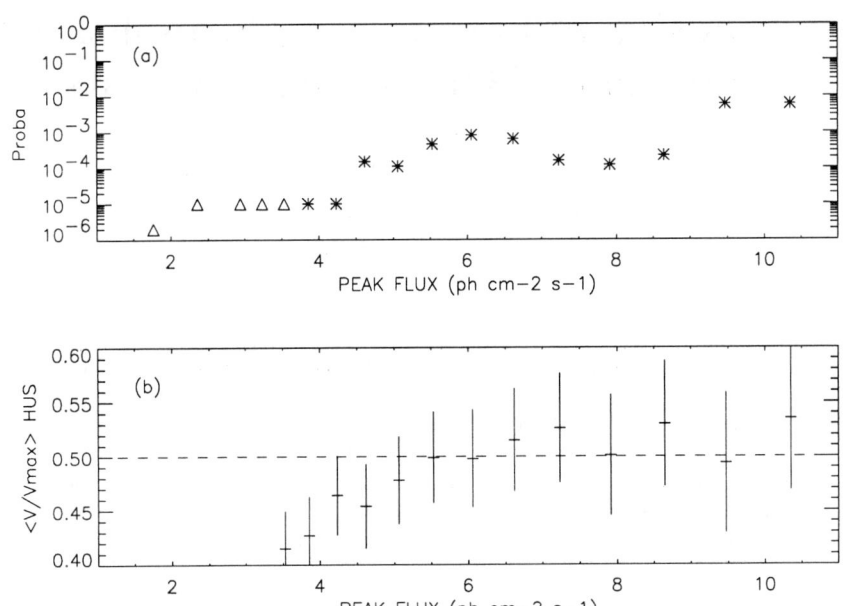

FIG. 2. Significance of HIC (a) (triangles are upper limits), and $\langle V/V_{\max} \rangle$ (b) vs the burst intensity.

REFERENCES

1. B. M. Belli, in AIP Conf. Proc. **265**, eds. W.S. Paciesas and G.J. Fishman, 100 (AIP, New York, 1992).
2. J.-L. Atteia, C. Barat, Jourdain, et al., Proc. of the 22nd ICRC Conference, Dublin, Ireland, Paper OG 2.11, 93 (1991).
3. I. Mitrofanov, A. Pozanenko, J.-L. Atteia, et al., in: Gamma-Ray Bursts, eds. C. Ho, R.I. Epstein and E.E. Fenimore, Cambridge University Press, 203 (Cambridge U Press, 1992).
4. J.-L. Atteia, C. Barat, Boer, et al., A&A **288**, 213 (1994).
5. W. S. Paciesas, G. N. Pendleton, C. Kouveliotou, et al., in AIP Conf. Proc. **265**, eds. W.S. Paciesas and G.J. Fishman, 190, (AIP, New York, 1992).
6. R. J. Nemiroff, J. Norris, J. T. Bonnell, et al., ApJ **435**, L133 (1994).
7. R. S. Mallozzi, W. S. Paciesas, G. N. Pendleton, et al., ApJ, Dec 1st issue (1995).
8. J.-L. Atteia in The Gamma Ray Sky with Compton GRO and SIGMA, eds. M. Signore et al., Kluwer Academic Publishers, 369-380 (1995).
9. E. E. Fenimore, R. I. Epstein, C. Ho, et al., Nature **366**, 40 (1993).
10. C. Barat, F. Cotin, M. Niel, et al., AIP Conf. Proc. **170**, 395 (AIP, New York, 1988).
11. K. Hurley, M. Sommer, J.-L. Atteia, et al., A&AS **02**, 401 (1992).
12. J.-P. Dezalay, J-L. Atteia, C. Barat, et al. A&A, in preparation (1996).

Luminosity Function and Cosmological Evolution of Gamma-Ray Bursts

Walid J. Azzam and Vahé Petrosian

Center for Space Science and Astrophysics
Stanford University
Stanford, CA 94305-4060

The so-called log N–log S relations of GRBs are analyzed in terms of the logarithmic slope of the cumulative counts using a new method which accounts for selection biases. The results are compared with the predictions of cosmological models with a finite range, power-law luminosity function and with pure density and luminosity evolution laws that scale as a power of $(1+z)$. We discuss the limits that may be placed on the range of luminosities and on the evolution.

INTRODUCTION

The source counts or the so-called log N–log S relations are the primary data that constrain the spatial distribution of sources such as gamma-ray bursts (GRBs) whose distances are unknown. However, application of the log N–log S analysis to GRBs is complicated since they have a variable threshold for detection, and suffer from several selection biases. We use a new statistical technique that takes proper account of the variable nature of the triggering threshold to analyze the BATSE 3B catalog. The purpose of this paper is to investigate the constraints that may be placed on simple cosmological models and/or on the luminosity function, and its evolution. A brief description of the method is provided, followed by the comparison of the data to models, and a discussion of the goodness of fits along with the constraints that may be obtained.

METHOD

For each burst characterized by peak flux, $f_{p,i}$, as given in the BATSE catalog, we define, as described by Lee and Petrosian (1), a limiting flux, $f_{\lim,i} \equiv f_{p,i}(\frac{C_{\lim,i}}{C_{p,i}})$, where $C_{p,i}$ and $C_{\lim,i}$ are the peak and threshold photon count rates respectively. We define an associated unbiased number of bursts, M_i, contained in the box with $f_p > f_{p,i}$ and $f_{\lim} \leq f_{p,i}$. The cumulative distribution $N(f_p)$ is then obtained in a stepwise fashion: $\delta \ln N(f_{p,i}) = \ln(1 + M_i^{-1})$. Note that before carrying out the analysis, we correct f_p for duration bias (1).

© 1996 American Institute of Physics

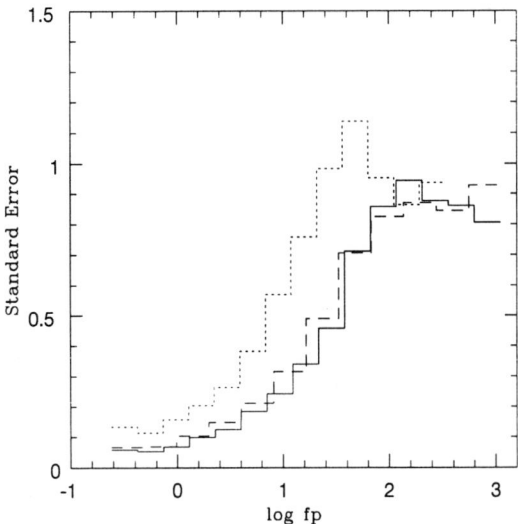

FIG. 1. Standard error of the logarithmic slope, obtained through Monte Carlo simulations, as a function of peak flux. The three sets of simulations, each consisting of a 100 runs, represent respectively 250 (dots), 600 (dashes), and 1000 (solid) (f_p, f_{lim}) pairs randomly drawn: f_p from a power law distribution with a differential slope of -2, and f_{lim} from a uniform distribution.

The logarithmic slope of the counts, $s(f_p) = -\frac{d \log N}{d \log f_p}$, known as the hazard rate in statistic's literature (2) is a more useful quantity and is also directly related to $\ln(1 + M_i^{-1})$. It is this quantity which we use when comparing the data to models.

We carried out Monte Carlo simulations to test the accuracy of determining $s(f_p)$. Figure 1 shows the standard errors, for the slopes determined by our method, obtained for three sets of simulations, each consisting of a 100 runs. The range and values of f_p and f_{lim} were chosen to correspond to those of the GRBs observed with BATSE. The three different sets consist respectively of 250, 600, and 1000 (f_p, f_{lim}) pairs randomly drawn: f_p from a power law distribution with a differential slope of -2, and f_{lim} from a uniform ditribution. We also carried out simulations for a power law distribution with a differential distribution of $-5/2$ and for a broken power law distribution. The results obtained in the latter cases are qualitatively similar to what is shown in Figure 1. Basically, as f_p increases so do the error bars, indicating that bursts with large f_p will not play a significant role when it comes to fitting models.

COMPARISON WITH MODELS AND FITS

Using $s(f_p)$ we compare the data to $\Omega = 0$ and $\Omega = 1$ cosmological models ($\Lambda = 0$) and with various forms for the distribution of peak luminosity and

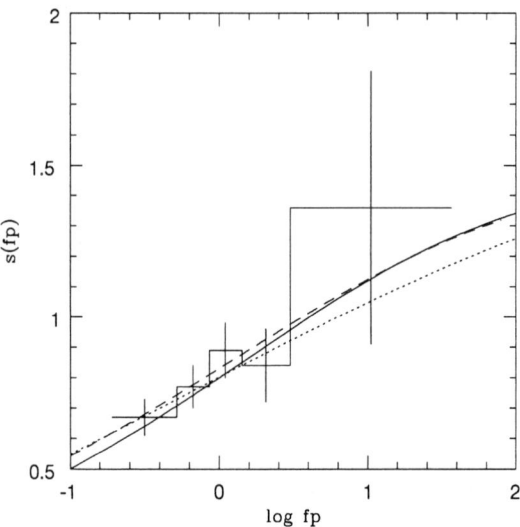

FIG. 2. Logarithmic slope as a function of peak flux. The histogram refers to BATSE 3B data (1024msec). The curves show finite range power-law luminosity function models, $\Psi \propto L^{-\beta}$ with $\beta = 2$, for $\Omega = 1$, with $\frac{L_{max}}{L_{min}} = 1$ (solid), 10 (dashes), and 100 (dots).

its evolution $\Psi(L_p, z)$. We consider three cases:

(1) A non-evolving finite range power-law luminosity function, $\Psi(L_p, z) = \rho_o(L_p/L_o)^{-\beta}$ for $L_o \leq L_p \leq L_{max}$. As an example, we show in Figure 2 the case for $\Omega = 1$ with $\beta = 2$, and $\frac{L_{max}}{L_o} = 1$ (solid), 10 (dashes), and 100 (dots), where L_o (in units of 10^{51} ergs/s) = 0.62, 3.1, and 0.62 respectively. Notice that the curves are quite similar, especially at low f_p where the fitting matters most.

(2) A standard candle luminosity function, $\Psi(L_p, z) = \rho(z)\delta(L_p - L_o)$, with density evolution, $\rho(z) = \rho_o(1+z)^\gamma$.

(3) A standard candle luminosity function, $\Psi(L_p, z) = \rho_o\delta(L_p - L(z))$, with luminosity evolution, $L_p(z) = L_o(1+z)^k$.

A maximum-likelihood technique, referred to as the hazard-rate statistic in statistic's literature (3), was used to fit models with different values of the indices, β, γ, and k to the data. This method permits us to determine which models give a better fit to the data.

Figure 3 shows the results for case (1), or the no-evolution models. A smaller value for the hazard-rate statistic indicates a better fit. The models, here and in what follows, were shifted horizontally (i.e along f_p axis) with respect to the data in order to obtain the best fits. For an assumed spectral index this determines the value of L_o. The horizontal lines (solid for $\Omega = 0$ and dahed for $\Omega = 1$) represent the standard candle values. We note first that the hazard-rate statistic increases with increasing width of the luminosity function, and

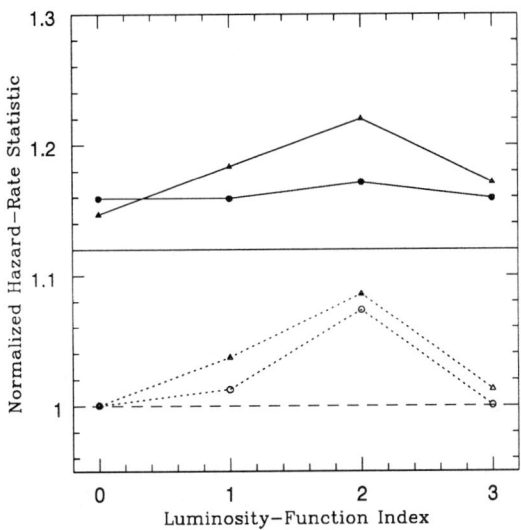

FIG. 3. Results of fitting 3B data to power-law luminosity function models with finite luminosity range, $\frac{L_{max}}{L_{min}} = 10$ (circles), and 100 (triangles) for $\Omega = 0$ (solid) and $\Omega = 1$ (open). Smaller values of the normalized hazard-rate statistic indicate better fits. The horizontal lines give the corresponding values for standard candles – solid line for $\Omega = 0$, and dashed line for $\Omega = 1$.

TABLE 1. Results for Density and Luminosity Evolution for $\Omega = 1$

Figure of Merit	$\gamma = k = 0$	$\gamma = 1$	$\gamma = 2$	$k = 1$	$k = 1.5$
Hazard-Rate Stat.	1.00	1.09	1.97	1.1	1.04
Reduced Chi-Square	0.33	0.30	5.38	0.36	1.2
L_o (10^{51} ergs/s)	0.62	3.1	15.5	0.98	1.2

that as expected for small or large values of β, we get the same result as the standard candle. The fits seem to be insensitive to the luminosity range and moderately sensitive to β, with the $\Omega = 1$ models giving a better fit. The interpretation of this statistic is an elaborate statistical problem (3). Thus, to get a feeling for the confidence levels, we have also calculated the reduced chi-square values, which are a more approximate measure of the goodness of fit in this case. The reduced chi-square values, which closely mimic the trends of the hazard-rate statistic, range from 0.38 to 0.60 for $\Omega = 0$, and from 0.24 to 0.39 for $\Omega = 1$, indicating acceptable fits for both models, with the difference between them consistent with the results from the hazard-rate statistic.

Table 1 shows the results of fitting the density and luminosity evolution models for $\Omega = 1$. Moderate, or no evolution, seem to give better fits.

We also investigated more complicated models that involve both density and luminosity evolution. As an example, we consider the case for $\Omega = 0$ with $k = 1$, and $\gamma = -2, -2.5$, and -3. All three models gave good fits with a normalized hazard-rate statistic of: 1.06, 1.22, and 1.12; and with a reduced chi-square of: 0.34, 0.43, and 0.52 for $\gamma = -2, -2.5$, and -3, respectively, as compared to a normalized hazard-rate statistic of 1.12 and a reduced chi-square of 0.38 for a standard candle $\Omega = 0$ model.

CONCLUSIONS

1) Although recent work by other authors (4) based on log N–log f_p fits, suggests that 90 % of detectable GRBs have L_p values within a range of 10, our results indicate that it is difficult to constrain the luminosity range for luminosity function models, primarily because points with high f_p do not play an important role when fitting, due to the large error bars at high f_p.
2) Models with moderate or no density and luminosity evolution fit the data well, with $\Omega = 1$ models giving, in general, better fits than the $\Omega = 0$ models. This is in agreement with recent work by other authors (5) in which time dilation effects were incorporated with log N–log f_p relations to obtain better cosmological model fits.
3) We also touched on more complicated models involving both density and luminosity evolution, and showed that, as expected, they can improve the fits to the data.

REFERENCES

1. T. T. Lee & V. Petrosian, these proceedings (1996).
2. B. Efron & V. Petrosian, JASA **80**, 452 (1994).
3. B. Efron, private communication (1996).
4. A. Ulmer, R.A.J.M. Wijers, & E.E. Fenimore, ApJ Letters **440**, L9 (1995).
5. A. Mészáros, & P. Mészáros, preprint (1995).

Repeater Models and Sky Coverage

David L. Band

CASS, University of California, San Diego, La Jolla, CA 92093

The BATSE data does not prove that burst sources repeat, but also does not rule out such repetitions. However, the implications of the observed constraints depend on the repetition model. I find that the repetition content is very weakly dependent on the sky coverage (fraction of the sky observed on average) for models where bursts occur at an average rate, and is primarily a function of the number of bursts in a sample. A given observed repetition fraction (fraction of events which originate from sources which are observed to burst more than once) can be produced by a small number of sources which burst frequently or from a larger number which burst less frequently.

INTRODUCTION

Whether burst sources repeat is of great current interest because most cosmological models destroy the source in producing the burst. The BATSE burst database has been searched for spatial (1) and spatial-temporal clustering (2,3). A repeater signal appears to be absent in the 3B catalogue (4), but the existence of repeaters is not ruled out by the observations. Because of the uncertainties in burst positions, the existence of repeaters or constraints on their presence are based on statistical tests, and not on the attribution of individual bursts to specific sources. Explicit or implicit in analyses of the repeater content of a burst sample are assumed models of source repetition. Here I show the importance of the assumed repetition model. Specifically, limits on the repeater content of a given burst database must be translated into limits on a physical repetition model, particularly to compare different burst samples.

As examples, I construct two models where sources burst at a constant average rate. In the first model bursts occur stochastically, and Poisson statistics are used to describe the number of observed repetitions during the period the source was observed. In the second there are a fixed number of bursts during the period of the detector's operation, and the number of bursts actually observed is described by the binomial distribution. These models are presented in greater detail elsewhere (5).

A crucial part of modeling is the clear definition of the fundamental quantities. Thus we assume that over a period ΔT our detector has a livetime τ averaged over the sky; the sky coverage f_s is the ratio of these times, $f_s = \tau/\Delta T$. The sky coverage is assumed here to be constant over the sky, while in fact it varies with spatial position. Given the rate at which bursts occur from

© 1996 American Institute of Physics

repeating and nonrepeating sources, the expected number of observed bursts $\langle N_B \rangle$ is proportional to the livetime τ. Of the observed bursts, $\langle N_{B,r} \rangle$ is the expected number which originate from sources which are observed to burst more than once. Thus the apparent repeater fraction f_r is the fraction of bursts from sources observed to repeat: $f_r = \langle N_{B,r} \rangle / \langle N_B \rangle$. Clearly f_r tends asymptotically to F_r, the fraction of all bursts, observed or not, produced by repeating sources. Of the N_S repeating sources distributed over the sky, only $\langle N_{S,\text{obs}} \rangle$ will be observed to repeat; asymptotically $\langle N_{S,\text{obs}} \rangle$ tends to N_S. Finally $\langle n_{\text{obs}} \rangle \geq 2$ is the number of events observed from sources observed to repeat.

STOCHASTIC REPETITIONS

In this model the bursts from the ith source occur at an average rate r_i; the probability of a burst during any time interval is constant and uncorrelated with the size of the previous burst. Therefore if a source is observed for a total livetime τ, the number of events observed from the ith source, n_i, will have a Poisson distribution with an average of $r_i \tau$:

$$P(n_i) = \frac{(r_i \tau)^{n_i}}{n_i!} \exp[-r_i \tau] \qquad (1)$$

It does not matter whether the livetime is continuous (i.e., large sky coverage fraction), or chopped into short intervals (i.e., low coverage fraction).

Figure 1 shows the dependence of observables such as the apparent repeater fraction f_r on the average number of observed bursts per source $r\tau$ for a model where all sources burst at the same rate r.

BURSTS AT A CONSTANT RATE

Here I assume that the ith source bursts at a constant rate r_i, and therefore in the period ΔT during which the detector operated there are a total of $m_i = r_i \Delta T$ bursts (m_i is assumed to be an integer) which may or may not be observed. The sky coverage f_s is the probability that a burst is observed; the actual number observed n_i has a binomial distribution:

$$P(n_i) = \frac{m_i!}{n_i!(m_i - n_i)!} f_s^{n_i} (1 - f_s)^{m_i - n_i} \; . \qquad (2)$$

Figure 1 also shows the dependence of observable quantities on the average number of observed events per source mf_s for a model where all sources burst m times during ΔT. Note that mf_s in this model is the same as $r\tau$ in the stochastic model. Integral values of $m \geq 2$ are plotted for $f_s = 1/3$ and $f_s = 1/4$, approximately the sky coverage for the 1B and 2B–1B catalogues, respectively. If $m = 1$ there are no repetitions.

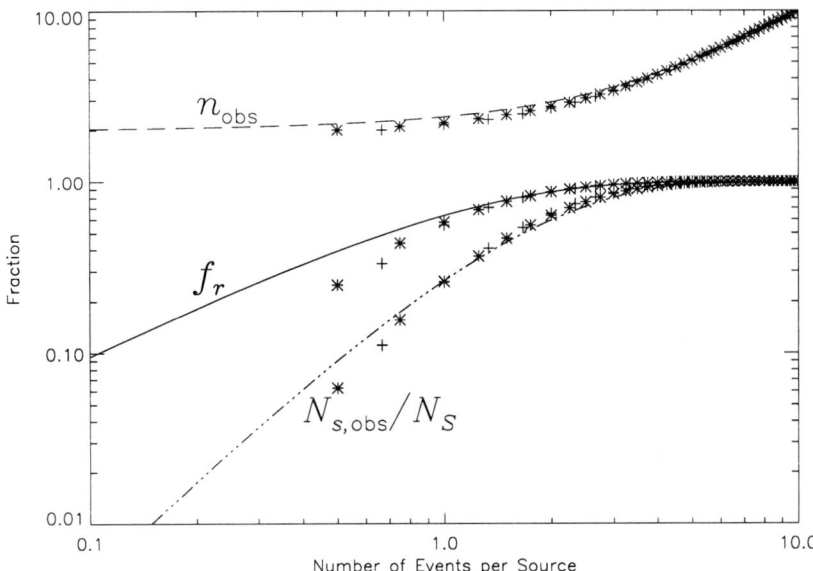

FIG. 1. Repetition quantities as a function of the mean number of observed bursts per source ($r\tau$ for the stochastic model and mf_s for the constant rate model). The curves are for the stochastic model and the points for the constant rate model. All sources burst at the same rate, and only repeating sources are included. The number of observed bursts from sources observed to repeat (i.e., from which there are two or more observed bursts) is $\langle n_{obs} \rangle$ (long dashes), while $\langle N_{S,obs} \rangle / N_S$ (dots and dashes) is the fraction of the sources from which repetitions are observed. The fraction of the observed bursts from sources with two or more observed bursts is f_r (solid). Points (the number of bursts per source is an integer) show these quantities for $f_s = 1/4$ (asterisks) and $f_s = 1/3$ (pluses) for the constant rate model.

AVERAGE RATE MODELS

If a source bursts at an average rate, then the number of bursts observed from the source will depend on the livetime, unless there are correlations between the sky coverage and the repetition pattern. The character of a burst sample's repeater content depends on whether the average number of observed bursts from a repeating source is greater than 1, which is proportional to the number of bursts in a sample. Thus the sky coverage is felt primarily through the livetime, which determines the sample size. This can be seen from Figure 1 where the curve for the stochastic model and the points for the constant rate model coincide, particularly when the average number of bursts per source is greater than 1.

Thus comparable size samples should be compared, otherwise allowances need to be made for the differences in the expected repeater content. Note that varying the burst detection threshold changes the burst rate for the resulting samples: bursts will be detected at a greater rate if weaker bursts can be detected.

Complications arise when the repetition pattern has a time scale commensurate with a characteristic time of the sky coverage. For example, if a repetition tends to occur ~45 min (i.e., half an orbit) after an initial burst, the Earth may occult one of the bursts for a detector in low Earth orbit.

WHY AVERAGE RATE MODELS DO NOT SHOW A STRONGER f_s DEPENDENCE

One might have expected a stronger dependence on the sky coverage f_s which is the probability of observing a given burst. Although the probability of seeing both members of a pair of bursts from a repeating source is f_s^2, the quantity of interest is the conditional probability of observing repetitions from the source of a previously observed burst. Assuming a repeating source bursts twice during the observation period ΔT, the conditional probability of detecting a repetition of an observed burst is f_s. The sample size accumulated is proportional to $\tau = f_s \Delta T$. Decreasing f_s does reduce the probability of observing a repeater, but it also decreases the number of bursts in the sample. To maintain the sample size, ΔT must be increased as f_s is decreased. But a key assumption was that there were only two bursts during ΔT; if ΔT increases then the effective burst rate $r = 2/\Delta T$ decreases.

Therefore for f_s to be the probability of observing the repetition of a given burst in a burst sample of a specified size, the effective burst rate must be proportional to the inverse of ΔT. Clearly this is not the case for any repetition model where the rate is constant.

ACTIVE PHASE MODELS

If the source goes through active bursting phases shorter than ΔT, and the average separation between the active phases is longer then ΔT, then the effective burst rate is inversely proportional to ΔT, $r \propto \Delta T^{-1}$. This is the class of models which searches for temporal-spatial clustering would find (2). For example, if a source bursts twice within five days, and then does not burst again for a decade, then $r = 2/\Delta T$ for $\Delta T \sim 1$ yr.

Similarly, studies of the sky coverage dependence of observing repetitions from sources with a fixed number of bursts within the observation period ΔT implicitly assume the burst rate is inversely proportional to ΔT. This assumption excludes fixed rate models, but permits active phase models.

As ΔT increases other repeating sources should become active. Therefore a burst sample with a smaller sky coverage but longer observation period should have more repeating sources with fewer observed bursts from each source than a sample of the same size accumulated over a shorter time but with a larger sky coverage. Thus the observed repeater content is clearly affected by the sky coverage if the sources burst in short active phases.

CONSTRAINING PHYSICAL MODELS

Models where there are repeating and nonrepeating sources are defined by many parameters, thus a single observable such as f_r from a burst sample does not fully determine the model parameters. Many observed repetitions (large $r\tau$ or mf_s) from a small number of repeating sources or few observed repetitions from many repeaters can result in the same observed repeater fraction f_r. With more observables the parameters can be determined; for the stochastic model, the average number of events observed from sources observed to repeat $\langle n_{\rm obs} \rangle$ determines $r\tau$, although it may be difficult to determine $r\tau \leq 1$ since $\langle n_{\rm obs} \rangle$ is very nearly constant at a value of 2 for $r\tau$ in this range. Similarly, the dependencies of the observables on $r\tau$ (or its equivalent) are very nearly the same for the two models developed here, and probably for most average rate models; it may be very difficult to identify the repetition pattern from observations, particularly if the sky coverage f_s is low. Of course, the repetition pattern can be determined if specific bursts can be attributed to the same source.

To prove the existence of repeating burst sources the observed repeater fraction f_r must be shown to be nonzero. On the other hand, to constrain the allowed repeater population, limits on the actual repeater fraction F_r must be derived from the data.

Acknowledgments. This research was supported by NASA contract NAS8-36081.

REFERENCES

1. J. M. Quashnock and D. Q. Lamb, M.N.R.A.S **265**, L59 (1993).
2. V. C. Wang and R. E. Lingenfelter, Ap. J. **416**, L13 (1993), **441**, 747 (1995).
3. V. Petrosian and B. Efron, Ap. J. Lett. **441**, L37 (1995).
4. C. A. Meegan, *et al.*, Ap. J. Supp., submitted (1995). Also available from grossc.gsfc.nasa.gov, username gronews.
5. D. L. Band, Ap. J., in press (1996).

Analysis of the Space Distribution of the Gamma-Ray Bursts in the BATSE 3B Catalog

B. M. Belli

Istituto di Astrofisica Spaziale, CNR, CP 67, 00044 Frascati, Italy

---

We extended our earlier analysis of the intensity and frequency distribution of Gamma-Ray Bursts to the BATSE 3B catalog. We considered the presence of two classes of events and studied them separately. We selected the total fluence of an event to represent the intensity of the burst. Our results indicate an isotropic and homogeneous spatial distribution and the presence of a possible luminosity spread within each GRB class. Considering possible biases due to the threshold effects, no disagreement between log N-log S relation and space distribution seems to be present.

INTRODUCTION

We consider for the present analysis the GRBs for which the BATSE 3B catalog (6) reports their hardness ratio HR, (i.e., the ratio of the fluences recorded in the two energy channels 100-300 keV and 50-100 keV), and their duration T90, (i.e., the time during which the integral counting rate goes from the 5% to the 95% of the total) (7). These parameters represent the intrinsic characteristics of the events and are essentially independent of their (unknown) distance. If we plot these events in the plane T90-HR they form two distinct groups (Fig. 1). We indicate the group at the right in Figure 1 as class I, (Type I bursts) and the group at the left, as class II, (Type II bursts). We draw in the plane log (T90)- log (HR) the straight line "d", defined by:

$$HR = 2(T_{90})^{1/2} \qquad (1)$$

For each HR this line indicates the minimum limit to the event duration for class I and the maximum limit for class II. Class I contains 570 events, while class II contains 229 events. Type II bursts have on average higher HRs than Type I bursts, and even if they are on average shorter than the other ones, they are not characterized by short duration and can reach in some cases as long a duration as the events of Type I, and vice versa (3,4).

Two classes in normal GRBs have been already suggested by other authors (5,8,7) on the basis of their temporal and spectral behaviour. Comparison between the previous classes and the present ones shows that they are substantially the same or at least that the previous ones represent a first step in our class definition (5,7).

© 1996 American Institute of Physics

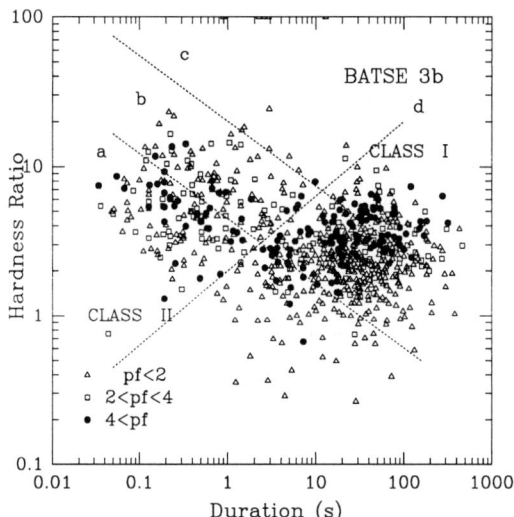

FIG. 1. 799 events of the BATSE 3B catalog. Pf is the peak count rate.

DATA ANALYSIS

To study the space and intensity distributions of the two classes we divide the bursts of each class in subsets of events with equal luminosity in order to separate spatial effects from luminosity effects (1,2). In previous papers we suggested to use the peak count rate as a measure of intensity of the events and the corresponding peak photon energy spectral hardness as the parameter to divide their luminosities. The BATSE 3B catalog does not report the photon spectral hardness, but only the energy spectral hardness. For this reason we use here as burst intensity the total fluence of the events, (erg/cm^2). In principle the intrinsic total energy of a burst increases with hardness and duration. To select subsets of equal intrinsic luminosity we divide the bursts with the straight lines drawn in the plane $\log T_{90}$-$\log HR$ (Fig. 1). These lines represent simple functions according to which an increase of the duration corresponds to a decrease of the hardness. We estimate that the three subsets "a", "b", "c" defined in Fig. 1 each consist of events of equal intrinsic luminosity. The slopes of the division lines are given by the position of the majority of the strongest peak events, and by the position of the majority of minimum HR events (Fig. 1).

LOGN–LOGS

Fig. 2a shows the $\log N$-$\log S$ curves for burst Types I and II. The curve relative to class I follows on a the power law with exponent -3/2, which represents the law of homogeneity and isotropy. The slope changes at low fluences. The curve relative to class I shows a similar behaviour, shifted to higher fluences.

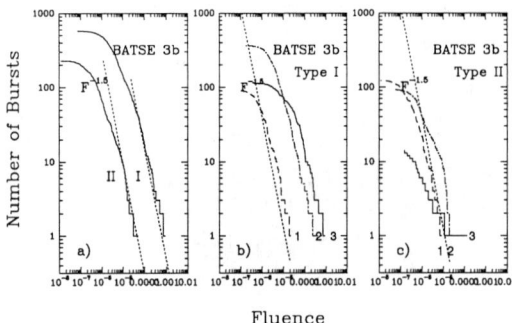

FIG. 2. a) the log$N$-log$S$ curve for Type I and II GRBs b) the log$N$-log$S$ curve for three groups of equal intrinsic luminosity for the Type I GRBs c) the log$N$-log$S$ curve for three groups of equal luminosity for the type II GRBs.

Fig. 2b shows the three curves relative to the three subsets of class I with equal intrinsic luminosity: it is possible to fit the first parts of these curves in a differential form with the -5/2 power law with good chi-square values. All three curves depart from this law for a flatter behaviour, each at a different value of fluence. In the third curve, probably constituted by the highest intrinsic luminosity and furthest events, this behaviour is more evident. It is possible that this result is due to a change in the geometry of the source space distrubution, toghether with the contribution of the threshold effects, that in this case might be more relevant.

Fig. 2c shows the three log$N$-log$S$ curves for class II bursts: the first two follow the -3/2 power law with the presence of instrumental and statistical effects. The third curve exhibits a different behaviour. We think that this different behaviour is principally due to threshold effects, without excluding a contribution due to a change of geometry. The threshold, triggered by a count rate of a fixed number of sigma over the background level in a selected temporal bin, cuts on the basis of the peak count rate. For the events of third groups, characterized by very high energy photon spectra hardness and longer duration, the peak count rates might be very low.

Let us consider all the events, without dividing them in two classes. Fig. 3 shows the log$N$-log$S$ curves for the three subsets of events "a", "b" and "c", separated by the lines defined in Fig. 1. It appears very clearly that for the group "b", numbered "2" in the figure, which is the central and richest in events (456), exhibits a log$N$-log$S$ curve strictly following a power law with exponent -3/2. The group "c" with the highest values of HR shows a

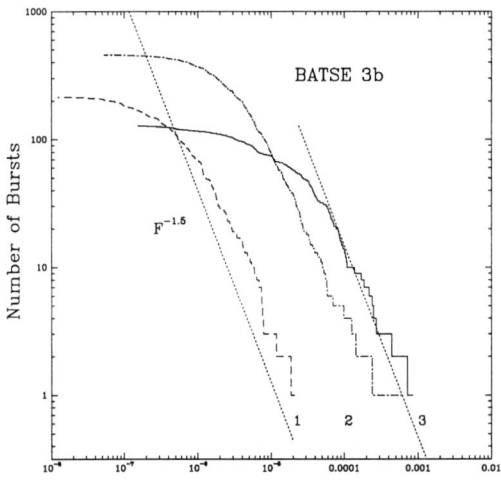

FIG. 3. Log$N$-log$S$ curves of the GRBs divided in three groups of equal intrinsic luminosity.

log$N$-log$S$ curve, which deviates from a -3/2 power law much more evidently. Considering the presence of the two classes, Figures 2b and 2c allow the separation of the log$N$-log$S$ curve "3" in Fig. 3, into the contributions of the Classes I and II. The two classes, having events of different duration and hardnesses, and presumably of different shapes of light curves and luminosity laws, suffer different biases, as it appears in the figures. The log$N$-log$S$ curve "2" of Fig. 3, which represents a homogeneous and isotropic event distribution, can also be the result of the addition of the two classes of events with different luminosity laws, which are both homogeneously and isotropically distributed in space. The spatial distribution of the two classes is consistent with an isotropic distribution (Fig. 4a and 4b).

CONCLUSIONS

Differing intrinsic luminosities and instrumental biases are for us the principal reasons of the deviation of the total log$N$-log$S$ from the -3/2 power law for Classes I and II. The spatial distribution of the two classes is consistent with an isotropic distribution. We think that these results might preferably suggest a galactic origin for the GRBs, but it is at the moment impossible to say if we are in presence of events close to us in the galactic disk or of events in the galactic halo. No disagreement between the log$N$-log$S$ and the spatial source distribution of GRBs has been found.

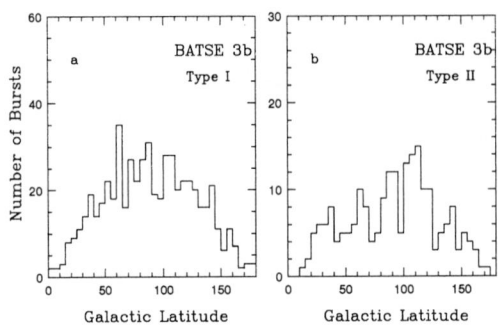

FIG. 4. Distribution of the galactic latitudes for a) Type I and b) Type II GRBs.

REFERENCES

1. B. M. Belli, in High Transients in Astrophysics, ed. S.E. Woosley, AIP Conf. Proc. **115**, 426 (AIP, New York, 1984).
2. B. M. Belli, in Gamma-Ray Bursts, eds W. S. Paciesas and G. J. Fishman, AIP Conf. Proc. **265**, 100 (AIP, New York, 1991).
3. B. M. Belli, in Towards the Source of Gamma-Ray Bursts, eds K. Bennet and C. Winkler, 29th ESLAB Symposium, ESTEC, Noordwijk, The Netherlands, Astr. Spa. Sci. **231**, 43 (1995).
4. B. M. Belli and M. N. Cinti, 24th ICRC Conf. Proc. OG. **2**, 69 (1995).
5. Dezalay J. P. et al. in Gamma-Ray Bursts, eds W. S. Paciesas and G. J. Fishman, AIP Conf. Proc. **265** 304 (1991).
6. G. J. Fishman, in preparation.
7. C. Kouveliotou et al., ApJ **413**, L101 (1993).
8. D. Q. Lamb, C. Graziani and I. A. Smith, ApJ **413**, L11 (1993).
9. R. E. Lingenfelter and J. C. Higdon, in High velocity neutron stars and GRBs, AIP Conf. Proceedings **366**, 164 (1995).

THE CORRECTED LOG N-LOG FLUENCE DISTRIBUTION OF COSMOLOGICAL γ-RAY BURSTS

Joshua S. Bloom[1,2], Edward E. Fenimore[2], Jean in 't Zand[2,3]

[1] *Harvard-Smithsonian Center for Astrophysics, Cambridge, MA 02138*
[2] *Los Alamos National Laboratory, Los Alamos, NM 87544*
[3] *Goddard Space Flight Center, Greenbelt, MD 20771*

Recent analysis of relativistically expanding shells of cosmological γ-ray bursts has shown that if the bursts are cosmological, then most likely total energy (E_0) is standard and not peak luminosity (L_0). Assuming a flat Friedmann cosmology ($q_o = 1/2$, $\Lambda = 0$) and constant rate density (ρ_0) of bursting sources, we fit a standard candle energy to a uniformly selected log N-log S in the BATSE 3B catalog correcting for fluence efficiency and averaging over 48 observed spectral shapes. We find the data consistent with $E_0 = 7.3^{+0.7}_{-1.0} \times 10^{51}$ ergs and discuss implications of this energy for cosmological models of γ-ray bursts.

INTRODUCTION

On the basis of strong threshold effects of detectors, Klebesadel, Fenimore, and Laros (7) concluded that GRB fluence tests were largely inconclusive. As a result, nearly all subsequent number-brightness tests have used peak flux (P) rather than fluence (S). However, the standard candle peak luminosity assumption that is required by log N-log P studies is unphysical. If, for instance, bursts originate at cosmological distances and are produced by colliding neutron stars then one might expect that total energy would be standard and not peak luminosity. Moreover, recent analysis of relativistically expanding shell models has cast doubt on the standard L_0 assumption (9).

In this paper, we seek to eliminate the large threshold effects present in log N-log S studies by correcting the observed number of bursts at a given fluence by the trigger efficiency of the detector.

PVO CONSISTENCY CHECK

The Pioneer Venus Orbiter (PVO) had a peak flux trigger sampled on 0.25, 1.0, and 4.0 sec timescales and was sensitive to bursts down to fluxes of 5×10^{-6} erg cm^{-2}. Despite a substantially lower fluence trigger sensitivity range, PVO saw hundreds more bright bursts than BATSE due the relatively long on-time and large sky-coverage of PVO. As the bright region of the BATSE log N-log

© 1996 American Institute of Physics

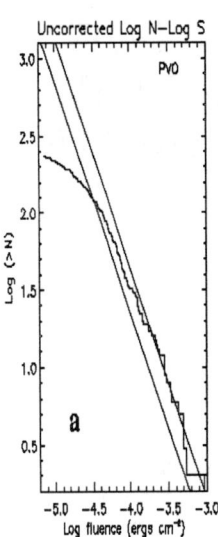 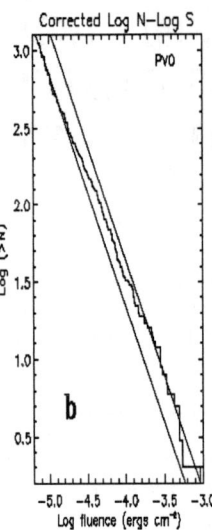

FIG. 1 The log N-log S curve for PVO. a) The uncorrected curve for 293 events in the PVO catalogue which shows significant departure ($P_{KS} \simeq 0.0$) from the $-3/2$ power law (shown as solid line) expected from BATSE observations. b) The corrected curve ($P_{KS} = 0.40$ at $\epsilon(S) = 0.50$) using the PVO trigger efficiency (6). Any local deviation from a $-3/2$ power-law at low fluences we attribute to an incomplete understanding of the trigger efficiency.

S curve seems to fit a $-3/2$ power law well, we would expect that the entire log N-log S curve of PVO should show a similar behavior.

Using the PVO trigger efficiency, $\epsilon(S)$, (6) for each burst i with fluence S_i in the PVO catalogue we take the expected number of bursts with pre-detection fluence to be $N_{i,\text{true}} = N_{i,\text{obs}}/\epsilon(S_i)$. We then compare the derived log $N(>S)$-log S curve with a $-3/2$ power law with arbitrary S-intercept as seen in figure (1b). Although at lower fluence there appears to be a deviation from $-3/2$, the fit is good: with a Kolmogorov-Smirnov (KS) probability of 40% that the corrected distribution comes from a $-3/2$ power law. We derive this KS statistic by finding the maximum distance between the corrected and $-3/2$ distributions (in linear space) down to $S = 10^{-4.5}$ erg cm^{-2}, the fluence at which $\epsilon(S)$ falls to 50%. Note that although the fit is acceptable, a $-3/2$ slope is not necessarily required by the corrected data. We thus conclude that the trigger efficiency determination algorithm (6) is sound, at least in PVO.

DERIVING THE LOG N-LOG S CURVE

BATSE Trigger Efficiency

One subtly worth noting is that BATSE trigger efficiencies are model dependent, i.e., they depend on the choice of E_0 and cosmology, since it is necessary to know *a priori* the true underlying distribution of bursts that passes by the detector. In fact, the derivation of the PVO $\epsilon(S)$ assumed an underlying $-3/2$ distribution. Petrosian and Lee (11) have constructed trigger efficiencies using bivariate correlation. While this method does not assume a particular cosmology, it does require that GRB brightness and duration are inherently uncoupled. Our method does not have this requirement and we make no as-

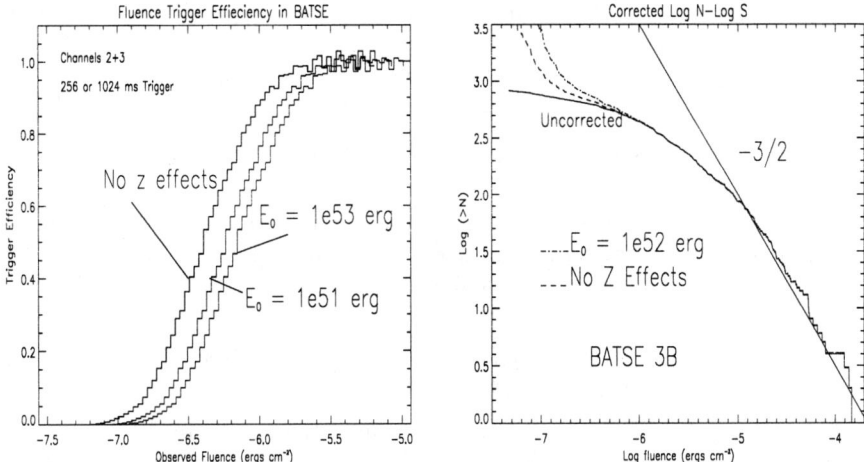

FIG. 2 a) The BATSE trigger efficiency for different E_0 and b) the corrected log N-log S curve for 830 BATSE bursts. Along with the uncorrected log N-log S curve, we depict a corrected curve corresponding to an assume standard candle energy of E_0 of 10^{52} ergs (dot-dash line) and a corrected curve where the effect of redshift on the baseline spectra is removed (dash line).

sumptions about the bursts other then they are cosmological in origin. The BATSE trigger efficiencies could be calculated for any E_0 ($q_0 = 1/2$, $\Lambda = 0$) and two are depicted in figure (2a). Note that the efficiency is nearly unity for the several orders of magnitude in fluence. The corrected log N-log S curve for BATSE is depicted in figure (2b) for two values of E_0. Interestingly, the two distributions are nearly identical for most of the fluence range. In addition, it is clear that the bend from $-3/2$ in log N-log S is true.

Standard Candle Energy Fits

The observed fluence of a source depends strongly on the spectrum, and since the observed spectral shape depends on the distance to the object, the intrinsic spectrum of a GRB object must be used. In addition, the normalization and the spectral shape vary over the duration of the burst, adding to the uncertainty in analysis.

Following a similar analysis as in Fenimore and Bloom (2), we take as our baseline spectra averages over the GRB spectra fit by Band et al. (1). Each such baseline burst has associated with it an observed fluence, S_i, and an observed spectral shape, $\phi_i(E)$. Since each Band et al. (1) burst spectrum is averaged over the burst duration, we assume that the spectral shape is constant, that is, $\phi(E, t_s) \simeq N(t_s)\phi_i(E)$. The fluences, S_i [ergs cm^{-2}], are available for 48 of the Band et al. (1) bursts in BATSE 3B (10).

The observed spectral shape, $\phi_i(E)$, will not necessarily come from a burst at $z \sim 0$ especially if E_0 is large. Therefore, for a given E_0, S_i, and $\phi_i(E)$ we first solve for the redshifts, z_i, of the baseline events associated with each spectral shape. The standard candle energy, E_0, is given by,

$$E_0 = 4\pi R_{i,z}^2 \int_0^\infty N(t_s) dt_s \int_{30}^{2000} E\phi_i\left(\frac{E}{1+z_i}\right) dE \qquad (1)$$

where $N(t_s)$ is the normalization of the spectrum (ergs keV^{-1}) at time t_s at the source. The comoving distance, $R_{i,z}$, is defined in eq. [2] of ref. (2).

The observed ith baseline burst fluence in the energy range 50–300 keV is,

$$S_i = \int_0^\infty N(t_{\rm obs}) dt_{\rm obs} \int_{50}^{300} E\phi_i\left[\frac{1+z_r}{1+z_i}E\right] dE, \qquad (2)$$

where $N(t_{\rm obs})$ is the observed normalization of the spectrum.

For a given standard candle energy, E_0, we numerically determine the redshift $(1+z_i)$ of the ith baseline burst using eqs. (1, 2) and letting $z_r = z_i$. Note that $(1+z_i) \int N(t_s) dt_s = \int N(t_{\rm obs}) dt_{\rm obs}$.

Instead of assuming a spectral shape at the source, we use an average over baseline spectra to compute the number of expected observed bursts, $\Delta N_{\rm exp}[S_j \text{ to } S_{j+1}]$ in some fluence range $[S_j, S_{j+1}]$:

$$\Delta N_{\rm exp}[S_j \text{ to } S_{j+1}] = \frac{4\pi}{N_{\rm BAND}} \sum_{i=1}^{N_{\rm BAND}} \int_{R(S_j)}^{R(S_{j+1})} \epsilon[S_i(r)] \frac{\rho_0}{1+z_r} r^2 dr. \qquad (3)$$

where $N_{\rm BAND} = 48$ is the number of baseline spectra used and ρ_o is the rate density of bursts per comoving volume. The quantity $S_i(r)$ is the predicted fluence (using eqs. [1, 2]) of the ith baseline burst if it was at a distance r. This distance corresponds to a redshift $1+z_r$.

We construct 11 fluence bins (in BATSE channels 2+3 corresponding to approximately 50–300 keV) of roughly equal number of bursts. We select bursts with $C_{\rm min}/C_{\rm max} > 1$ on either the 256 or 1024 ms timescale, then find a minimized χ^2 between the number of predicted bursts and observed by varying E_0. For 9 degrees of freedom we find an acceptable $\chi^2 = 14.7$ corresponding to a standard candle $E_0 = 7.3^{+0.7}_{-1.0} \times 10^{51}$ ergs. Table (1) gives the bin ranges, number of observed bursts per bin, number of predicted bursts for the best fit energy, and their implied redshifts.

CONCLUSIONS

Our fit of $E_0 = 7.0^{+0.7}_{-1.0} \times 10^{51}$ [30–2000 keV] ergs seems a plausible number on the basis that GRBs last on the average 10 sec and $L_0 = 4.6 \times 10^{50}$ erg s^{-1} from log N-log P studies (2). However, this E_0 implies a rather large efficiency of energy conversion to γ-rays ($\sim 10\%$) if the bursting mechanism is colliding neutron stars ($M_{\rm total} \simeq 2.8 M_\odot$). Nevertheless, this result would seem to help resolve the "no-host" problem (cf. ref (3)). Interestingly, that

TABLE 1. Best Fit Distribution of $E_0 = 7.0 \times 10^{51}$ ergs

Bin Number(j)	Fluence Ranges[a] (50–300 keV) S_j	S_{j+1}	$\Delta N[S_j$ to $S_{j+1}]$ Observed[b]	Predicted	$1 + z_j$
1	2.16e-07	3.82e-07	51	38.4	3.88
2	3.82e-07	5.85e-07	42	50.5	3.24
3	5.85e-07	7.55e-07	42	36.3	2.84
4	7.55e-07	1.13e-06	46	63.0	2.64
5	1.13e-06	1.43e-06	37	36.8	2.36
6	1.43e-06	2.00e-06	48	49.7	2.22
7	2.00e-06	2.80e-06	39	44.3	2.04
8	2.80e-06	4.05e-06	44	41.1	1.89
9	4.05e-06	6.20e-06	37	37.2	1.74
10	6.20e-06	1.36e-05	40	44.7	1.60
11	1.36e-05	6.60e-05	41	32.3	1.41

[a] In ergs cm^{-2}
[b] Bursts with $C_{min}/C_{max} > 1$ on the 256 or 1024 ms timescale in BATSE 3B.

the dimmest bursts ($S \simeq 5 \times 10^{-8}$ erg cm^{-2}) are required to be at a redshift of $1 + z \simeq 6.4$ given this E_0, would seem to rule out several cosmological models that require GRB progenitors to be within galaxies (although see reference (8)). This surprisingly high redshift is due to the correct blueshifting of the baseline spectra back to the source in eq. (1). If we neglect this factor, we obtain a smaller, more tenable redshift of the dimmest bursts ($1 + z = 5.2$).

Whatever the conclusion about the models, we note two important results. First, the bend in the log N-log S curve in BATSE is real, not an artifact of strong threshold effects. This implies that we are seeing either a truncated spatial distribution of GRBs (as in Galactic models) or an effect due to the expansion of the universe. The bend might also be caused by a combination of rate density or number density evolution, and a study of their possible effects is certainly warranted. Secondly, with the availability of Monte Carlo modeling of trigger efficiencies, log N-log S tests need no longer be inconclusive.

REFERENCES

1. D. L. Band, et al., Ap.J. **413**, 201 (1993).
2. E. E. Fenimore, and J. S. Bloom, Ap.J. **235**, 29 (1995).
3. E. E. Fenimore, et al., Nature **366**, 40 (1993).
4. E. E. Fenimore, et al., in preparation (1996).
5. J. C. Higdon, R. E. Lingenfelter, Ap.J. **307**, 197 (1986).
6. J. in t' Zand, and E. E. Fenimore, these proceedings (1996).
7. R. Klebesadel, E. Fenimore, and J. Laros, in AIP Conf. Proc. **115**, 429 (1982).
8. L. Lu, W. W. Sargent, D. S. Womble, T. A. Barlow, Ap.J. **457**, L1 (1996).
9. C. Madras, E. E. Fenimore, in preparation (1996).
10. C. A. Meegan, et al., Ap.J.S., in press (1996).
11. V. Petrosian, and T. Lee, these proceedings (1996).

On the distribution of BATSE Gamma-Ray Bursts

Majid Borumand, W. Kluźniak

University of Wisconsin-Madison
Department of physics
Madison, Wisconsin 53706

Models in which GRB sources are distributed in a disk or in intersecting planes are inconsistent at the 8σ level with the mean value of $C_{\min}/C_{\max}$ determined for 411 BATSE GRBs. This conclusion does not depend on the position of GRBs in the sky.

We have also analyzed the directional data on 585 BATSE GRBs to test earlier claims of departure from a random distribution in galactic longitude. Using the Kolmogorov-Smirnov test, we found no evidence for any departure from isotropy.

INTRODUCTION

Different distance scales for GRBs require different energy outputs which in turn call for drastically different physical processes and astronomical scenarios. The lack of known counterparts to GRBs forces us to use statistical methods to obtain clues to their distance. Any nonuniformity in the distribution of bursts on the celestial sphere would have severe implications for distance scale. In this regard we test the BATSE 3B sample against earlier claims of nonuniformity in galactic longitude. Those claims were interpreted by some (4) as evidence that bursts are residing in intersecting planes. We test this hypothesis directly by examining the $C_{\min}/C_{\max}$ distribution.

RULING OUT DISK DISTRIBUTIONS, $C_{\mathrm{MIN}}/C_{\mathrm{MAX}}$ TEST

A uniform disk distribution (or a combination of intersecting disk distributions) of GRBs on the sky corresponds to (3):

$$< C_{\min}/C_{\max} > \geq 0.5$$

The value we found for 411 GRB's of the 3B catalogue is:

$$< C_{\min}/C_{\max} > = 0.42$$

Since the dispersion for a uniform planar distribution is $\sigma \approx 0.01$ we conclude that disk distributions can be ruled out at the 8σ level (Fig. 1). In computing

© 1996 American Institute of Physics

the mean we have chosen the maximum value of $C_{\min}/C_{\max}$ among the three given for each burst.

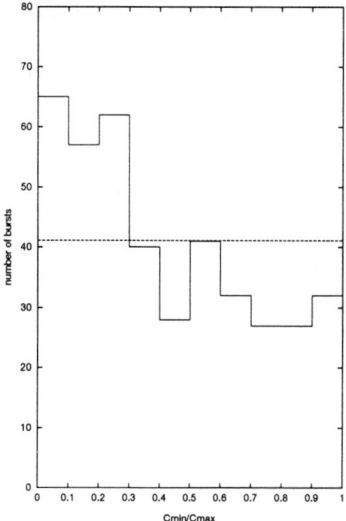

FIG. 1. A histogram of C_{min}/C_{max} for 411 bursts from the 3B Catalogue. The dashed line corresponds to a uniform distribution in a plane. The observed excess number of strong bursts rules out disk distributions at the 8σ level.

UNIFORMITY IN GALACTIC LONGITUDE
THE KOLMOGOROV-SMIRNOV TEST

We compared the cumulative distribution function in longitude $S_{\text{BATSE}}(l)$ of 585 BATSE bursts with the cumulative distribution function of a uniform distribution in Galactic longitude l (Fig. 2). We find the maximum value of the absolute difference between the two cumulative distribution functions to be:

$D_{\text{observed}} = \max | S_{\text{BATSE}}(l) - l/360° | = 2.89 * 10^{-2}$

For a random uniform distribution, the probability of D being greater than this observed value is (2):

$P(D \geq 2.89 * 10^{-2}) = 0.70$

Thus we find no evidence for departure from isotropy in the distribution of gamma-ray bursts in Galactic longitude.

STATUS OF SYSTEMATIC EFFECTS.

In an earlier publication we have pointed out that when the earliest BATSE bursts were binned in Galactic longitude the resulting distribution was not the one expected for a random and isotropic distribution of sources. We ascribed this anomaly to systematic effects (6). As is well known, other authors have later reported similar anomalies and have tried to interpret them in terms of peculiarities of the actual distribution of GRBs.

We have now repeated the chi-square test described in (6) on the larger and revised sample of the 2B bursts. As expected, we have found no departures from isotropy. At this conference we have learned that the BATSE team has reanalysed GRB positions. With the release of the 3B catalogue (1) the unnecessary controversy regarding peculiarities in the distribution of GRBs has been put to rest.

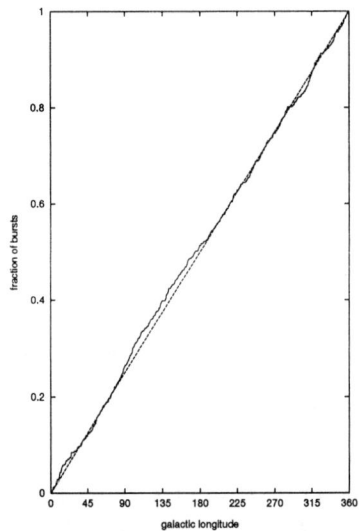

FIG. 2. Comparison between the cumulative distribution functions in galactic longitude of 585 BATSE bursts (solid line) and of a uniform distribution (dashed line) in galactic longitude. The two distributions are compatible–BATSE GRBs are consistent with a sample randomly drawn from a uniform parent distribution.

CONCLUSIONS

1. Models in which GRBs are distributed in a disk or in intersecting disks are ruled out at the 8σ level by the count statistics (C_{min}/C_{max}) alone.

2. We find no evidence for departure of the gamma-ray burst distribution from isotropy.

REFERENCES

1. C. Meegan, et al., APJ Suppl., in press (1996).
2. P.R. Bevington and D. Robinson, Data Reduction and Error Analysis For The Physical Sciences, (McGraw-Hill, New York 1992).

3. W. Kluźniak, in Gamma-ray Bursts, eds. W.S. Paciesas & G.J.Fishman, AIP Conf. Proc. **265**, 105 (AIP, New York 1992).
4. E. Maoz, Preprint, (1994).
5. W. H. Press, S.A. Teukolsky, W. T. Veterling, and B.P. Flannery, in Numerical Recipes In Fortran, 614 (Cambridge University Press 1992).
6. W. Kluzniak, in Compton Gamma-Ray Observatory, eds. M. Friedlander, N.Gehrels, O. J. Macomb, AIP Conf. Proc. **280**, 724 (AIP, NewYork 1993).

Comparison of Two Burst Repetition Tests

J. J. Brainerd

University of Alabama in Huntsville

I demonstrate that the two-point correlation function is always a superior test of burst repetition than the version of the nearest neighbor test invented by Quashnock and Lamb. Because no evidence of burst repetition appears in a two-point correlation function analysis of the 1B BATSE burst catalog, the signal from the nearest neighbor test must be a statistical fluctuation.

TWO REPETITION TESTS

Knowledge of whether gamma-ray burst sources produce more than one outburst on a time scale of days or longer would provide strong constraints on theory. But the large errors inherent to burst locations from individual experiments make the direct association of two gamma-ray bursts with a single source impossible. For this reason, in the largest gamma-ray burst data bases, burst repetition can only be found through a statistical analysis of small scale clustering associated with burst repetition. Two clustering tests were applied to the 1B catalog of the Burst and Transient Source Experiment (BATSE) (4) immediately after its release: the two-point correlation function, which has been applied to catalogs from earlier experiments; and Quashnock and Lamb's variation of the nearest neighbor test. Applied to the BATSE 1B catalog, these tests produce different results. The two-point correlation function finds no evidence of burst repetition (1), but the nearest neighbor test finds a deviation of 0.02 significance, and a subset of bursts exhibit a deviation of 10^{-4} significance (6). In this article, I show that the two-point correlation function is a superior test of burst repetition, and that the deviation found through the nearest neighbor test must be a statistical fluke.

An advantage of the two-point correlation function is that one knows from the error distribution function the range of angles over which the burst repetition signal appears (3,5). A Fisher distribution function with a characteristic location error of $\theta_e \approx 4.73°$ provides a reasonable description of the distribution of θ_σ, the combined statistical and systematic location error, for the BATSE 3B catalog burst locations. This distribution function is

$$\frac{dP^\sigma(\theta_\sigma)}{d\Omega} = \frac{1}{2\pi\mu(1-e^{-2/\mu})} e^{-\frac{1-\cos\theta_\sigma}{\mu}}, \quad (1)$$

where $\mu = 1 - \cos\theta_e$. If n_s sources produce more than one observed burst in a sample of N gamma-ray bursts, with each repeating source i ($1 \leq i \leq n_s$)

© 1996 American Institute of Physics

producing ν_i outbursts, then the two-point correlation function is (2)

$$w(\theta) = S\left[\frac{2\sqrt{2}\,e^{-\frac{2}{\mu}}\sinh\left(\frac{\sqrt{2+2\cos\theta}}{\mu}\right)}{\mu\left(1-e^{-2/\mu}\right)^2\sqrt{1+\cos\theta}} - 1\right], \qquad (2)$$

where

$$S = \frac{\sum_{i=1}^{n_s}\nu_i(\nu_i - 1)}{N(N-1)}. \qquad (3)$$

From these equations, one sees that the burst model only appears in the amplitude of $w(\theta)$, and that the shape of $w(\theta)$ is solely determined by the average location error. This enables one to choose a range of angles for averaging $w(\theta)$ that maximizes the signal over random fluctuations. For the purposes of this study, I integrate over the angles for which $w(\theta)$ is positive. For $\mu \ll 1$, this is given by $0 < 1 - \cos\theta < -2\mu\ln\mu$.

The nearest neighbor test is applied by constructing a cumulative distribution of nearest neighbor values and deriving the maximum deviation D of this distribution from the nearest neighbor distribution produced by a uniform distribution of burst locations. One then constructs the Kolmogorov-Smirnov statistic $D\sqrt{N}$ and determines the significance of the value through numerical simulation—this is required because the set of nearest neighbors is partially correlated.

COMPARISONS OF TEST LIMITS

Earth blockage and the occasional disabling of the burst trigger on BATSE give a sky exposure that is between 0.25 and 0.40 of the full sky exposure, creating a distinction between the physical model of burst repetition and the observed burst repetition. For models in which each repeating source produces a small number of outbursts (e.g., 2 or 3 repetitions), reducing the sky exposure reduces the number of sources that produce at least two observed outbursts. For models in which a large number of outbursts occur, reducing the sky exposure reduces the number of observed outburst per repeater, but it does not significantly lower the number of sources producing two or more observed outbursts.

Figures 1 through 3 show the limits that the two tests place on various physical models of burst repetition. In each model, only a fraction of the M bursts are emitted by repeating sources. The number of repeating sources is m_s, and each repeating source produces κ outbursts. The total number of bursts M is chosen to give a particular average number of observed bursts $\langle N \rangle$ for a given sky exposure. The sky exposures and the values of $\langle N \rangle$ in Figures 1 through 3 are those for the 1B, 2B, and 3B BATSE catalogs. The figures are calculated under the assumption that each repeating source produces all of its outbursts on a time scale much shorter than the CGRO orbital precession time scale of 50 days, although one finds essentially identical results when the

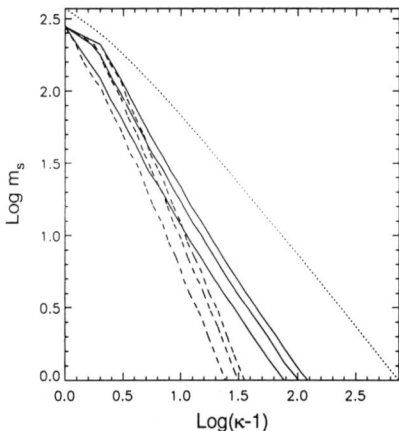

FIG. 1. Limitations on burst repetition models for the 1B BATSE catalog sample size. The total number of bursts is $M = 743$, the average number of observed bursts is $\langle N \rangle = 260$, and the sky exposure is 0.35. The y-axis variable m_s is the number of repeating sources in the model, and the x-axis variable κ is the number of outbursts produced by a repeating source. Contours for significances of 10% (lowermost), 1%, and 0.1% (uppermost) are plotted as solid lines for the nearest neighbor test and as dashed lines for the two-point correlation function. The dotted line gives $m_s = M/\kappa$.

burst repetition time scale is longer than the precession time scale. For each model, 10^3 random burst distributions are generated, and the average value of the statistic from each of the repetition tests is calculated. The significance is then derived for this average value.

Two effects are apparent in these figures. First, the significance contours for the two-point correlation function run parallel to those for the nearest neighbor test in the upper left of each figure, where the number of outbursts per repeating source is small. The sky exposure makes it unlikely that more than two bursts are observed from each repeating source, so increasing the number of repetitions only increases the number of sources that are observed to repeat. Because both tests are proportional to the number of repeating sources, their significance curves behave similarly in this regime. One also notes that in this regime, the limits on repetition fraction are very weak. Second, as the repetition rate increases, one enters the regime where more than two bursts are observed from each repeating source. As one sees from equation (3), the signal in the two-point correlation function increases as $\nu_i(\nu_i - 1)$, so the significance curves become more constraining as the number of outbursts per repeating source increases. The repetition signal in the nearest neighbor test, however, is proportional to $(\nu_i - 1)^\alpha$, where $1.1 \leq \alpha \leq 1.3$ (2), making it less sensitive to the repetition rate and causing its significance contours to diverge from the two-point correlation function contours. The two-point correlation function is always superior to the nearest neighbor test, and if the latter finds a strong repetition signal, the former should find a stronger repetition signal.

The effect of changing the sky exposure between catalogs is weak. For the two-point correlation function, the contours are identical relative to the maximum repetition rates given by the dotted lines. The contours from the nearest neighbor test are affected slightly by the exposure, being strongest for the 3B simulation, which has the largest sky exposure, and weakest for the 2B simulation, which has the smallest sky exposure. For a model in which both κ and m_s increase proportionally with time, a 2σ signal in the 1B catalog

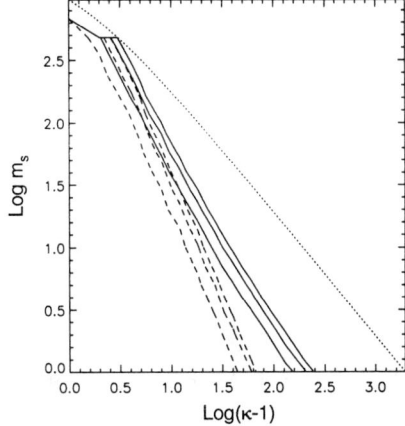 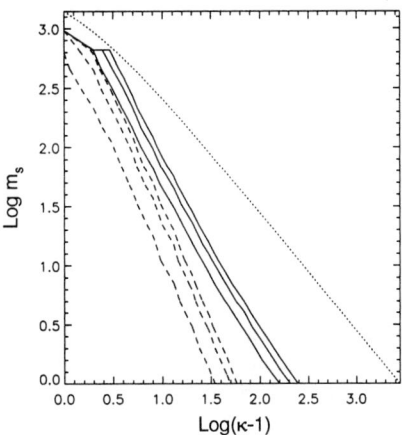

FIG. 2. Limitations on burst repetition models for the 2B BATSE catalog sample size. Same as Figure 1, but with $M = 1940$, $\langle N \rangle = 485$, and a sky exposure of 0.25.

FIG. 3. Limitations on burst repetition models for the 3B BATSE catalog sample size. Same as Figure 1, but with $M = 2805$, $\langle N \rangle = 1122$, and a sky exposure of 0.40.

should appear as a $1.5\,\sigma$ signal in the 2B catalog and as a $3\,\sigma$ signal in the 3B catalog. If a strong burst repetition signature appears in one catalog, it should appear in all catalogs.

REPETITION IN THE 1B CATALOG

Quashnock and Lamb (6) find that the 1B catalog deviates from isotropy with a significance of 0.017 ($2.5\,\sigma$) in the nearest neighbor test. The test applied to the subset of bursts with location errors less than $9°$ produces an anisotropy of 1.1×10^{-4} ($4.0\,\sigma$) significance. In contrast, the two-point correlation function for the full 1B catalog deviates from isotropy by $0.4\,\sigma$ between $0° < \theta < 11.3°$, the range over which $w(\theta)$ is positive. If the deviation seen in the nearest neighbor test is a measure of burst repetition, one expects a positive deviation of $> 2.5\,\sigma$ significance. If smaller intervals are used, reflecting the shape of $w(\theta)$, one finds deviations from isotropy of $2.7\,\sigma$ for separations of $0° < \theta < 5.7°$ and of $-2.5\,\sigma$ for $5.7° < \theta < 8.1°$. But if burst repetition is present, one expects a deviation that is $> 3.6\,\sigma$ significance for the first bin and $> 1.7\,\sigma$ significance for the second bin. For bursts with errors $< 9°$, the deviation from isotropy of the two-point correlation function is $0.6\,\sigma$ between $0° < \theta < 11.3°$. Over smaller intervals the deviations are $3.1\,\sigma$ for $0° < \theta < 5.7°$ and $-3.4\,\sigma$ for $5.7° < \theta < 8.1°$; if the nearest neighbor deviation is from repetition, one expects the first bin to have a significance $> 5.8\,\sigma$ and the second bin to have a significance $> 2.8\,\sigma$. These results suggest that the strong signal seen by Quashnock and Lamb in the nearest neighbor test of the

1B catalog has a statistical origin.

REFERENCES

1. G. R. Blumenthal, D. H. Hartmann, & E. V. Linder, in Gamma-Ray Bursts, AIP Conf. Proc. **307**, ed. G. J. Fishman, J. J. Brainerd, & K. Hurley (New York: AIP), 117 (1994).
2. J. J. Brainerd, Astrophys. J., submitted (1996).
3. J. J. Brainerd, C. A. Meegan, M. S. Briggs, G. N. Pendleton, & M. N. Brock, Astrophys. J. **441**, L39 (1995).
4. G. J. Fishman, C. A. Meegan, R. B. Wilson, M. N. Brock, J. M. Horack, C. Kouveliotou, S. Howard, W. S. Paciesas, M. S. Briggs, G. N. Pendleton, T. M. Koshut, R. S. Mallozzi, M. Stollberg, & J. P. Lestrade, Astrophys. J. Supp. **92**, 229 (1992).
5. C. A. Meegan, D. H. Hartmann, J. J. Brainerd, M. S. Briggs, W. S. Paciesas, G. N. Pendleton, C. Kouveliotou, G. Fishman, G. Blumenthal, & M. Brock, Astrophys. J. **446**, L15 (1995).
6. J. M. Quashnock, & D. Q. Lamb, MNRAS **265**, L59 (1993).

Testing the Dipole and Quadrupole Moments of Galactic Models

Michael S. Briggs[1], William S. Paciesas[1], Geoffrey N. Pendleton[1], Charles A. Meegan[2], Gerald J. Fishman[2], John M. Horack[2], Chryssa Kouveliotou[3], Dieter H. Hartmann[4], Jon Hakkila[5]

[1] *Department of Physics*
University of Alabama in Huntsville, Huntsville, AL 35899
[2] *Space Sciences Laboratory*
NASA/Marshall Space Flight Center, Huntsville, AL 35812
[3] *Universities Space Research Association,*
NASA/Marshall Space Flight Center, Huntsville, AL 35812
[4] *Department of Physics and Astronomy*
Clemson University, Clemson, SC 29634
[5] *Department of Physics and Astronomy*
Mankato State University, Mankato, MN 56002

INTRODUCTION

If gamma-ray bursts originate from a galactic source population, then at some level a galactic pattern must exist in their locations. Expected patterns for galactic sources are a concentration towards the galactic center, measured by the mean dipole moment of the locations towards the center, $\langle \cos \theta \rangle$, where the θ_i are the angles between the burst locations and the galactic center, or a concentration towards the galactic plane, measured by the mean quadrupole moment about the plane, $\langle \sin^2 b - \frac{1}{3} \rangle$, where the b_i are the galactic latitudes of the locations (23,2). To date, neither pattern has been found in the BATSE data: the values $\langle \cos \theta \rangle = 0.011 \pm 0.017$ and $\langle \sin^2 b - \frac{1}{3} \rangle = 0.002 \pm 0.009$ (these values have been corrected for BATSE's nonuniform sky exposure) for the 1122 bursts of the 3B catalog are both consistent with zero and thus with isotropy (21). The dominant uncertainty in these values is due to the finite sample size (4). What galactic signatures could be hidden under these uncertainties?

The tight limits on the quadrupole moment, in conjunction with the fall-off in the number of faint sources, rule out a disk origin for the majority of the sources (20,9,3,4). Galactic models which remain under discussion either consist of a extended halo or of multiple components. A halo consistent with the data must be much larger than the solar galactocentric distance of $R_o = 8.5$ kpc so that the dipole moment will be sufficiently small. We can determine the typical scale by considering a very simple halo: a galactocentric shell of radius R_{shell}. Such a shell has a dipole moment (13):

© 1996 American Institute of Physics

$$\langle \cos\theta \rangle = \frac{2}{3} \frac{R_o}{R_{\text{shell}}}. \quad (1)$$

Using the dipole moment of the 3B catalog (above), we obtain a 2σ lower-limit for R_{shell} of 120 kpc. Any GRBs inside of this radius will have to be balanced with sources located farther away.

In the remainder of this paper we compare galactic models which have published moments with the observed moments of the 3B catalog. The procedures and the models are discussed in greater detail in an earlier work, which used a smaller sample of GRBs (4).

GALACTIC MODELS

The models with quantitative moments that we are aware of appear in Table 1. For each model we list in the table all moments meaningfully different from zero. Each model listed has one or two such moments. In some cases the parameters of the published model are based on fits to the then existing GRB sample; conversely the parameters of some models are not based upon fits but are (presumably very good) examples of the model. We have merely extracted the model moments from the publications–we have made no effort to reoptimize the models. Since in most cases the models have free parameters and were created when the BATSE GRB sample was smaller, rechoosing the parameters might improve agreement with the data. In some cases, a full refitting might worsen agreement because of the tightened constraints on the brightness distribution, logN-logP.

From the table it is apparent that while some galactic models are quite inconsistent with the observed moments, others agree well. The best (and most recent) model (5) is within 0.3σ of the observed dipole moment. The model (19) most distant from the data deviates by 4.0σ from the observed dipole moment and 6.9σ from the observed quadrupole moment. There is an approximate trend for the moments of the more recent models to be smaller, as additional data from BATSE has indicated that the moments of the first post-BATSE models were too large.

Of the models including a disk component (19,27,14,28), the one that best matches the data is the Dark Matter Halo/Disk model of Smith & Lamb (27), which has a 2.7σ deviation in $\langle \cos\theta \rangle$. The largest moment in this model is the dipole moment of its halo component, since only 20% of the bursts originate from the disk component.

The remaining models all assume that GRBs originate from an extended halo. These models fall into two classes: arbitrary models which postulate a source radial distribution and high-velocity neutron star (HVNS) models which assume that the halo consists of HVNS ejected from the disk. The HVNS models have the advantage of being based upon plausible sources and incorporating more physics, but have potential difficulties explaining why only HVNS burst and whether there are sufficient HVNS to produce the observed burst rate.

TABLE 1. Moments of Galactic Model Compared with the Observations

Model	Statistic	Prediction [a]	Dev. [b]
Eichler & Silk (7)	$\langle\cos\theta\rangle$	0.05	2.3
Hartmann (12)	$\langle\sin^2 b - \frac{1}{3}\rangle$	−0.05	5.8
Li & Dermer (16)	$\langle\cos\theta\rangle$	0.048	2.2
Lingenfelter & Higdon (19)	$\langle\cos\theta\rangle$	0.08	4.0
	$\langle\sin^2 b - \frac{1}{3}\rangle$	−0.06	6.9
Fabian & Podsiadlowski (8)	$\langle\cos\theta_{\text{LMC}}\rangle$ [c]	0.038	1.4
Smith & Lamb (27): Disk/Gaussian Shell Halo	$\langle\sin^2 b - \frac{1}{3}\rangle$	−0.027	3.2
Smith & Lamb (27): Dark Matter Halo/Disk	$\langle\cos\theta\rangle$	0.057	2.7
Higdon & Ling. (14): $R_{\text{core}} = 7.5$ kpc, 25% disk	$\langle\cos\theta\rangle$	0.088	4.5
Higdon & Ling. (14): $R_{\text{core}} = 15$ kpc, 20% disk	$\langle\cos\theta\rangle$	0.073	3.6
Higdon & Ling. (14): $R_{\text{core}} = 30$ kpc, 8% disk	$\langle\cos\theta\rangle$	0.060	2.9
Li, Duncan & Thompson (17) [d]	$\langle\sin^2 b - \frac{1}{3}\rangle$	−0.084	1.8
	$\langle\cos^2\theta - \frac{1}{3}\rangle$ [e]	0.073	2.6
Podsiadlowski, Rees & Ruderman (26): Fig. 5a	$\langle\cos\theta\rangle$	0.043	1.9
	$\langle\sin^2 b - \frac{1}{3}\rangle$	−0.019	2.3
Podsiadlowski, Rees & Ruderman (26): Fig. 5b	$\langle\cos\theta\rangle$	0.054	2.5
	$\langle\sin^2 b - \frac{1}{3}\rangle$	−0.024	2.9
Smith (28)	$\langle\cos\theta\rangle$	0.050	2.3
	$\langle\sin^2 b - \frac{1}{3}\rangle$	−0.023	2.8
Bulik & Lamb (5)	$\langle\cos\theta\rangle$	0.016	0.3

[a] Not corrected for BATSE's nonuniform sky exposure.
[b] Deviation, in σ, of the prediction from the value observed for the 1122 bursts of the 3B catalog (expect 109 bursts for Li et al. (17)). Includes correction for sky exposure.
[c] Statistic is the dipole moment to the Large Magellanic Cloud; the observed value is −0.010 and the predicted sky exposure bias is −0.024.
[d] The predictions are for bursts with 1024 ms peak flux > 3.45 γ s^{-1} cm^{-2}, of which there are 109 in the 3B catalog.
[e] The observed value of this quadrupole moment is −0.005 and the sky exposure predicted value is −0.004.

The models (7,12,14) which postulate a source distribution usually assume a dark matter halo form, following the example of Paczyński (24). To match the data, these models are driven to very large core radii, larger than assumed in dark matter models of the galactic rotation curve and larger than any observed galactic component (4). Also suggested are a Gaussian shell halo (27) and a exponential halo (27,28), both of which differ from any known galactic population.

The first HVNS model (16) is still in acceptable agreement with the data, probably because of the 1000 km s^{-1} velocity assumed for all of the bursting sources, a higher value than used by more recent versions of this model. The most recent HVNS model (5) closely matches the data. The unusual model of Fabian and Podsiadlowski (8) is also in good agreement with the data despite using an unusually low source velocity, 400 km s^{-1}. This is achieved by the unique assumption that GRB sources are born only in the Magellanic Clouds, so that the sources are born at halo distances and can easily escape their birth site. However, if sources are born in the disk of the Milky Way at even a small fraction of the Magellanic Cloud birth rate per mass, a strong disk signature would be produced.

Since the uncertainties on the observed moments decrease as $1/\sqrt{N_B}$, where N_B is the number of burst locations in the sample, further progress in testing the moments of galactic models will be slow. Collecting additional data still has several important benefits: it will tighten the constraints on galactic models and aid in analyzing suggested sub-classes of bursts (1,15,25). The tightened limits on the properties of a hypothetical Milky Way halo will aid in interpreting proposed observations of the corresponding halo of M31, observations which are intended to distinguish between halo and cosmological distance scales (6,10,11,18,22).

REFERENCES

1. B. M. Belli, Ap&SS, **231**, 43 (1995).
2. M. S. Briggs, ApJ **407**, 126 (1993).
3. M. S. Briggs, Ap&SS, **231**, 3 (1995).
4. M. S. Briggs, W. S. Paciesas, G. N. Pendleton, C. A. Meegan, G. J. Fishman, J. M. Horack, M. N. Brock, C. Kouveliotou, D. Hartmann & J. Hakkila, ApJ **459**, 40 (1996).
5. T. Bulik & D. Q. Lamb, these proceedings (1996).
6. T. Bulik, P. S. Coppi & D. Q. Lamb, these proceedings (1996).
7. D. Eichler & J. Silk, Science **257**, 937 (1992).
8. A. C. Fabian & P. Podsiadlowski, MNRAS **263**, 49 (1993).
9. J. Hakkila, C. A. Meegan, G. N. Pendleton, G. J. Fishman, R. B. Wilson, W. S. Paciesas, M. N. Brock & J. M. Horack, ApJ **422**, 659 (1994).
10. F. A. Harrison & S. E. Thorsett, ApJ **460**, L99 (1996).
11. F. A. Harrison, J. E. Grindlay, N. Gehrels, C. J. Hailey, W. A. Mahoney, T. A. Prince, B. D. Ramsey, P. Ubertini, G. K. Skinner & M. C. Weisskopf, these proceedings (1996)
12. D. Hartmann, Comments Astrophys. **16**, 231 (1992).

13. D. Hartmann, L. E. Brown, L.-S. The, E. V. Linder, V. Petrosian, G. Blumenthal & K. Hurley, ApJS **90**, 893 (1994).
14. J. C. Higdon & R. E. Lingenfelter, ApJ **434**, 552 (1994).
15. C. Kouveliotou, T. Koshut, M. S. Briggs, G. N. Pendleton, C. A. Meegan, G. J. Fishman & J. P. Lestrade, these proceedings (1996).
16. H. Li & C. D. Dermer, Nature **359**, 514 (1992).
17. H. Li, R. Duncan & C. Thompson, in AIP Conf. Proc. **307**: Gamma-Ray Bursts, ed. G. J. Fishman, J. J. Brainerd, K. Hurley, (New York: AIP), 600 (1994).
18. H. Li & E. Liang, these proceedings (1996).
19. R. E. Lingenfelter & J. C. Higdon, Nature **356**, 132 (1992).
20. C. A. Meegan, G. J. Fishman, R. B. Wilson, W. S. Paciesas, G. N. Pendleton, J. M. Horack, M. N. Brock & C. Kouveliotou, Nature **355**, 143 (1992).
21. C. A. Meegan, G. N. Pendleton, M. S. Briggs, C. K. Kouveliotou, K. M. Koshut, J. P. Lestrade, W. S. Paciesas, M. L. McCollough, J. J. Brainerd, J. M. Horack, J. H. Hakkila, W. Henze, R. D. Preece, R. S. Mallozzi & G. J. Fishman, ApJ, in press (1996).
22. C. A. Meegan, G. J. Fishman, B. A. Harmon, J. M. Horack, R. B. Wilson, J. J. Brainerd, M. S. Briggs, W. S. Paciesas, G. N. Pendleton, C. Kouveliotou & J. Hakkila, these proceedings (1996).
23. B. Paczyński, ApJ **348**, 485 (1990).
24. B. Paczyński, Acta Astron. **41**, 157 (1991).
25. G. N. Pendleton, W. S. Paciesas, R. D. Preece, M. S. Briggs, C. Kouveliotou & C. A. Meegan, these proceedings (1996).
26. P. Podsiadlowski, M. J. Rees & M. Ruderman, MNRAS **273**, 755 (1995).
27. I. A. Smith & D. Q. Lamb, ApJ **410**, L23 (1993).
28. I. A. Smith, ApJ **444**, 686 (1995).

Detection of Gamma-Ray Bursts from Andromeda

Tomasz Bulik[1,2], Paolo S. Coppi[3] and Donald Q. Lamb[1]

[1] *Department of Astronomy and Astrophysics*
University of Chicago, Chicago, Illinois 60637

[2] *Nicolaus Copernicus Astronomical Center*
Bartycka 18, 00-716 Warsaw, Poland

[3] *Department of Astronomy, Yale University,*
P.O. Box 208101, New Haven CT 06520

If gamma-ray bursts originate in a corona around the Milky Way, it should also be possible to detect them from a similar corona around Andromeda. Adopting a simple model of high velocity neutron star corona, we evaluate the ability of instruments on existing missions to detect an excess of bursts toward Andromeda. We also calculate the optimal properties of an instrument designed to detect such an excess. We find that if the bursts radiate isotropically, an experiment with a sampling distance $d_{\max} \gtrsim 500$ kpc could detect a significant excess of bursts in the direction of Andromeda in a few years of observation. If the radiation is beamed along the neutron star's direction of motion, an experiment with $d_{\max} \gtrsim 800$ kpc would detect such an excess in a similar amount of time, provided that the width of the beam is greater than $10°$. Lack of an excess toward Andromeda would therefore be compelling evidence that the bursts are cosmological in origin if made by an instrument at least 50 times more sensitive than BATSE, given current constraints on Galactic corona models. Comparisons with detailed dynamical calculations of the spatial distribution of high velocity neutron stars in the coronae around the Milky Way and Andromeda confirm these conclusions.

I. INTRODUCTION

In the current discussion on origin of gamma-ray bursts (GRBs) (1,2) it has been suggested that detection (or lack thereof) of gamma-ray bursts from our nearest neighbor galaxy M31 would finally settle the debate. Detection of bursts from M31 has been discussed in the past (3–5), while the constraints due to the presence of M31 are summarized in (6).

In section 2 we estimate a sampling distance for a simple detector, in section 3 we estimate a bursting rate expected for a given detector in the framework of the isotropic emission model and the beaming model (7).

© 1996 American Institute of Physics

II. DESCRIPTION OF A DETECTOR

Let us consider a non-imaging, collimated detector with an area A and collimation half angle θ_d. and estimate the limiting flux, or sampling distance for a source of given luminosity, of such a detector. Given the diffuse X-ray background rate n_B and the particle background rate n_P [cm^{-2}s^{-1}] the background count rate in the detector is

$$N_{BG} = n_B A F(\theta_d) + n_P A = A n_B \left(F(\theta_d) + \frac{n_P}{n_B} \right) \ [\text{s}^{-1}]. \quad (1)$$

where $F(\theta_d)$ is the collimation factor. Let a burst with the luminosity of L[phot s-1] go off at the distance d at the center of the field of view of the detector. The signal count rate is $N_{SIG} = LA(4\pi d^2)^{-1}$, and this source will be detected if the signal is k times larger than the variance in the background (for BATSE $k = 5.5$), $N_{SIG}T \geq k \times \sqrt{N_{BG}T}$, where T is the triggering time. Thus, assuming that T is smaller than the live time of the source, we obtain the sampling distance for the detector

$$d_{\max} = \left[\frac{L^2 A T}{(4\pi)^2 k^2 n_B F'(\theta_d)} \right]^{1/4}, \quad (2)$$

where $F'(\theta_d) = F(\theta_d) + \frac{n_P}{n_B}$. For BATSE LAD detectors $n_P = 0.25$ cm^{-2}s^{-1} and $n_B = 1.23$ cm^{-2}s^{-1}. Thus $F'(\theta_d) = F(\theta_d) + 0.20$, but in the following we have assumed that the particle background can be reduced by a factor of 4.

Thus the minimum detectable flux can be improved by: 1. increasing the area, 2. collimating the detector, 3. optimizing the triggering timescale, 4. varying the triggering criterion, and 5. optimizing the energy response. In the case of an imaging detector on one hand the effective area is reduced but on the other hand the triggering criterion can be relaxed.

III. ESTIMATE OF THE BURSTING RATE

We assume that gamma-ray bursts come from high velocity neutron stars in the galactic corona. We ignore the effects due to potential of our Galaxy and the M31, and assume that the neutron stars have high enough velocities to stream along straight lines. Moreover, we assume that the bursting rate is constant and that the bursting never stops and we ignore the distance between the Sun and the center of the Galaxy. We assume that the bursting rate in M31 is identical to that in the Milky Way. The rate-density of the sources in the isotropic emission model is

$$n(r) = \mathcal{N} \left[r^{-2} + (r^2 - 2r r_{M31}\mu + r_{M31}^2)^{-2} \right] \quad (3)$$

where $\mu = \cos(\vec{r}, \vec{r}_{M31})$ The constant $\mathcal{N}$ can be estimated from the rate of events observed by BATSE. The number of bursts seen by BATSE per year

Isotropic emission

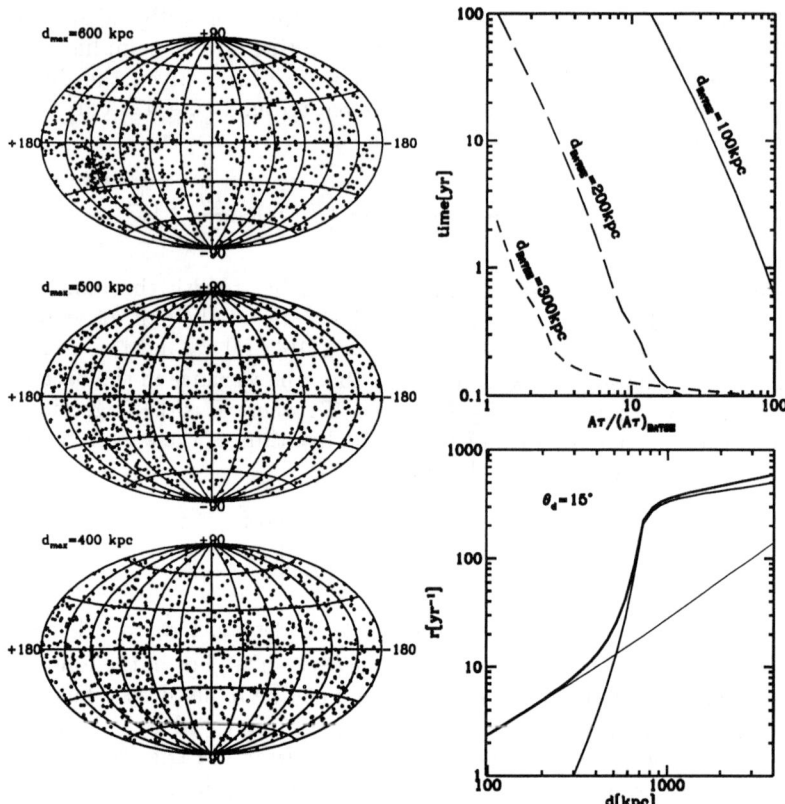

FIG. 1. This figure describes the model with isotropic emission. The left panels show skymaps for three sampling distances: 400, 500, and 600 kpc. Andromeda becomes visible after reaching 500 kpc. The thick line in the right lower panel shows the expected bursting rate as a function of sampling distance when looking towards M31 with a detector collimated to $\theta_d = 15°$ assuming that the current BATSE sampling distance is 200 kpc. The thin lines show contributions of bursts from the Milky Way and M31. The upper right panel shows the time required to detect M31 as a function of a detector area assuming that the current BATSE sampling distance is 100, 200, and 300 kpc.

can be estimated as $N_{\rm BATSE} = 4\pi \int_0^{d_{\rm BATSE}} dr\, n(r)\, r^2$ and ignoring the bursts from M31 is given by $N_{\rm BATSE} = 4\pi \mathcal{N} d_{\rm BATSE}$. In the isotropic emission model the burst rate when looking when at M31 and in the opposite direction can be found by straightforward integration of $n(r)$ when the turn-on time is neglected.

The beaming model of bursting is based on assumption that bursts are emitted in a cone with the half width θ_b around the direction of initial kick velocity. Bursts can be observed only if the observer happens to be in such a cone. The rate of bursting rate in the beaming model becomes

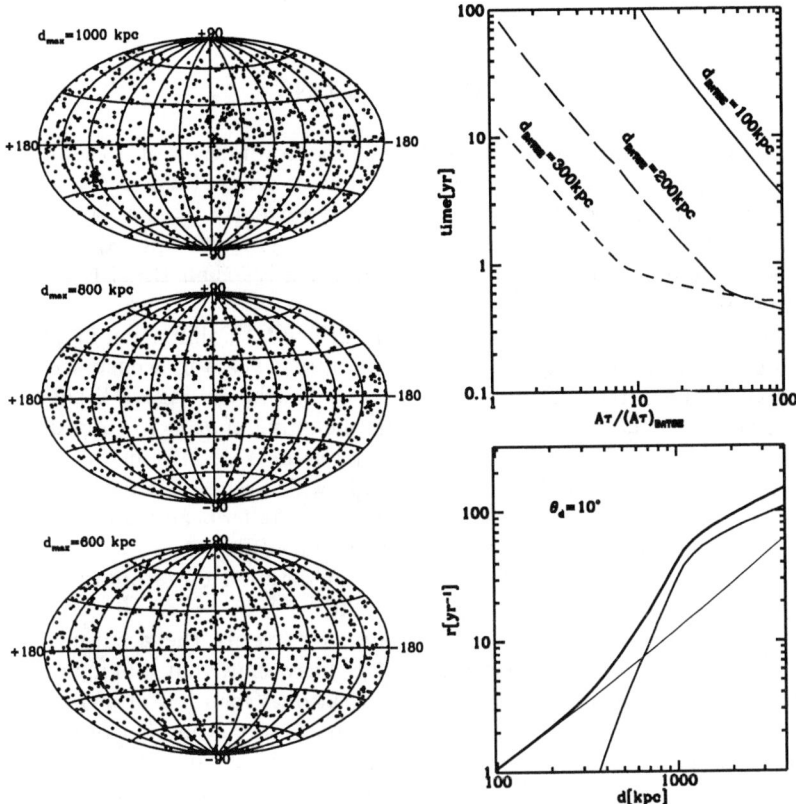

FIG. 2. This figure describes the beaming model with $\theta_b = 20°$. The left panels show skymaps for three sampling distances: 600, 800, and 1000 kpc. Andromeda becomes visible after reaching 800 kpc. The thick line in the right lower panel shows the expected bursting rate as a function of sampling distance when looking towards M31 with a detector collimated to $\theta_d = 10°$ assuming that the current BATSE sampling distance is 200 kpc. The thin lines show contributions of bursts from the Milky Way and M31. The upper right panel shows the time required to detect M31 as a function of a detector area.

$$n(r) = \mathcal{N}\left[r^{-2} + \Theta(|\vec{n}(\vec{n} - \vec{n}_{M31})| - \mu_b)(r^2 - 2rr_{M31}\mu + r_{M31}^2)^{-2}\right] \quad (4)$$

where $\Theta(x)$ is the Heavyside function, $\vec{n}$ is a unit vector, and $\vec{n}_{M31}$ is the unit vector pointing in the direction of M31.

We calculate the bursting rate looking at M31 and in the opposite direction and estimate the time required to detect M31 with significance of 5σ. In the limit when the bursting rates are similar, and the signal is small we need to wait to accumulate enough counts so that the excess is significant. If the signal is strong it is enough to wait for a detection of 14 bursts looking at M31 while not seeing anything in the other direction. However, in such a case we would like the experiment to detect at least one burst over its lifetime, so that the experiment can be tested, which provides another constraint on the timescale.

IV. CONCLUSIONS

Isotropic emission: an experiment with the collimation of $15°$, and area $A = 5A_{\text{BATSE}}$ (the BATSE effective area at trigger is ≈ 1300 cm^2) , and well tuned triggering timescale should see M31, if currently $d_{\max} > 200$ kpc,(see Figure 1.) The current limit on the distance is $d_{\max} > 130$ kpc (8). Therefore we need a detector at least 20 times more sensitive than BATSE to detect gamma-ray bursts from M31.

The beaming model: detection is more difficult. Collimation of $5°$ to $7°$ seems optimal, however an observing time of ≈ 3 years is needed with $A = 5A_{\text{BATSE}}$ if the current sampling distance is 200 kpc.. Figure 2 shows the results of a simulation of a model with the beaming of $\theta_b = 20°$ and detector collimation of $\theta_d = 10°$. In order to detect bursts from M31 in the beaming model we need reach the distance of 800 kpc which requires a detector about 50 times more sensitive than BATSE. Note that smaller beaming angle require much larger initial kick velocities of neutron stars (8)

We find that the optimal field of view for detction of gamma-ray bursts from M31 is between $10° \times 10°$ and $20° \times 20°$.

We have performed numerical simulations to verify the accuracy of the simple analytical model presented here (9). They show that looking opposite M31 may not be optimal because of gravitational focusing of sources from M31 behind us. Thus, a good choice of the background number of bursts for comparison with M31 might be e.g. looking perpendicular to the direction to M31. We have also found that the analytical expressions presented here are in agreement with the detailed numerical model.

REFERENCES

1. D. Q. Lamb., PASP, in press (1996).
2. B. Paczyński, PASP, in press (1996).
3. I. S. Shklovski, I. G. Mitrofonov, MNRAS **212**, 545 (1985).
4. J. L. Atteia, K. Hurley, Adv.Sp.Res. **6**, 39 (1986).
5. F. Harrison and S. E. Thorsett, ApJ **460**, L99 (1996).
6. P. S. Coppi, T. Bulik, D. Q. Lamb, these proceedings.
7. H. Li, et al., in Gamma-Ray Bursts, eds. G. J. Fishman, J. J. Brainerd and K. Hurley, AIP Conf. Proc. **307**, 600 (AIP, New York, 1993)
8. T. Bulik, D. Q. Lamb, these proceedings.
9. T. Bulik, P. S. Coppi, D. Q. Lamb, Ap.J. submitted, (1996).

Constraints on the Galactic Corona Models of Gamma-Ray Bursts From the 3B Catalogue

Tomasz Bulik[1,2] and Donald Q. Lamb[1]

[1] *Department of Astronomy and Astrophysics*
University of Chicago
5640 South Ellis Avenue, Chicago, IL 60637

[2] *Nicolaus Copernicus Astronomical Center*
Bartycka 18, 00-716 Warsaw, Poland

We investigate the viability of Galactic corona models of gamma-ray bursts by calculating the spatial distribution expected for a population of high-velocity neutron stars born in the Galactic disk and moving in a gravitational potential that includes the Galactic bulge, disk, and a dark matter halo. We consider models in which the bursts radiate isotropically and in which the radiation is beamed. We place constraints on the models by comparing the resulting brightness and angular distributions with the data in the BATSE 3B catalog. We find that, if the burst sources radiate isotropically, the Galactic corona model can reproduce the BATSE peak flux and angular distributions for neutron star kick velocities $\gtrsim 800$ km s^{-1}, source turn-on ages $\gtrsim 20$ Myrs, and BATSE sampling distances 130 kpc $\lesssim d_{\max} \lesssim 350$ kpc. If the radiation is beamed, no turn-on age is required and agreement with the BATSE data can be found provided that the width of the beam is $\lesssim 20°$.

INTRODUCTION

Gamma-ray bursts (GRB's) continue to confound astrophysicists nearly a quarter century after their discovery (1). Before the launch of CGRO, most scientists thought that GRB's came from magnetic neutron stars residing in a thick disk (having a scale height of up to ~ 2 kpc) in the Milky Way (2). The data gathered by BATSE showed the existence of a rollover in the cumulative brightness distribution of GRB's and that the sky distribution of even faint GRB's is consistent with isotropy (3).

Galactic models attribute the bursts primarily to high-velocity neutron stars in an extended Galactic halo, which must reach one fourth or more of the distance to M31 ($d_{\text{M31}} \sim 690$ kpc) in order to avoid any discernible anisotropy (4–6). Cosmological models place the GRB sources at distances $d \sim 1-3$ Gpc, corresponding to redshifts $z \sim 0.3 - 1$; a source population at such large

© 1996 American Institute of Physics

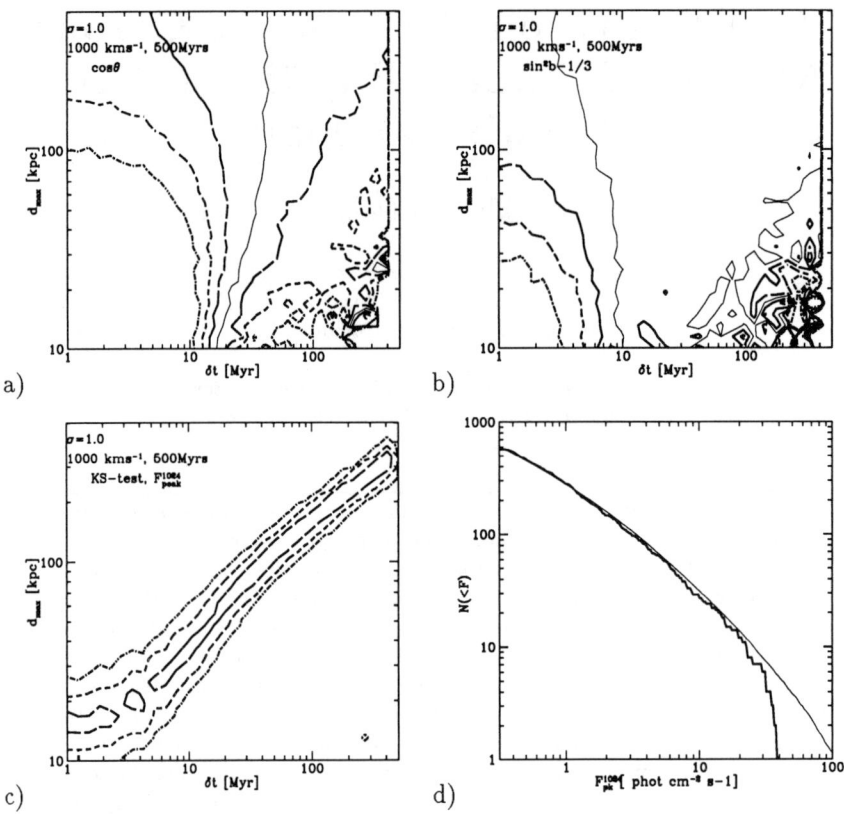

FIG. 1. Comparison of a Galactic halo model in which neutron stars are born with a kick velocity of 1000 km s^{-1}, have a burst-active phase lasting $\Delta t = 5 \times 10^8$ years, and a luminosity function with width $\sigma = 1.0$, with a self-consistent sample of 570 bursts from the BATSE 3B catalogue. Panels (a) and (b) show the contours in the (δt, $d_{\max}$)-plane along which the Galactic dipole and quadrupole moments of the model differ from those of the data by $\pm 1\sigma$ (solid lines), $\pm 2\sigma$ (dashed line), and $\pm 3\sigma$ (short-dashed line) where σ is the model variance; the thin line in panels (a) and (b) show the contour where the dipole and quadrupole moments for the model equals that for the data. Panel (c) shows the contours in the (δt, $d_{\max}$)-plane along which 32%, 5%, and 0.4% of simulations of the cumulative distribution of 570 bursts drawn from the peak flux distribution of the model have KS deviations D larger than that of the data. Panel (d) shows brightness distribution of a model with $\delta t = 30$ Myrs and $d_{\max} = 80$ kpc and the BATSE data.

distances naturally produces an isotropic distribution of bursts on the sky. In addition, studies show that the expansion of the universe can reproduce the observed rollover in the cumulative brightness distribution e.g., (7).

Recent studies (8,9) have revolutionized our understanding of the birth velocities of radio pulsars. They show that a substantial fraction of neutron stars have velocities that are high enough to produce an extended halo around the Milky Way like that required by Galactic halo models of GRBs (10).

Detailed studies of the spatial distribution expected for high-velocity neutron stars born in the Galactic disk (6,11) show that there is a large region in the parameter space where galactic models are consistent with the data from the 2B catalogue. The aim of this work is to re-evaluate this models in the light of the BATSE 3B catalogue.

MODELS

We have calculated detailed models of the spatial distribution expected for a population of high-velocity neutron stars born in the Galactic disk and moving in a Galactic potential that includes the bulge, disk, and a dark matter halo.

We use the mass distribution and potential (12) which includes a doubly exponential disk, a bulge, and a dark matter halo (6). The circular velocity v_c and the Galactic disk lead to characteristic angular anisotropies as a function of burst brightness which provide a signature, and therefore a test, of high-velocity neutron star models.

We assume that the neutron stars are born with the local circular velocity $v_c \approx 220$ km s^{-1} of the Galactic disk and an isotropic distribution of initial kick velocities v_{kick} ranging from 200 to 1200 km s^{-1}. We follow the resulting orbits for up to 3×10^9 years. Given that current knowledge of v_{kick} is poor, we adopt a Green's-function approach: we calculate the spatial distribution of neutron stars for a set of kick velocities (e.g., $v_{\text{kick}} = 200, 400, ..., 1200$ km s^{-1}).

We consider a model in which the bursts are standard candles, i.e. $L = \delta(L - L_0)$, and also models with the log-normal luminosity function with some width, $P(L, L + dL) \approx \exp(-(\log(L/L_{\text{av}})/\sigma)^2)dL/L$. We denote d_{av} the distance to an average burst. If the width of the luminosity function is small than d_{av} tends to d_{max} - the sampling distance. We parametrize the burst-active phase by a turn-on age δt and a turn-off time Δt, and assume that the rate of bursting is constant throughout the burst-active phase, i.e., the burst rate $r = $ const for $\delta t \leq t \leq \Delta t$ and 0 otherwise. The high-velocity neutron star model then has the following parameters: v_{kick}, δt, Δt, BATSE sampling depth d_{max}, and the width of the luminosity function σ. We also consider a beaming mode (10) in which bursting occurs in a cone with width θ_b along the initial kick velocity of a neutron star. Such models do not require a turn-on delay.

COMPARISON BETWEEN MODELS AND DATA

We compare the models with a carefully-selected data set that is self-consistent (6). We use only bursts that a) trigger on the 1024 ms timescale, and have $t_{90} > 1024$ ms, b) have F_{pk}^{1024}, the peak flux in 1024 ms, since we adopt it as the brightness measure. c) have $F_{\text{pk}}^{1024} \geq 0.35$ photons cm^{-2} s^{-1} in order to avoid threshold effects (7,13). We also exclude overwriting bursts and MAXBC bursts. The 3B catalogue contains 570 bursts satisfying the above criteria; this set of bursts has Galactic dipole and quadrupole moments $\langle \cos\theta \rangle = 0.018 \pm 0.0241$, and $\langle \sin^2 b - \frac{1}{3} \rangle = -0.011 \pm 0.012$, and

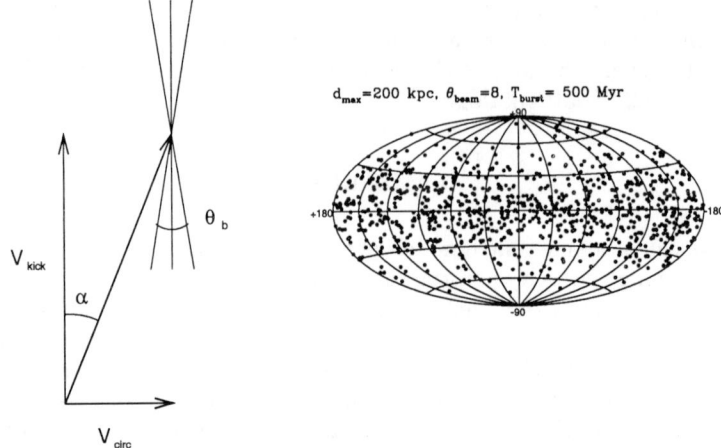

FIG. 2. The left panel shows the lower limit on the beaming angle θ_b. A star with the kick velocity V_{kick} perpendicular to the disk plane also has a galactic circular velocity V_{circ} in the plane, however, it is bursting along the direction of V_{kick}. If the angle α is larger than θb the bursts will never be seen by an observer in the Galaxy, relatively close to the birthplace of the neutron star. The right panel shows a skymap for $v_{kick} = 10^3$ kms^{-1}, and $\theta_b = 8°$.

$\langle \cos \theta_{M31} \rangle = 0.0078 \pm 0.0241$.

We test the viability of Galactic halo models by comparing the Galactic dipole and quadrupole moments, $\langle \cos \theta \rangle$ and $\langle \sin^2 b - 1/3 \rangle$, of the angular distribution of bursts for the model with those for the above set of bursts, using χ^2. We have also compared the peak flux distribution for the model with that for the above set of bursts, using the KS test.

These comparisons do *not* provide estimates of model parameters (i.e., they do not yield parameter confidence regions), but are meant only to be a rough "goodness-of-fit" guide to models which should be tested using a more rigorous approach like the maximum likelihood method.

In Figure 1 the results for a Galactic halo model in which neutron stars are born with a uniform single velocity 1000 km s^{-1} turn-off time $\Delta t = 5 \times 10^8$ yrs, and a log-normal luminosity function with $\sigma = 1.0$. As an example, in the model with $\delta t = 100$ Myrs and $d_{av} = 170$ kpc the expected dipole and quadrupole moments are $\langle \cos \theta \rangle = 0.0033$ and $\langle \sin^2 b - 1/3 \rangle = -0.0046$, after correcting for the BATSE exposure. This values are extremely close to those expected for isotropy.

The beaming model (10) fits the data when the beaming angle $\theta_b \approx 20°$, and the constraints on the BATSE sampling distance are similar to the case of isotropic emission. An interesting feature of the beaming model is that θ_b is bounded on both sides. The upper limit is due to the galactic anisotropy seen for young neutron stars because of their birth places. The lower limit is due to the galactic rotational velocity which leads to lack of bursts in galactic polar directions when the beaming angle is too small. The lower limit is approximately $\tan \theta_b > v_c/v_{kick}$, see Figure 2.

Comparisons of this kind show that the high-velocity neutron star model can reproduce the peak flux and angular distributions of the bursts in the BATSE 3B catalogue for neutron star kick velocities $v_{\text{kick}} \gtrsim 800$ km s^{-1}, burst turn-on ages $\delta t \gtrsim 20$ million years, and BATSE sampling depths 130 kpc $\lesssim d_{\text{max}} \lesssim 350$ kpc. It is clear from this comparison that global isotropy comparisons will not yield the answer to the question of the origin of gamma-ray bursts.

In high-velocity neutron star models, the slope of the cumulative peak flux distribution for the brightest BATSE bursts and the PVO bursts reflects the space density of the relatively small fraction of burst sources in the solar neighborhood. The standard candle models reproduce the BATSE peak flux distribution but they are unable to explain the $-\frac{3}{2}$ slope observed for the PVO bursts. The luminosity function "fills" the void near us with distant bright bursts which appear as isotropic, nearby bursts. Moreover, increasing the width of the luminosity function allows to relax the constraint on the burst turn-on time δt.

M31 provides a strong constraint on the BATSE sampling distance d_{max} (4). We have investigated the effects of M31 within the framework of the high-velocity neutron star model described above by including the distortion of the Galactic halo potential due to M31, as well as the spatial distribution of burst sources emanating from M31. The results of this work are summarized in these proceedings (14).

REFERENCES

1. R.W. Klebesadel, I.B. Strong, and R.A. Olson, ApJ **182**, L85 (1973).
2. J. C. Higdon and R. E. Lingenfelter, Ann. Rev. Astron. Astrophys. **28**, 401 (1990).
3. M. S. Briggs, et al., ApJ **459**, 40 (1996).
4. J. Hakkila, et al., ApJ **422**, 659 (1994).
5. D. Hartmann, et al., ApJ Suppl. **90**, 893 (1994).
6. T. Bulik and D. Q. Lamb, A & Sp. Sc., in press (1995).
7. E. E. Fenimore, et al., Nature **366**, 40 (1993).
8. A. G. Lyne and D. R. Lorimer, Nature **369**, 127 (1994).
9. D. A. Frail, W. W. Goss and J. B. Whiteoak, ApJ **437**, 781 (1994).
10. H. Li and C. Dermer, Nature **359**, 514 (1992).
11. Ph. Podsiadlowski M. J. Rees, and M. Ruderman, MNRAS **237**, 755 (1995).
12. K. Kuijken and G. Gilmore, MNRAS **239**, 571 (1989).
13. J. J. M. in 't Zand and E. Fenimore, in Gamma-Ray Bursts, eds. G.J. Fishman, J.J. Brainerd, and K. Hurley, AIP Conf. Proc. **307**, 692 (AIP, New York, 1994).
14. P. S. Coppi, T. Bulik, and D. Q. Lamb, these proceedings (1996).

Search for Repeating Classical Bursts with the Interplanetary Network

T. L. Cline[1], K. C. Hurley[2], M. Boer[3], M. Niel[3], M. Sommer[4], C. Kouveliotou[5,6], G. J. Fishman[5], and C. A. Meegan[5]

[1] *Code 660, NASA Goddard Space Flight Center, Greenbelt, MD 20771, USA*
[2] *University of California, Berkeley, CA USA*
[3] *Centre d'Etude Spatiale des Rayonnements, Toulouse, France*
[4] *Max Planck Institut fur Extraterrestrische Physik, Garching, FRG*
[5] *NASA Marshall Space Flight Center, Huntsville, AL USA*
[6] *Universities Science Research Association at Huntsville, AL USA*

> The comparison of gamma-ray burst observations from Ulysses and from CGRO BATSE has provided several hundred source regions that consist of large-diameter but narrow annuli. Although awkward for use in most applications, the one-dimensional accuracies of these ring segments can be exploited in the search for classical repeaters (as evidenced by the necessary crossing of multiple annuli at the same locations). The catalog of Ulysses-to-BATSE source loci (1) and the BATSE 3B catalog (2) have been used to search for source rings from the interplanetary network that intersect within or close to their own 3B source fields, with calculations done varying the BATSE field size. The regions determined in this manner are consistent with randomicity or can be accounted for as systematic to this technique. Thus, no evidence for the repetition of classical GRBs has been found. Extensions of this work could be possible using event catalogues from other missions, e.g., GGS-Wind, and from the earlier networks.

JUSTIFICATION

One peculiar characteristic of cosmic gamma ray bursts (GRBs) is that events have not been observed to repeat from the same source, as opposed to the well-known soft gamma repeaters (SGRs) that were found because they do. Clearly, the possibility of event repetition is of great interest, since models that require the destruction or radical alteration of the system responsible for the burst event would be consistent only with the impossibility of repetition. Alternatively, evidence for classical event repetition might indicate the possibility of a relationship between classical events and SGRs; this could support the point of view that all the various gamma ray transients, GRBs, SGRs, and the 79 Mar 5 event, might be made to be compatible with some kind of unified model.

Before the Compton GRO era, the data available to place an upper limit to

© 1996 American Institute of Physics

the classical GRB repetition rate consisted of a few tens of well-determined source fields and an additional variety of other source loci of irregular sizes and shapes. The typical repetition interval, reflecting in part the short observational time span, could not be stretched beyond one to a few years (3). The possible existence of patterns suggestive of event clustering could be inferred in the early CGRO BATSE data, promoting a hope that additional data would provide concrete evidence of repetition; however, these indications have not been supported by analyses using the revised 3B data. It is fair to state that prior to the present time, there existed no strong evidence of any kind for the repetition of classical GRBs.

ANALYSIS

Several hundred network GRB source fields are now available as a catalog supplement (1). These consist mostly of the thin (up to a few arc-minutes wide) rings that result from comparing Ulysses and BATSE measurements. The one-dimensional precision comes from the multi-astronomical-unit separation of these spacecraft since the launch of CGRO, several months after Ulysses began its initial travel towards Jupiter. A small fraction of the entries, for those events observed with three or more spacecraft, give smaller source fields of varying extents. The majority, consisting of lengthy, annular-segment regions, are obviously impractical or useless for many applications, although, e.g., a candidate source direction can be eliminated for any given event. Similarly, the possibility of source commonality can be easily examined by searching for intersections of the source annuli and checking for those intersections that are consistent with their respective BATSE source regions, using the recently released 3B catalog.

The over-250 annuli in the Ulysses catalog supplement were used with simple computerized calculations to find the two-annulus intersections (about 34,000) and to study the separations of these locations from their respective 3B positions. The BATSE source fields, keeping in mind their relative coarseness and their statistical nature (4), were also rms-augmented in recalculations with 1-degree steps. For runs up to a 3-degree BATSE augmentation, in fact, only several two-event intersections were found. The 3-degree trial produced the first and only multiple intersection, that for BATSE events # 451, 2286 and 2619. The location of this candidate is at RA = 198.8 and Dec = -3.8 (J2000), using an error of 0.25 degrees. It is assumed to be an entirely random occurrence. When the errors were increased, a curious result with events clustering in time was found; this can be explained as a systematic effect peculiar to this technique, however. (The source annuli are determined with the direction from BATSE to Ulysses as the center of each arc; this celestial location moves slowly as a function of time, being nearly static for closely spaced events. Thus, for these, the possibility of nearly tangent loci is greatly increased.)

CONCLUSIONS

The patterns of two-event BATSE-consistent interplanetary-network source intersections are found to be randomly distributed in the sky, with a rate that is considered to be also within statistical necessity. When the BATSE source field is given a 3-degree rms augmentation, only one multiple intersection is found; no others were found for an additional, moderate increase. This result is also considered to be accidental coincidence. We conclude that there is no evidence here for classical GRB repetition. This result is, of course, conditioned by the brevity of the 4-year time span of the data considered; calculations of this kind need to include the historical annuli from earlier interplanetary networks and the more recent observations, such as from GGS-Wind and other new missions. Thus, the available data are not yet exhausted.

REFERENCES

1. K. Hurley, et al., these proceedings (1996).
2. C.A. Meegan, et al., these proceedings (1996).
3. B.E. Schaefer, & T.L. Cline, ApJ **289**, 490 (1985).
4. C. Gratziani, & D.Q. Lamb, these proceedings (1996).

A Simultaneous Spectral Invariant Analysis of the GRB Count Distribution and Time Dilation

Ehud Cohen and Tsvi Piran

Racah Institute of Physics, The Hebrew University, Jerusalem, Israel 91904

The analysis of the BATSE's count distribution within cosmological models suffers from observational uncertainties due to the variability of the bursts' spectra: when BATSE observes bursts from different redshifts at a fixed energy band it detects photons from different energy bands at the source. This adds a spectral dependence to the count distribution $N(C)$. Similarly, variation of the duration as a function of energy (1) at the source complicates the time dilation analysis. It has even been suggested that these methods lead to inconsistent estimates of the redshift from which the bursts are observed (2). Clearly it would be best to combine the estimates and to perform a joint analysis of the strength and the duration of the bursts. But for this we have to eliminate first the spectral dependence problem.

We describe here a new statistical formalism that performs the required "blue shifting" of the count number and the burst duration in a statistical manner. This formalism allows us to perform a combined best fit (maximal likelihood) to the count distribution, $N(C)$, and the duration distribution simultaneously. The outcome of this analysis is a single best fit value for the redshift of the observed bursts.

INTRODUCTION

When BATSE observes bursts from different redshifts at a fixed energy band it detects photons from different energy bands at the source. This spectral dependence complicates the interpretation of the peak-flux and time-dilation distributions. So far several attempts have been made to overcome this problem by modeling the spectral shape of the bursts. We suggest here a different method, which is based on the availability of multi channel data in different energy bands. The basic idea beyond our scheme is that we "view" all bursts at the same intrinsic energy band independently of their redshift by scanning over the different channels until we find the most likely redshift and using this value to "blue-shift" back the observed spectrum to the initial spectrum at the source. Now we look at the same energy band at the source for all bursts, avoiding the issue of the spectral shape.

© 1996 American Institute of Physics

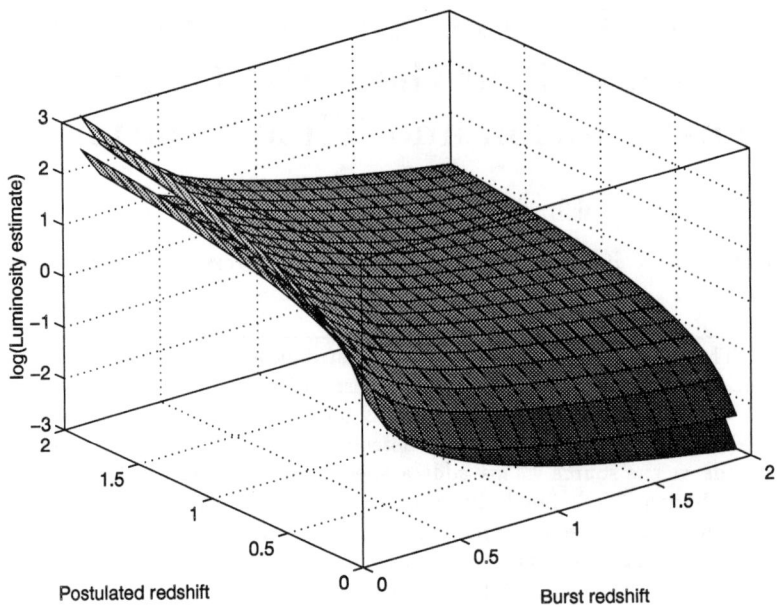

FIG. 1. The "blue-shifted" luminosity of an observed burst with an intrinsic luminosity $L = 1$, with power-law parameter $\alpha = 2.5$ (upper sheet) and $\alpha = 1$ (lower sheet) in an $\Omega = 1$ cosmology, for different assumed z values. The correct luminosity $L = 1$ is obtained only when the chosen z is the same as the real one.

STANDARD CANDLE SOURCES.

Consider first a population of standard candle sources that have a fixed luminosity, L, at a given energy band centered around an energy E. For each burst we have a set of measured peak fluxes $C_i(E_i)$ at different energy bands. We determine the redshift of a particular burst by solving the equation:

$$L = L(z) = C(\frac{E}{1+z})d_L^2(z)(1+z)^{-2} , \quad (1)$$

where d_L is the luminosity distance and $C(\frac{E}{1+z})$ is the observed peak-flux at the energy $\frac{E}{1+z}$.

One may wonder whether there will be multiple solutions to this equation. The answer is surprisingly no, provided that the spectrum is softer than $N(E)dE \propto E^{-1}dE$. As an example we consider a simulated source with a spectral shape: $N(E)dE \propto E^{-\alpha}dE$. Figure 1 shows the luminosity as calculated by this method for simulated sources with $\alpha = 2.5$ (upper sheet) and $\alpha = 1$ (lower sheet) in an $\Omega = 1$ cosmology. The only z which solves the luminosity equation is the actual redshift.

Once we obtain the redshift for each burst we proceed to compare the redshift distribution with the one predicted by different cosmological models.

This is done, for example, by using the maximum likelihood method. We check whether the method is consistent by repeating the whole process for different energy channels (at the source). If there is no noise the different channels should give the same redshift distribution.

SOURCES WITH VARIABLE LUMINOSITY.

In reality we don't have standard candle sources. Current estimates suggest that the GRB luminosity function is quite narrow (with variation of peak flux of no more than one order of magnitude) but it is unlikely that it is a delta function. Thus, the luminosity of a burst is not known a priori, and the redshift can not be deduced uniquely from the peak flux. However, this method can be used in a statistical manner even when faced with this uncertainty.

We assume that the bursts have a known luminosity function $\phi(L)$ at a standard energy band at the source. In fact all we need to assume is that $\phi(L)$ has a given functional shape characterized by a few parameters which will be determined by the analysis. As in the standard candle case we blue-shift the peak flux in each of the energy channels by a suitable factor $(1+z)$ to our canonical energy at the source and we calculate the corresponding luminosity: $L(z)$, using equation 1. We then estimate the likelihood that the burst is at redshift z as:

$$h(z) = \phi(L(z))\frac{dL}{dz} \equiv \tilde{\phi}(L(z)) \ . \tag{2}$$

For standard candles $\phi(L) = \delta(L-L_0)$ and $h(z) = \delta(z-z_0)$). Figure 2 depicts the spread in z due to a power-law luminosity distribution:

$$\phi(L) = \begin{cases} \left(\frac{L}{L_0}\right)^3 & L < L_0 \\ \left(\frac{L}{L_0}\right)^{-3} & L > L_0 \end{cases} \tag{3}$$

A given cosmological model, with a given evolution model (that is number density of bursts per co-moving as a function of cosmological time) determines $n(z)dz$, the expected number of bursts from a dz interval centered around z, per observer unit time. With a given luminosity function at a fixed energy band at the source we calculate now the likelihood of a burst, given the set of data $C_i(E_i)$:

$$P(C_{1...n}) = \int n(z)h(z)dz \\ = \sum_i \tilde{\phi}\left[C_i(\tfrac{E}{1+z_i})d_L(z_i)^2(1+z_i)^{-2}\right]n(z_i) \ . \tag{4}$$

Once we have this estimate for the likelihood of all bursts we proceed to write the likelihood function of the whole set of data. Then we use the maximal likelihood method to estimate the most likely cosmological parameters and the luminosity function parameters.

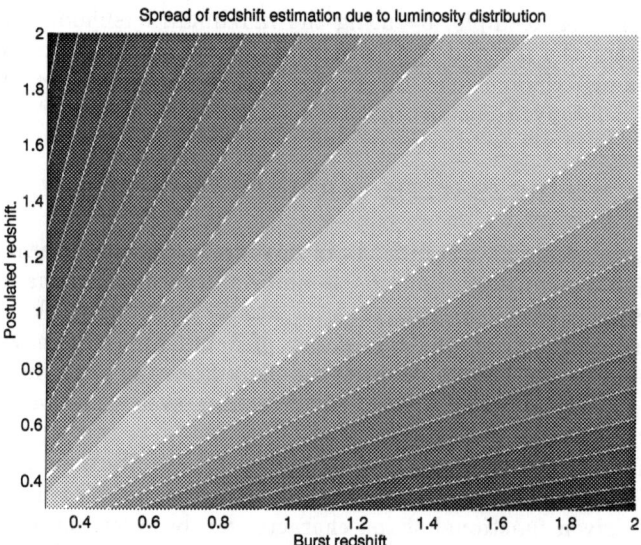

FIG. 2. Figure 2: Likelihood of a burst from redshift z to be detected as a nominal burst from a postulated redshift. Standard candle sources would make the graph non zero only on the main diagonal. The spread is due to a power-law luminosity distribution given by Eq. 3. The lines represents 33%, 10%, 3.3%... likelihood.

It is apparent that as the luminosity distribution widens, the constraint on the redshift weakens, and the information about cosmology is vaguer. The difference between reasonable spectra is not large. For example to confuse a redshift $z = 0.5$ bursts and a redshift $z = 1$ with $\alpha = 2$ requires a luminosity factor of 4.5 while $\alpha = 1$ requires a luminosity factor of 3.4. Naturally, as the amount of data increases, the confidence in the determination of z increases. Once more, self consistency requires that similar results are obtained when using different energy channels.

A JOINT ANALYSIS OF TIME-DILATION.

The time-dilation analysis should be combined with the peak-flux estimate. The combined analysis provide a self consistent method which estimates the redshift of each burst using both sets of data. This should be more accurate than the standard analysis (3–5) in which the bursts are grouped into large groups of bright, dim and dimmest bursts.

Using the spectral invariant method we view all bursts in the same intrinsic energy, and in this way we eliminate the energy dependence of the bursts' duration. Now we would like to combine information about the duration of a burst in the different energy channels, $\delta t_i(E_i)$ (this could be Δt_{50} or alternatively the autocorrelation characteristic time) with the set of peak-fluxes $C_i(E_i)$. We estimate the likelihood of the burst as:

$$P(C_{1...n}, \delta t_{1...n}) = \int n(z)h(z)\Psi\left[\frac{\delta t(\frac{E}{1+z})}{1+z}, E\right] dz =$$
$$\sum_i n(z_i)\tilde{\phi}\left[C(\frac{E}{1+z_i})d_L(z_i)^2(1+z_i)^{-2}\right]\Psi\left[\frac{\delta t_i(\frac{E}{1+z_i})}{1+z_i}, E\right] \quad (5)$$

Where $\Psi(t, E)$ is the intrinsic energy dependent duration distribution function. Using the maximum likelihood method we can now estimate the best-fit parameters of the luminosity function, $\tilde{\phi}$, the duration distribution, $\Psi(t, E)$, and the maximal redshift from which the bursts are observed.

DISCUSSION

The spectrum invariant method is the only way to have a complete usage of all the information available in the multi-spectral channel detectors in situations in which the spectra are being redshifted differently from burst to burst. It also uses the time-dilation and the peak-luminosity in a way that more than just brings a consistent answer, it uses the whole data to enlarge the statistical confidence of the results. A luminosity distribution influences considerably the confidence of the redshift determination of a single burst origin. However, it has a minor influence when using a large number of bursts. Similarly, the duration distribution makes it impossible to determine the redshift of a single burst, however one can determine the parameters of the duration distribution and find whether the redshift determined from the time dilation analysis is consistent with the one determined from the peak-flux distribution.

REFERENCES

1. E.E. Fenimore, et al., ApJ Letters **448**, L101 (1995).
2. E.E. Fenimore, & J.S. Bloom, ApJ **453**, 25 (1995).
3. J.P. Norris, et al., ApJ **424**, 540 (1994).
4. D.L. Band, preprint (1994).
5. R.A.M.J. Wijers, and B. Paczyński, ApJ Letters **437**, L107 (1994).

Contraints on Galactic Corona Models of Gamma-Ray Bursts Imposed by Andromeda and the BATSE 3B Catalog

Paolo Coppi[1], Tomasz Bulik[2,3] and Donald Q. Lamb[2]

[1] *Department of Astronomy, Yale University,*
P.O. Box 208101, New Haven CT 06520

[2] *Department of Astronomy and Astrophysics, University of Chicago*
5640 South Ellis Avenue, Chicago, IL 60637

[3] *Nicolaus Copernicus Astronomical Center*
Bartycka 18, 00-716 Warsaw, Poland

> If our galaxy posseses a corona of neutron stars responsible for most or all currently observed gamma-ray bursts, then Andromeda (M31) should possess a similar corona. Bursts from the M31 corona are detectable by a sufficiently sensitive detector, and this places important constraints on Galactic corona models, in particular on the maximum distance out to which a detector like BATSE can see. We investigate the viability of coronal burst models by calculating the spatial distribution of bursts expected for a population of high-velocity neutron stars born in the disks of the Milky Way and Andromeda, and moving in a gravitational potential that includes the bulges, disks, and dark matter halos of both galaxies. We consider two burst emission scenarios, one in which the emission is isotropic and one in which it is beamed along the neutron star kick velocity. We constrain the models by comparing the resulting burst brightness and angular distributions with those of the BATSE 3B catalog. If bursts radiate isostropically, we find that the Galactic corona model can reproduce the BATSE data for kick velocities $\gtrsim 800$ km s^{-1}, source turn-on ages $\gtrsim 20$ Myrs, and BATSE sampling distances 130kpc $\lesssim d_{\max} \lesssim$ 400kpc. If bursts are instead beamed, kick velocities $\gtrsim 800$ km s^{-1} and BATSE sampling distance 100 kpc $\lesssim d_{\max} \lesssim$ 400 kpc are still required, but a characteristic burst turn-on age is not.

INTRODUCTION

Recent studies showing that a significant fraction of neutron stars acquire large kick velocities during their birth (3,2) have rekindled interest in models where gamma-ray bursts originate in an extended halo surrounding our galaxy.

© 1996 American Institute of Physics

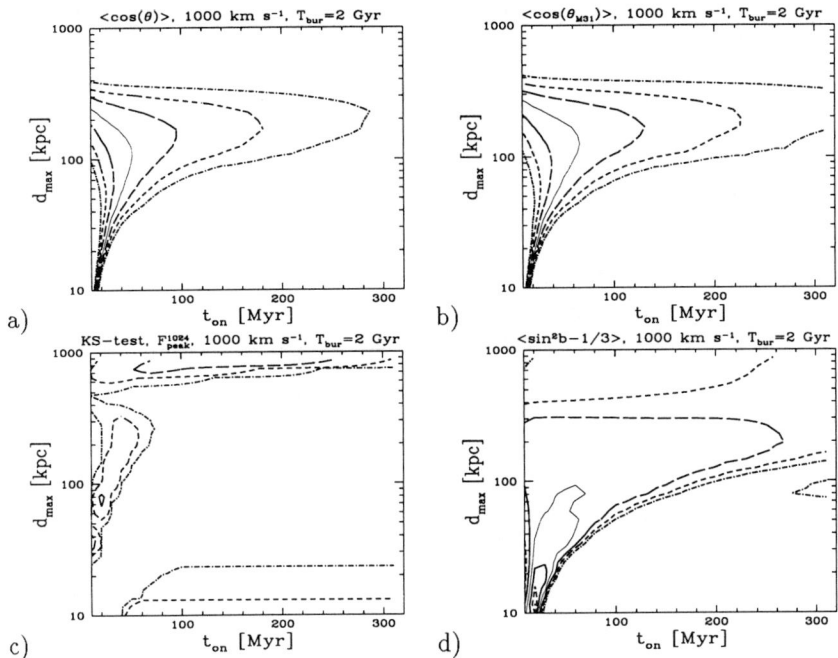

FIG. 1. Comparison of a neutron star halo model that includes M31 with a self-consistent sample of 570 bursts from the BATSE 3B catalog. The neutron stars are born with a kick velocity of 1000 km s^{-1} and have a burst-active phase that starts after a characteristic (sharp) turn-on time t_{on} and lasts $\Delta t = 2 \times 10^9$ years. Their burst emission is *isotropic* and has a luminosity function a factor of ten wide. Panels (a), (b), and (d) show respectively the contours in the (t_{on}, d_{max})-plane along which the Galactic dipole, the dipole in the direction of M31, and the Galactic quadrupole moment differ from those of the data by $\pm 1\sigma$ (solid lines), $\pm 2\sigma$ (dashed line), and $\pm 3\sigma$ (short-dashed line) where σ is the model variance; the thin solid lines show the contour where the expected model value equals the data value. Panel (c) shows similarly the contours along which the model peak flux distribution of bursts is consistent with the data distribution at the 1,2, and 3 σ levels as determined by the Kolmogorov-Smirnov test.

Given a model for the spatial distribution and kick velocities of newly born neutron stars along with a model for the gravitation potential of our galaxy, one can calculate explicitly the spatial distribution of neutron stars in the halo of our galaxy. Remarkably, the distribution of gamma-ray bursts expected from a halo of fast ($v \gtrsim 800$ km s^{-1}) neutron stars can match both the BATSE isotropy and brightness distribution data under fairly reasonable assumptions (5,4,1). In fact, the model parameters can be tuned to give a sufficiently isotropic burst sky distribution that many years of BATSE observations would be required to rule it out (6).

Although BATSE may never discriminate between galactic corona and cosmological models for this reason, a future, sufficiently sensitive detector should be able to. If our galaxy posseses a neutron star halo, it is reasonable for the galaxy Andromeda (M31) to also have one. A detector that can see out to M31 should thus find a significant excess of bursts around M31. The detec-

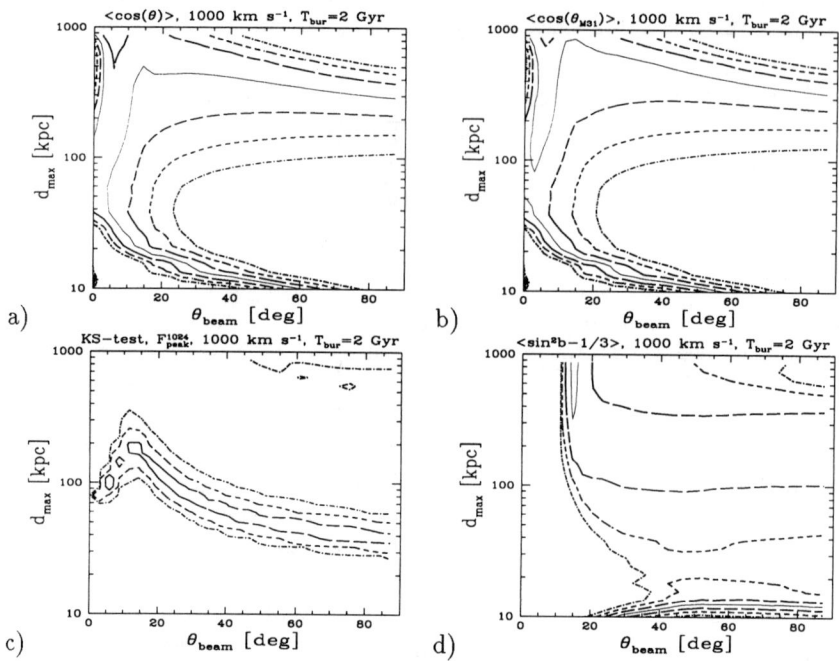

FIG. 2. The same as in Figure 1, except that the burst emission is *beamed* into a cone of half-opening angle θ_{beam}, the neutron star burst-active phase begins immediately ($t_{on} = 0$), bursts have constant luminosity (zero-width luminosity function), and contours are now plotted in the (θ_{beam}, d_{max}) plane.

tor characteristics and observing time required to detect such an excess are discussed in (7). For clarity, the results presented there relied on a simple analytic model of the coronas surrounding our galaxy and M31. The aim of this paper is to: (i) calculate numerically the exact space distribution of bursts expected from a galactic corona model when the presence of M31 is taken into account in a fully self-consistent manner, and (ii) more accurately assess the constraints imposed on galactic corona models by the latest BATSE data. Of particular interest are the limits on the current BATSE sampling distance since the detector collection area required to confidently test a model via an M31 excess goes as the fourth power of that distance. The work presented is an extension of that in (6) and is based on the same numerical techniques and assumptions about the gravitational potential of our galaxy, and the birth distributions and bursting properties of neutron stars. For the calculations shown, M31 is modeled as a galaxy with statistically identical properties to our own, and neutron stars are followed as they move in the combined gravitational potential of *both* galaxies.

CALCULATIONS AND COMPARISON WITH BATSE 3B DATA

The inclusion of M31 in a gamma-ray burst corona model has two main effects. First, M31 represents an additional source of bursts that originate

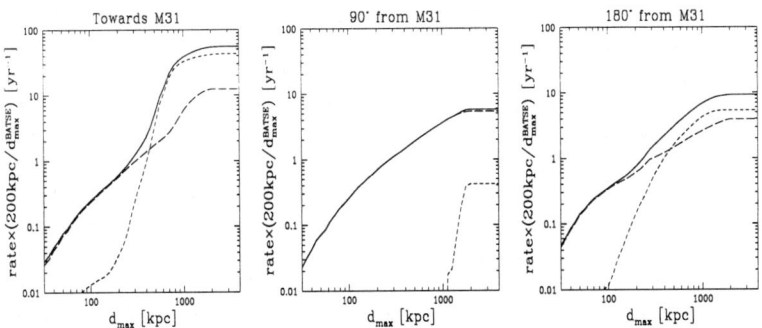

FIG. 3. The burst rates seen by a slab detector with 5° collimation looking in the directions of M31, 90° to M31, and 180° from M31 as a function of d_{max}, the characteristic detector sampling distance–see (7) in these proceedings for more detail on the assumed detector characteristics. The underlying neutron star halo model is the beamed emission model of Figure 2. The rates are scaled in terms of a BATSE sampling distance, d_{max}^{BATSE}, of 200 kpc. The heavy solid lines show the total observed rates; the short-dashed and long-dashed lines show respectively the contributions from bursts originating in M31 and in our galaxy. Note the significant number of M31 bursts seen at 180° from M31.

from the same point on the sky. If BATSE or another detector were to see sufficiently far towards M31, an excess of bursts about M31 should become readily apparent, e.g., as in Fig. 1 of (7). For isotropic burst emission, the lack of such an excess in the BATSE data sets a firm upper limit on the the current BATSE sampling distance, d_{max}; as seen in Fig. 1 here, the expected sky distribution becomes highly anisotropic once a detector sees beyond $\sim$ 400 kpc, about half the distance to M31 ($d_{M31} = 640$ kpc).

The only way to prevent detection of M31 bursts by a detector seeing beyond 400 kpc is to postulate that M31 gamma-ray bursts are somehow different from those in our galaxy. One moderately plausible reason (and perhaps the only one) why this might be the case is that emission from a burst could be beamed along a preferred direction, the neutron star's kick velocity (5). In this case, most bursts escaping our galaxy eventually become visible while only a small fraction of bursts from M31, $\sim \Delta\Omega_{beam}/4\pi$ where $\Delta\Omega_{beam}$ is the beam solid angle, can ever be seen. As discussed in (6), however, the beam solid angle cannot be made arbitrarily small. In Fig. 2, the the quadrupole anisotropy due to galactic bursts increases rapidly below $\theta_{b,min} \sim 20°$. Above $\theta_{b,min}$, the anisotropy from M31 bursts increases so that the most difficult corona to detect is the one with beaming angle $\theta_{b,min}$. A detector capable of seeing to ~ 800 km s^{-1} would be required in this case. In general, $\theta_{b,min} \propto v_{kick}^{-1}$ where v_{kick} is the typical kick velocity. Note that in the beamed burst corona model, it is not anisotropy but rather the shape of the burst brightness distribution (Fig. 2d) that limits the BATSE sampling distance d_{max} to $\lesssim 400$ kpc.

The second, more subtle effect of M31 is to distort the neutron star corona structure and kinematics. Stars from our galaxy which pass near M31 are strongly accelerated and deflected towards M31 as they approach M31 and then decelerated once they have passed M31 and begin to climb out of its potential well. As noted in (4), this leads to an excess or "hot spot" of bursts from our galaxy located behind M31, and by symmetry, a similar hot spot of bursts from M31 located behind our galaxy, in the direction opposite M31. This effect can be clearly seen in Figure 3, which shows the burst rate as a function of detector sampling distance in several directions. A search for an M31 excess based on burst counts towards and away from M31 should therefore look at 90° to M31 to determine the burst rate away from M31 and *not* at 180° to M31. Also, because M31 is comparable in mass to our galaxy, the outer halo surrounding the combined M31-Milky Way system is significantly more compact than in the Milkly Way-only case, i.e., M31 changes significantly the overall log N-log S distribution for bursts located $\gtrsim$ 400 kpc from us. Finally, as noted in (4), the inclusion of M31 breaks any spherical or circular symmetries in the overall gravitational potential and leads to bound neutron star orbits that are typically not periodic and closed. In particular, we have verified that bound stars returning to our galaxy at late times do not show a strong disk-like signature.

While M31 has significant dynamical effects on the neutron star halo, in general these effects do *not* play an important role if if one restricts attention to sampling distances consistent with BATSE, i.e., $\lesssim$ 400 kpc. Thus, the conclusions presented in (6), e.g., the lower bounds on the $d_{\max}$, remain largely unaltered by the inclusion of M31. The one exception is the case where the burst active phase lasts longer than $\sim$ 1 Gyr. This is enough time for bound neutron stars to begin returning and for neutron stars from M31 to reach us. The number of such stars inside our galaxy is small and in the beamed burst case, their effect is negligible because few of the stars are beamed at us. In the non-beamed case, however, enough of these stars can be present to significantly alter the density of nearby bursts, i.e., change the bright end of the burst brightness distribution. This effect may be seen by comparing Fig. 1d with Fig. 1c of (6). Note that even in this case, parameters consistent with BATSE data can still be found and except for the value of t_{on}, they are not very different from those obtained when the bursts have short active phases. In fact, the the presence of old, fast neutron stars in our galaxy may be part of the explanation for why bright bursts currently do not appear to show a strong galactic disk signature.

REFERENCES

1. T. Bulik, and D. Q. Lamb, A& Sp.Sc., in press (1995).
2. D. A. Frail, W. M. Goss, and J. B. Whiteoak, ApJ **437**, 781 (1994).
3. A. G. Lyne, and D. R. Lorimer, Nature **369**, 127 (1994)
4. Ph. Podsiadlowski, M. J. Rees, and M. Ruderman, MNRAS **273**, 755 (1995).
5. H. Li, et al., in Gamma-Ray Bursts, eds. G. J. Fishman, J. J. Brainerd and K. Hurley, AIP Conf. Proc. **307**, 600, (AIP, New York, 1994).
6. T. Bulik, D. Q. Lamb, these proceedings.
7. T. Bulik, P. S. Coppi, D. Q. Lamb, these proceedings.

A Wide Ranging Search For Correlations Among Burst Properties

Samuel E. Dyson and Bradley E. Schaefer

Yale University, Department of Physics, New Haven CT 06520-8121

One primary reason that Gamma Ray Bursts (GRBs) have remained such a mystery is that there are virtually no known significant, non-definitional correlations or bimodalities among the burst properties. The best we have at the present time is a weak hardness/duration correlation, a disputed duration/intensity correlation, a problem plagued hardness/intensity correlation, and a duration bimodality. But these have not provided the sort of break-through similar to those provided by the H-R diagram or the PopI/PopII results in stellar astronomy. The physical processes which relate burst properties to one another are masked by the lack of correlations and bimodalities. In the past, searches for correlations and bimodalities in the data have been made for a relatively small set of properties, usually with no regard for the light curve shape.

We present our results of a systematic search for statistically significant correlations and bimodalities from a set of 49 burst properties, with particular emphasis on the inclusion of light curve characteristics. We have used the data base of the 260 bursts in the BATSE 1B catalog. We keep an accurate account of the number of trials for which we test the data, and then fold this into our detection threshold. Our method recovered the known correlation between the log of the duration and the spectral hardness, with a significance of 9.3×10^{-5} after correcting for trials. We also recovered the expected risetime/decay time correlation and the bimodality of the logarithm of duration. We found no original, significant, non-definitional correlations or bimodalities.

INTRODUCTION

Correlations between GRB properties are among the most basic of pieces in the Gamma Ray Burst puzzle. Most of our ignorance regarding the nature of GRBs can be ascribed to the lack of low-energy counterparts and the lack of significant correlations between observables (1).

Correlations are useful and many have immediate implications. For example, if we were to observe a correlation between burst duration and galactic latitude, we could conclude that GRBs are sensitive to the location of the Galaxy, and that they are therefore likely to be of Galactic origin. The same would be true if any GRB property were to show a correlation with any galactic coordinate. Similarly, a correlation between observed duration and peak

© 1996 American Institute of Physics

flux might imply a cosmological time dilation.

Historically speaking, a modern astrophysicist working without GRB correlations is akin to a stellar astronomer working before the discovery of the H-R diagram or Wien's Law, or like a cosmologist working before the discovery of the Hubble Law. In all these cases, a lack of a critical correlation would deny the researcher an insight into fundamental physical relations. The field of Gamma Ray Burst studies desperately needs a believable correlation on which to build a sound physical model.

Given the vital role that property correlations play, many previous searches have been made. Hardness/hardness correlations have been sought, in analogy with the z-diagrams of x-ray quasi-periodic oscillators (2,3). Duration/intensity and hardness/intensity relations have been claimed as evidence of cosmological time dilation and red shift (4–7). A bimodality of observed burst durations is found to correlate with burst hardness (8, and references therein). A study of two parameters characterizing the structure of the light curve reveals neither significant correlations nor bimodalities (9).

Yet, there are limitations to these previous searches. First, the research has been directed toward likely pairs of potentially correlated properties, so that only a small and haphazard array of potential correlations has been tested. This means that some striking but unexpected correlation might yet be found by a systematic survey. Second, these tests generally do not involve properties of GRB light curve shapes, despite their rich store of information. Thus, we have made a wide ranging systematic search for correlations and bimodalities among many burst properties, with emphasis on the light curve shapes.

METHODS

The research project presented here differed from most previous investigations in three ways. First, it included many properties of GRB light curves in correlation and bimodality searches. Secondly, it involved *systematic* searches for correlations and bimodalities. Thirdly, it used correct formalisms for evaluating the statistical significance of observed correlations.

We have defined a total of 49 properties and have measured them for all 260 bursts in (10). Some properties are taken directly from the 1B catalog: date, fluences* in channels 1, 2, 3, 4, and 1–4, peak fluxes* in 64 ms, 256 ms, and 1024 ms time bins, T50*, T90*, galactic latitude and longitude, and the angle from the galactic center. We also used two definitions for deriving the number of peaks in a light curve, the rise and decay time of the main peak, the rise and decay time of the whole burst*, the number of well separated episodes, two definitions for the slope of the peak tops, two definitions for the rms of the peak times, the location of the main peak with respect to the first and last flux, two definitions of the isolation of the main peak, two definitions of the symmetry of the burst, and three hardness ratios. A dozen properties (indicated above by an asterisk) were also used with their logarithm as a separate property.

The correlations between all pairs of properties can be presented as a matrix with elements equal to the correlation coefficient, r. We have correlation tables for all bursts as well as 8 other correlation tables involving 8 burst subsets. These subsets include long bursts, short bursts, single peak bursts, multiple peak bursts, and various combinations.

The closer r is to zero, the more likely that the observed degree of correlation could result from random chance. A r value near $+1$ implies a strong correlation, while a r value near -1 implies a strong anti-correlation. After accounting for our total number of trials, we adopt a conservative threshold of $r = 0.35$ for correlations involving all 260 bursts.

SIGNIFICANCE THRESHOLD

Since we were looking through several matrices, or tables, each with a large number of entries, we considered the statistical probability that the correlation coefficients were the result of random variations, rather than of intrinsic relationships. The appropriate integral over the linear correlation coefficient distribution is given in (11), and we have related this to a tabulated probability function. The probability Q_c that a given correlation coefficient exceeding r could have been generated by an uncorrelated pair of variables is

$$Q_c(r, \nu) = Q\{(1 - r^2)/(r^2\nu)|\nu, 1\}. \tag{1}$$

Here, $\nu = N - 2$ is the number of degrees of freedom and N is the number of GRBs in the sample. Q is the F-(variance-ratio) distribution function (12).

To get the final probability, we must know the number of trials N_t, where

$$N_t = \Sigma_m[(N_p^2 - N_p)/2], \tag{2}$$

where N_p is the number of properties in a matrix and the summation is over all matrices used in the search. Given that the main correlation table as well as four of the subset tables contain 49 properties while the other four subset tables contain 37 properties, we find a total number of trials $N_t = 8544$.

In the cases of interest with a small Q_c value, the final probability that any one trial will result in a correlation coefficient with a magnitude of r or larger is

$$P_{\text{final}} = Q_c(r, \nu) \times N_t. \tag{3}$$

We have chosen a conservative acceptance criterion of $P_{\text{final}} < 0.0001$, which corresponds to a 3.9-sigma confidence level. Note that this is after accounting for the number of trials. For $N = 260$ and $N_t = 8544$, the magnitude of r must be greater than or equal to 0.35 for a correlation to be significant. So our threshold r magnitude is $r_{\text{th}} = 0.35$. For some pairs of properties, N was smaller than 260 so that our significance threshold was greater in magnitude than 0.35.

When noise or scatter is added to the values of two perfectly correlated variables A and B, the observed correlation coefficient r is degraded. We define a figure of merit M such that

$$M = [(\sigma_A/\rho_A)^2 + (\sigma_B/\rho_B)^2]^{0.5}, \tag{4}$$

where the σ values are the rms scatter for variables A and B, and the ρ values are the associated dynamic ranges of variables A and B. The correlation coefficient r varies with M as $r = 1 - 2M^2$, as derived from Monte Carlo simulations.

We now consider how good a correlation would have to be to exceed our significance threshold. For $r_{\rm th} = 0.35$, M must be 0.57 or smaller. Therefore, a significant correlation can be found if the observed rms scatter in both variables is 40% of the total ranges. Alternatively, if all the scatter is in one variable, we would detect a correlation as significant if the variable's scatter was 57% of its total range. It is heartening to see that our technique will prove such a weak correlation.

In the event of the discovery of a previously unobserved, statistically significant correlation in the 1B Catalog data, we planned to extend the present correlation search to the 2B Catalog where we could search only for expected correlations, thereby greatly reducing the number of trials and increasing the statistical significance of a result. By using an independent set of bursts from the 2B Catalog, we would be able to confirm the existence of correlations that appeared to be significant in the 1B Catalog.

RESULTS

We recovered the non-definitional correlation observed by (8) between spectral hardness (the ratio of fluences in channels 3 and 2) and the logarithm of duration (for T50 and T90). Given $r = -0.38$, $N = 212$, and $N_t = 8544$, the probability that this correlation is the result of random fluctuation is 9.3×10^{-5}. This probability is degraded from the 10^{-8} Kouveliotou result because of the large number of trials which our systematic search involved.

Another recovered correlation is the relation between the rise and decay times of the main peak. This property is not surprising and indicates that the physical mechanism of the early and late light emission are connected. To our knowledge this correlation was first explicitly stated by (13).

Our systematic search recovered the bimodality in duration presented on a logarithmic scale reported previously by many groups from Cline and Desai (14) to Kouveliotou et al. (8). We note however, that when the duration is plotted on a linear scale, this bimodality apparently vanishes. This is disturbing as it implies that bimodalities are dependent on the chosen scale.

Our systematic search did not discover any other correlations or bimodalities. This includes a lack of any correlation between duration and intensity (4,6) or between hardness and intensity (5,7).

Our systematic search for linear correlations has examined 8544 pairs of GRB properties. Despite these many trials, our methods are sensitive to relations where the scatter is $\sim 50\%$ of the dynamic range of data. Indeed, we have recovered the previously known correlations between hardness and intensity as well as between rise and decay times. In addition we have recovered the duration bimodality.

Our search was systematic (so that unexpected correlations could have been found) and we considered many light curve shape properties (virtually ignored in the past). Thus, we were surprised and disappointed that our search did not discover any new relations.

REFERENCES

1. B. Paczynski, in Compton Gamma Ray Observatory, eds. M. Friedlander, N. Gehrels, and D. J. Macomb, AIP Conf. Proc. **280**, 981 (AIP, New York, 1993).
2. W. S. Paciesas et al., in Gamma Ray Bursts, eds. W. S. Paciesas and G. J. Fishman, AIP Conf. Proc. **265**, 190 (AIP, New York, 1992).
3. C. Kouveliotou, Ap. J. Supp. **92**, 637 (1994).
4. J. P. Norris et al., Ap. J. **424**, 540 (1994).
5. R. J. Nemiroff et al., Ap. J. **435**, L133 (1994).
6. I. Mitrofanov et al., in Gamma Ray Bursts, eds. G. J. Fishman, J. J. Brainerd, and K. Hurley, AIP Conf. Proc. **307** 187 (AIP, New York, 1994).
7. B. E. Schaefer, Ap. J. **404**, L87 (1993).
8. C. Kouveliotou et al., Ap. J. **413**, L101 (1993).
9. J. P. Lestrade et al., in Gamma Ray Bursts, eds. G. J. Fishman, J. J. Brainerd, and K. Hurley, AIP Conf. Proc. **307** 212 (AIP, New York, 1994).
10. G. J. Fishman et al., Ap. J. Supp. **92**, 229 (1994).
11. P. R. Bevington, Data Reduction and Error Analysis for the Physical Sciences, New York: McGraw-Hill (1969).
12. M. Abramowitz and I. A. Stegun, Handbook of Mathematical Functions, Washington: National Bureau of Standards (1964).
13. C. Barat et al., Ap. J. **285**, 791 (1984).
14. T. L. Cline and U. D. Desai, in Proc. 9th ESLAB Symp., Noordwijk: ESRO, p. 37 (1974).

Continuing search of the EGRET data for high-energy gamma-ray microsecond bursts

C. E. Fichtel[1], D. L. Bertsch[1], R. C. Hartman[1], S. D. Hunter[1], C. von Montigny[1], D. J. Thompson[1], B. L. Dingus[2], J. A. Esposito[2], R. Mukherjee[2], P. Sreekumar[2], G. Kanbach[3], H. A. Mayer-Hasselwander[3], D. A. Kniffen[4], Y. C. Lin[5], P. F. Michelson[5], P. L. Nolan[5], L. McDonald[6], and E. J. Schneid[7]

[1] *NASA/Goddard Space Flight Center, Greenbelt, MD 20771*
[2] *Universities Space Research Association, NASA/GSFC, Greenbelt, MD 20771*
[3] *Max-Planck Institut fur Extraterrestrische Physik, 85748 Garching, Germany*
[4] *Hampden-Sydney College, P. O. Box 862, Hampden-Sydney, VA 23943*
[5] *HEPL, Stanford University, Stanford, CA 94305*
[6] *Hughes STX, NASA/GSFC, Greenbelt, MD 20771*
[7] *Northrop Grumman Corporation, Mail Stop A01-26, Bethpage, L.I., NY 11714*

In the mid 1970's, Hawking (1) and Page and Hawking (2), investigated theoretically the possibility of detecting high-energy gamma rays produced by the quantum-mechanical decay of a small black hole created in the early universe. They concluded that, at the end of the life of the small black hole, it would radiate a burst of gamma rays peaked near 250 MeV with a total energy of about 10^{34} ergs in the order of a microsecond or less if certain details of the theory were true. The characteristics of a black hole are determined by laws of physics beyond the range of current particle accelerators; hence, the search for these short bursts of high-energy gamma rays provides at least the possibility of detecting directly the gamma rays from such bursts, and a search of the EGRET data has led to an upper limit below 5×10^{-2} black hole decays per pc^3 yr^{-1}, placing constraints on this and other theories predicting microsecond high-energy gamma-ray bursts.

INTRODUCTION

There has long been an interest in making general relativity compatible with quantum mechanics. One attempt to achieve this marriage was made by Hawking (1). A very specific prediction arises from this work that can be tested or at least constrained by high energy gamma-ray observations. In the early Universe, small black holes may be created if the Universe was chaotic or had a soft equation of state (3, 4, 5). They slowly evaporate and the temperature increases until the black hole will begin radiating particles of

© 1996 American Institute of Physics

higher rest mass. Using the statistical bootstrap approach of (6), the number of species of particles increases exponentially with energy; therefore, the black hole will lose its energy very quickly when it reaches the Hagedorn limiting temperature of 160 MeV. The heavy hadrons emitted by the black hole would decay rapidly and about 10 to 30 percent of their energy would appear as a burst of gamma rays with energies primarily between 10^2 and 10^3 MeV. The total energy in the form of gamma rays would be about 10^{34} ergs and the time of emission would be 10^{-7}s according to (2). It should be noted that the exponential increase of the number of particles with energy is not predicted by the "Standard Model", which is based on a finite number of quarks, leptons, and gauge bosons. If the number of states is not infinite, then the GeV emission from the evaporation of a primary black hole will occur much more slowly. (See, e.g., (7,8) and references therein). In this alternative, although a burst of the order of seconds is expected for a threshold energy, $E(t)$, of 10 TeV, the length of the burst in this model is proportional to $E(t)^{-3}$; so a group of gamma rays in the 10^2 to 10^4 MeV range would be so long in time as to in effect not be a burst. Cline and Hong (9) have suggested that these bursts might be much shorter, a small fraction of a second. Cline and Hong (10) have proposed still a third alternative leading to millisecond gamma-ray bursters which might be seen by BATSE on the Compton Observatory.

Attempts have been made to see bursts of two of these types using ground based telescopes which can record secondaries produced in the atmosphere. For the former 10^{-7} s 10^2 MeV type burst, Porter and Weekes (11) used a ground-based system, in their case a large Cerenkov telescope, to attempt to detect the low energy photons resulting from an air shower that would be produced by such a high energy gamma-ray burst. They set an upper limit of 4 x 10^{-2} decays pc^{-3} yr^{-1}. Alexandreas, et al. (12), using an ultra high energy extensive air shower array, have set an upper limit for the proposed TeV bursts of a fraction of a second to a few seconds. Connaughton et al. (13) also have set a similar limit for TeV bursts.

In an earlier paper by Fichtel et al. (14), based on data from the Energetic Gamma Ray Experiment Telescope (EGRET) on the Compton Gamma Ray Observatory, an upper limit for primordial black hole (PBH) decays of 5 x 10^{-2} pc^{-3} yr^{-1} was determined based on direct measurements of gamma rays with energies above 30 MeV. In this work, results from data in phases 1 to 4, (April 1991 to October 3, 1995), or approximately four and a half years, are reported.

THE HIGH-ENERGY GAMMA-RAY TELESCOPE

EGRET has the standard elements of a high energy gamma-ray telescope, specifically an anticoincidence scintillator dome to discriminate against charged particles, a particle track detector consisting of spark chambers with interspersed high z material to convert the gamma rays into electron pairs, a trigger telescope which detects the presence of the pair and determines that the particles have the correct direction of motion, and an energy measurement device which in the case of EGRET is a NaI (T1) crystal. A description of the instrument and its general capabilities is given by (15). The results of the instrument calibration, both before and after launch, are given by (16). The telescope covers the energy range from about 20 MeV to 30 GeV. The effective area is about 1.5×10^3 cm^2 from 0.2 to 1.0 GeV, and lower outside of this range. The field of view of EGRET is about one-half of a steradian, with the effective area dropping to about one-half the on axis value at 20° from the axis and one-sixth at 30°. The instrument is designed to be free of internal background, and the calibration tests have verified that it is at least an order of magnitude below extragalactic gamma radiation.

Hence, the only significant radiation besides the sources themselves is the diffuse galactic and extragalactic radiation. Because EGRET can detect several gamma rays in one event, it has the capability of detecting a submicrosecond high energy gamma-ray burst of the type predicted by Hawking and determining its direction. The subsequent sections will describe how the detection occurs, the results, and the implications for the theory.

THE SEARCH FOR MICROSECOND BURSTS IN THE EGRET DATA

It is possible to search for high energy gamma ray microsecond bursts in the EGRET data because there is a time interval of approximately 6×10^{-7} seconds from the passage of the initial gamma ray in a group to the triggering of the spark chamber system. Hence, unless the event is vetoed by the anticoincidence system, all gamma rays converting in the active volume of the telescope will be recorded and one can look for events in which there is more than one gamma ray from a given direction in one event. The approach to processing the gamma-ray events is described in detail in (16). A review of the data indicates that a multiple gamma ray event would almost certainly be flagged as a "questionable gamma ray" event since in the great majority of the cases a gamma ray is found in the automatic program, but there are more than the minimum allowable "unstructured sparks", that is sparks not associated with the recognized gamma ray. These events are examined on a graphics unit screen by data analysts. When a multiple gamma ray event is found by a data analyst it is flagged regardless of the nature of the gamma rays detected. These selected events are then examined by an EGRET scientist. Even these potential multiple gamma ray events were in fact rare. When they were examined in detail, none were found to be multiple gamma ray events coming from the same direction in the sky.

Most were consistent with their coming from a common point inside the telescope, usually the pressure shell under the anticoincidence scintillator dome, presumably as the result of the interaction of a neutron there. The very small number for the given time is approximately consistent with what might be expected, although an exact calculation is difficult. The rest were of several types, most frequently appearing to be a separate secondary gamma ray directed back to one of the secondaries from a primary gamma ray interaction.

Hence, there was then no evidence for any multilple gamma ray microsecond burst. It should be noted that in the instrument calibration, multiple gamma ray events were seen from the same direction as expected since there were some multiple gamma ray pulses.

In the paper by (14), it was shown that the probability of missing a multiple gamma ray event was small. That paper also describes the method for calculating the upper limit for the black hole emission density for a black hole with total energy E. The energy spectrum of the gamma rays is that given by (2), which is peaked at 250 MeV. For $E = 10^{34}$ ergs, the energy suggested by (2), the upper limit is 3×10^{-2} pc^{-3} yr^{-1}. Since approximately 85 percent of the effective collecting life of EGRET has now been expended, this limit will not change significantly, unless, of course, a short multiple gamma-ray event is observed. Even then, it might be due to something else. The relevant distance range is from about 20 to 100 pc. Below 20 pc the instrument would be overwhelmed and not record the event; beyond 100 pc, the probability of detection becomes quite small. For $E = 10^{35}$ ergs, the upper limit is 1×10^{-3} pc^{-3} yr^{-1}, and the distance range is somewhat larger.

DISCUSSION AND CONCLUSION

There remains no evidence within the work reported here for the existence of a microsecond burst of high-energy gamma rays, and a severe upper limit has been set. In the work related to the final burst of gamma radiation from a primordial black hole by (2), they indicated that "It would radiate away all its energy in a time of about 10^{-7}s, giving a burst of gamma-rays peaked around 250 MeV with a total energy about 10^{34} ergs". As noted in the last section, the upper limit of primordial black holes decays is then 3×10^{-2} pc^{-3} yr^{-1}. Also, for 10^{34} ergs, the black holes that could have been detected would have been predominantly within 100 pc, and hence within the local region of our galaxy. Page and Hawking note that – "one might expect that the [PBH] would be concentrated in the gravitational potential wells of galaxies". Hence, this limit, in view of the distant range, is particularly relevant in terms of this model.

REFERENCES

1. S. W. Hawking, Nature **248**, 30 (1974).
2. D. N. Page and S. P. Hawking, ApJ **206**, 1 (1976).
3. S. W.Hawking, MNRAS **152**, 75 (1971).
4. B. J. Carr and S. W. Hawking, MNRAS **168**, 243 (1974).
5. B. J. Carr, ApJ **201**, 1 (1975).
6. R. Hagedorn, Nuovo Cim. **269**, 1027 (1968).
7. J. H. MacGibbon and B. R. Webber, Phys. Rev. **D41**, 3052 (1990).
8. F. Halzen, E. Zas, J. H. MacGibbon, and T. C. Weekes, Nature **353**, 807 (1991).
9. D. G. Cline and W. Hong, in Gamma Ray Bursts, eds. G. J. Fishman, J. J. Brainerd, & K. Hurley, AIP Conf. Proc. **307**, 577 (AIP, New York, 1994).

10. D. G. Cline and W. Hong, ApJ **401**, L57 (1992).
11. N. A. Porter and T. C. Weekes, MNRAS **183**, 205 (1978).
12. D. E. Alexandreas, Phys. Rev. Letters **71**, 2524 (1993).
13. V. Connaughton, et al., in Gamma Ray Bursts, eds. G. J. Fishman, J. J. Brainerd, & K. Hurley, AIP Conf. Proc. **307**, 470 (AIP, New York, 1994).
14. C. E. Fichtel, et al., ApJ **434**, 557 (1994).
15. G. Kanbach, et al., Space Science Reviews **46**, 69 (1988).
16. D. J. Thompson, et al., ApJ Suppl. **86**, 629 (1993).

The GRB Rate At High Photon Energies

Burkhardt Funk[a], Karl Mannheim[b], Dieter Hartmann[c]

[a] Universität Wuppertal, Fachbereich Physik, D-42097 Wuppertal, Germany
[b] Universitäts-Sternwarte, Geismarlandstrasse 11, D-37803 Göttingen, Germany
[c] Dept. of Physics and Astronomy, Clemson University, Clemson, SC 29634, USA

> We calculate how many bursts can reasonably be expected at a gamma ray energy E for a cosmological distribution of bursts satisfying the observed apparent brightness distribution. The crucial point is that the gamma ray absorption by pair production in the intergalactic diffuse radiation field eliminates bursts from beyond the energy-dependent *gamma ray horizon* $\tau_{\gamma\gamma} \sim 1$, thus drastically reducing the number of bursts at high energies. Our results are consistent with current experimental upper limits.

INTRODUCTION

The lack of a significant large scale anisotropy in the angular distribution of GRBs argues in favor of their cosmological origin. Assuming they are all standard candles, the observed number of bursts at a given flux directly relates to a number of sources at a given redshift. The maximum redshift sampled by BATSE under these assumptions is $z_{\max} \sim 2$ (1). As a direct consequence absorption of the high energy gamma rays by diffuse background radiation becomes important at photon energies >30 GeV.

In this work we calculate the expected burst rate as a function of photon energy accounting for cosmic absorption and compare the result with experimental limits. We will start with the computation of the gamma ray horizon, which in a second step will be used to determine the number of bursts above a given detector energy threshold. Finally, we determine the experimental sensitivity required for the detection of all GRBs within the gamma ray horizon at a given threshold energy and compare this with the sensitivity of current experiments.

GAMMA RAY ABSORPTION FROM GEV TO 100 TEV

Gamma rays of energy E propagating from a distant source at redshift z towards a terrestrial observer can be absorbed by inelastic interactions with low energy photons of present-day energy ϵ from an isotropic diffuse background radiation field. The dominant process is pair creation $\gamma_E + \gamma_\epsilon \to e^+ + e^-$ with a threshold energy $\epsilon_{\rm th} = 2m_e^2 c^4 / E(1-\mu)(1+z)^2$ where $\mu = \cos\theta$ denotes the

© 1996 American Institute of Physics

cosine of the scattering Angle. According to Peebles (2) for a cosmological model with $\Omega = 1$ and $\Lambda = 0$ we obtain the optical depth (3)

$$\tau_{\gamma\gamma}(E, z) = \frac{c}{H_\circ} \int_0^z dz' \, (1+z')^{1/2} \int_0^2 dx \, \frac{x}{2} \int_{\epsilon_{th}}^\infty d\epsilon \, n_b(\epsilon) \sigma_{\gamma\gamma}(E, \epsilon, x, z') \quad (1)$$

where $n_b(\epsilon)$ is the non-evolving present-day background density and $\sigma_{\gamma\gamma}$ is the pair creation cross section. The shape of the diffuse background density $n_b(\epsilon)$ is obtained by averaging over various galaxy formation models presented by MacMinn & Primack (4). We multiplied this shape by a small factor to obtain agreement with the background density (including the contribution from the 2.7 K microwave background) estimated by Beichman & Helou (5) in the FIR (for a modest galaxy luminosity or density evolution $\gamma = 2$), and by Madau & Phinney (6) in the NIR through UV range (Fig. 1).

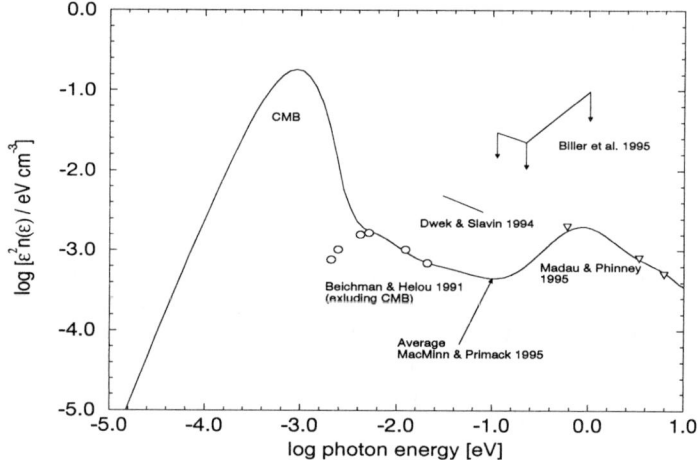

FIG. 1. *Solid line:* FIR-to-UV background adopted in the present work. *Straight line segment:* FIR background photon density inferred by Dwek & Slavin (7) assuming TeV absorption for Mrk421. *Limits:* Experimental upper limits on the background IR density obtained by Biller *et al.* (8)

We numerically integrate the optical depth function and solve for the gamma ray horizon $\tau_{\gamma\gamma}(E, z) = 1$. Results are shown in Fig. 2 for two values of the Hubble constant.

EXPECTED TOTAL NUMBER OF BURSTS

The number of bursts above the detector threshold energy E is given by the number of bursts in the volume enclosed by the hyperplane $\tau_{\gamma\gamma}(z, E) = 1$. It has been shown (3) that the burst rate is given by

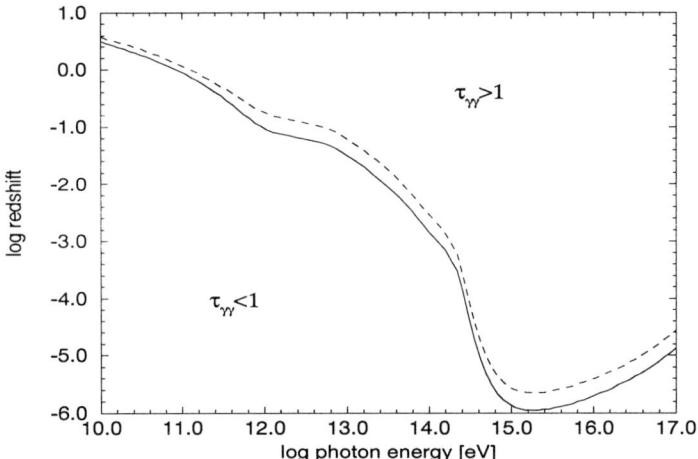

FIG. 2. Gamma ray horizon for the diffuse background radiation shown in Fig.1. *Solid line: $h = 0.5$. Dashed line: $h = 1.0$.*

$$N(\geq E) = 16\pi n_o \left(\frac{c}{H_o}\right)^3 \left(\frac{1}{6} + \frac{1}{2(1+z)^2} - \frac{2}{3(1+z)^{3/2}}\right). \quad (2)$$

where $n_o = 6.9 \times 10^{-9} h^3$ Mpc^{-3} yr^{-1} denotes the volume burst rate assumed in this work. A dependence on the Hubble constant enters Eq. 2 through the value of $z = z(E, H_o)$. We point out that it is therefore in principle possible to measure the Hubble constant with number counts of GRBs, if the diffuse background photon density is known. In Fig. 3 we show the number of bursts vs. detection energy threshold, demonstrating the rapid shrinking of the observable Universe with increasing E due to pair absorption. At energies of ~10 TeV, the standard candle scenario predicts $2-4$ ($h = 0.5-0.75$) bursts per year.

LIMITING SENSITIVITY

The number of bursts which can be *detected* above a given energy threshold also depends on the particular instrumental sensitivity F_{det}, the spectral shape and the duration of the bursts. In order to detect *all* GRBs within the gamma ray horizon the necessary flux sensitivity must obey the relation (3) $F_{\text{det}} \leq F_o$ where F_o is given by

$$F_o = \frac{L[\geq E(1+z)]}{16\pi (c/H_o)^2 [1 - (1+z)^{-1/2}]^2 (1+z)^2} \quad (3)$$

Using the standard candle luminosity as inferred from BATSE measurements ($L_o = 7 \times 10^{50} h^{-2}$ ergs/s) and assuming a simple power law spectrum above the break energy E_b (~1 MeV) with a spectral index α yields

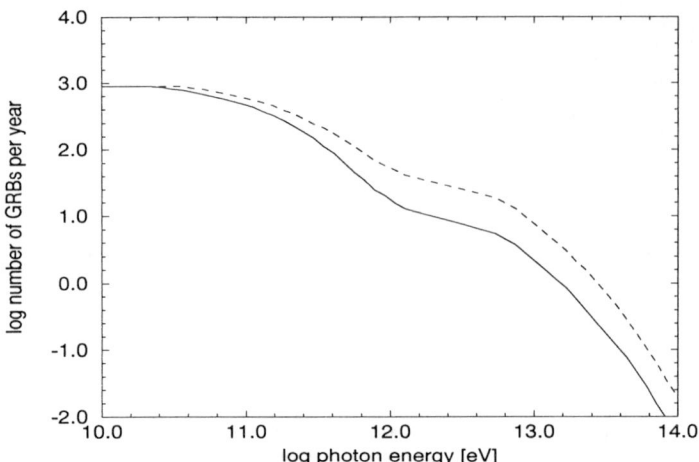

FIG. 3. Number of GRBs at a given detection energy threshold E. *Solid line:* $h = 0.5$. *Dashed line:* $h = 1.0$.

$$F_0 = 1.7 \times 10^{-7} \frac{(1+z_{\max})^{\alpha-2}(E/E_b)^{2-\alpha}}{(1+z)^{\alpha-1}(\sqrt{1+z}-1)^2} \left(\frac{\delta t}{1\text{ s}}\right)^{-1} \text{ erg cm}^{-2}\text{ s}^{-1} \quad (4)$$

for the source flux at the distance of the gamma ray horizon. Although the burst luminosity in the standard candle picture is fixed ($\delta t \sim 1$ s), we leave the possibility open that the high energy tails persist longer than the typical BATSE bursts. Hurley et al. (9) indicate spectral slopes $\alpha = 2.2 - 3.7$ and extended or delayed durations of 30 s—90 minutes for EGRET detected bursts. In Fig. 4 we plot the limiting sensitivity for $\alpha = 2.2$ and tail durations of 1s and 100s. The apparently paradoxical increase of the limiting flux with energy reflects the rapid shrinking of the gamma ray horizon.

CONCLUSIONS

We estimated the expected gamma ray burst rate to be about 20—40 bursts per year in the TeV range. The sensitivities of current experiments are sufficient to detect bursts if their spectra are not steeper than $\alpha \sim 2.7$. Taking into account sky exposure and triggering efficiency, we rule out the possibility of cosmological burst detection by CYGNUS, EAS-TOP and CASA-MIA while MILAGRO could at most detect ~ 10 events per year. The Whipple Observatory, the Tibet air shower array and HEGRA have very low expected detection rates (0.01-0.1 events per year).

Acknowledgements. This work was supported in part by NASA under grant NAGW 5-1578. DH and KM are grateful for hospitality and support (under NSF grant PHY94-07194) during their stay at the ITP, where this

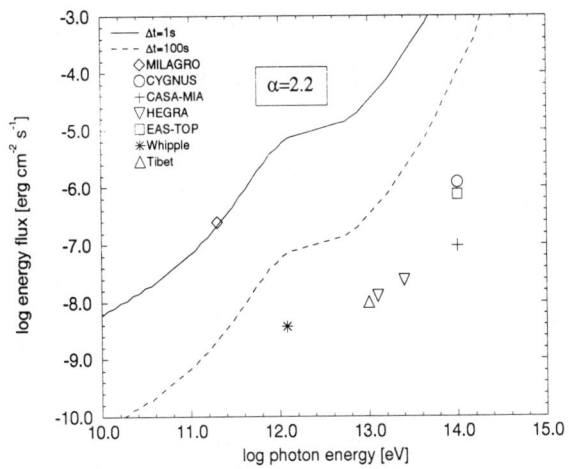

FIG. 4. *Solid lines:* limiting sensitivity required for detection of all GRBs at a given threshold energy ($h = 0.5$) assuming a high enery tail of spectral index α. *Dashed lines:* same for high energy tails with extended durations relative to the BATSE bursts by a factor of 100. *Symbols:* current experimental flux limits for burst detection (typically assuming $\delta t \sim 30$ s)

collaboration began. BF is supported by the BMBF, Germany, under contract 05 2WT164. KM acknowledges travel support by the DFG under grant Ma 1545/5-1.

REFERENCES

1. E. Cohen, T. Piran, ApJ 444, L25 (1995).
2. P.J.E. Peebles, *Principles of Physical Cosmology*, Princeton Univ. Press (1993).
3. K. Mannheim, D. Hartmann, B. Funk, ApJ, submitted (1995).
4. D. MacMinn, J. Primack, Proc. of the Heidelberg Workshop on TeV Astrophysics, Eds. H. Völk and F. Aharonian, Kluwer, in press (1995).
5. C.A. Beichman, G. Helou, ApJ 370, L1 (1991).
6. P. Madau, E.S. Phinney, ApJ, in press (1995).
7. E. Dwek, J. Slavin, ApJ **436**, 696 (1994).
8. S.D. Biller, C.W. Akerlof, J. Buckley, *et al.*, ApJ 445, 227 (1995).
9. K. Hurley, B.L. Dingus, R. Muherjee, *et al.*, Nature 372, 652 (1994).

GRBs OBSERVED WITH WATCH AND BATSE (3B CATALOGUE)

J. Gorosabel[1], A.J. Castro-Tirado[1], N. Lund[2],
S. Brandt[3]*, O. Terekhov[4], R. Sunyaev[4]

[1] *LAEFF-INTA, P.O. Box 50727, 28080 Madrid, Spain*
[2] *DSRI, Gl. Lundtoftevej 7, DK-2800 Lyngby, Denmark*
[3] *Astrophysics Division, ESTEC, Noordwijk, The Netherlands*
[4] *Space Research Institute, Profsoyuznaya 84/32, Moscow, Russia*

We have correlated the positions of 44 GRBs observed with *WATCH* with the 1122 bursts given in the *BATSE* 3B Catalogue. We conclude that there is no indication of recurrent activity of *WATCH* bursts in the *BATSE* data and constrained the number of *WATCH* repeaters to ≤ 4.

INTRODUCTION

One of the clues for determining the nature of GRBs is based on studying the possible repeating behavior of their sources. Therefore, analysis of the *BATSE* data has been performed, focusing on the possibility of time and spatial clustering (1–3). Some techniques make use of the two-point correlation function (4) and the nearest neighbor statistics, and some constraints have been reported in the literature (5). Among other techniques are the fuzzy angular correlation (6), the pair-matching statistic (7), and the network synthesis method (8–10).

A commonly accepted result is that there is no longer evidence of repeaters in the 2B catalogue (4), although an excess of pairs of GRBs clustered in both time and space was found (1,11). Recent studies have concluded that the 3B data set contains no excess of matched burst pairs (7). It is obvious that an improvement of the GRB error boxes will better constrain the number of repeaters (6,12).

We have studied GRBs observed with *WATCH* with error boxes with $\sim 1°$ 3σ error radii, including systematic errors. The *WATCH* all-sky monitor on both *GRANAT* and *EURECA* discovered more than 70 GRBs between 1990 and 1993 (13–17), with one of its main advantages being the capability of locating 44 bursts to relatively small error-boxes. The distribution of the 44 GRBs is consistent with isotropy (17). Thirteen of these bursts were also observed with *BATSE*. In the following section we study the possible repetitions of the 44 *WATCH* bursts in the 3B catalogue.

---

*present address: Los Alamos National Laboratory, MS D436, NM 87545, USA.

© 1996 American Institute of Physics

METHOD

The 3B catalogue contains 1109 bursts, excluding the bursts seen simultaneously by the two instruments. Hereafter, we describe our 4-step method:

1. The overlaps between the *WATCH* and *BATSE* positions given in both catalogues were calculated. Gaussian normalized error-distribution functions for the error boxes were assumed. Only for the case of overlap between the *BATSE* and *WATCH* GRBs, (considered at 3σ for both of them) the following integral extended up to the 4σ *WATCH* error box was calculated.

$$c_{ij} \equiv \iint B_i(l,b) W_j(l,b) \, ds$$

   with $B_i(l,b)$ and $W_j(l,b)$ denoting the probability distributions in galactic coordinates (l^{II}, b^{II}), of a *BATSE* GRB, b_i, $i = 1,..44$ and a *WATCH* GRB, w_j, $j = 1,..1109$ respectively. This integral gives us an indication of the probability that *BATSE* and *WATCH* bursts (b_i and w_j respectively) arise from the same region in the sky. The probability that the 3B catalogue contains the *WATCH* bursts w_j is: $c_j \equiv \sum_{i=1}^{n_j} c_{ij}$, where n_j is the number of overlaps of a particular *WATCH* burst w_j in the *3B* catalogue.

   Thus, $C \equiv \sum_{j=1}^{44} c_j$ can be considered as a measure of the correlation between *BATSE* and *WATCH* catalogues. A value of 64.9 is calculated for C. The mean value of the overlap for the *WATCH* and the 1109 *3B* bursts is therefore: $\langle c_{\text{nsim}} \rangle \equiv 64.9/44 = 1.47$

2. We also obtained 100 random catalogues containing 44 GRBs each. All of them show isotropic distribution of bursts. Similar methods have been used in the past (2,3).

   Let $R_j(l,b)$ be the probability distribution of a particular GRB belonging to one of these catalogues. As in 1, we proceeded to calculate correlations between the *3B* and each of those 100 catalogues. We determined $q_{ij} \equiv \iint B_i(l,b) R_j(l,b) \, ds$, $q_j \equiv \sum_{i=1}^{n_j} q_{ij}$, and $Q \equiv \sum_{j=1}^{44} q_j$. These Q values for the 100 catalogues show a mean value, $\langle Q \rangle \equiv 66.9$ and a $\sigma \equiv 8.8$.

3. For the thirteen GRBs detected simultaneously with *WATCH* and *BATSE*, the values c_j were calculated, with a mean value $\langle c_{\text{sim}} \rangle \equiv 4.4$, that can be considered as indicative of the expected correlation for bursts arising from the same source.

4. Taking in account $\langle c_{\text{sim}} \rangle$ (step 3), we compared $\langle Q \rangle$ (step 2) with C (step1), thus obtaining a functional relation between the number of related sources, N_r, and the correlation between catalogues Q.

RESULTS

The results of our study can be better understood when viewing Fig. 1. The mean value of the correlation for 100 simulated catalogues, $\langle Q \rangle = 66.9$, is even larger than for the *WATCH* catalogue $C = 64.9$.

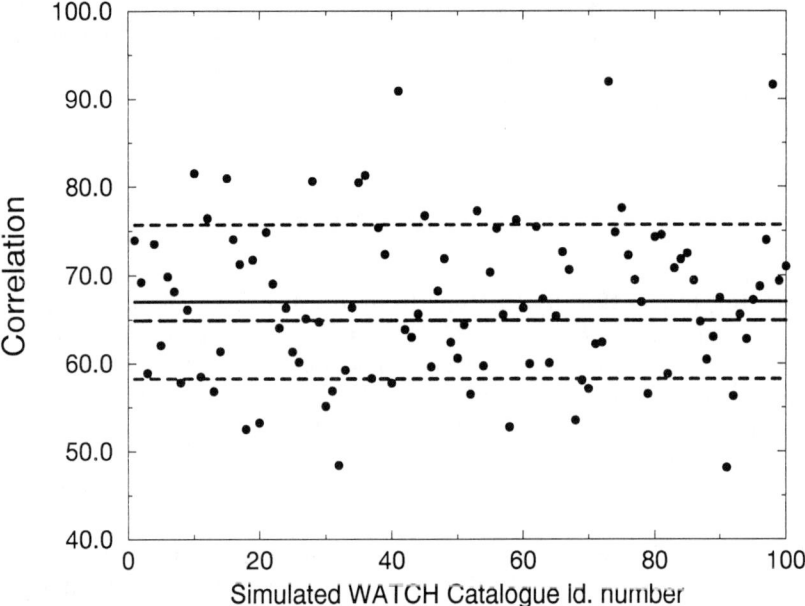

FIG. 1. The continuous straight line represents the $\langle Q \rangle$ for the random catalogues. The short-dashed lines are the $\pm 1\sigma$ limits. The long-dashed line is the correlation C between the *WATCH* and the *3B* bursts

Therefore the correlation between two catalogues Q, and the numbers of repeaters GRBs N_r, could be estimated through the following formula:

$$Q = \langle c_{\text{sim}} \rangle \ N_r + \langle c_{\text{nsim}} \rangle \ (44 - N_r)$$

Assuming the correlation Q like the dependent variable, we get:

$$N_r = \frac{Q - 44 \langle c_{\text{sim}} \rangle}{\langle c_{\text{sim}} \rangle - \langle c_{\text{nsim}} \rangle}$$

This formula gives us an estimation of the repeaters for a known correlation Q. In the case of our concern, taking the maximum value of Q inside the

$\langle Q \rangle \pm 1\sigma$ region, we would expect an upper limit of ~ 4 accidental overlaps, but we would not be able to distinguish whether they are indeed related to repeaters. This empirical formula has been checked by calculating the correlation with simulated catalogues. One should notice that this formula is only an approximation, and cannot be used in the region close to the mean value, where the random fluctuations are very high. Our study is in good agreement with the result obtained by Strohmayer et al. (12), who deduced that if the 10%-15% *BATSE* GRBs were repeaters none of them would be noticeable, and with the analyses carried out by Bennett et al. (7) and Tegmark et al. (18). Both of them use a bigger sample and therefore impose more severe constraints than ours.

CONCLUSION

This paper is aimed to get correlations among different instruments. Similar studies have been performed for other experiments. (19). We conclude there is no indication of recurrent activity of *WATCH* bursts in the *BATSE* data and derive an upper limit of ≤ 4 repeaters for the *WATCH* GRBs, although we would like to point out that we have not used the exposure maps of *WATCH* and *BATSE*.

REFERENCES

1. V. Petrosian, E. Bradley, ApJ **441**, L37 (1995).
2. J. Hakkila, Alabama University Research Reports (1990).
3. J. Hakkila, et al., ApJ **442**, 659 (1994).
4. J.J. Brainerd, ApJ **441**, L39B (1995).
5. V.C. Vo, et al., BAAS **184**, 1005 (1994).
6. D. Hartmann, H. Dieter, ApJ **367**, 186 (1991).
7. D.P. Bennett, S.H. Rhie, submitted to ApJ Letters (1996).
8. K. Hurley, Adv. Sp. Res. **15**, 127 (1995).
9. K. Hurley, ApJ **431**, 31 (1994).
10. K. Hurley, BAAS **185**, 509 (1994).
11. V. Wang, R.E. Lingenfelter, ApJ **441**, 747 (1995).
12. T.E. Strohmayer, E.E. Fenimore, J.A. Miralles, ApJ **432**, 665 (1994).
13. A.J. Castro-Tirado, et al., in Gamma-Ray Bursts, eds. G.J. Fishman, J.J. Brainerd & K. Hurley, AIP Conf. Proc. **307**, 17 (AIP, New York, 1994).
14. S. Brandt, Ph. D., University of Copenhagen (1994).
15. N. Lund, in press (1996).
16. S. Brandt, N. Lund, A.J. Castro-Tirado, in Gamma-Ray Bursts, eds. G.J. Fishman, J.J. Brainerd & K. Hurley, AIP Conf. Proc. **307**, 13 (AIP, New York, 1994).
17. A.J. Castro-Tirado, Ph. D., University of Copenhagen (1994).
18. M. Tegmark, D.H. Hartmann, M. Briggs, J. Hakkila, C.A. Meegan, submitted to ApJ Letters (1996).
19. I.G. Mitrofanov, et al., Adv. Sp. Res. **15**, 131 (1995).

Analysis of the Systematic Errors in the Positions of BATSE Catalog Bursts

C. Graziani[1] and D. Q. Lamb[2]

[1] NASA Goddard Space Flight Center, Greenbelt, MD 20771
[2] Dept. of Astronomy and Astrophysics, University of Chicago, Chicago, IL 60637

> We analyze the systematic errors in the positions of bursts in the BATSE 1B, 2B and 3B catalogs, using a likelihood approach. We use the BATSE data in conjunction with 196 single IPN arcs. We assume circular Gaussian errors, and that the total error is the sum in quadrature of the systematic error σ_{sys} and statistical error σ_{stat}, as prescribed by the BATSE catalog. We find that the 3B burst positions are inconsistent with the value $\sigma_{\text{sys}} = 1.6°$ stated in the BATSE 3B catalog.

INTRODUCTION

The stated systematic error in the BATSE 3B catalog (1) burst locations is $\sigma_{\text{sys}} = 1.6°$. This value was estimated by taking the RMS deviation of BATSE positions from known positions of 36 bursts, determined by the IPN, WATCH, and COMPTEL (2). Unfortunately, many of these same known positions were used to calibrate the BATSE burst location software, i.e., as a guide in determining what effects to include in the burst location algorithm. In order to calibrate properly the BATSE burst position errors, an independent set of burst locations is needed. Such a set exists, in the form of 196 bursts for which single IPN arcs exist (3). In this paper we analyze the systematic errors in the positions of bursts in the BATSE 1B, 2B and 3B catalogs using these 196 bursts. Our analysis is based upon the likelihood approach.

ANALYSIS

We assume that the systematic error σ_{sys} and statistical error σ_{stat} in position are circular Gaussians, to be added in quadrature, as prescribed by the BATSE catalog. The circular approximation should be good, since roughly $2/3 - 3/4$ of all BATSE bursts have χ^2 positional contours that are nearly circular (2), and the bursts with IPN arcs tend to be those with larger fluences. We further assume that the sky is flat (i.e., that the size of the total error in the BATSE burst position is not too large), and that the IPN arcs have zero width. The former is a good approximation, since the largest BATSE $\sigma_{\text{tot}} \approx 12° \ll 90°$ among the bursts with IPN arcs. The latter is a good approximation since the characteristic width of the 3-σ contours of the IPN arcs is a few arcminutes, which is much less than the smallest BATSE $\sigma_{\text{tot}} \approx 1°$.

© 1996 American Institute of Physics

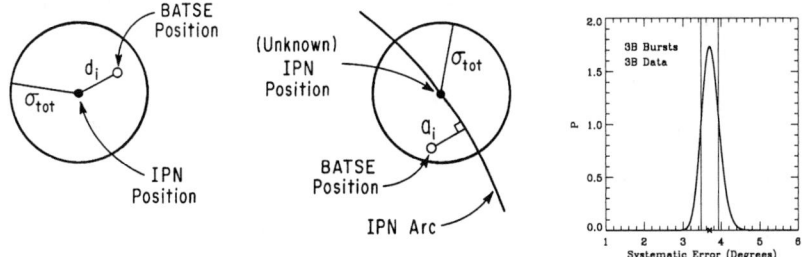

Figure 1. Configurations for a burst with a known position (left diagram) and for a burst with a single IPN arc (right diagram).

Figure 2. Probability density as a function of σ_{sys} for the ML constant σ_{sys} model. The vertical lines show the $\pm 1\text{-}\sigma$ interval.

The likelihood function for bursts with known positions (e.g., two intersecting IPN arcs) is given by

$$\mathcal{L}(\sigma_{\text{sys}}) = \prod_{i=1}^{N} L_i \equiv \prod_{i=1}^{N} \frac{1}{2\pi\mu_i^2} \exp\left(-\frac{1}{2}\frac{d_i^2}{\mu_i^2}\right), \quad (1)$$

whereas the likelihood function for bursts with single IPN arcs is given by

$$\mathcal{L}(\sigma_{\text{sys}}) = \prod_{i=1}^{N_{\text{arc}}} L_i \equiv \prod_{i=1}^{N_{\text{arc}}} \frac{1}{\sqrt{2\pi}\mu_i} \exp\left(-\frac{1}{2}\frac{a_i^2}{\mu_i^2}\right). \quad (2)$$

Here, d_i is the deviation of the BATSE position from the known position, a_i is the perpendicular deviation of the BATSE position from the IPN arc (see Figure 1), and $\mu_i^2 = 0.43\sigma_{\text{tot}}^2 = 0.43(\sigma_{\text{sys}}^2 + \sigma_{\text{stat}}^2)$.

The likelihood function $\mathcal{L}(\sigma_{\text{sys}})$ allows exploration of any model that we may have for σ_{sys}. Here we focus on three models: (1) $\sigma_{\text{sys}} = $ a constant, (2) $\sigma_{\text{sys}} = A(S/10^{-5} \text{ erg cm}^{-2})^\alpha$, and (3) $\sigma_{\text{sys}} = A(\sigma_{\text{stat}}/1°)^\alpha$. The first model has one free parameter (σ_{sys}); the second and third models have two free parameters (scale factor A and power law index α).

RESULTS

Fitting the constant σ_{sys} model to the 18 BATSE 3B bursts with IPN positions, we find a maximum likelihood (ML) value $\sigma_{\text{sys}}^{\text{ML}} = 1.85°^{+0.28°}_{-0.22°}$. This is consistent with the value of 1.6° quoted in the BATSE 3B catalog (1), which was found using these bursts and some others.

Fitting the constant σ_{sys} model to the 196 BATSE 3B bursts with IPN arcs, we find a ML value $\sigma_{\text{sys}}^{\text{ML}} = 3.7°^{+0.24°}_{-0.22°}$. This is inconsistent (at the 10σ level!)

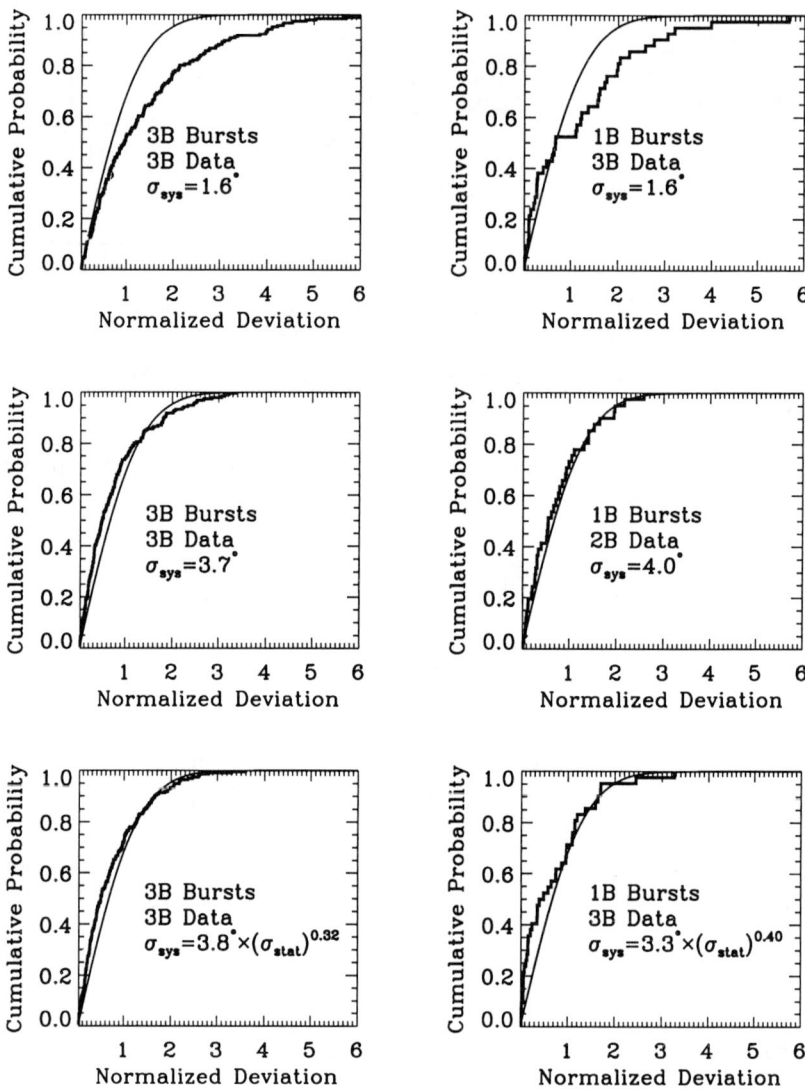

Figure 3. Comparison of expected and observed cumulative distributions of the normalized deviation $D = |a_i/\mu_i|$ for the burst samples and data as labeled, assuming the ML models of σ_{sys} as labeled.

with the value of 1.6° quoted in the BATSE 3B catalog, as shown by the probability distribution in σ_{sys} (see Figure 2) and comparison of the expected and observed distribution of normalized deviations $D = |a_i/\sigma^i_{\text{tot}}|$ assuming $\sigma_{\text{sys}} = 1.6°$ (see Figure 3).

Fitting the second model, in which $\sigma_{\text{sys}} = A(S/10^{-5} \text{ erg cm}^{-2})^\alpha$, we find

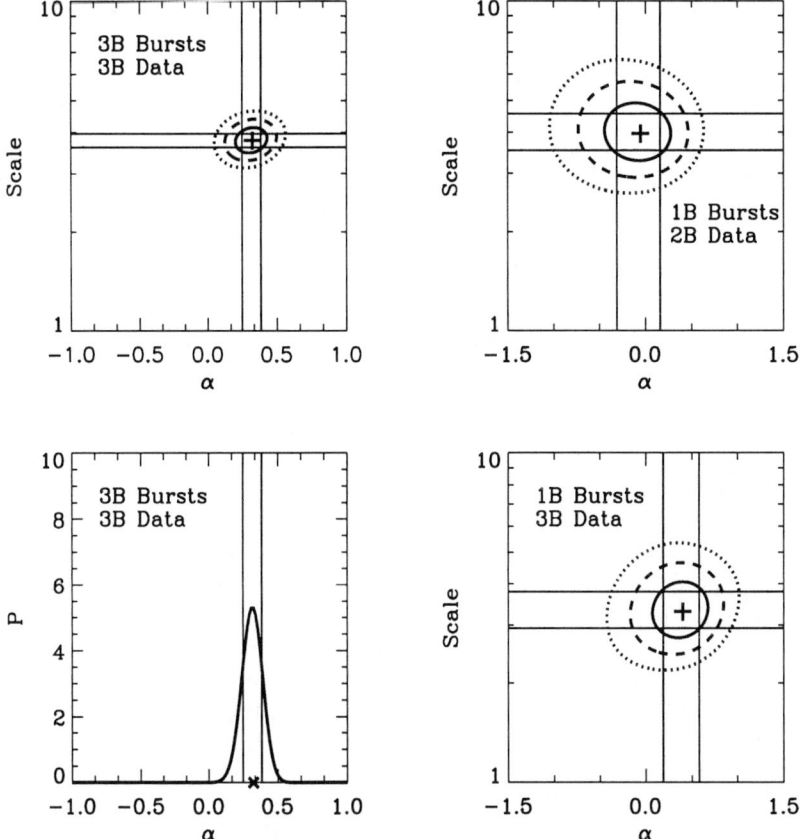

Figure 4. Upper left-hand and lower and upper right-hand panels: 1-σ, 2-σ, and 3-σ contours in the (A, α)-plane for the burst samples and data as labeled, assuming the σ_{stat}-dependent model. Lower left-hand panel: same as Figure 2, except for the σ_{stat}-dependent model.

ML values $A_{\text{ML}} = 3.8 \pm 0.3$ and $\alpha_{\text{ML}} = -0.18^{+0.06}_{-0.05}$. The probability $P(\alpha \geq 0) = 1.6 \times 10^{-3}$, showing that the data prefer the power-law model over the ML model with constant $\sigma_{\text{sys}} = 3.7°$.

Fitting the third model, in which $\sigma_{\text{sys}} = A\sigma_{\text{stat}}^{\alpha}$, we find ML values $A_{\text{ML}} = 3.8 \pm 0.2$ and $\alpha_{\text{ML}} = 0.32^{+0.06}_{-0.08}$ (see Figure 4). The probability $P(\alpha \leq 0) = 4.7 \times 10^{-5}$, showing that the data strongly prefer the power-law model over the ML model with constant $\sigma_{\text{sys}} = 3.7°$.

These results imply that σ_{sys} is strongly correlated with S and σ_{stat}. For example, $\sigma_{\text{sys}}(\sigma_{\text{stat}} = 0.1°) = 2°$ while $\sigma_{\text{sys}}(\sigma_{\text{stat}} = 10°) = 8°$. This is illustrated in Figure 5, which shows the ML S-dependent and σ_{stat}-dependent models.

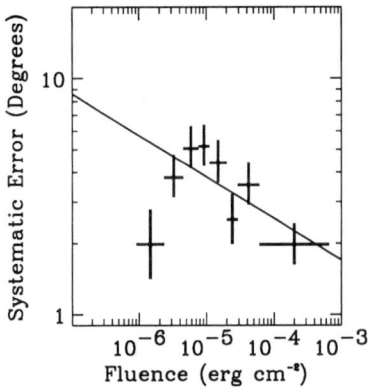 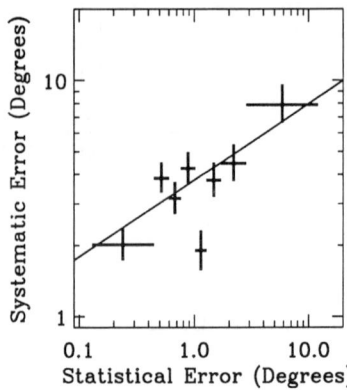

Figure 5. Left-hand panel: The ML S-dependent model. Right-hand panel: The ML σ_{stat}-dependent model. The points are *not* data points, but rather are the ML *constant* σ_{sys} model for the given interval in S or σ_{stat}.

It is also interesting to examine the character of σ_{sys} for subsets of the 3B catalog, in particular the 1B sample of bursts in which evidence for repeating was found. Fitting the three models to the "old" 1B positions, we find $\sigma_{\text{sys}} = 4.0°$ for the ML constant σ_{sys} model, and that the data do not request a more complicated model(see Figures 3 and 4). In contrast, fitting the 3B positions for the 1B sample of bursts, we find that the data request the σ_{stat}-dependent model at the 95% confidence level [e.g., $P(\alpha \leq 0) = 1.6 \times 10^{-2}$] (again see Figures 3 and 4).

These results imply that for the 1B sample of bursts, the new (3B) σ_{sys}'s are smaller than the old (1B) ones for bursts with $S > 4 \times 10^{-6}$ erg cm^{-2} or with $\sigma_{\text{stat}} < 1.6°$. However, for the $\approx 80\%$ of bursts with lower fluences or larger σ_{stat}'s, they imply that the new σ_{sys}'s are larger than the old ones.

Acknowledgments. We are grateful to Kevin Hurley for providing us with the 3rd catalog of IPN arcs, without which this analysis would not have been possible. We also wish to thank Geoff Pendleton and Michael Briggs for discussing BATSE burst location issues with us.

REFERENCES

1. C. A. Meegan, et al., electronic catalog (grossc.gsfc.nasa.gov).
2. G. N. Pendleton, private communication (1995).
3. K. Hurley, et al., Third IPN Catalog (1995).

Constraints on the Luminosities of Cosmological Gamma-Ray Bursts

Jon Hakkila*, Charles A. Meegan†, John M. Horack†,
Geoffrey N. Pendleton‡, Michael S. Briggs‡,
Robert S. Mallozzi‡, Thomas M. Koshut‡,
Robert D. Preece‡, and William S. Paciesas‡

*Dept. Physics and Astronomy, Mankato State Univ., Mankato, MN 56002
†NASA/MSFC, ES-84, Huntsville, AL 35812
‡Dept. Physics, Univ. Alabama in Huntsville, Huntsville, AL 35899

> Chi-squared analysis of the combined BATSE/PVO peak flux distribution places constraints on the luminosity function of gamma-ray bursts in a $\Lambda = 0$, $\Omega_o = 1$ Friedmann cosmology with a non-evolving density distribution. The analysis indicates that the minimum burst luminosity, the maximum burst luminosity, or both, are constrained, with the opposite end of the luminosity function often unconstrained. For this cosmological model, a non-zero width of the intrinsic luminosity function is preferred, but luminosity functions are forbidden that let many bursts from an extremely broad luminosity range be observed.

INTRODUCTION

A cosmological origin (17) naturally explains the gamma-ray burst angular and peak flux distributions (15,1,18), and would naturally explain any time dilation effects seen in burst time histories. The BATSE 3B catalog (16) used both alone and in tandem with the PVO data (4) indicates that complexities such as non-standard cosmologies, density or luminosity evolution, and/or luminosity functions of non-zero width are preferred but not required for cosmological models (5,9).

The role of the luminosity function in cosmological gamma-ray burst scenarios has only been examined by a few authors (3,2,11), and exhaustive studies have been impeded by the large number of free parameters available. One limitation to understanding cosmological luminosity constraints has resulted from terminology pertaining to the luminosity function. Several analyses (14,8,3,20,21) have described properties of the *observed* luminosity function, describing luminosities of detected bursts. This distribution is narrow, with 90% of observed bursts spanning less than one order of magnitude and 80% spanning less than a factor of five; exceptions are allowed only in accelerating cosmologies (3). These limitations are not necessarily the same as those present on the intrinsic luminosity range.

© 1996 American Institute of Physics

COSMOLOGICAL MODEL

In a Friedmann cosmology with density parameter $\Omega_o = 1$, cosmological constant $\Lambda = 0$, and no density evolution, the rate at which bursts are detected is

$$N(>F) = 4\pi \int_{L_{\min}}^{L_{\max}} \Phi(L) dL \int_0^{z_{\max}(L,F)} \frac{n(z)[r(z)]^2}{(1+z)} \left(\frac{dr}{dz}\right) dz, \qquad (1)$$

where $n(z)$ is the rate density in the comoving frame, r is the radial coordinate, $\Phi(L)$ is the intrinsic luminosity function, and $z_{\max}$ is the maximum redshift z at which a source of luminosity L (erg sec^{-1}) and peak flux F (erg cm^{-2} sec^{-1}) could be detected. We assume that the comoving frame rate density is constant: $n(z) = n_0$, and that the luminosity function $\Phi(L)$ is given by

$$\Phi(L) \propto L^{-\beta}; L_{\min} < L \leq L_{\max}, \qquad (2)$$

and we refer to the luminosity range by $\Re = L_{\max}/L_{\min}$.

We further assume that burst energy distributions may be modeled by power-law spectra with a common energy spectral index of 0, which appears to be a reasonable approximation for an ensemble of bursts (13).

Using these assumptions, we find $z_{\max}(L,F)$ from

$$F = \left[\frac{L}{4\pi(\frac{c}{H_0})^2}\right] \left(\frac{1}{\sqrt{1+z_{\max}}-1}\right)^2. \qquad (3)$$

DATA

The dataset used in this analysis is the BATSE 3B Catalog containing bursts in the observed energy range 50 to 300 keV with peak fluxes measured on the 1024 ms timescale. In order to avoid sampling incompleteness due to BATSE's decreased burst trigger sensitivity near the detection threshold, the sample is restricted to the 645 bursts with photon peak fluxes greater than 0.42 photons cm^{-2} sec^{-1}. The peak fluxes are then converted to energy peak fluxes (F in units of erg cm^{-2} sec^{-1}) by assuming a constant spectrum.

These bursts are combined with 144 PVO bursts (4) to create a combined BATSE/PVO integral rate vs. peak flux curve. PVO peak fluxes have been converted to energy fluxes by correcting for the PVO triggering bandpass, and the resulting integral rate curve has been normalized to the BATSE peak flux curve and detection rate (N in detected bursts year^{-1}). This curve has a $-3/2$ logarithmic slope for approximately 1.5 to 2 decades in F, as expected for a homogeneous distribution, but changes slope significantly within a decade, indicating a rapid depletion of distant (low peak flux) sources. The curve reaches a logarithmic slope of -0.8 at F $\approx 8.45 \times 10^{-8}$ erg cm^{-2} sec^{-1}.

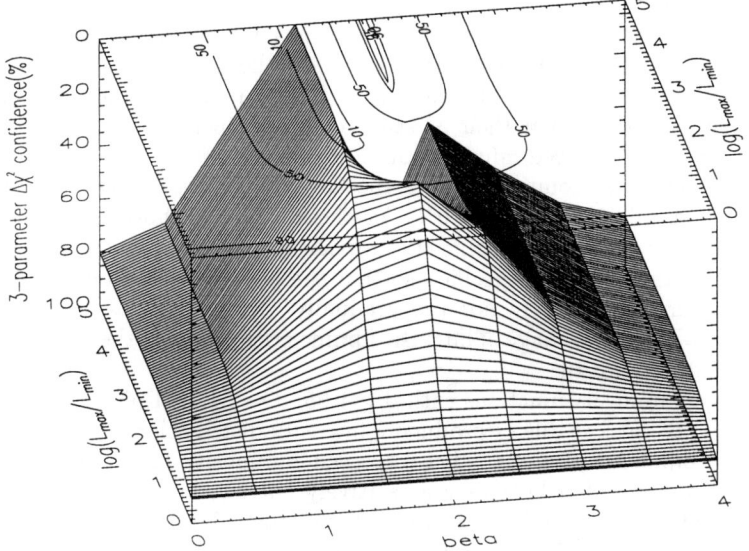

FIG. 1. Luminosity parameter space.

The combined rate is compared to theoretical curves representing different models in the $(\beta, \Re)$ parameter space using a χ^2 test (19). The rate data have been binned differentially, maintaining at least eight bursts per bin. The binned rate data contains 14 points, and stresses the relative importance of the $-3/2$ portion of the curve sampled by PVO.

ANALYSIS

The best-fit model, optimized over the three parameters β, L_{max}, and $\Re$ has $\chi^2_{min} = 6.49$ for 10 degrees of freedom. The probability of obtaining this value or larger assuming that the model is correct is 77.3%, which does not contradict the assumed model. We therefore assume that this model is correct and estimate the parameter ranges of the luminosity function that are statistically consistent with the data.

Parameters are estimated via $\Delta\chi^2 = \chi^2 - \chi^2_{min}$ for the three free parameters in question (12,19). Since this study's goal is to determine constraints on the luminosity function shape, the best-fit $\Delta\chi^2$ confidence regions are projected on the β and $\log\Re$ axis and are shown in Fig. 1. Three solution types are identified (6,7,10): L_{max}-dominated, L_{min}-dominated, and range-dominated.

L_{max} dominated solutions are those in which the observations of the luminosity function/radial distribution are biased heavily in favor of the most luminous bursts ($\beta < 1.95$). In this case the L_{min} bursts are so undersampled that their presence is of no consequence to the observed distribution — they have peak fluxes below the detection threshold. L_{max}-dominated solutions can have infinite luminosity ranges.

L_{min}-dominated solutions are those in which the observations of the luminosity function/radial distribution are biased heavily in favor of the least luminous bursts ($\beta > 2.1$). In contrast to the previous case, the L_{max} bursts are so few in number that their presence is inconsequential. L_{min}-dominated solutions also are allowed infinite luminosity ranges.

Range-dominated solutions within the 90% confidence region have $1.95 \leq \beta \leq 2.1$ and $1.6 \leq \Re \leq 1.6 \times 10^3$. For these solutions, both the high- and low-luminosity end of the luminosity function are observed. In this case, the luminosity range must be specified so that the peak flux distribution change slope neither too abruptly nor too gradually. As is the case for Galactic models, the narrowest range of intrinsic luminosities produces the widest range of observed luminosities.

Intrinsic luminosity functions with $\beta \geq 1.95$, $\beta \leq 2.1$, or $1.95 < \beta < 2.1$ and $1.6 \leq \Re \leq 10^3$ are capable of producing intensity distributions that lie within the 90% confidence contour. The observed luminosity functions produced by these intrinsic distributions are relatively narrow, but at the same time slightly wider than standard-candle distributions. The optimal fits to the data are produced by L_{max}-dominated solutions near $\beta = 1.5$ with $\Re \geq 100$.

The effective redshifts (the redshift of the most luminous bursts identifiable in the sample) of the best-fit models identified in Fig. 1 are shown as functions of β and $\Re$ in Fig. 2. The effective redshifts are large (in excess of 3) for L_{max}-dominated and range-dominated models and small (less than 2) for L_{min}-dominated models. The latter is consistent with time-dilation measurements of bursts whereas the former is not. This is not a definitive statement; there are good-fit L_{max}-dominated models with smaller effective redshifts and good-fit L_{min}-dominated models with larger effective redshifts. However, it does suggest that the gamma-ray burst luminosity function might be L_{min}-dominated.

CONCLUSIONS

Few constraints exist on the luminosities of cosmological gamma-ray bursts. Standard-candle luminosities are not favored, but are only marginally unacceptable. Luminosity functions with large ranges and power-law indices near 2 are not favored, but few other constraints are strongly identified. The effective redshift of the best-fit solutions, coupled with the relatively small amount of time dilation observed in gamma-ray burst time histories, might indicate that the gamma-ray burst luminosity function is L_{min}-dominated.

REFERENCES

1. M. S. Briggs, et al., ApJ **459**, 40 (1996).
2. E. Cohen, and T. Piran, ApJ, in press (1995).
3. A. G. Emslie and J. M. Horack, ApJ **435**, 16 (1994).

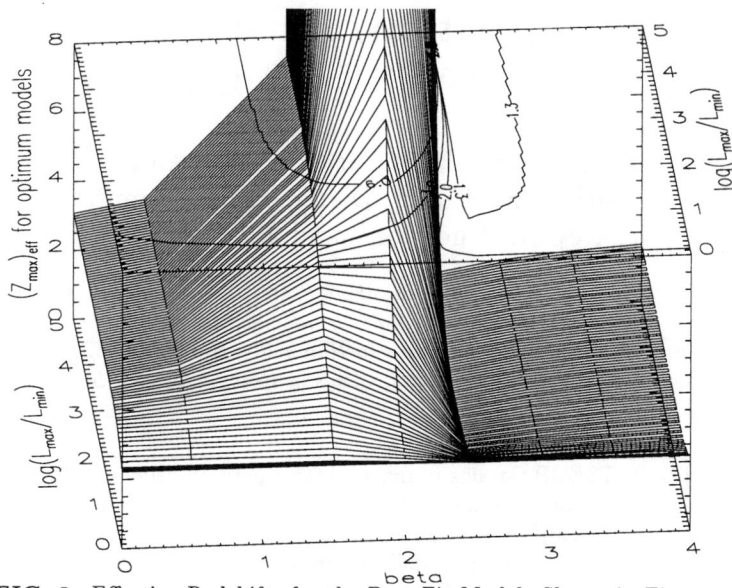

FIG. 2. Effective Redshifts for the Best-Fit Models Shown in Fig. 1.

4. E. E. Fenimore, et al., Nature **366**, 40 (1993).
5. E. E. Fenimore, these proceedings (1996).
6. J. Hakkila, et al., ApJ **454**, in press (1995).
7. J. Hakkila, et al., Ap. Sp. Sci., in press (1996).
8. J. M. Horack, A. G. Emslie, and C. A. Meegan, ApJ Letters **426**, L5 (1994).
9. J. M. Horack, et al., submitted (1996).
10. J. M. Horack, et al., ApJ in press (1996).
11. I. Horváth, P. Mészáros, and A. Mészáros, ApJ in press (1996).
12. M. Lampton, B. Margon, and S. Bowyer, ApJ **208**, 177 (1976).
13. R. S. Mallozzi, et al., these proceedings (1996).
14. S. Mao and B. Paczynski, ApJ **388**, L45 (1992).
15. C. A. Meegan, et al., Nature **355**, 143 (1992).
16. C. A. Meegan, et al., ApJ Supp., in press (1996).
17. B. Paczyński, ApJ Letters **308**, L43 (1986).
18. G. N. Pendleton, et al., ApJ submitted (1996).
19. W. H. Press, B. P. Flannery, S. A. Teukolsky, and W. T. Vetterling, in Numerical Recipes, (New York, Cambridge University Press), (1989).
20. A. Ulmer and R.A.M.J. Wijers, ApJ **439**, 303 (1995).
21. A. Ulmer, R.A.M.J. Wijers, and E. E. Fenimore, ApJ Letters **440**, L9 (1995).

Repetition/Clustering in the BATSE 3B Catalog

Jon Hakkila*, Charles A. Meegan†,
Michael S. Briggs‡, Dieter H. Hartmann§,
Geoffrey N. Pendleton‡, and John M. Horack†

*Dept. Physics and Astronomy, Mankato State Univ., Mankato, MN 56002
†NASA/MSFC, ES-84, Huntsville, AL 35812
‡Dept. Physics, Univ. Alabama in Huntsville, Huntsville, AL 35899
§Dept. Physics and Astronomy, Clemson Univ., Clemson, SC 29634

The BATSE 3B Catalog is analyzed for repetition using a technique that simultaneously accounts for the instrumental effects of location errors and incomplete sky exposure. A repetition model is assumed in which bursts are emitted sporadically from burst sources. Upper limits are found on the repetition rates in the entire 3B Catalog, as well as on the 1B, 2B−1B, and 3B−2B components of the catalog. Additional anisotropy variations are predicted based upon the distribution of close burst pairs. Solutions are favored in which no repetition of the type studied is present.

INTRODUCTION

Gamma-ray burst repetition has been suspected (14,18,16,19) from angular distribution analyses of the BATSE 1B catalog (5). This interpretation has been disputed, as other authors claim that an excess number of close burst pairs in the 1B catalog is due to instrumental effects (13); primarily those arising from statistical variations due to burst localization errors (9). Subsequent analyses (3,11) (7) of the BATSE 2B catalog (10) find no evidence of repetition, which has been attributed (upon removal of less accurately localized MAXBC bursts) to decreased sky exposure (15) (8). Indeed, poor sky exposure can influence the ability of the experiment to detect burst repetition, although the amount by which sky exposure affects the experiment is model-dependent (6,1).

The ability of BATSE to detect repetition thus depends on (a) appropriate modeling of the instrumental effects of localization errors and incomplete sky exposure, (b) a repetition model, and (c) an analysis technique sensitive to the presence of a repetition signal. One such technique analyzes the close burst pair distribution (6), which is the distribution of burst pairs close enough that they might be repetitions from the same bursting source.

The BATSE 3B catalog (12) is the largest gamma-ray burst catalog to date (1122 bursts) for which location errors and sky exposure are known. This

© 1996 American Institute of Physics

catalog and its published subsets (the 1B, containing the first 262 bursts; the 2B−1B, containing the next 323 bursts; and the 3B−2B, containing the remaining 526 bursts) provide an ideal database for an analysis of this type.

ANALYSIS TECHNIQUE

The analysis proceeds as follows: (a) Bursts are selected via Monte Carlo sampling from a burst source distribution using a specified repetition model, (b) Localization errors are assigned to each burst, and a model of BATSE sky exposure is used to determine whether or not the Monte Carlo bursts are detected by the mock experiment, (c) The angular distribution of detected bursts is analyzed in terms of the Two-Point Angular Correlation Function (TPACF) and the dipole and quadrupole moments of the close burst pair distribution, and (d) The Monte Carlo results are used to determine the probability that the BATSE distribution is drawn from the repetition model.

Repetition Model

Bursts are selected randomly from an isotropic distribution containing a predefined number of sources. The number of bursts detected corresponds to the number BATSE observed during the timespan of the catalog in question. The distribution of repetitions source^{-1} year^{-1} in each Monte Carlo catalog is thus described by Poisson statistics. Repetition is a feature of the model, since the number of sources is less than or equal to the number of bursts detected. Each source has an equal probability of repeating.

Instrumental Effects

Each burst position is reassigned in order to mimic BATSE's inability to perfectly localize bursts. Burst localization errors (recalculated for the entire 3B catalog) are described by the sum in quadrature of a systematic error $\sigma_{sys} = 1.5°$ and a statistical error σ_{stat} that depends on the detected burst brightness. A reasonable fit to the statistical error is given by $\sigma_{stat} = 2.805 F_{peak}^{-0.74}$, where the peak photon flux F_{peak} is given in units of photons cm^{-2} sec^{-1}. Because the statistical error depends on burst brightness, Monte Carlo bursts are assigned peak fluxes as modeled from the BATSE 3B peak flux distribution, which includes all bursts brighter than BATSE's minimum detection threshold $F_{det} = 0.05$ photons cm^{-2} sec^{-1}.

BATSE is not able to perfectly sample the gamma-ray burst sky. BATSE's sky exposure (12) differs for each data subset used in this study, and is 34% for the 1B catalog, 26% for the 2B−1B catalog, 41% for the 3B−2B catalog, and 38% for the entire 3B catalog. Furthermore, sky exposure is a function of declination (5), which has not changed during the experiment's lifetime.

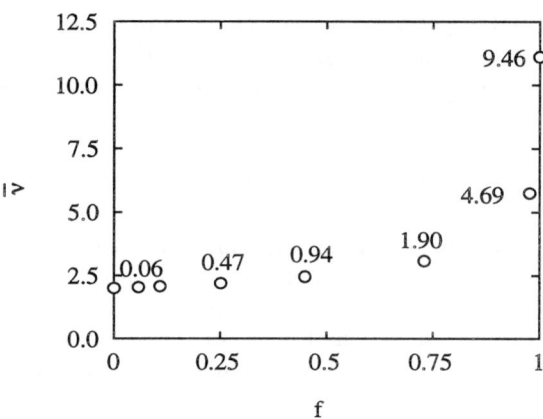

FIG. 1. Repetition rate R as a function of the parameters f and $\bar{\nu}$.

The model produced for each catalog can be characterized in terms of a mean repetition rate R (bursts source^{-1} year^{-1} brighter than minimum detection threshold), from which the sky exposure detracts to produce a rate of repetition from the *detected* bursts (bursts source^{-1} year^{-1}). We compare the modeled quantity R to two generic (model-independent) theoretical quantities: the fraction of sources f that produce more than one detected burst, and the mean number of detected repetitions $\bar{\nu}$ per source producing more than one detected burst. Note that these definitions are similar but not identical to those of Meegan *et al.* (11). Poisson statistics define a characteristic relationship between f and $\bar{\nu}$ for different mean repetition rates (6,1); this is demonstrated in Fig. 1. The quantities f and $\bar{\nu}$ are based on the ability of a model to associate each burst with the source from which it emanates; in testing the model this information is lost due to localization uncertainties.

Repetition Tests

The TPACF is used to test burst clustering, as it is more sensitive than the Nearest Neighbor test for detecting burst repetition (2). BATSE's TPACF angular resolution is dependent on how the repetition model is convolved with selection effects. The angular resolution of the 3B catalog is found to be 7.2°, based on Monte Carlo analysis of the model in question. This means that a measurable signal associated with repetition can be seen on angular scales as large as 7.2°, so exclusion of bursts separated by angles smaller than this degrades any repetition signal.

Additional isotropy variations are observed in the large-scale distribution of close burst pairs (6). BATSE burst detection is favored in the directions of the celestial poles, particularly the North Celestial Pole, due to anisotropic

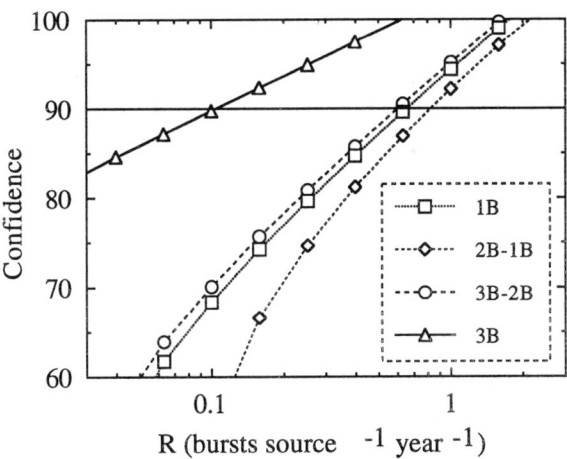

FIG. 2. Statistical limits on repetition rates for catalogs in question.

sky exposure. These enhancements are magnified for repetitions, since the probability of observing many repetitions from a source sensitively depends on the sky exposure in that region. Large repetition rates produce well-populated clusters containing many burst pairs within 7.2° of one another. Thus, random close burst pairs are distributed anisotopically due to sky exposure, and close burst pairs resulting from repetition are distributed even more anisotropically, with the degree of anisotropy dependent upon the mean repetition rate R. Since repetitions cannot be identified *a priori* from random close bursts, all burst pairs separated by less than 7.2° must be considered in the analysis. Dipole and quadrupole anisotropies can be detected for repetition models by applying the Rayleigh-Watson and Bingham statistics (4) to the close burst pair distribution.

ANALYSIS AND CONCLUSIONS

We analyze the 1B, 2B−1B, 3B−2B, and 3B catalogs to determine upper limits on the burst repetition rate. The close burst pair TPACF is found to vary strongly with R, while the Rayleigh-Watson and Bingham statistics vary more weakly with R. In fact, the close burst pair Rayleigh-Watson and Bingham statistics vary too slowly over the range of R in question to be valuable for parameter estimation. These statistics would provide more meaningful constraints on R (a) for repetition models producing larger detected clusters (e.g. models where a small number of sources bursts often while the remainder burst only once), and/or (b) under the condition that the sky exposure was both more complete and more anisotropic.

Limits on the mean repetition rate are obtained from the TPACF statistic by examining the probability that $\Delta\chi^2$ is exceeded for one free parameter. Ta-

TABLE 1. 90% confidence limits on maximum allowed burst repetition rates. R is the mean repetition rate (repetitions source^{-1} year^{-1}), f is the fraction of sources that produce more than one detected burst, and $\bar{\nu}$ is the mean number of detected repetitions per source producing more than one detected burst.

Catalog	R	f	$\bar{\nu}$
1B	0.65	0.30	2.29
2B−1B	0.85	0.39	2.37
3B−2B	0.60	0.33	2.30
3B	0.10	0.07	2.05

ble 1 and Fig. 3 summarize the average repetition rate limits for the catalogs in question. Although repetition rates as high as $R = 0.85$ bursts source^{-1} year^{-1} are allowed in the published subcatalogs, the 3B catalog can only have repetition rates $R \leq 0.10$ bursts source^{-1}year^{-1}. This indicates that the selection effects do not detract from the ability of catalogs as large as the 3B to provide more stringent repetition limits. The 3B upper limit corresponds to an observed repeater fraction of 7% for 2.05 bursts per repeating source. These results are similar to (but not as stringent as) results found using the very different technique of spherical harmonic analysis (17). We conclude that repetition is not significantly present in the BATSE 3B catalog.

REFERENCES

1. D. L. Band, ApJ, in press (1996).
2. J. J. Brainerd, ApJ, submitted (1996).
3. J. J. Brainerd, et al., ApJ **441**, L39, (1995).
4. M. S. Briggs, ApJ **407**, 126, (1993).
5. G. J. Fishman, et al., ApJS **92**, 229 (1994).
6. J. Hakkila, et al., Ap. Space Sci. **231**, 23 (1995).
7. D. H. Hartmann, et al., ApJ, submitted (1995).
8. D. Q. Lamb and J. Quashnock, Ap. Space Sci. **231**, 19 (1995).
9. C. A. Meegan, et al., in Gamma-Ray Bursts, eds. G. J. Fishman, J. J. Brainerd, and K. Hurley, AIP Conf. Proc. **307**, 3 (AIP, New York, 1994).
10. C. A. Meegan, et al., electronic catalog: (grossc.gsfc.nasa.gov), (1994).
11. C. A. Meegan, et al., ApJ **446**, L15 (1995).
12. C. A. Meegan, et al., ApJS, in press (1996).
13. R. Narayan and T. Piran, MNRAS **265**, L65 (1993).
14. J. M. Quashnock and D. Q. Lamb, MNRAS **265**, L45 (1993).
15. J. M. Quashnock, Ap. Space Sci. **231**, 35 (1995).
16. T. E. Strohmayer, et al., ApJ **432**, 665 (1994).
17. M. Tegmark, et al., ApJ, submitted, (1996).
18. V. Wang and R. E. Lingenfelter, ApJ **416**, L13 (1993).
19. V. Wang and R. E. Lingenfelter, ApJ **441**, 747 (1995).

Search for Supergalactic Anisotropies in the 3B Catalog

Dieter H. Hartmann[1], Michael S. Briggs[2], Karl Mannheim[3]

[1] Dept. of Physics and Astronomy, Clemson University, Clemson, SC 29634
[2] Dept. of Physics, University of Alabama, Huntsville, AL 35899
[3] Universitäts-Sternwarte Göttingen, Geismarlandstrasse 11, D-37803 Göttingen

> The angular distribution of GRBs is isotropic, while the brightness distribution of bursts shows a reduced number of faint events. These observations favor a cosmological burst origin. If GRBs are indeed at cosmological distances and if they trace luminous matter, we must eventually find an anisotropic distribution of bright bursts. If a significant number of bursts originate at redshifts less than $z \sim 0.1$, the concentration of nearby galaxies towards the supergalactic plane is pronounced enough that we could discover the corresponding clustering of burst locations. We used the 3B catalog to search for a pattern visible in supergalactic coordinates. No compelling evidence for anisotropies was found. The absence of anisotropies in SG coordinates implies a minimum sampling distance of 200 h^{-1} Mpc.

INTRODUCTION

The isotropic angular distribution of γ-ray bursts (GRBs) argues in favor of their cosmological origin. However, this link only applies if bursts sample a significant fraction of the universe. On "small" scales the universe is known to be lumpy, and we expect local anisotropies in the angular pattern of GRB positions, but only if a significant number of bursts have occured close enough to trace these spatial inhomogeneities do we expect to find significant deviations from isotropy. The brightness distribution of bursts in the BATSE sample (8,12) suggests a maximum sampling redshift of $z_{max} \sim 1$, while the nearest bursts would originate at a redshift $z_{min} \sim 0.1$ (4,9,27). While the probability of observing bursts from distances less than a few 100 Mpc is thus apparently small, neither the extent of the local inhomogeneities nor the actual minimum sampling distance of BATSE are well know. It is thus useful to test the data for anisotropies of bright bursts.

If the burster brightness distribution indeed implies a sampling range $z \sim 0.1 - 1$, there would be too few bursts closer than ~ 100 Mpc, and we would not be able to detect local cosmic inhomogeneities. There are, however, a few indications that some bursts may have been much closer than the 100 Mpc limit suggested by their brightnesn distribution. Weak evidence for an association of ultra-high energy photons (~ 100 TeV) with a GRB has been

© 1996 American Institute of Physics

claimed for the bright burst GRB910511 (15). At distances in excess of D $\sim$ 300 Mpc such ultra-high energy photons or particles from bright bursts would not reach the Earth without excessive degradation through interaction with the cosmic microwave background or the intergalactic IR photons. The second indication is the possible association of GRBs with ultra-high energy cosmic rays (UHECRs) at E $\sim 10^{20}$ eV (13; but see 24-26). At 3 10^{20} eV the particle horizon is less than 50 Mpc (7). While not conclusive, these observations suggest a closer origin of at least some GRBs, which motivates a search for nearby cosmic patterns.

On scales $\sim$ 100 Mpc the luminous universe is not uniform, but shows a well known concentration towards the supergalactic (SG) plane (5,6,18,19). Analysis of UHECR arrival directions suggests a concentration towards this plane (20), which may suggest that sources in the local universe are indeed responsible for some or all of the cosmic rays above $\sim 10^{19-20}$ eV (see also 1). We investigate the possibility that a significant number of GRBs may also originate in this inhomogeneous volume of the local universe. In any cosmological model where GRBs trace luminous matter, the anisotropies of the nearby universe must eventually be revealed.

OBSERVATIONAL CONSTRAINTS

We use the 1122 improved BATSE positions of the 3B catalog (12) to test the angular distribution for a dipole or quadrupole moment in the supergalactic frame. We compile a subset of N = 867 bursts with measured peak fluxes. We calculate the dipole moment as the average

$$D_{SG} = \langle \cos \theta_{SG} \rangle \quad (1)$$

where θ_{SG} is the angle between the direction to the GRB and the supergalactic center direction. Similarly, the quadrupole is

$$Q_{SG} = \langle \sin^2 SGB \rangle - 1/3 \quad (2)$$

where SGB is the supergalactic latitude. If bursts are isotropic, the moments should average to zero with individual samples fluctuating according to a Gaussian distribution with $\sigma_D = 1/\sqrt{3N}$ and $\sigma_Q = 2/\sqrt{45N}$ (2). The non-uniform sky exposure induces artificial moments (3), which for D_{SG} cause a bias of 0.022 and for Q_{SG} a bias of 0.010. We normalize the difference between the observed values and the expected values (from isotropy plus uneven sampling) to the 1σ statistical uncertainties and quote these ratios as the intrinsic dipole or quadrupole deviation, Δ_D and Δ_Q, in Table 1.

SG anisotropies are expected for nearby, bright sources. We assume a simple Friedmann universe with zero cosmological constant and standard candle GRBs. The fraction of bursts in the sample with redshifts less than z is f(z) = I(z,Ω)/I(z_{max},Ω), where

$$I(z, \Omega) = \int_0^z dz \, y(z)^2 \, E(z)^{-1} \, (1+z)^{-1} \quad (3)$$

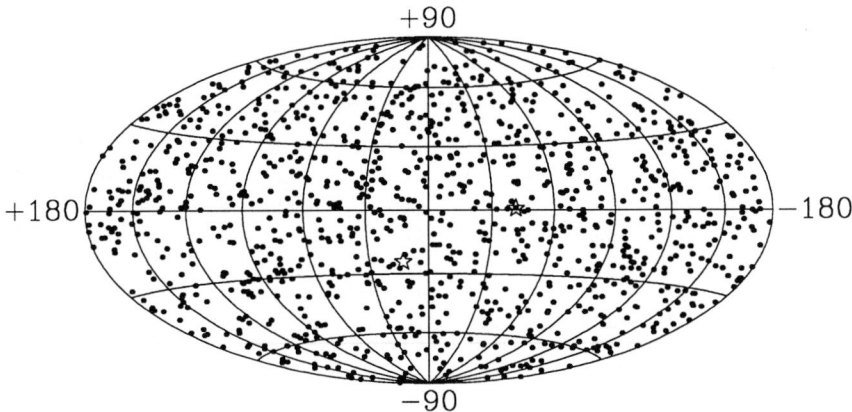

FIG. 1. Distribution of 3B bursts in supergalactic coordinates. Stars indicate two UHECRs (see text). The map is consistent with isotropy.

with E(z) and y(z) defined in Peebles (14). To what redshift do we expect supergalactic anisotropies to be evident in the angular distribution of sources? The large scale structure of nearby radio sources (18,19) suggests strong SG concentrations to $z \sim 0.02$, corresponding to $D \sim 60 \ h^{-1}$ Mpc, where h is the Hubble constant in units of 100 km s^{-1} Mpc^{-1}. We shall assume $\Omega = 1$ hereafter. The maximum sampling distance is a crucial parameter. The standard interpretation of the log N − log P curve (9,27) yields $z_{max} \sim 1$. We thus expect local (within 100 Mpc) extragalactic structure to be detectable from a fraction $f \sim 1.4 \times 10^{-4}$ of the sample. It is thus very surprising that there may have already been coincidences of UHECRs with bright GRBs.

However, the sampling redshift of the BATSE detectors is not very well determined. Cohen & Piran (4) find that the maximum redshift could be as small as $z_{max} \sim 0.7$. Rutledge et al. (17) argue that that the maximum redshift could be as small as $z_{max} = 0.8$, while the most likely range is $1 - 2$. If we assume an extreme case of $z_{max} = 0.5$, the fraction of bursts originating within 100 Mpc increases to 5×10^{-4}, corresponding to 0.4 bursts in the reduced 3B sample. We could also argue that the supergalactic concentrations may extend to much larger distances, say 300 Mpc (a redshift of ~ 0.1). As pointed out by Shaver & Pierre (18), the distribution of optically selected rich Abell clusters provides evidence for super-galactic structures beyond $z = 0.02$, and an extension to $z = 0.1$ appears to be possible (21−23). Galaxy surveys of the nearby ($z \leq 0.2$) universe clearly show large scale power around 100 h^{-1} Mpc 11 and references therein). For the extremely favorable case of such an extended structure combined with a small BATSE volume of $z_{max} = 0.5$ we obtain a fraction of 1.6%, or 14 bursts. It is clear that even an extremely favorable combination of parameters results in a small number of GRB sources that would originate from within the structured volume.

We investigate the flux-ordered sample (3B') of 867 bursts, by selecting the brightest 1/64 (N=14) of these bursts, the top 1/32 (N=27), top 1/16 (N=54),

TABLE 1. Supergalactic statistic of GRBs

Dataset	N	F_p^1	D_{SG}	Q_{SG}	Δ_D	Δ_Q	B_{SG}	Δ_B
3B	1122	–	+0.010	−0.003	−0.7	+0.8	+0.008	−0.9
3B'	867	0.05	−0.003	+0.006	−1.3	+1.6	+0.006	−1.0
top 1/2	433	0.07	−0.000	+0.031	−0.8	+2.8	−0.011	−1.9
top 1/4	217	1.65	−0.013	+0.021	−0.9	+1.5	+0.004	−0.6
top 1/8	108	3.45	−0.046	+0.050	−1.2	+2.1	−0.017	−1.1
top 1/16	54	6.56	−0.136	+0.017	−2.0	+0.7	−0.020	−0.9
top 1/32	27	12.2	−0.154	+0.028	−1.6	+0.7	−0.026	−0.7
top 1/64	14	21.0	−0.228	+0.000	−1.6	+0.1	+0.048	+0.4

top 1/8 (N=108), top 1/4 (N=217), and top 1/2 (N=433). The total 3B sample only shows deviations of less than 1σ statistical significance. The reduced sample of 867 bursts with known fluxes shows comparable deviations. Going from the 1/2 sample to the 1/64 sample the quadrupole starts out to be significantly enhanced (2.8 σ) and then reduces to a value that is consistent with isotropy. The dipole moment for the 1/2 sample is consistent with isotropy and by the time we arrive at the bright samples small anisotropes (1–2 σ) develop. If these deviations were to reflect the underlying spatial distribution of galaxies located in the superstructure, the quadrupole and dipole moments should be correlated. That this is not the case suggests that the deviations are random and not related to the geometry of the local universe.

If the sources are not only concentrated near the (0,0) direction, but also in the antipodal direction, the statistic of $\cos\theta_{SG}$ could be diluted because two excesses along opposite directions may partially compensate. The distribution of galaxies within ~ 100 Mpc resembles this kind of geometry (10). To test the hypothesis that some part of these known structures induces a bipolar anisotropy we use the quadrupole statistic

$$B_{SG} = \langle \cos^2\theta_{SG} \rangle - \frac{1}{3} \qquad (4)$$

The statistical error in this measure is identical to σ_Q. The bias in B expected from uneven sampling of the sky is +0.016. There is no evidence for a bipolar distribution of GRBs, using $(0,0)_{SG}$ as symmetry axis (Table 1).

The absence of a quadrupole moment in the supergalactic frame can be used to derive a lower limit on the sampling distance (redshift) of BATSE. If we simply model the concentration of sources towards the plane as a Watson distribution with $\kappa = -3$ (which places 1/2 of the bursts within 15.7 degrees of the SG plane; 3), detection of the quadrupole at the 2 σ level requires that more than $\sim 10\%$ of all sources are located within z_0. For standard candle, non-evolving sources this implies a lower limit on the sampling redshift of $z_{max} \sim 0.065$, i.e., detection to a minimum distances of 200 h^{-1} Mpc.

CONCLUSIONS

We conclude that there is currently no compelling evidence for a statistically significant concentration of sources in super-galactic coordinates, consistent with expectations for a cosmological burst origin. While BATSE has probably detected giant explosions near the "edge" of the universe, it is not clear how many events may have come from "local" galaxies. The arguably most promising "smoking gun" of cosmological burst models that trace the light is supergalactic anisotropy. So far we have not detected it, but we also did not yet expect it to be detectable (see also 16).

Acknowledgements. This work was supported in part by NASA under grant NAG 5-1578, and by the Deutsche Forschungsgemeinschaft under grant Ma95/3-1. We dedicate this publication to the memory of Gérard de Vaucouleurs.

REFERENCES

1. P. L. Biermann, Nucl. Phys B **43**, 221 (1995).
2. M. S. Briggs, ApJ **407**, 126 (1993).
3. M. S. Briggs, et al., ApJ **459**, 40 (1996).
4. E. Cohen & T. Piran, ApJ **444**, L25 (1995).
5. G. de Vaucouleurs, Vistas in Astr. **2**, 1584 (1956).
6. G. de Vaucouleurs, A. de Vaucoleurs, H. G. Corwin, R. J. Buta, G. Paturel, & S. Fouqué, *The Third Reference Catalog of Bright Galaxies*, Springer (1991).
7. J. W. Elbert, & P. Sommers, ApJ **441**, 151 (1995).
8. G. J. Fishman, et al., ApJS **92**, 229 (1994).
9. E. E. Fenimore, et al., Nature **366**, 40 (1993).
10. T. Kolatt, A. Dekel, & O. Lahav, MNRAS **275**, 797 (1995).
11. S. D. Landy, et al. 1996, ApJ **456**, L1
12. C. A. Meegan, et al., ApJ, submitted (1995).
13. M. Milgrom, & V. Usov, ApJ **449**, L37 (1995).
14. P. J. E. Peebles, *Principles of Physical Cosmology* Princeton Univ. Press (1993).
15. S. P. Plunkett, et al., ApSS **231**, 271 (1995).
16. J. M. Quashnock, these proceedings (1996).
17. R. E. Rutledge, L. Hui, & Lewin, W. H. G., MNRAS **276**, 753 (1995).
18. P. A. Shaver, & M. Pierre, A&A **220**, 35 (1989).
19. P. A. Shaver, Aust. J. Phys. **44**, 759 (1991).
20. T. Stanev, et al., Phys. Rev. **75**, 3056 (1995).
21. R. B. Tully, ApJ **303**, 25 (1986).
22. R. B. Tully, ApJ **323**, 1 (1987).
23. R. B. Tully, ApJ **388**, 9 (1992).
24. M. Vietri, ApJ **453**, 883 (1995).
25. E. Waxman, Phys. Rev. Lett. **75**, 386 (1995).
26. E. Waxman, ApJ **452**, L1 (1995).
27. W. Wickramasinghe, et al. ApJ **411**, L55 (1993).

BATSE Detection Biases Against "Slow Rising" Gamma-Ray Bursts

J. C. Higdon* and R. E. Lingenfelter[†]

*Joint Sciences Department
Claremont-McKenna College, Claremont, California 91711
[†]Center for Astrophysics & Space Sciences
University of California, San Diego, La Jolla, California 92093

We investigate quantitatively the effect of a selection bias against the detection of slowly rising gamma-ray bursts. Weak bursts may not be detected when the slowly rising flux from the burst contributes to the background, increasing the threshold flux needed to trigger detection. To quantify the efficacy of this selection bias we employ Monte Carlo simulations, using the light curves of the most intense bursts from the BATSE catalog as burst templates. We find that this detection bias signifcantly affects the observed distribution of gamma-ray bursts.

INTRODUCTION

The gamma-ray bursts detected by BATSE, which constitute a much larger data sample than any previously available, have provided much new information on the properties of bursts that can constrain models of the origin of the bursts. But in order to understand the nature of gamma-ray bursts and their sources, and apply meaningful constraints on burst models, both Galactic and cosmological, it is essential that we consider all burst selection effects which bias the observed properties of the bursts. Some of these selection biases have been discussed in conjunction with the BATSE measurements, but the effects of other very important biases have not yet been investigated. Here, we briefly discuss the selection bias against "slow rising" bursts which are missed because untriggered flux from the bursts is counted as "background", raising the trigger threshold. We find that these missed bursts significantly affect the observed flux distribution.

Problems of selection effects in gamma ray burst samples were first raised by Cline and Schmidt (1), who pointed out that flux triggers dramatically distort the fluence size-frequency distributions then being used to study the spatial distribution. We subsequently made a quantitative analysis of both spectral and temporal selection effects on the KONUS burst sample (3,4) resulting from the wide range of variations in the spectra and durations of individual bursts. We showed (4) that the observed deviations of the KONUS cumulative size-frequency distributions (7) of burst peak power and fluence from the $-3/2$ power law, expected for a uniform Euclidian spatial distribution, resulted

© 1996 American Institute of Physics

entirely from the burst spectral and duration variations.

Because of these problems, Schmidt, Higdon and Hueter (10) suggested the use of the $\langle V/V_{\max}\rangle$ as the least biased test of spatial uniformity, where $V/V_{\max} = (C_{\max}/C_{\min})^{-3/2}$ and $C_{\max}/C_{\min}$ is the peak detector count rate relative to the trigger threshold. Applying the $\langle V/V_{\max}\rangle$ test to the KONUS bursts showed (5) that these burst were consistent with a uniform distribution. The subsequent BATSE measurements (8), which sample bursts at roughly 4 times lower threshold, now show, however, that at twice the distance the burst sources are not uniformly distributed in space, since the observed $\langle V/V_{\max}\rangle$ of 0.329 ± 0.012 is significantly less than the value of 0.5, expected for a uniform distribution. The $\langle V/V_{\max}\rangle$ test, however, assumes a complete sample above the detection threshold and can not compensate for missed slow risers or other incompleteness.

"SLOW RISER" SELECTION BIAS

A burst triggers (8) the BATSE detector array when the count rate "excess" above the background levels in at least two detectors simultaneously exceed a threshold value of 5.5σ above the detector background count rates in the same energy band and trigger time interval, averaged over the background in consecutive 17.408-second intervals.

The "slow riser" bias results from the problems of recognizing the "excess" counts from bursts. The recognition of such an excess depends strongly on the shape of the burst light curve, or time history, because the detectors measure only the sum of the burst and detector background fluxes and the BATSE on-board trigger system *assumes* that all counts are from the background until the system triggers on a burst. Thus, there is a strong bias against the detection of "slow rising" bursts, because burst triggers do not occur when the burst count rate exceeds 5.5σ above the actual detector background, as they would in an unbiased system, but instead they occur only when the burst count rate rose fast enough compared to the 17.408 second background sampling time.

Otherwise, if the burst flux rises "slowly" enough that the burst does not trigger before burst counts appear in the 17.408-sec background interval, the next interval includes burst signal in addition to the background, raising the apparent background and hence the trigger threshold, making it harder for the burst to trigger, and if the burst flux continues to rise, but still doesn't rise fast enough to trigger, the background interval advances in 17.408-sec steps further adding even more signal and raising the threshold even higher. In this way bursts that are relatively "slow" risers can fail entirely to trigger the system and be missed, even if their peak fluxes are well above the threshold for the actual detector background. This is particularly important since 17 sec also happens to be the median duration (T90) of the triggered bursts (8).

A clear example of such missed bursts is the first, and most intense, peak of the multipeaked burst GRB911208 (trigger no. 1152) whose combined time history for the two brightest detectors is shown in Figure 1 from the First

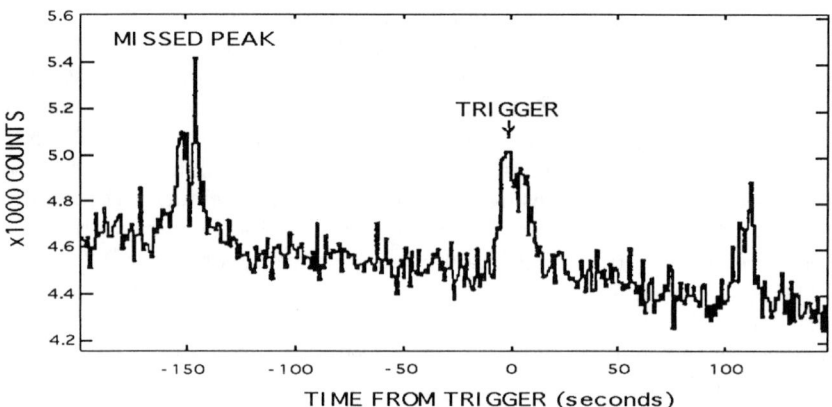

FIG. 1. An example of missed "slow" rising bursts. The combined time history of the multipeaked burst GRB911208 (trigger no. 1152) from the First BATSE Catalog shows that the on-board trigger missed in the "slow" rising first, and brightest peak, occurring 144 seconds prior to triggering $t = 0$ on the faster rising, but weaker second peak.

BATSE Catalog (2). As can be seen, BATSE only triggered ($t = 0$) on the second peak of the burst which rose more rapidly than the brighter first peak at $t = -144$ seconds. We found that this burst peak would in fact trigger ($C_{max}/C_{min} > 1$) the BATSE array only 65% of the time and be missed ($C_{max}/C_{min} < 1$) the remaining 35% of the time. We also found that there was a large variation in the apparent C_{max}/C_{min} of the burst peak ranging over a factor of 1.6 from 0.88 to 1.44. This in turn gives an arbitrary selection-dependent variation of a factor of 2 in the apparent V/V_{max}, depending solely on the chance occurrence time of the burst peak with respect to the background integration time. Since the third peak would also not trigger, this burst was detected only because the rise time in the second peak was short compared to the 17.408-second background integration time. There are, of course, many untriggered bursts that did not have this second chance.

Thus, we see that there is a strong selection bias against "slow rising" bursts which are missed because untriggered flux from the bursts is counted as "background" raising the trigger threshold, and that these missed bursts significantly affect the observed flux distribution.

To make a quantitative determination of the effects of the "slow riser" trigger bias, we are analyzing the raw count data of the gamma-ray bursts which have been obtained by the BATSE detectors. In particular, we use the continuous 1024 ms averaged DISCLA data in the four energy channels from the triggering (second-brightest) detectors for each burst, because these are the only continuous data on any of the trigger time scales that cover a significant time before the trigger. In't Zand & Fenimore (6) have studied the Malmquist bias using with 64 ms averaged data after the trigger, but that data does not cover the necessary pretrigger history to test the slow riser bias.

We have taken a sample of the flux time histories of the brightest, and hence best defined, BATSE bursts as templates for those of all bursts. These time histories are determined from the total count rates by subtracting background models fit to the count rates well before and after the bursts. We then draw a large random set of time histories from this sample, each of which we normalize to peak count rates, $C_{\max}$, drawn from both homogeneous and nonhomogeneous burst model populations, for values that exceed the threshold value of $C_{\min}$ determined from the model fit of the actual detector background. We then apply the BATSE onboard trigger criteria, including truncations and round-ups (Chip Meegan, private communication), to determine whether or not, each of the bursts in this set, which should all be detectable were it not for the slow riser bias, would in fact actually trigger the onboard system.

In this way we can quantitatively measure the missed burst fraction and its dependence on $C_{\max}/C_{\min}$. Because the time history sample was itself drawn from bursts already selected using a biased trigger, we have also reversed the count histories to provide an alternative test sample that should set a limit on the sample bias itself. In addition, we have included the effects of the Malmquist bias. We can also stretch the count histories to include time dilation that is expected in cosmological models, although we have not done so in the preliminary analysis discussed here. Time stretching, nonetheless, is particularly important for the slow riser biases because it leads to increased losses of bursts by slowing all of the rise times.

PRELIMINARY RESULTS

From a preliminary analysis of the slow riser bias using the DISCLA data for a sample of the brightest 1B Catalog bursts (2), we find that $\sim 19 \pm 6\%$ of the bursts above the actual background threshold are missed because of the slow riser bias. In the BATSE team's off-line search for missed bursts in a 40 day sample period (9) also found $13 \pm 10\%$ missed bursts during the regular (non-readout) operating time, which is consistent with our preliminary analysis. This effect, however, was not included in the corrections applied to the peak flux Log$N$–Log$P$ differential distributions shown in Figs. 11 to 13 of the 1B catalog (2). The latter figure for the 1024 ms trigger, is reproduced here as Fig. 2, showing the BATSE data as crosses, the BATSE angular corrections as open diamonds, and our additional preliminary slow riser and Malmquist bias corrections as black diamonds.

As can be seen these additional corrections significantly increase the number of bursts with fluxes below 1 photon cm^{-2}s^{-1}. We see that the corrected data might suggest that the weaker bursts between 0.2 and 2.0 photon cm^{-2}s^{-1} may actually be homogeneously distributed with a $-3/2$ power, and that the nonhomogeneity may result from an "excess" of bright bursts rather than a "lack" of weak bursts as it has been assumed. Clearly we need to carry out a more thorough analysis of this effect.

This preliminary analysis shows the corrections may significantly alter the

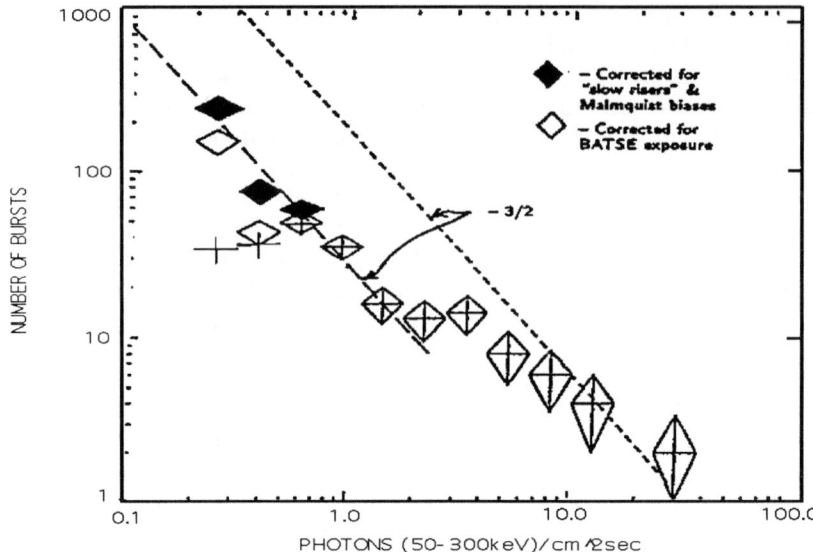

FIG. 2. Differential distribution of BATSE peak flux $\text{Log}N$–$\text{Log}P$ for the 1024 ms trigger from the 1B Catalog with our preliminary corrections for the "slow riser" and Malmquist biases, suggesting most burst are from a uniform, possibly nearby, population.

constraints on nonuniform models, such as Galactic halo and nearby disk populations, and they also have important consequences for cosmological models.

Acknowledgements. We thank NASA for support under grants NAG 5-1597 (R.E.L.) and NAG 5-2010 (J.C.H.). J.C.H. also thanks Todd Richmond for technical assistance and encouragement.

REFERENCES

1. T. Cline & W. Schmidt, Nature **266**, 749 (1977).
2. G. Fishman, et al., Astrophys. J. Supp. **92**, 229 (1994).
3. J. C. Higdon & R. E. Lingenfelter, Proc. 19th ICRC **1**, 37 (1985).
4. J. C. Higdon & R. E. Lingenfelter, Astrophys. J. **307**, 197 (1986).
5. J. C. Higdon & M. Schmidt, Astrophys. J. **355**, 13 (1990).
6. J. in't Zand, & E. Fenimore, in Gamma-Ray Bursts, eds. G. J. Fishman, J. J. Brainerd and K. Hurley, AIP Conf. Proc. **307**, 692 (AIP, New York, 1994).
7. E. Mazets, et al., Astrophys. Space Sci. **80**, 3 (1981).
8. C. Meegan, et al., Second BATSE Burst Catalog, on-line at GROSSC. (1994).
9. B. Rubin, et al., in Compton Gamma-Ray Observatory, eds. M. Friedlander, N. Gehrels and D. J. Macomb, AIP Conf. Proc **280**, 719 (AIP, New York, 1993).
10. M. Schmidt, J. C. Higdon, & G. Hueter, Astrophys. J. **329**, L85 (1988).

Reconciling GRB Time Dilation Measurements to the Brightness Distribution in Standard Cosmology

John M. Horack[†], Robert S. Mallozzi[‡], Thomas M. Koshut[‡]

[†]*NASA/MSFC, ES-84, Huntsville, AL, 35812*
[‡]*Department of Physics, University of Alabama in Huntsville, Huntsville, AL, 35899*

We present a means to assess the consistency of time dilation and energy shifting measurements with both the brightness distribution of BATSE and various Friedmann cosmological models. We utilize the methodology to assess the compatibility of a recent claim (1) of a time dilation factor of 2.25 from bursts separated by a factor of 21 in brightness. We show that these results are *not* compatible with the brightness distribution of BATSE, assuming a non-evolving population of standard-candle bursts embedded in an Einstein-de Sitter universe. Recent revisions of this value (2) downward to 1.75, which help to alleviate the discrepancy, however are still somewhat marginal in light of the energy-width corrections that need to be applied.

LUMINOSITY DISTANCE AND TIME DILATION EQUATIONS

Peak Flux Ratios

Assuming that bursts are standard candles, one can easily determine the ratio of observed peak fluxes for two sets of bursts at redshifts z_{P_1} and z_{P_2} ($z_{P_1} > z_{P_2}$):

$$\frac{P_2}{P_1} = \left(\frac{r(z_{P_1})}{r(z_{P_2})}\right)^2 \left(\frac{1+z_{P_1}}{1+z_{P_2}}\right) \frac{L(z_{P_2})}{L(z_{P_1})}. \tag{1}$$

Evaluation of (1) requires that a model cosmology be adopted to determine $r(z)$ and that a spectral form be utilized in evaluating $L(z)$, the amount of the source's bolometric luminosity that is accessible to the detector with finite energy window E_1 to E_2.

For the cosmological model, we adopt the familiar Einstein-de Sitter universe. For the spectral form of the burst, we utilize the function

$$\phi(E) = A_\circ E^{-1} \exp(-E/kT) \tag{2}$$

used for bursts elsewhere in the literature (3–5) where E is the energy in keV. Utilizing (2), the luminosity can be expressed as

© 1996 American Institute of Physics

$$L(z) = A_\circ \left(\ln(6) + \sum_{j=1}^{\infty} \frac{(-1)^j}{j\, j!} (kT)^{-j}(1+z)^j (300^j - 50^j) \right), \qquad (3)$$

where we have expanded the exponential in an infinite series and performed the integration term by term using the BATSE energy window of 50–300 keV.

Time Dilation Equations

In addition to having different peak fluxes, these two sets of bursts will have different average durations. Neglecting effects other than pure time dilation, the ratio of the average burst durations from each sample can be expressed simply as

$$\tau_c \equiv \frac{\langle T(P_1) \rangle}{\langle T(P_2) \rangle} = \frac{1 + z_{P_1}}{1 + z_{P_2}}, \qquad (4)$$

where we have assumed no evolution with cosmic time in the intrinsic duration distribution of bursts. In practice, the situation is more complicated. The intrinsic dependence of burst duration on energy results in measured duration ratios that are less than expected from cosmology alone, and must be corrected for in order to infer the proper redshifts for the burst sources (3). It has been shown (4) that the relation between the amount of *observed* temporal stretching τ_m and the value expected from cosmology alone τ_c in (4) has the simple form

$$\tau_m = (\tau_c)^{1-w} \equiv \left(\frac{1 + z_{P_1}}{1 + z_{P_2}} \right)^{1-w}, \qquad (5)$$

with $w \approx 2/5$ in these previous analyses (3,6).

Whether one chooses to incorporate this "width correction" ($w \neq 0$) or not ($w = 0$), we have obtained a means to determine the unique set of redshifts z_{P_1} and z_{P_2} which result from an observed peak flux ratio P_2/P_1 given by (1) and a measured temporal stretching factor τ_m from (5).

SOLUTIONS TO THE EQUATIONS

The solutions to the simultaneous equations for the peak flux ratio and time dilation or energy shifting factor are illustrated graphically in Figure 1. We concentrate first on Figure 1a. On the x-axis of this figure, we label the values of the redshift for the dim set of bursts z_{P_1}. The y-axis denotes the values of redshift for the brighter set of bursts z_{P_2}. There are two sets of curves presented in this figure.

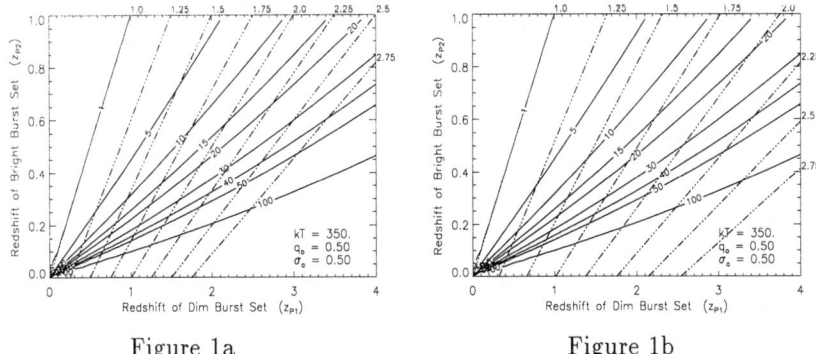

Figure 1a   Figure 1b

The solid curves are related to Equation (1). These curves denote the values of z_{P_1} and z_{P_2} which yield a peak flux ratio P_2/P_1 whose value is indicated by the label associated with the line. The dot-dashed curves are related to Equation (5). These curves indicate the values of redshift (z_{P_1}, z_{P_2}) which yield the value τ_m indicated by the label at the top and right hand side of the figure. Although generated using a curved spectral form (2), the figures are nearly identical to those generated using a power-law spectral form with index -1.8.

The differences between Figures 1a and 1b are the values of w that are assumed for the energy-width correction to the time dilation measurement. In Figure 1a, the value of $w = 0$, i.e. there is no width correction applied. This figure may also therefore be used for energy shifting measurements (6) which require no correction for the energy-dependent widths of burst profiles. Figure 1b employs a correction of $w = 0.2$.

A related work (7) investigates in significantly greater detail the effects of varying kT, the spectral form, and the cosmological model.

OBSERVATIONS

We now consider one claim of time dilation (1) subsequently referred to as N95. A value $\tau_m \sim 2.25$ is found between bright and dim bursts, with the intensity measured in instrumental units of counts s^{-1}. This value has subsequently been revised downward (2) to ~ 1.75. Either value is suitable for illustrating the comparison, however different conclusions will be obtained depending on the value used. We shall examine both values, and assess the compatibility of the resulting redshifts with the brightness distribution seen by BATSE assuming a non-evolving, standard candle population.

The bursts in N95 were divided into three intensity groups: bright, dim, and dimmest. We have obtained the peak fluxes in units of photons cm^{-2} s^{-1} for these three groups of bursts, computed on a timescale of 0.256 s. We have computed the average peak fluxes for each of the three intensity groupings, as well as the combination "dim + dimmest" group of bursts. These are shown in Table 1, along with the minimum P, maximum P, and the error in the mean. We note that most of the "bright" bursts from N95 possess peak fluxes which place them on or very near the portion of the BATSE intensity distribution which displays a consistency with a $-3/2$ logarithmic slope, (8) indicative of a homogeneous distribution of sources in Euclidean (nearby) space. This is a key result for the analyses to follow.

Table 1 – Peak Flux data of bursts used in N95 photons cm^{-2} s^{-1} in 0.256 s intervals					
	Bursts	$\langle P \rangle$	$P_{\min}$	$P_{\max}$	$\sigma_{\langle P \rangle}$
Bright	43	16.20	4.21	58.28	1.93
Dim	59	0.93	0.33	2.82	0.04
Dimmest	51	0.61	0.38	0.95	0.02
Dim + Dimmest	110	0.78	0.33	2.82	0.03

In Table 2 of N95, measured time dilation factors are shown for the two methods used and three different energy regimes. These factors represent the ratio of the average duration found for the combined "dim + dimmest" sample to the average duration of the "bright" sample. The average of the time dilation factors in Table 2 of N95 is 2.259 ± 0.144. The uncertainty reflects the spread in measured T values computed simply as $\sqrt{\langle T^2 \rangle - \langle T \rangle^2}$.

From Table 1, the peak flux ratio between the "dim + dimmest" bursts and the "bright" bursts is given by 20.7 ± 2.5. Combining this value with the derived $\tau_m = 2.259 \pm 0.144$, we can examine Figure 1a (first assuming no width correction) to assess the compatibility of the redshift values in the Einstein-de Sitter cosmology. We observe that a peak flux separation of ~ 21 and $\tau_m = 2.26$ results in redshifts of $z_{0.78} \approx 2.65$ and $z_{16.2} \approx 0.62$.

These redshift values are inconsistent with the observed BATSE peak flux distribution assuming non-evolving, standard candle bursts (7). One cannot easily reconcile the near-Euclidean homogeneity displayed by the bright bursts (8) with a redshift of 0.62, which itself results in a strong deviation from $-3/2$ in the slope of the integral distribution. Indeed, in one consistent model (7) a redshift of $z = 0.62$ is consistent with that of bursts possessing a peak flux of ~ 1.75 photons cm^{-2} s^{-1}, significantly below our value of 16.2. Only by modifying *both* the measured time dilation factor and peak flux ratio by nearly 3 error-bars can one then obtain redshifts that are approximately consistent with the observed brightness distribution of non-evolving bursts.

The revised (2) τ_m value of 1.75 can also be examined using Figure 1a, resulting in redshifts of $z_{0.78} \approx 1.35$ and $z_{16.2} \approx 0.35$. These are less inconsistent with the observed brightness distribution assuming no evolution than the values obtained with a $\tau_m = 2.25$, however there still is a slight discrepancy, especially at the dim end. In our consistent model, (7) a redshift of 0.35 corresponds to a peak flux of approximately 6, still below the measured value of 16.2 for these bursts. However, one may argue that these redshifts are consistent within the statistics of the newly-revised τ_m value of 1.75.

We now address the possibility of energy-width corrections (the w term in Equation 5). Performing a similar comparison as was done above, however this time utilizing Figure 1b (which incorporates a w value of 0.2) we see that the discrepancy noted before is enhanced. Indeed the redshifts compatible with a time dilation measurement of 2.25 obtained from sets of bursts separated by a factor of ~20 in peak flux are exceptionally large (3). A non-zero value for w only heightens the discrepancy between redshifts obtained from the brightness distribution and those found using the time-dilation measurement and peak flux ratios. The larger the width correction, the more discrepant the redshift values become. With a minimal energy-width correction, the revised τ_m value of 1.75 is now clearly at odds with the non-evolving burst brightness distribution.

SUMMARY

Evolution in the comoving rate density of bursts as strong as $\sim (1+z)^2$ can serve as one simple explanation of the incompatibility (7). Substantial energy corrections eventually require the incorporation of a finely-tuned, contrived evolutionary form to generate a $-3/2$ logarithmic slope in the brightness distribution over a large range of brightnesses and redshifts which would therefore have no relation to a homogeneous distribution of bursts in nearby space.

There are other possibilities which can explain the apparent inconsistency, among them the presence of an inverse luminosity-duration correlation. If such a correlation is present, one need only attribute $\sim 30\%$ of the measured τ_m to this cause, leaving the remaining $\sim 70\%$ to cosmological time dilation, in order to obtain a consistent result with a non-evolving population (7).

If the cosmological time-dilation results (1,2) are substantially correct, one can draw an important conclusion regarding the nature of GRBs. Taken in conjunction with the observed brightness distribution, these time dilation results indicate that bursts likely have evolution in their comoving rate density and/or luminosity function at least as strong as $(1+z)^2$, or a systematic increase in the measured time dilation factor (e.g., from an inverse luminosity-duration correlation) is present at a level not exceeding $\sim 30\%$.

REFERENCES

1. J. P. Norris, et al., ApJ. **439**, 542 (1995).
2. J. P. Norris, These Proceedings, (1996)
3. E. E. Fenimore & J. S. Bloom, ApJ. **453**, 25 (1995)
4. D. L. Band, et al., ApJ. **413**, 281 (1993)
5. E. E. Fenimore, et al., Nature **366**, 40 (1993)
6. P. Meszaros & A. Meszaros, ApJ. **449**, 9 (1995).
7. J. M. Horack, R. S. Mallozzi, & T. M. Koshut, ApJ. **466**, in press, (1996)
8. C. A. Meegan, et al., Nature **351**, 143 (1992)

Analytic Constraints on Gamma–Ray Burst Luminosity Functions

John M. Horack[†], Charles A. Meegan[†],
Jon Hakkila[‡], A. Gordon Emslie[*]

[†] *NASA/MSFC, ES-84, Huntsville, AL, 35812*
[‡] *Department of Physics, Mankato State University, Mankato, MN, 56002*
[*] *Department of Physics, University of Alabama in Huntsville, Huntsville, AL, 35899*

The brightness distribution of detected gamma–ray bursts enables constraints to be placed on their luminosities; specifically, analyses by previous authors have demonstrated that the range of luminosity from which approximately 90% of the observed bursts are drawn is very likely less than an order of magnitude. It has also been demonstrated that when power–law functional forms with infinite ranges of luminosity (a.) from 0 to L_{max} or (b.) from L_{min} to ∞ are employed, it is necessary (but perhaps not sufficient) to restrict the power–law index $\beta < 1.8$ and $\beta > 2.5$, respectively, in order to obtain consistency with the observed brightness distribution. Similarly, these previous works have demonstrated that when $1.8 < \beta < 2.5$, the range of the burst luminosity function must be restricted and cannot be arbitrarily large.

We present here analytic derivations of these results valid for *any* spatial density function and for both Euclidean and cosmological scenarios. This analysis shows that the limiting values of β are simply the maximum and minimum absolute value logarithmic slopes of the brightness distribution. We also demonstrate that this result is consistent with those requiring a narrow range in luminosity for a majority of the detected bursts.

ANALYSIS

L_{min}–dominated Luminosity Functions

We begin by exploring L_{min}–dominated luminosity functions (1) of the form

$$\phi(L) = A_o L^{-\beta} \text{ for } L_{min} \leq L \leq \infty, \tag{1}$$

and are zero outside this range. The integral brightness distribution $\mathcal{N}(>P)$ can be written

$$\mathcal{N}(>P) = 4\pi \int_0^\infty n(r) \, r^2 \int_{4\pi r^2 P}^\infty \phi(L) \, dL \, dr. \tag{2}$$

© 1996 American Institute of Physics

Inserting the luminosity function in (1), we now differentiate equation (2) twice with respect to P, obtaining

$$\frac{d\ln N(P)}{d\ln P} + \beta = \frac{A_\circ P^{-5/2}}{2\sqrt{4\pi}N(P)} L_{\min}^{5/2-\beta} n\left(\sqrt{\frac{L_{\min}}{4\pi P}}\right). \qquad (3)$$

The right hand side of the above equation is non-negative for all P. Consequently,

$$\beta \geq -\frac{d\ln N(P)}{d\ln P}. \qquad (4)$$

for all P.

From the BATSE differential brightness distribution $N(P)$, the steepest logarithmic slope (in absolute value) is $\sim 5/2$, occurring at large values of P where the distribution is consistent with a homogeneous source population (2). At all other values of P, the absolute value of the logarithmic slope is less than (or equal to) $5/2$. The above equation therefore shows that in order for an $L_{\min}$-dominated luminosity function with a formally infinite luminosity range to possibly be consistent with the observed brightness distribution, the value of β must be ≥ 2.5.

$L_{\max}$-dominated Luminosity Functions

In the $L_{\max}$-dominated case, the luminosity function has the form

$$\phi(L) = A_\circ L^{-\beta} \text{ for } 0 \leq L \leq L_{\max} \qquad (5)$$

and is zero outside this range. With (the relatively flat) luminosity functions of this form that are consistent with the brightness distribution of BATSE, dim bursts are undetectable because of the small volume of space within which they must be situated in order to have a brightness above the BATSE threshold. Most of the observed bursts are drawn from a portion of $\phi(L)$ where $L \approx L_{\max}$, thus the nomenclature "$L_{\max}$-dominated" (1). Incorporating this luminosity function into (2) and again differentiating twice with respect to P, we obtain

$$\frac{d\ln N(P)}{d\ln P} + \beta = \frac{-A_\circ P^{-5/2}}{2\sqrt{4\pi}\,N(P)} L_{\max}^{5/2-\beta} n\left(\sqrt{\frac{L_{\max}}{4\pi P}}\right). \qquad (6)$$

Since the right hand side of this equation is negative definite, we must have

$$\beta \leq -\frac{d\ln N(P)}{d\ln P} \qquad (7)$$

for a luminosity function of this form to be consistent with the brightness distribution. The minimum absolute value of the logarithmic slope in the observed $N(P)$ is approximately 1.8, occurring at low values of P (2). We therefore conclude that consistent $L_{\max}$-dominated intrinsic luminosity functions with formally infinite ranges $L_{\max}/L_{\min}$ must have $\beta \leq 1.8$.

Range–dominated Luminosity Functions

The remaining luminosity function we examine is the so-called "range–dominated" case (1) where both $L_{\max}$ and $L_{\min}$ are specified, i.e.

$$\phi(L) = A_o L^{-\beta} \text{ for } L_{\min} \leq L \leq L_{\max}. \tag{8}$$

From the previous two cases that have been examined, it is apparent that one is required to specify an $L_{\max}$ and $L_{\min}$ only for luminosity functions with β between 1.8 and 2.5. Otherwise the range can be formally infinite.

Performing the analysis as before, utilizing the range–dominated $\phi(L)$, we obtain the expression

$$\frac{d\ln N(P)}{d\ln P} + \beta = \frac{A_o P^{-5/2}}{2\sqrt{4\pi}N(P)} \left[L_{\min}^{5/2-\beta} n\left(\sqrt{\frac{L_{\min}}{4\pi P}}\right) - L_{\max}^{5/2-\beta} n\left(\sqrt{\frac{L_{\max}}{4\pi P}}\right) \right]. \tag{9}$$

The sign of the right–hand side is no longer obvious, and indeed depends on the form of the radial distribution $n(r)$. Since in the previous sections we noted that the range of luminosity can be formally infinite for β outside the interval 1.8 to 2.5, we choose an intermediate value of 2 to illustrate the behavior in this regime. With $\beta = 2$, the left–hand side can be either positive or negative depending on the value of P at which the equation is evaluated. The sign of the right–hand side is determined by the sign of the quantity in the braces. In this expression, the density function is evaluated at a distance $\sqrt{L_{\min}/4\pi P} \equiv r_{\rm sm}$ and at the larger distance $\sqrt{L_{\max}/4\pi P} \equiv r_{\rm lg}$.

For low–P, where the left–hand side is approximately $-1.8 + 2 = 0.2$, with any density function $n(r)$ that decreases with distance, the requirement that the bracketed quantity sometimes be positive limits the possible values of $L_{\max}$ relative to $L_{\min}$. If $L_{\max}$ is too large relative to $L_{\min}$, the bracketed expression will be negative even for low P. Therefore, for this intermediate value of $\beta = 2$, the range of luminosity cannot be arbitrarily large. Similar arguments hold for *any* value of β between 1.8 and 2.5, since the right–hand side must be both positive and negative (for different values of P) for these values of β. Hence the range of gamma–ray burst luminosity is constrained for any power–law in the range $1.8 < \beta < 2.5$.

It is perhaps important to emphasize that this restriction is levied on the range of luminosity that bursts may possess *regardless of their detectability*, whereas the restriction of previous analyses (3,4,5) confining 90% of the detected bursts to a factor of 10 in luminosity applies (self–evidently) to those bursts that are actually detected. Furthermore, it is clear that a restriction on the range of luminosity that bursts may possess *de facto* constrains the range of luminosity that can be found in any detected set of bursts.

The numerical value of the upper–limit obtained here depends on both how rapidly the left–hand side of the previous equation changes sign as P changes, and on the functional form of $n(r)$. To illustrate this, again assume $\beta = 2$. At low P, from observations of the brightness distribution, we know that

$$\frac{d\ln N(P)}{d\ln P} + 2 > 0. \tag{10}$$

Therefore it must also follow that

$$K^{0.5} \leq \frac{n(r_{\rm sm}(P))}{n(r_{\rm lg}(P))}. \tag{11}$$

We observe above the interplay between the drop–off in the number density $n(r)$ and the range of luminosity $K \equiv L_{\max}/L_{\min}$. If the radial density function exhibits a steep drop–off, i.e. $n(r_{\rm sm}) \gg n(r_{\rm lg})$, the range of luminosity can, in general, be larger and still preserve the inequality. Conversely, if the drop–off in the radial distribution is more gradual, the range of luminosity K is more severely constrained. The maximum range in luminosity is permitted for the most rapid drop–off in the radial distribution, and the minimum range in luminosity is found where the depletion of sources with increasing distance is least severe.

These two factors, combine to produce the break from one logarithmic slope to the other that is observed in the BATSE brightness distribution. The break they produce occurs over a range of brightness that is of order 10. We therefore observe here the basis for the independently–found conclusions (3,4,5) that the range of luminosity for a significant majority of the observed bursts cannot exceed a factor of ~ 10.

Application to Cosmological Scenarios

In this section, we demonstrate that the previous conclusions can also be obtained for cosmological distributions of bursts. We analyze the $L_{\min}$-dominated luminosity funciton. Utilizing this $\phi(L)$, it can be shown[3] that the differential brightness distribution $N(P)$ can be written

$$N(P) = (4\pi)^2 A_\circ P^{-\beta} \times$$
$$\int_{z_*(P)}^{\infty} \frac{n_c(z) r(z)^2}{(1+z)\sqrt{1-kr(z)^2}} \left|\frac{dr}{dz}\right| \left(4\pi d_{\rm l}(z)^2 (1+z)^{\alpha-2}\right)^{-\beta} d_{\rm l}(z)^2 dz \tag{12}$$

where $d_{\rm l}(z)$ is the redshift–dependent luminosity distance, k is the curvature of the cosmological model, and $n_c(z)$ is the co–moving rate density of bursts. The quantity $z_*(P)$ is the redshift at which a burst with luminosity L and power–law spectral index α will produce a measured brightness P.

If we again differentiate this expression with respect to P, we obtain a similar result as before, however with a more complicated positive definite expression on the right–hand side of the equation. This leads to the result

$$\beta \geq -\frac{d\ln N(P)}{d\ln P}. \tag{13}$$

We have noted before that the largest absolute value of the logarithmic slope observed in the BATSE brightness distribution is 2.5. Based on this, we conclude that for $L_{\min}$-dominated luminosity functions with formally infinite ranges of luminosity, β must be larger than 2.5, a conclusion identical to that obtained in the Euclidean case. This result is independent of the cosmological model or the comoving rate density of bursts $n_c(z)$. Similar mathematics applied to the $L_{\max}$-dominated $\phi(L)$ yield the result that $\beta \leq 1.8$ in these cases.

Analyses in the cosmological scenario involve more complicated expressions, which make deduction of conclusions in the range–dominated case more difficult. However, since the range of burst luminosity is not necessarily limited for $\beta \leq 1.8$ or $\beta \geq 2.5$, if the range of luminosities is limited, it will be for β between 1.8 and 2.5.

CONCLUSIONS

Each of these results have been found in previous works, which utilized various different computational techniques and methodologies. What we have demonstrated here is that these results can be straightforwardly derived from the mathematical expression of the GRB brightness distribution, and that the constraints are a direct consequence of the maximum and minimum logarithmic slope of the observed $N(P)$ distribution. These conclusions are furthermore independent of the specific form of the bursts' radial distribution and do not depend on whether GRBs are embedded in a Euclidean or cosmological geometry. This both lessens the requirement for extensive computer simulations in obtaining these general constraints and enhances our physical understanding of their origins.

REFERENCES

1. J. Hakkila, et al., ApJ in press, (1996).
2. C. A. Meegan, et al., Nature **355**, 143 (1992).
3. J. M. Horack, et al., ApJ in press, (1995).
4. A. Ulmer & R.A.M.J. Wijers, ApJ **439**, 303 (1995).
5. A. Ulmer, R.A.M.J. Wijers, & E.E. Fenimore, ApJ **440**, L9 (1995).

The Compatibility of Friedmann Models With Observed Properties of GRBs and a Large Hubble Constant

John M. Horack[†], A. Gordon Emslie[‡], Thomas M. Koshut[‡], Robert S. Mallozzi[‡], Charles A. Meegan[†]

[†] NASA/MSFC, ES-84, Huntsville, AL, 35812
[‡] Department of Physics, University of Alabama in Huntsville, Huntsville, AL, 35899

> In this work, we investigate the implications of the constraints provided by both measurements of the Hubble Constant H_o and by the brightness distribution of GRBs. We show that the range of cosmological models that can be consistent with the GRB brightness distribution, H_o near 80 km s^{-1} Mpc^{-1}, and age of the universe near 15 Gyr is constrained significantly, largely independent of a wide range of assumptions regarding the evolutionary nature of the burst population. Specifically, low–density, $\Lambda > 0$ cosmological models with deceleration parameter in the range $-1 < q_o < 0$ and density parameter σ_o in the range ≈ 0.05–0.15 ($\Omega_o \approx 0.1 - 0.3$) are strongly favored.

ANALYSIS

For Friedmann dust cosmologies that admit a non–zero cosmological constant Λ, the age of the universe τ_o can be written as

$$y(\sigma_o, q_o) \equiv \tau_o H_o = \int_0^\infty \frac{(1+z')^{-1}\, dz'}{\sqrt{2\sigma_o(1+z')^3 + (1+q_o-3\sigma_o)(1+z')^2 + (\sigma_o - q_o)}}, \quad (1)$$

where σ_o is the density parameter and q_o is the deceleration parameter.

Given a reliable lower limit to τ_o, we can use Equation 1 to obtain the region in the (σ_o, q_o) plane that is consistent with both the estimated age of the universe τ_o and the measured value of H_o. As in the work of Leonard & Lake (1), we incorporate the Hubble Constant measurements of Pierce et al. (2) (87 ± 7 km s^{-1} Mpc^{-1}) and Freedman et al. (3) (80 ± 17 km s^{-1} Mpc^{-1}), as well as the age measurements of Chaboyer (4) ($\tau_o = 15.5 \pm 4$ Gyr) and VandenBergh (5)($\tau_o = 16.5 \pm 2$ Gyr).

Utilizing the inverse–variance–weighted means for both H_o and age of the universe, we find that the best estimate of the dimensionless product $y_o = \tau_o H_o$ is

© 1996 American Institute of Physics

$$y_o = 1.43 \pm 0.19. \tag{2}$$

We note that the only cosmological models consistent with this value of y_o are accelerating ($q_o < 0$). This result, which applies for all $y > 1$, is well-known. What remains to be determined is whether the brightness distribution of cosmological gamma–ray bursts observed by BATSE is also consistent with any of the models that correspond to the measured value of y_o.

It is straightforward to construct a model differential brightness distribution $N_M(P)$ from a set of cosmological parameters (σ_o, q_o), spectral index α, burst comoving rate density function $n_c(z)$, and limiting redshift $z_{0.5}$. This model distribution can be tested against the observed brightness distribution using the χ^2 test. One thereby obtains both a value for the reduced χ^2 statistic and a probability $P(>\chi^2)$ that a random sample taken from the model would result in a distribution of equal or greater disparity than the observed data. Low values of $P(>\chi^2)$ therefore indicate the model is unlikely to be correct, as it fails to reproduce a distribution similar to the observed $N(P)$.

COMBINING THE MEASUREMENTS

We previously noted that given the observations of H_o and estimates of the age of the universe, their product is measured to be $y_o \pm \sigma_{y_o} = 1.43 \pm 0.19$. Assuming a (minimal–information) Gaussian distribution for the distribution variable y', it is straightforward to compute for each point in the (σ_o, q_o) plane a probability value

$$\mathcal{P}(|y' - y_o| > |y - y_o|) = \mathrm{erfc}\left(\frac{y - y_o}{\sigma_{y_o}}\right), \tag{3}$$

which represents the probability that the actual value of y' found in nature is equally or more discrepant from the measured value y_o than the calculated value $y(\sigma_o, q_o)$. Points in the (σ_o, q_o) plane which yield very low probabilities $\mathcal{P}$ are unlikely to represent reality as they predict a value y that is inconsistent with the measurement y_o at a high level.

We adopt the reasonable assumption that the measurements of $\tau_o H_o$ obtained in part from observations carried out with the HST are independent of the measured brightness distribution of gamma–ray bursts obtained by BATSE on the GRO. A joint probability P_j can therefore be obtained for each point in the (σ_o, q_o) plane by simply multiplying $P(>\chi^2)$ and $\mathcal{P}$, i.e.

$$P_j = P(>\chi^2) \times \mathrm{erfc}\left(\frac{y - y_o}{\sigma_{y_o}}\right). \tag{4}$$

The quantity P_j is the probability that a random sampling from the cosmological model in question would return *both* a χ^2 value more discrepant than the one obtained comparing the model to the BATSE brightness distribution *and* an actual value of y' that is farther from the measured value $y_o = 1.43 \pm 0.19$ than the y calculated from Equation 1.

We employ P_j as our statistic to assess a model's capability to reproduce observations of both the brightness distribution and value of y. As such, it is more demanding than either of the individual estimators $[P(> \chi^2)$ or $\mathcal{P}]$ alone. We stress that P_j is used here only to reject models and *cannot* be interpreted as the probability that a model is correct. Nevertheless, models which yield low values of P_j are unlikely to be the actual cosmological configuration found in nature, as they predict a brightness distribution and/or value of y that is significantly different than those observed. Conversely, models yielding large values of P_j cannot be rejected, as they reproduce both a brightness distribution and value of y consistent with observation.

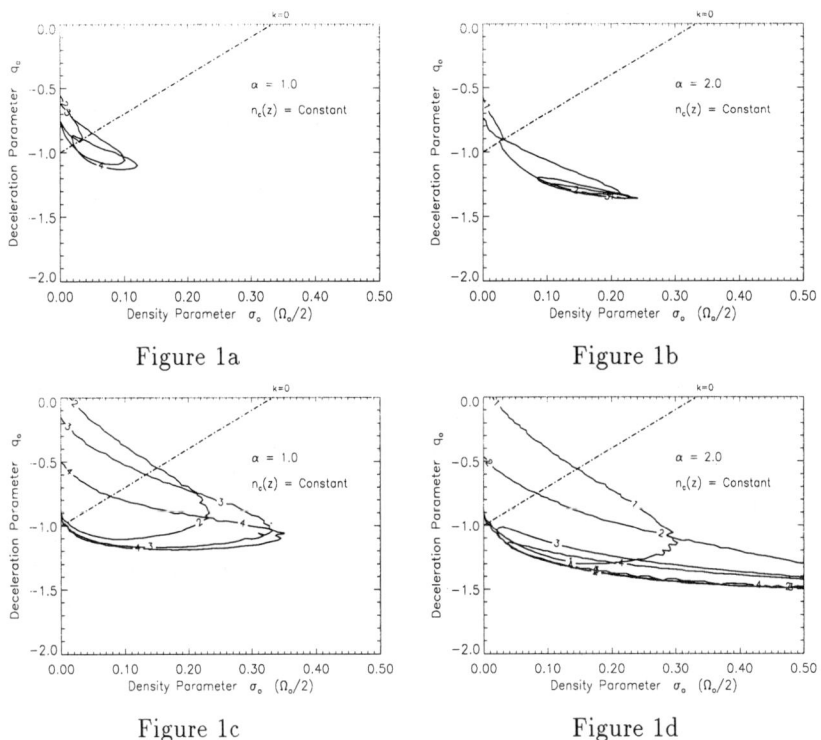

Figure 1a

Figure 1b

Figure 1c

Figure 1d

RESULTS OF COMBINING MEASUREMENTS

Figures 1a, b, c, and d show the results of combining these two probabilities for a variety of assumed cosmological models. In Figure 1a we show the level contours in the (σ_o, q_o) plane for which $P_j = 0.32$, assuming a non-evolving burst population with photon spectral index $\alpha = 1.0$. Each different limiting redshift $z_{0.5}$ that is assumed produces a different region in the plane where $P_j = 0.32$. The contour labels ('1' through '4') denote the values of these limiting redshifts.

We have also indicated with a dashed line the positions of spatially flat ($k = 0$) cosmological models, which have $(3\sigma_o - q_o - 1) = 0$. To better illustrate the regions of constant P_j, we have also truncated the x-axis at a value of $\sigma_o = \Omega_o/2 = 0.5$, the critical density for $\Lambda = 0$. All models shown are therefore not only accelerating, but have a sub-critical density. Figure 1b presents the same information for cosmological models in which $\alpha = 2$. In Figures 1c and 1d, the value of P_j is lowered to 0.01 for $\alpha = 1$ and $\alpha = 2$, respectively. As expected, the contours of $P_j = 0.01$ are significantly larger than those for $P_j = 0.32$. For a given point in the (σ_o, q_o) plane, we observe that higher limiting redshifts are required to obtain a particular value P_j for the $\alpha = 1$ spectral form than are required for $\alpha = 2$.

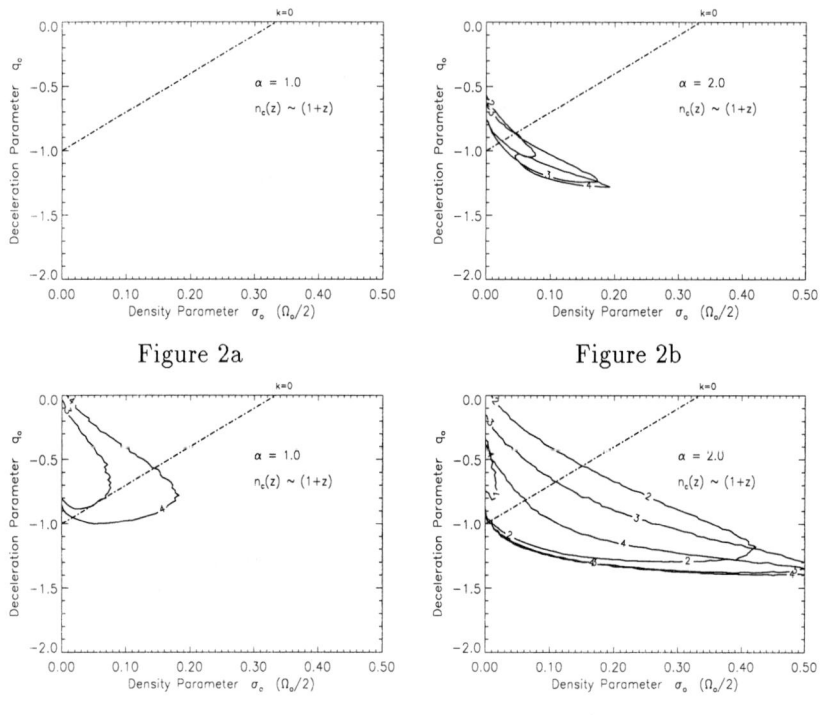

Figure 2a   Figure 2b

Figure 2c   Figure 2d

Figures 2a, b, c, and d present similar information to Figures 1, but for an evolutionary comoving rate density function $n_c(z) \sim (1 + z)$. The general forms of the figures are similar, however there is one significant difference, namely that a given value of P_j is found in the non-evolving case with lower redshifts than in the evolving case.

By inspection of Figures 1 and 2, we observe that virtually *any* cosmological model (σ_o, q_o) with a density in excess of the critical density $(\sigma_o = \Omega_o/2 > 0.5)$ will result in a joint probability P_j that is less than 0.01, *regardless of the assumptions about the burst population* (evolution, spectral index, limiting redshift, etc.). Both the observed value of y_o and the BATSE brightness distribution $N(P)$ are considerably more disparate from those that might be expected from a random sampling of these theoretical models. Exceptions are those rapidly accelerating ($q_o << 1$) models with steep burst spectra ($\alpha \sim 2$), no evolution, and limiting redshifts in excess of $z_{0.5} > 2$.

Because of this, we conclude that the combined measurements of H_o, τ_o, and the BATSE brightness distribution make cosmological models with the density parameter σ_o near 0.5 ($\Omega_o \approx 1$) extremely unlikely. A significantly more detailed account of this work can be found elsewhere (6).

CONCLUSIONS

We find that in the context of a large Hubble Constant (~ 80 km s^{-1} Mpc^{-1}) and an age of the universe near 15 Gyr, cosmological models which possess a density near $\sigma_o = \Omega_o/2 = 0.5$ are unlikely representations of reality, as they fail to produce both values of $y_o \equiv \tau_o H_o$ and a burst brightness distribution $N(P)$ which agree with the observational data. This conclusion is independent of a wide range of assumptions regarding the burst evolutionary rate density, spectral index, or limiting redshift.

Furthermore, examination of the best-fit models, i.e. those which most accurately reproduce the observed $\tau_o H_o$ value and BATSE brightness distribution, each lie in a region of parameter space where $\sigma_o = \Omega_o/2 \approx 0.05 - 0.15$, significantly less than the $\Lambda = 0$ critical density. Members of this family of models are all low density, accelerating models with a positive cosmological constant Λ, and minimal spatial curvature ($|3\sigma_o - q_o - 1|$ near zero).

Coles & Ellis (7) summarize the observational evidence favoring a low-density universe, and Ostriker & Steinhardt (8) have further argued, that not only is Ω_o in the vicinity of 0.1–0.3, but the universe also is likely to have minimal spatial curvature and a positive cosmological constant. If recent measurements of H_o are substantially correct and our estimates of the age of the universe are fairly accurate, we find that these cosmological models also best explain jointly the observed value of y_o and the BATSE brightness distribution.

REFERENCES

1. S. Leonard, & K. Lake, ApJ **441**, L55, (1995).
2. M. J. Pierce, et al., Nature **371**, 385, (1994).
3. W. L. Freedman, et al., Nature **371**, 757, (1994).
4. B. Chaboyer, 1994, private communication cited in Leonard & Lake, ApJ **441**, L55, (1995).
5. D. A. VandenBergh, in ASP Conf. Ser. 13, The Formation and Evolution of Star Clusters, ed. K. Janes (San Francisco: ASP), 183, (1991).
6. J. M. Horack, et al., ApJ, submitted, (1996).
7. P. Coles, & G. Ellis, Nature **370**, 609, (1994).
8. J. P. Ostriker, & P. J. Steinhardt, Nature **377**, 600, (1995).

The Ulysses Supplement to the BATSE 3B Catalog

K. Hurley*, T. Cline†, G. Fishman‡, C. Kouveliotou‡,
C. Meegan‡, M. Sommer*, M. Boer§, M. Niel§

*University of California, Space Sciences Laboratory, Berkeley, CA 94720-7450
†NASA Goddard Space Flight Center, Greenbelt, MD 20771
‡NASA Marshall Space Flight Center, Huntsville, AL 35812
*Max-Planck Institut für Extraterrestrische Physik, 85740 Garching, Germany
§Centre d'Etude Spatiale des Rayonnements, 31029 Toulouse Cedex, France

> The gamma-ray burst experiment aboard Ulysses detects about one event every three days. Most of these are BATSE bursts, which are found in the real-time (untriggered) data by searching the Ulysses-crossing time window. The resulting annuli of arrival directions generally intersect the BATSE error circles to define an area which is about two orders of magnitude smaller than the area of the BATSE error circle. These localizations are useful for counterpart searches, studies of burst repetition, and confirmation of the BATSE location accuracies.

INTRODUCTION

The Ulysses GRB detector (1) consists of two hemispherical CsI scintillators with a projected area of about 20 cm^2 in any direction (or $\sim$1/100 the area of a BATSE LAD). It has an unobstructed view of the full sky and operates continuously; over 97% of the data is recovered. Because it is in interplanetary space, the background is stable, and, with the exception of solar particle events, there are practically no data gaps. The instrument has both triggered modes (burst time histories with 8 or 32 ms resolution) and real-time modes (0.25, 0.5, 1 or 2 s resolution). It detects about one event every three days. The majority of them are BATSE bursts, which are found in the real-time data by searching a Ulysses-crossing time window of several hundred seconds duration; this is generated for every BATSE trigger using the BATSE burst location (routinely communicated to UC Berkeley from Huntsville with a delay of a day or so) and the position of the Ulysses spacecraft.

Using the arrival time analysis method, localizations have been generated for all the bursts detected through February 1994, and for many, but not yet all bursts between March 1994 and the present date. We summarize the results here. The complete list of localizations can be e-mailed to anyone interested; request it from khurley@sunspot.ssl.berkeley.edu.

© 1996 American Institute of Physics

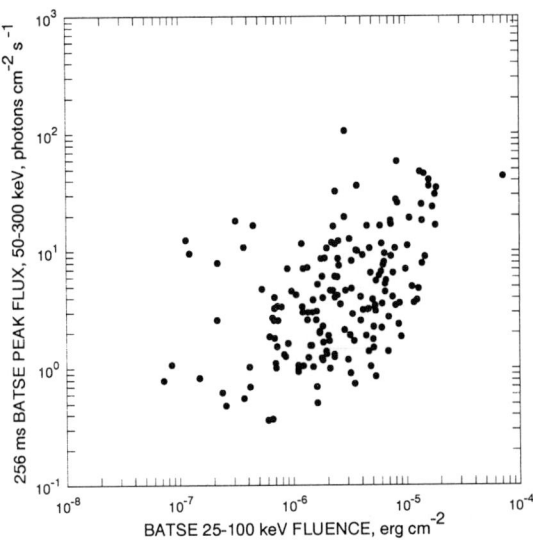

FIG. 1. BATSE peak fluxes and fluences of 230 events detected with Ulysses.

SUMMARY OF RESULTS

Number of Events

Ulysses detected approximately 230 BATSE bursts between 21 April 1991 (the first BATSE burst) and 28 February 1994. During this time, BATSE detected 921 events. Thus Ulysses detects one out of every 3.8 BATSE bursts on the average.

Sensitivity

The BATSE fluence and peak flux of the bursts detected with Ulysses are shown in figure 1. The weakest event detected with Ulysses GRB had a fluence of 7.25×10^{-8} erg cm^{-2}. This was BATSE burst No. 2513 on 3 September 1993. (Bursts this weak would probably never be reliably detected without knowledge of the crossing window).

Comparison of BATSE and Ulysses Locations

For those BATSE events detected only with Ulysses (i.e., for which there is a single annulus of location), a minimum distance between the annulus and the center of the BATSE error circle may be computed. The distribution is shown in figure 2. The average minimum separation is $\sim 2.4°$.

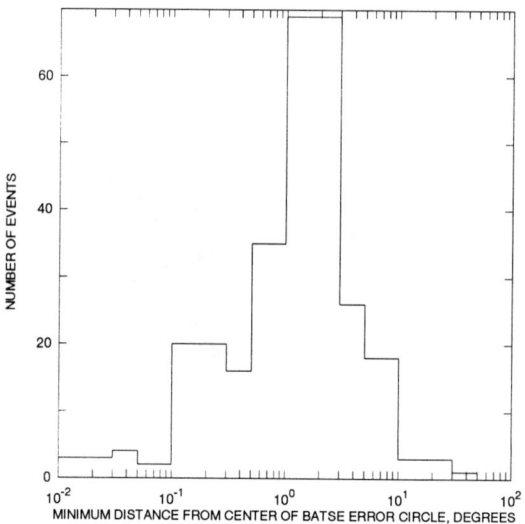

FIG. 2. Distribution of the minimum distance between Ulysses/BATSE annuli and the centers of the corresponding BATSE error circles for 197 events.

Annulus Widths

The distribution of annulus widths is shown in figure 3. The average width is ∼5.7′. Note that many annuli are still preliminary. The emphasis has usually been placed on speed of location rather than accuracy. Widths have been overestimated so as not to mislead workers conducting counterpart searches. Improvements of a factor of several can be expected in many cases.

Reduction of Location Areas

Assuming a systematic error of 1.6° added in quadrature to the statistical error, the average 1σ area of the BATSE error circles for which there are Ulysses/BATSE annuli is 22.5 square degrees. If we take an average annulus width of 5.7′ and assume that the annulus passes through the center of the BATSE error circle, then it reduces the location area by a factor of about 40. However, since the annuli do not generally pass through the centers of the error circles, the reduction is actually larger, and roughly two orders of magnitude.

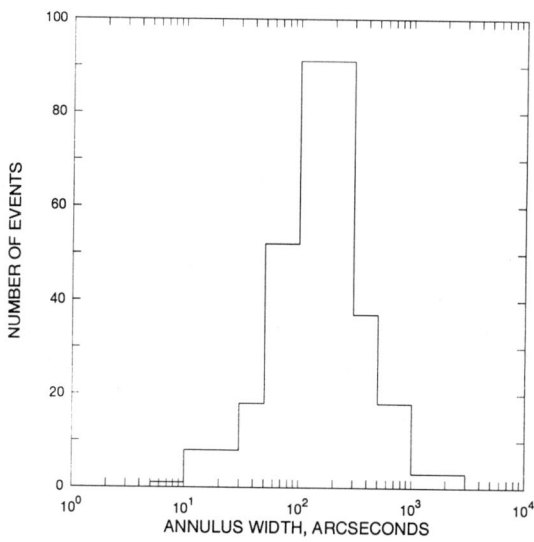

FIG. 3. Distribution of annulus widths for 229 events.

Electronic Version

The complete event list is too long to present here, but it is available by e-mail. The format is:
22, 'JAN', 91, 15, 13, 59, 0, 0, 0, 0, 275.666, -27.379, 44.907, 0.055, 308.4, -21.524, 49.597, 0.021, 297.988, -70.875, 1, 0, 0, 0, 0, 0, 0, 0, 0, 0, 0, 0, 0
8, 'FEB', 91, 9, 0, 15, 0, 0, 0, 0, 137.321, 19.205, 82.746, 0.016, 0, 0, 0, 0, 0, 0, 0, 0, 0, 0, 0, 0, 0, 0, 0, 0, 0, 0, 0
19, 'FEB', 91, 11, 45, 26, 0, 0, 0, 0, 98.01, 26.17, 79.947, 0.025, 141.559, 17.886, 65.594, 0.017, 213.512, 58.767, 1, 0, 0, 0, 0, 0, 0, 0, 0, 0, 0, 0, 0
etc.

Each line in the file contains all the known localization information for a burst. The columns contain the following information:
day, month, year, hours, minutes seconds, BATSE No., BATSE RA, BATSE Dec, BATSE Error Circle Radius*, RA of IPN annulus No. 1, Dec of IPN annulus No. 1, Radius of IPN annulus No. 1, Width of IPN annulus No. 1, RA of IPN annulus No. 2, Dec of IPN annulus No. 2, Radius of IPN annulus No. 2, Width of IPN annulus No. 2, RA of WATCH error circle, Dec of WATCH error circle, Radius of WATCH error circle, RA of COMPTEL error circle, Dec of COMPTEL error circle, Radius of COMPTEL error circle†, RA of EGRET error circle, Dec of EGRET error circle, radius of EGRET error circle, RA of PHEBUS error circle, Dec of PHEBUS error circle, Radius of PHEBUS error circle†, RA of SIGMA error circle, Dec of SIGMA error circle,

Radius of SIGMA error circle†.

Notes:
All coordinates are J2000. Zeros indicate that there are no data for the event from the instrument in question.

*The error circle radius follows the convention in the 3B catalog. It represents only the statistical errors, which must be added in quadrature to the systematic error. Where a number >99 appears, this means that the localization error could not be estimated due to the BATSE data type.

†These error circles generally have complex shapes. The radius given here is only an approximation to the true region.

Elsewhere in these proceedings, the data in this table have been analyzed for evidence of repeating bursts (2), and to independently assess the BATSE statistical and systematic location errors (3).

Acknowledgments: Ulysses data reduction is carried out under JPL Contract 958056. Analysis in conjunction with BATSE data is supported by CGRO guest investigator grant NAG5-1560. We are grateful to M. Briggs for comments.

REFERENCES

1. K. Hurley, et al., Astron. Astrophys. Suppl. Ser. **92(2)**, 401 (1992).
2. T. Cline, et al., these proceedings (1996).
3. C. Graziani and D. Lamb, these proceedings (1996).

BACODINE/3rd Interplanetary Network Burst Localization

K. Hurley*, S. Barthelmy[†], P. Butterworth[†], T. Cline[†],
M. Sommer[‡], M. Boer*, M. Niel*, C. Kouveliotou[§], G. Fishman[§],
C. Meegan[§]

*University of California
Space Sciences Laboratory
Berkeley CA 94720-7450
[†]NASA Goddard Space Flight Center
Greenbelt, MD 20771
[‡]Max-Planck Institut für Extraterrestrische Physik
85740 Garching, Germany
*Centre d'Etude Spatiale des Rayonnements
31029 Toulouse Cedex, France
[§]NASA Marshall Space Flight Center
Huntsville, AL 35812

Even with only two widely separated spacecraft (Ulysses and GRO), 3rd Interplanetary Network (IPN) localizations can reduce the areas of BATSE error circles by two orders of magnitude. Therefore it is useful to disseminate them as quickly as possible following BATSE bursts. We have implemented a system which transmits the light curves of BACODINE/BATSE bursts directly by e-mail to UC Berkeley immediately after detection. An automatic e-mail parser at Berkeley watches for these notices, determines the Ulysses crossing time window, and initiates a search for the burst data on the JPL computer as they are received. In ideal cases, it is possible to retrieve the Ulysses data within a few hours of a burst, generate an annulus of arrival directions, and e-mail it out to the astronomical community by local nightfall. Human operators remain in this loop, but we are developing a fully automated routine which should remove them, at least for intense events, and reduce turn-around times to an absolute minimum. We explain the current operations, the data types used, and the speed/accuracy tradeoffs.

INTRODUCTION

The BATSE Coordinates Distribution Network (BACODINE) (1) has made it possible to search for flaring and fading counterparts to gamma-ray bursts. Prior to BACODINE, such searches could only be done serendipitously; virtually all dedicated counterpart searches were for quiescent counterparts, and these were done at best weeks after the bursts. BACODINE data illustrate

© 1996 American Institute of Physics

well the classic trade-off between speed and accuracy: they are available in many cases while the burst is in progress, but the error circles have radii of several degrees, restricting multi-wavelength counterpart searches to wide-field instruments with only moderate sensitivities. Here we describe an enhancement to the BACODINE procedure which reduces the search area by about two orders of magnitude. The ever-present speed/accuracy tradeoff is in effect here, too, since these refined positions are not available in less than several hours after a burst at best. There are, however, several unique circumstances under which this delay is not critical. For example, it may occur during daylight hours on some continents, so that optical telescopes could not begin searches in any case. Or, since large BACODINE error circles cannot be examined by some types of radio telescopes, and since delays in the radio emission from a GRB source are expected in some models, a 1 day waiting period for a smaller error box is consistent with search strategies.

ULYSSES OPERATIONS

To understand the inherent limitations in BACODINE/IPN localizations, it is necessary to consider the operation of the Ulysses mission. Ulysses is not a "real-time" mission. Data are received every day during an 8 hour tracking pass from one of three Deep Space Network stations (Canberra, Madrid, or Goldstone). However, data are only processed Monday through Friday at JPL during working hours. Thus a burst which occurs during a tracking pass on a weekday might be processed within hours. But a burst which occurs on a Friday afternoon might not be processed until the following Monday or Tuesday, depending on the backlog of data from the weekend. Note too that a burst arriving from the Earth-to-Ulysses direction could arrive at Ulysses up to 50 minutes after triggering BATSE, and the light-travel time from Ulysses back to Earth doubles this delay.

Independent of BACODINE, we search for every BATSE burst in the Ulysses data. To do this, it is necessary to know the approximate arrival direction of the burst so that a Ulysses-crossing window can be generated. (Otherwise, with Ulysses to Earth distances of up to 3000 light-seconds, the amount of data to be searched becomes too large to permit the reliable detection of very weak events.) Thus information on each BATSE burst is transmitted from Huntsville to Berkeley within a day or so of the BATSE trigger. If the burst is detected in the Ulysses data, we request the BATSE high time resolution (usually 64 ms) digital data from Huntsville. There is a minimum delay of 24 hours for receipt of this data at Huntsville; thus it occasionally happens that the Ulysses data are received prior to the BATSE data. The time to process the data in a non-urgent way and transmit it to Berkeley can be days to weeks. Once the data have been received, the arrival direction can be calculated within $\sim$20 minutes, but in general, in this non-urgent, routine processing mode, the data may not actually be processed for several days. A typical timeline for the above procedure is shown in Figure 1.

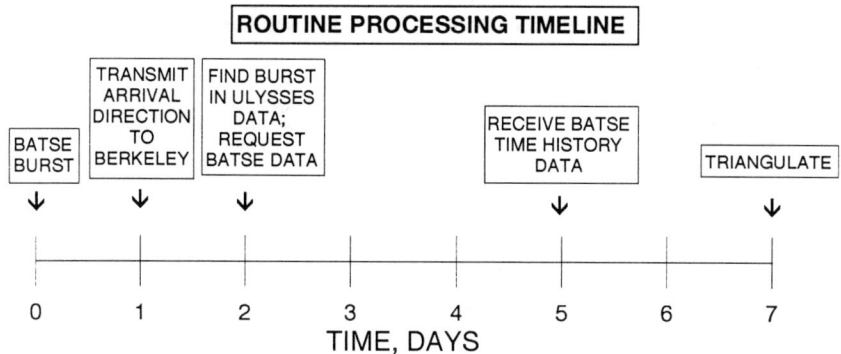

FIG. 1. Routine (non-urgent) processing of BATSE/IPN bursts.

BACODINE/IPN OPERATIONS

When a BACODINE burst occurs, the arrival direction is known within seconds. The burst time history can also be retrieved almost instantaneously from the BACODINE data, but in constrast to the BATSE 64 ms time resolution, BACODINE light curves have 1.024 s resolution. Thus two speed/accuracy trade-offs are involved: the first due to the BACODINE localization, which is less accurate than the BATSE one, and the second due to the coarser time resolution, which causes the IPN annulus to be wider. The BACODINE data are transmitted to a special e-mail address at Berkeley whose mailbox is examined every few minutes. When a new message arrives, the data are automatically processed to plot the time history, generate a Ulysses-crossing window, and retrieve and plot the Ulysses data. At this point, human operators enter the procedure. The Ulysses data are scanned to search for the burst; if it is found, the data are cross-correlated with the BACODINE light curve, and the time difference is used to generate an annulus of arrival directions. The data for the annulus are e-mailed to Goddard, where they are put into a standard format and forwarded to BACODINE recipients. The timeline for this procedure is shown in Figure 2. When the high time resolution BATSE data are received, follow-up notices are often sent out.

To date, approximately 50 bursts have been localized using BACODINE data; the fastest time from detection to notification was 14 hours.

FUTURE IMPROVEMENTS

Except for very special periods, Ulysses is not tracked for over 8 hours/day, and data are not processed outside normal working hours. Thus improvements to the timeline of Figure 2 must come from the removal of human operators. This is particularly important for bursts which arrive outside of normal working hours, or when people are on travel or on vacation. It is quite feasible to do this for bright bursts. For example, automatic scanning of the

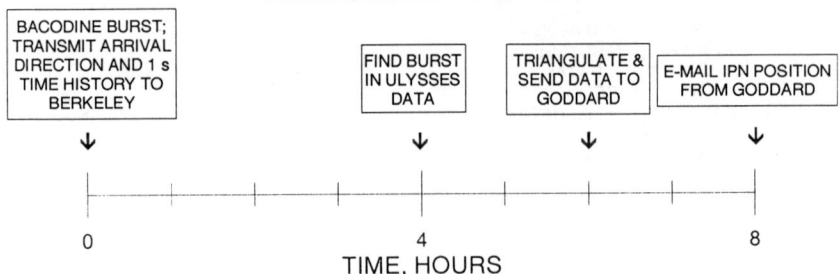

FIG. 2. A typical timeline for triangulation using BACODINE data under optimum conditions; it is assumed that the burst occurs on a weekday.

Ulysses data to find bursts is already being done; human intervention is only for the purpose of verification, and this is generally not necessary if the burst is intense. Cross-correlating the time histories, too, is routine enough to be automated for bright bursts, for which the correlation coefficient is high. Finally, sending the data to Goddard for final checks and formatting is easily done automatically. Over the coming months, these improvements will be implemented, reducing the total time to hours in the best cases.

Acknowledgments. Ulysses data reduction is carried out under JPL Contract 958056. Analysis in conjunction with BATSE data is supported by CGRO guest investigator grant NAG5-1560.

REFERENCES

1. S. Barthelmy, et al., in Gamma-Ray Bursts, Second Workshop, AIP Conf. Proc. 307 (AIP: New York), Eds. G. Fishman, J. Brainerd, and K. Hurley, 643 (1994).

Threshold effects in GRB brightness distributions

J.J.M. in 't Zand[1]

NASA Goddard Space Flight Center, Code 661, Greenbelt, MD 20771

E.E. Fenimore

Los Alamos National Laboratory, MS D436, Los Alamos, NM 87545

Brightness distributions (such as Log N-Log P or Log N-Log S) can be used to estimate the distance of gamma-ray bursts (GRBs). For low brightness, one must take into account threshold effects, for instance that a number of bursts were missed because they did not fulfill the trigger criteria. A suggestive method of correction for these threshold effects is the use of the trigger efficiency. Although the trigger efficiency indicates how severe threshold effects are, it does not provide complete information to correct for them. A more sophisticated tool is the trigger response matrix because it preserves the essential distinction between true and observed brightness. We discuss concept issues of how to take threshold effects into account, and illustrate this with calculations specific to BATSE.

INTRODUCTION

The purpose of an analysis of a brightness distribution of instrument-triggering gamma-ray bursts (GRBs) is to find constraints on the GRB distance scale. Such an analysis should include an evaluation of instrumental effects. The instrumental effects are important for dim bursts and are statistical in nature. If the brightness is a type of peak intensity (or flux), there are three effects: 1) the dimmer the burst, the smaller the probability that the burst triggers the instrument; 2) the dimmer the burst, the more the peak intensity estimate becomes susceptible to noise; 3) the dimmer the burst, the more the peak intensity estimate becomes biased to high values due to selection of upward statistical fluctuations (this is the so-called peak count rate bias: the peak intensity will systematically select upward statistical fluctuations). Instrumental effects are usually avoided by simply restricting the analysis to those parts of the brightness distribution where the effects are known to be negligible. This may imply disregarding half of the bursts (e.g., (3)) or more and, therefore, a less sensitive analysis. Obviously, including

---

[1] NAS/NRC research associate

© 1996 American Institute of Physics

these bursts and taking into account the instrumental effects can give added value to the analysis.

BRIGHTNESS DISTRIBUTION ANALYSES

An analysis of an observed brightness distribution of instrument-triggering bursts involves finding constraints on a GRB population model. Formally, one should follow the following steps to find those constraints: 1) propose a model; 2) predict the response of the instrument to this model; 3) assess the likelihood of the model given the observations; 4) repeat steps 2 and 3 to find those model constraints which maximize the likelihood. This procedure is quite similar to spectral analyses (see, e.g., ref. (2)) when bursts are thought of as the analogue of photons. However, a brightness analysis seems more complex than a spectral analysis, particularly because bursts need to be described by more parameters than photons need to.

The central and most nontrivial part in this analysis is step 2. Analogous to the detector response function in spectral analyses one would want to have a tool that eases the calculation of the instrument response to any model: the model-independent trigger response function (TRF). The TRF describes what the probability is that a burst with a certain set of values for sky location, incident time history, spectrum and brightness fulfills a set of instrument trigger conditions and has a certain observed value for that brightness. Each different set of instrument trigger conditions and each different type of brightness defines a different TRF. When the problem is translated into a discrete form the function is an n-dimensional matrix where n depends on how one parameterizes time history and spectrum (which is a non-trivial problem particularly for the time profile). Step 2) involves the multiplication of this n-dimensional matrix with an $n-1$ matrix which describes the model GRB population. The result is a 1-dimensional array specifying the expectation for the distribution of observed brightnesses.

Unfortunately, the TRF is elaborate to calculate because of its high dimensionality. Also, running each trial model through a Monte Carlo simulation of the instrument and trigger algorithm involves many computations. Therefore, one might want to resort to less extensive simplified tools and methods. It can be imagined that this is possible without loosing much accuracy if one employs a brightness measure which is closely related to the brightness measure used in the trigger algorithm (e.g., 50–300 keV peak count rate over 64, 256 or 1024 ms time scales for BATSE). Obviously, in this case the trigger condition does not depend on any other burst parameter[2]. The same applies to the peak flux instead of peak count rate in the same bandpass and over the same time scale, as long as the energy response of the detector is flat enough.

How does such a simplified TRF look like? We propose a matrix which describes the probability that a burst in a certain true brightness interval and

---

[2] we assume that GRBs are isotropically distributed so that the location does not matter

in a certain observed brightness interval satisfies the set of trigger conditions. We call this the trigger response matrix (TRM).

The 2-dimensional TRM is a sophisticated version of the 1-dimensional trigger efficiency (TE) (4). The TRM leaves free the *observed* brightness and *true* brightness, whereas the TE only leaves free the latter. Although informative, the TE is not sufficient to do an analysis because the link between *observed* brightnesses and efficiency is missing (due to the peak count rate bias effect the true brightness can be very different from the observed brightness, even in the average value). In a previous paper ((5), paper I) we employed a slightly different definition for the trigger efficiency to be able to link observed brightnesses to trigger efficiencies. However, this approach suffers from systematic errors which are only avoided by using the TRM instead.

If one would like to analyze a distribution of a brightness for which the bandpass or time scale deviates from the values used in the trigger algorithm (in the extreme going from peak flux to fluence) the trigger condition will depend on burst spectral and/or timing parameters. Two situations can occur, depending on whether the burst spectral and timing parameters depend on the brightness or not. If not, one may expect that the brightness distribution will be the same as the one predicted for the trigger-based bandpass and time scale except for a scaling of the brightness, this situation can conveniently be dealt with in calculating a TRM. Otherwise, the distributions will change in shape also, this is difficult to analyze without full-scale Monte Carlo simulations.

ILLUSTRATION

We are able to calculate TRMs for BATSE and PVO essentially in the same manner as we did TEs in paper I, with similar assumptions for the instrument, trigger algorithms, and background conditions. The only basic difference is that we now specify the results as a function of both the observed and true brightness. For details we refer the reader to paper I. We here merely summarize: a Monte Carlo simulation of several million bursts is performed and the response of the instrument evaluated. The TRM is calculated by counting the relevant burst numbers as a function of relevant brightness measures. The bursts are characterized by a number of characteristics (time profile, spectrum, peak flux or fluence, sky location with respect to instruments) which are drawn from representative distributions (e.g., time profile from a set of 98 bright BATSE and PVO bursts, sky location random, and spectrum using Band et al.'s (1) findings). All burst characteristics are assumed to be either independent or dependent following a cosmological model in which case spectrum and time profile depend on brightness through a cosmological time dilation and redshift.

Scattering of gamma rays against the atmosphere of the Earth, relevant for BATSE, is not taken into account.

We illustrate some results of our calculations in Fig. 1 in the form of the efficiency of triggering BATSE on either 64, 256 or 1024 ms as a function of the true peak flux in 50 to 300 keV (a) and 100 to 2000 keV (b). Although

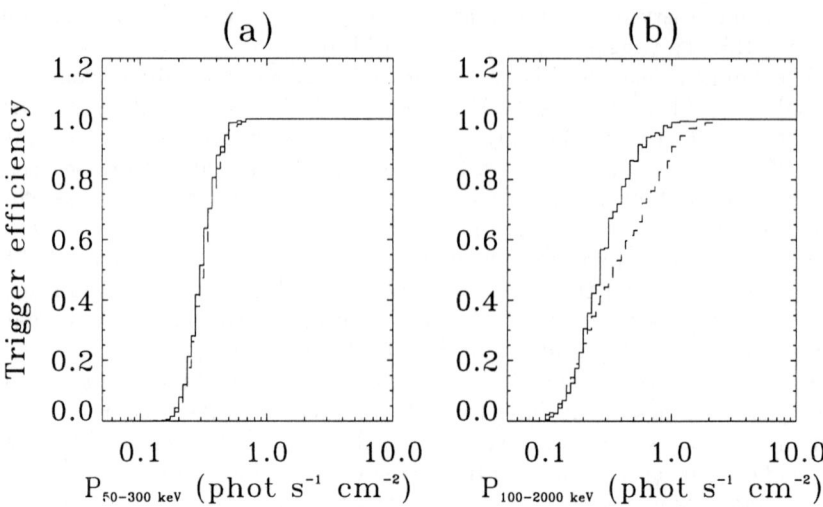

FIG. 1. Computed result for the efficiency of triggering BATSE on time scales of 64, 256 or 1024 ms, as a function of true peak flux measured over 256 ms in bandpasses 50 to 300 keV (a) and 100 to 2000 keV (b). In both bandpasses the results are given for 2 burst models: a cosmological model (dashed line) and a model where all burst parameters are independent (solid line) (see text). The sometimes noisy appearance is due to statistical fluctuations of a limited number of bursts

the TE is not the ultimate tool to use in an analysis of brightness distribution (and if one would want to choose it anyway it would be better to use observed instead of true peak flux as abscissa, see paper I), it lends itself well for illustration purposes here. Results are presented for 2 burst models where all burst parameters are either independent or dependent following a cosmological model with a standard candle bolometric energy of 10^{52} erg. Figure 1 shows that in the case of BATSE data the results for the 100 to 2000 keV peak flux are much more sensitive to the burst model than those for the 50 to 300 keV flux (which is the same energy bandpass as used in the BATSE trigger algorithm). This illustrates that the simplified tools discussed here do not work properly if applied to a brightness distribution of a brightness measure deviating substantially from the one employed in the trigger algorithm and one needs to investigate burst models with parameters dependent on the brightness.

It is noted that any subset of the actual trigger conditions can be employed, depending on what subset of the instrument-triggering bursts one would like to analyze; for instance Fishman et al. (4) give BATSE TE curves for triggering on exclusively one time scale, as compared to our illustration with bursts triggering on either one of three time scales (the latter set of trigger conditions actually concerns all BATSE-triggering bursts).

CONCLUSIONS

Given a few limitations it is possible to take threshold effects into account during analyses of GRB brightness distributions employing the trigger response matrix. The limitations are that one either investigates distributions of a brightness measure closely related to the one used in the trigger algorithm or one restricts the analysis to burst models where there is no relationship between burst spectrum nor timing behavior and brightness (making it difficult to apply to strong cosmological models).

For cosmological models (or other models where burst brightness is correlated with spectrum and/or timing behavior) it is difficult to analyze data from two instruments in the form of a single peak flux distribution because there is no easy way with TRMs to take into account threshold effects on the combined distribution for both instruments at the same time unless the bandpass and time scales are very similar. For such models one might need to do full-scale trigger response calculations for each model proposed.

Acknowledgments. We thank J. Bloom and A. Crider for useful discussions. Part of the research described in this paper was performed under the auspices of the Department of Energy.

REFERENCES

1. D. Band et al., Ap.J. **413**, 281 (1993)
2. M. S. Briggs, these proceedings (1996)
3. E.E. Fenimore et al., Nature **366**, 40 (1993)
4. G.J. Fishman et al., Ap. J. Sup. Ser. **92**, 229 (1994)
5. J.J.M. in 't Zand, E.E. Fenimore, in proceedings of 1993 Huntsville workshop on GRBs, eds. G.J. Fishman, J.J. Brainerd & K. Hurley, 692 (paper I, 1994)

The Angular Distribution of COMPTEL Gamma-Ray Bursts

R.M. Kippen[*], J. Ryan[*], A. Connors[*], M. McConnell[*],
D.H. Hartmann[†], C. Winkler[‡], L.O. Hanlon[‡], V. Schönfelder[§],
J. Greiner[§], M. Varendorff[§], W. Collmar[§], W. Hermsen[||]
and L. Kuiper[||]

[*] Space Science Center, University of New Hampshire, Durham, NH 03824
[†] Dept. of Physics and Astronomy, Clemson University, Clemson, SC 29634
[‡] Astrophysics Division, ESA/ESTEC, NL-2200 AG Noordwijk, NL
[§] Max-Planck-Institut für Extraterrestrische Physik, D-85748 Garching, FRG
[||] SRON-Utrecht, Sorbonnelaan 2, 3584 CA Utrecht, NL

> The superior burst location capability of the COMPTEL instrument aboard the *Compton* Gamma-Ray Observatory allows us to study the small-scale angular distribution of burst sources with good sensitivity even though the number of burst detections is small. We accumulate four years (April 1991 – April 1995) of observations to form a catalog of 27 burst locations whose mean 1σ uncertainty is $\sim 1°$. We find that the COMPTEL bursts are consistent with an isotropic distribution of sources, yet the spatial coincidence of two of the bursts within COMPTEL's angular resolution indicates the possibility of repetition. This possibility is studied using the two-point angular correlation function and the nearest neighbor statistic. Model dependent upper limits on the fraction of repeating sources are derived.

INTRODUCTION

The isotropic, inhomogeneous distribution of burst sources derived from BATSE data has led to the general acknowledgment that GRBs could be of cosmological origin. If bursts are associated with cosmologically distant galaxies, we do not expect (in most scenarios) to observe multiple events from the same direction. The small-scale angular distribution of burst sources thus plays a vital role in confirming or dispelling the cosmological hypothesis. Numerous investigations using BATSE data have yielded conflicting results due to BATSE's limited angular resolution of a few degrees at best. However, most recent analyses agree that the BATSE burst locations are consistent with isotropy even on small angular scales — indicating that burst recurrence is either rare or non-existent (1–3).

In the present study, we use the burst localization capabilities of The Imaging Compton Telescope (COMPTEL) to *independently* investigate the angular distribution of GRBs. Although it detects far fewer bursts, COMPTEL

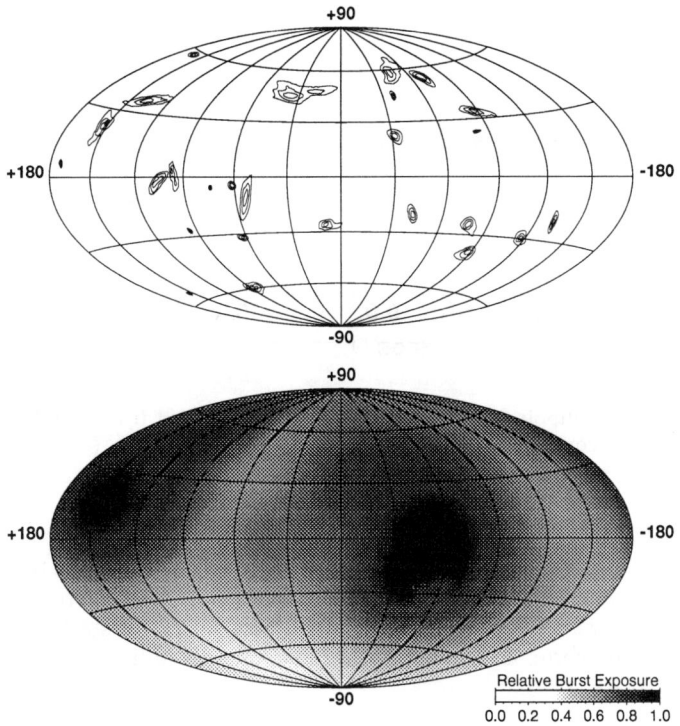

FIG. 1. Statistical location uncertainty contours (1,2,3σ confidence) of 27 bursts imaged by COMPTEL in four years of observations (top) and corresponding "exposure" map (bottom) plotted in Galactic coordinates.

has the advantage over BATSE that its burst localization accuracy is superior ($\sim 1°$). This makes even a modest sample of COMPTEL burst locations particularly valuable for investigating the small-scale angular distribution of burst sources.

OBSERVATIONS AND BURST LOCALIZATION

Gamma-ray bursts are observed regularly within the field-of-view ($\sim$1–3 sr depending on fluence) of the main COMPTEL "telescope" operating mode ($\sim$0.75–30 MeV). Lacking an on-board transient triggering system, these bursts are identified by searching for excess numbers of telescope "events" (4) at the times of all BATSE burst triggers. This technique is effective since BATSE is a far more sensitive detector at energies where bursts emit most of their power.

In the four year period from April 1991 through April 1995, 27 significant ($\gtrsim 4\sigma$) detections have been identified out of more than 230 candidate BATSE triggers (5). Each of these bursts has been localized through direct imaging of

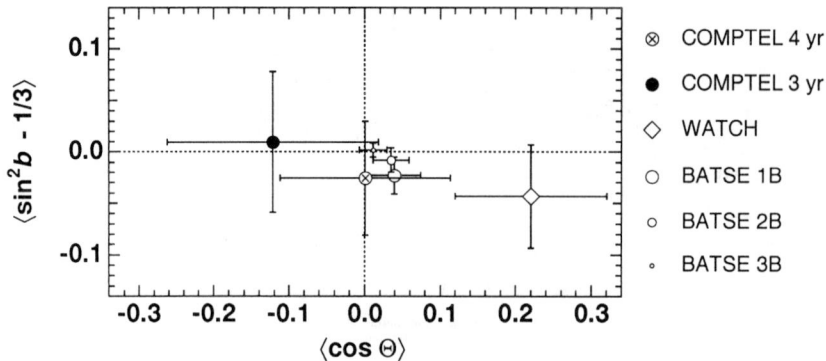

FIG. 2. Galactic dipole and quadrupole moments derived from COMPTEL burst localizations are compared to BATSE (2) and *Granat*–WATCH (8) results.

the MeV photon events using a maximum likelihood technique that provides quantitative constraints on the source direction and flux (6). Statistical burst location accuracy (1σ confidence) ranges from $<0.5°$ for the strongest events to $\sim 2°$ for the weakest, with a mean of $\sim 1°$ for all 27 bursts (see Figure 1). Simulations and comparisons with other instruments indicate that systematic location errors are $<0.5°$.

SPATIAL ANALYSIS

The angular distribution of COMPTEL burst directions is analyzed using two inclusive samples: the 27 bursts detected in four years of observations and the 18 bursts (7) of the first three years. Although the full sky has been observed by COMPTEL, several regions (most notably Virgo and the Galactic center) have received substantially (~ 30–50%) higher exposure. For each sample interval, a map of COMPTEL's non-uniform probability of detecting a burst on the sky (or "exposure") has been computed by accumulating data from the many pointed observation periods (Figure 1).

As shown in Figure 2, Galactic dipole and quadrupole moments corrected for the non-uniform exposure indicate that both the three year and four year burst location samples are consistent (on large scales) with an isotropic angular distribution of sources. Uncertainties in these quantities caused by burst location errors are negligible compared to the large statistical errors due to the small sample size.

Small-scale structure in the angular distribution is examined using the (now standard) two-point angular correlation function $w(\theta)$ and the nearest neighbor statistic $NN(\theta)$ (1). Both of these statistical tools are sensitive to small-scale clustering, which could indicate repeated bursts from individual sources. To account for burst location errors, observed $w(\theta)$ and $NN(\theta)$ distributions are evaluated by averaging over points sampled randomly from the COMP-

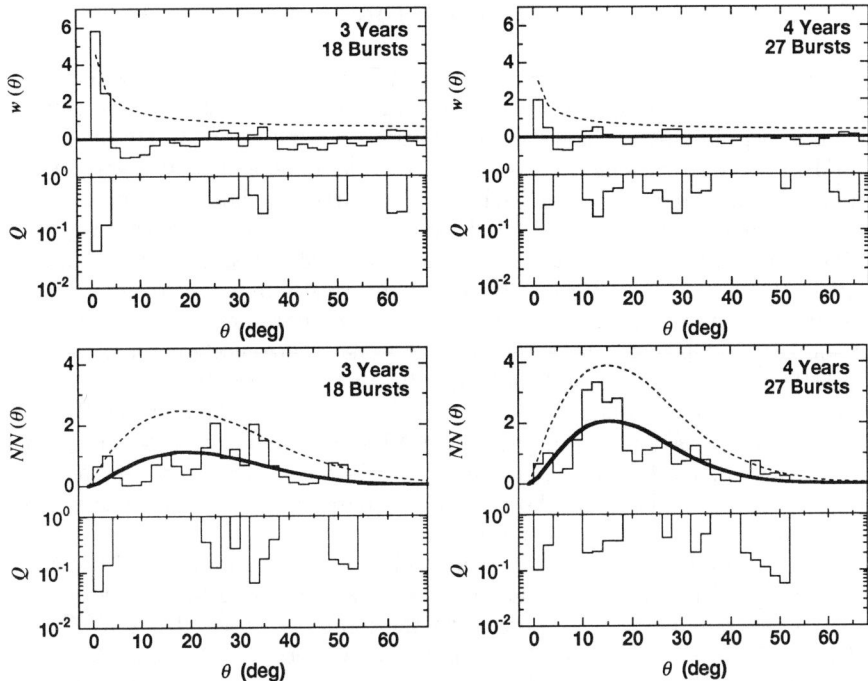

FIG. 3. Angular correlation functions (upper plots) and nearest neighbor distributions (lower plots) of two samples of COMPTEL burst locations. Each plot compares the average *observed* distribution (solid histogram) with the distributions (solid curve) and 1σ statistical deviations (dashed curve) *expected* from isotropy. The statistical significance Q of observed deviations from isotropy are indicated in the lower panels of each plot.

TEL maximum likelihood burst localizations (i.e., Figure 1). Each observed average distribution is compared to the null hypothesis of isotropy, weighted with COMPTEL's non-uniform sky exposure map, as shown in Figure 3. The significance Q of observed deviations in excess of the isotropic hypothesis is evaluated through Monte Carlo simulation.

In the three year sample of 18 GRBs, there is evidence of a marginally significant excess of bursts with small angular separations ($\theta \lesssim 2°$). The cause of this excess is that two of the COMPTEL bursts (GRB 930704 and GRB 940301) are localized to the same direction within their combined location errors. The probability of this coincidence occurring by chance based on either angular correlation or nearest neighbor analyses is $Q \lesssim 3\%$ — suggesting the possibility of burst recurrence (6). However, the small-angle excess is *not* observed in the full four year burst sample, indicating that no other coincidences have since been detected, thus diluting the significance of the GRB 930704/940301 overlap.

DISCUSSION

Given their limited numbers, the COMPTEL burst location samples provide little more than a consistency check on BATSE's unsurpassed measurements of the large-scale angular distribution of burst sources. However, the COMPTEL localizations do provide a valuable, independent measure of the small-scale properties of this distribution by virtue of their superior location accuracy. Results from the 27 burst sample presented here indicate that the COMPTEL burst locations are statistically consistent with an isotropic distribution of sources. If we accept this result, models of burst recurrence can be constrained. As an example, we fit a simple, one parameter (f_r, the fraction of sources that produce repeated bursts) burst recurrence model (9) to the COMPTEL observations and find that this particular model can be rejected with 95% confidence when $f_r > 23\%$. Similar analysis of the first 260 BATSE bursts was not as constraining due to BATSE's poor location accuracy (9). Unfortunately, models predicting fewer bursts per repeating source are not constrained by COMPTEL observations because we would not yet expect to see many recurrences in such small samples.

The preceding arguments still do not rule out the possibility that the coincident bursts GRB 930704 and GRB 940301 were produced by a single source. Nor can this possibility be excluded with the combined observations of COMPTEL, BATSE, EGRET and the Interplanetary Network. We can, however, conclude that if recurrence exists, it must be a relatively rare phenomenon. Continued observations may yet yield a more definitive answer.

Acknowledgements. This research was supported by the Compton Observatory Guest Investigator Program under NASA contract NAG5-2350. The COMPTEL project is supported by NASA under contract NAS5-26645, by the German government through DARA grant 50 QV 90968 and by the Netherlands Organization for Scientific Research (NWO). JG acknowledges the support of DARA under contract FKZ 50 OR 9201.

REFERENCES

1. C.A. Meegan et al., ApJ **446**, L15 (1995).
2. C.A. Meegan et al., ApJ in press.
3. M. Tegmark et al., ApJ in press.
4. V. Schönfelder et al., ApJS **86**, 657 (1993).
5. R.M. Kippen et al., ApJ in preparation.
6. R.M. Kippen et al., A&A **293**, L5 (1995).
7. R.M. Kippen et al., ApSS **231**, 231 (1995).
8. A.J. Castro-Tirado et al., in *Gamma-Ray Bursts*, ed. G.J. Fishman, J.J. Brainerd & K. Hurley (AIP **307**, New York, 1994), p. 17.
9. T.E. Strohmayer, E.E. Fenimore & J.A. Miralles, ApJ **432**, 665 (1994).

A Search For Untriggered Events In the BATSE Data Base

Jefferson M. Kommers[*], Walter H. G. Lewin[*],
Jan van Paradijs[†,‡], Chryssa Kouveliotou[§],
Gerald J. Fishman[¶], Michael S. Briggs[†]

[*]*Massachusetts Institute of Technology, Cambridge, MA 02139*
[†]*University of Alabama in Huntsville, Huntsville, AL 35800*
[‡]*University of Amsterdam, Amsterdam, The Netherlands*
[§]*Universities Space Research Association, Huntsville, AL 35812*
[¶]*NASA/MSFC, Huntsville, AL 35812*

> Between 13 January 1993 and 24 December 1993, BATSE detected 340 cataloged gamma-ray bursts and triggered six times on events from SGR 1806 − 20. The availability of (almost) continuous background data from the BATSE detectors permits a search for events that *did not* meet the on-board trigger criteria. As part of an on-going program, we have searched nearly one year of data both for possible fainter emission from soft gamma-ray repeaters or other sources (seen only in the low energy channel, 20–50 keV) and for untriggered cosmic gamma-ray bursts (seen in the burst energy channels, 50–300 keV).

The detection of gamma-ray bursts and other transient events by BATSE is controlled by the triggering algorithm running on-board the spacecraft. When the on-board computer signals a trigger, data are collected at high spectral and temporal resolution. Even when the trigger is not activated, however, the count rates in the detectors are recorded at lower resolution in the continuous background data types: CONT, DISCSP, and DISCLA (1). The availability of almost continuous background data permits a search for statistically significant transients that *did not* cause a trigger on-board the spacecraft. We have begun a comprehensive search of the archival background data for such "untriggered" events.

The motivation for our search is twofold. First, classical gamma-ray bursts (GRBs) are expected to escape the on-board trigger for any of several reasons—including low statistical significance, occurrence during the read-out period of a brighter burst (when the on-board trigger is partially disabled), or rising on a time scale longer than the on-board background averaging. A catalog of these events will include GRBs with peak fluxes a factor of two lower than those detected with BATSE's on-board threshold. Second, hard X-ray transients occurring primarily in channel 1 of the DISCLA data (20–50 keV) may have too few counts in the nominal gamma-ray burst channels (50–300 keV) to trigger on-board. Emission from soft gamma-ray repeaters (SGRs) or

© 1996 American Institute of Physics

other low-energy phenomena may be of this type; our search for these events is a continuation of previous work by van Paradijs et al. (2).

SEARCH ALGORITHM

For most of the mission, the on-board computer has been programmed to monitor the count rates in the 50–300 keV range. The average background rate is computed every 17 s and an on-board trigger is signaled when the count rate in the second brightest detector exceeds 5.5 times the uncertainty (σ) in the background rate on any of 3 time scales: 64 ms, 256 ms, and 1024 ms (1). A consequence of this strategy is that the on-board trigger is less sensitive to faint bursts with directions directly in front of one of the detectors than to ones with directions mid-way between two detector normals (3). For example, a 10σ event occurring directly in front of a detector may be reduced by the cosine response to 3.5σ in the second brightest detector and fail to trigger on-board. On the other hand, the same event occurring mid-way between two detector normals would register approximately 7.1σ in both detectors and comfortably cause an on-board trigger.

Our laboratory search procedure partially combats this detection anisotropy by triggering on the *average* significance of the two brightest detectors. We use a moving window on data before and after the time bin in question to find the estimated background count rate in each detector. To signal a "laboratory trigger", we demand first that the count rates in each of two contiguous detectors experience $\geq 2.5\sigma$ increases simultaneously. By itself, this criterion would produce far too many laboratory triggers per day to process. Thus we also demand that the summed significances of the fluctuations in the two detectors exceed 8.0σ; that is, the mean significance must exceed 4.0σ. With these laboratory trigger criteria, statistical fluctuations are expected to contribute approximately 0.08 events per day to our data base.

The results discussed in these proceedings are based on a search of the DISCLA data taken between 13 January 1993 and 24 December 1993. The DISCLA data type provides count rates for each large-area detector in 4 energy channels at 1.024 s time resolution. We search nine time series formed by binning the raw DISCLA data into three energy ranges: a search of channel 1 only (25–50 keV) provides sensitivity to SGR and other low-energy events; a search of the sum of channels 2 and 3 (50–300 keV) provides sensitivity to classical gamma-ray bursts; and a search of the sum of all three channels (25–300 keV) provides sensitivity to weaker events with high-energy tails, such as shot noise from Cyg X-1. Each of these energy ranges is searched at 1.024 s, 4.096 s, and 8.192 s time resolution (with correspondingly longer background averaging intervals) to catch slow-rising events.

We visually inspect each laboratory trigger to separate the data base into useful categories. The majority of laboratory triggers can be easily identified as magnetospheric or otherwise terrestrial in origin. Many events are identified as solar flares based on intensity, duration, spectral softness, and location. Events which appear to be neither terrestrial nor solar and which

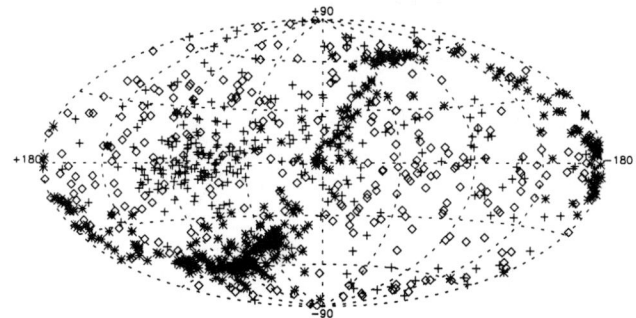

FIG. 1. All non-terrestrial untriggered events in Galactic Coordinates (see text).

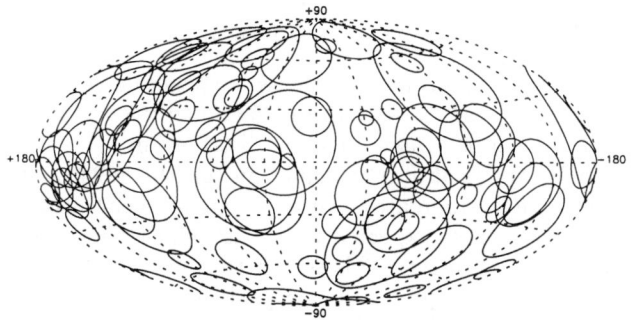

FIG. 2. Untriggered GRB candidates.

have sufficient spectral hardness to be seen in channels 2 and 3 are classified as GRB candidates. Events which do not make it into any of the previous three categories are classified as "unknown", a category which includes almost all the low-energy (channel 1 only) events. For solar flares, GRB candidates, and unknown events we derive burst directions using our port of the BATSE Locburst code. For the untriggered GRB candidates, we also find peak fluxes using the BATSE WINGSPAN spectral analysis package. Figure 1 shows a sky map which combines events from the GRB candidate ($\diamond$), solar flare ($*$), and unknown ($+$) categories. The concentration of solar events in the ecliptic plane is clearly visible, as is a concentration of low-energy events in the vicinity of Cyg X-1 ($\ell = 71.3°$, $b = 3.0°$).

GRB CANDIDATES AND OTHER EVENTS

Our search so far yields 92 untriggered GRB candidates. The durations of these events vary from less than 1.024 s to ~ 50 s with a variety of time profiles. The directions of these events are shown as 1.0σ error circles in Figure 2. To evaluate whether these events are consistent with isotropy requires knowledge of the the sky exposure and laboratory trigger efficiency as a function of burst direction. Since a detailed sky exposure and trigger efficiency study is not finished, rough estimates of the dipole and quadrupole statistics are available

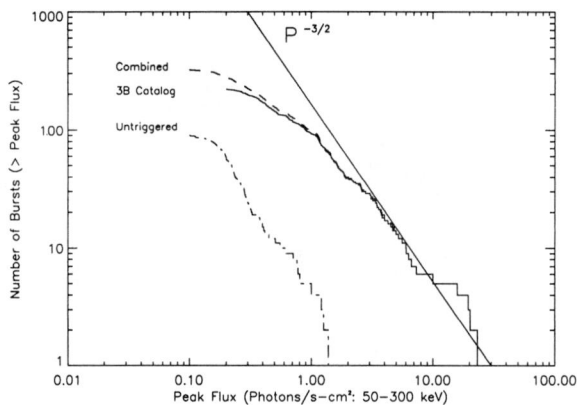

FIG. 3. Integral number vs. peak flux for untriggered GRB candidates.

by assuming (somewhat incorrectly) the sky exposure given in the 2B catalog: the resulting values $\langle \cos\theta \rangle = -0.015 \pm 0.05$ and $\langle \sin^2 b - 1/3 \rangle = -0.043 \pm 0.04$ are consistent with an isotropic direction distribution.

Figure 3 shows the peak flux distribution (on the 1024 ms time scale) for the untriggered GRB candidates. These events are concentrated at the faint end of the distribution, as expected for events which were generally too faint to cause an on-board trigger. This search thus extends the BATSE integral number versus peak flux plot to lower peak fluxes. When the untriggered bursts are added to the triggered bursts for the same same time period, the departure from a $-3/2$ power law remains evident. Interpreting the peak flux distribution in greater detail requires knowledge of the trigger efficiency, so Figure 3 shows raw data only. The completeness limits of our search will be determined when our detailed trigger efficiency study is finished.

We estimate that these 92 GRB candidates failed to trigger on-board the spacecraft for the following reasons: 8 occurred while the trigger was disabled, 72 were below the 5.5σ threshold in the second brightest detector, 6 occurred during the readout of a brighter event, and 6 had a slow rise that modified the on-board background estimate. The on-board trigger mechanism's bias against slow-rising GRBs has been discussed by Lingenfelter and Higdon (5). The 6 events we find that failed to trigger on-board the spacecraft *solely* because of the slow-rising effect constitute 1.7% of the total 352 GRBs that have been detected above the on-board threshold while the trigger was active. This is a lower fraction than estimated elsewhere (5); we note, however, that our search algorithm is biased against events which rise on time scales longer than ~ 40 s.

The "unknown" category of laboratory triggers includes all events which were not obviously of terrestrial or solar origin and which do not resemble a GRB. Most of the channel 1 only (low-energy, 25–50 keV) events fall into this category. The major problem with this class of events is that it is dominated by fluctuations from Cyg X-1: of 799 events in the unknown category, 689 are consistent with this source (although they may not *all* be from Cyg X-1);

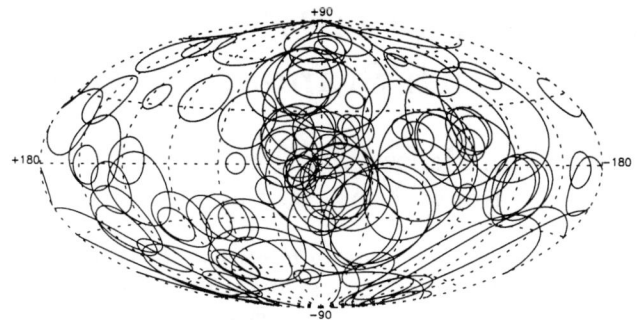

FIG. 4. Sky map of the directions of the unknown events.

see the clustering of events marked (+) in Figure 1. If we remove all the events consistent with Cyg X-1 we are left with the sky map shown in Figure 4, where events are plotted as their 1.0σ error circles. Although Figure 4 shows some general clustering toward the galactic center, there is no obvious clustering that would indicate the activity of any particular source. However, we find two events which can convincingly be attributed to SGR 1806−20 based on intensity, spectral softness, and location; both occur within one day of on-board triggered emission from SGR 1806 − 20 (4).

Our search for untriggered events has yielded a significant number of interesting transients. When the proper sky exposures and trigger efficiencies are calculated, we will be able to make a useful extension of the $\log N$-$\log P$ distribution of gamma-ray bursts, which will in turn provide more stringent constraints on models of the source distribution. We have demonstrated our sensitivity to untriggered emission from SGRs in that we have found two convincing examples well within our detection limits. We plan to expand our search to cover the entire BATSE data base to produce a catalog of untriggered gamma-ray bursts. We may also discover new source activity or new phenomena occurring in the lowest energy channel.

Acknowledgments. This work is supported by the NASA CGRO Guest Investigator Program under grant NAG5-2755 (JvP) and NAG5-2046 (WHGL). JMK acknowledges additional support from a National Science Foundation Graduate Research Fellowship.

REFERENCES

1. G. J. Fishman et al., Proc. of the Gamma Ray Observatory Workshop, GSFC April 10-12, ed. W. N. Johnson, 2-39 (1989).
2. J. van Paradijs et al., AIP Conf. Proc. **280**, 877 (1993).
3. M. N. Brock et al., AIP Conf. Proc. **265**, 399 (1991).
4. C. Kouveliotou et al., Nature **368**, 125 (1994).
5. R. E. Lingenfelter & J. C. Higdon, AIP Conf. Proc **366**, 164 (1996). See also these proceedings.

Consistency of GRB Durations and Spectra With Standard Cosmology

T. Koshut[†], R. Mallozzi[†], J. Horack[‡], W. Paciesas[†], C. Kouveliotou[*], R. Rutledge[**]

[†] Dept. of Physics, Univ. of Al. in Huntsville, Huntsville, AL, 35899
[‡] NASA/MSFC, ES-84, Huntsville, AL 35812
[*] Universities Space Research Association
[**] Dept. of Physics, Massachusetts Institute of Technology, Cambridge, MA 02139

> We present an analysis of the consistency of burst durations and spectra with a population of nonevolving sources at Gpc distances. The parameters T_{90} (T_{50}) and E_p are calculated using data obtained with BATSE, and are used to characterize burst durations and spectra, respectively. Effects consistent with cosmological redshift and time dilation have been reported in the E_p and T_{90} data, respectively. We simultaneously examine the distributions of these parameters for cosmological time dilation and redshift effects, being careful to avoid the introduction of selection effects. We show that the T_{90} (T_{50}) data are currently less constraining than E_p; the data do not require time dilation, but are consistent with the effect observed in the E_p data.

INTRODUCTION

The arguments over the distance scale for the burst source population are divided between an extended Galactic Halo, at a distance of $\sim$ 100 kpc, and a population at such large distances ($\sim$ Gpc) that cosmological effects resulting from the large-scale expansion of spacetime should be observable in the data. If burst sources are at appreciable redshifts z, then cosmological time dilation will stretch intrinsic burst temporal structures on all time scales by a factor of (1+z). In addition, all photons emitted at the source with energy E will undergo redshift and be lowered by a factor of (1+z). Both of these effects have the same underlying cause and would be simultaneously present. Standard cosmology predicts that, on average, the furthest bursts should have larger values of T_{90} and smaller values of E_p, relative to the nearer bursts.

Evidence consistent with cosmological time dilation has been presented by (1,2). They examined the intensity dependence of the BATSE T_{90} and T_{50} distributions (3), two parameters used to characterize burst durations, and reported time dilation factors $\approx$ 2. Mallozzi et al. (4) examined the intensity dependence of the BATSE distributions of E_p (the photon energy at which the best-fit νF_ν spectrum is a mamimum) and found evidence consistent with cosmological redshift. Because these two effects are not independent, we simul-

© 1996 American Institute of Physics

taneously examine the BATSE T_{90} (T_{50}) and E_p data sets for self-consistent cosmological effects.

DATA SET SELECTION

We will use the T_{90} and T_{50} data from the BATSE 3B Catalog (5) to search for effects caused by cosmological time dilation. The E_p data described in (4) are used to search for the effects of cosmological redshift on the burst photon spectra. As our distance indicator, we will use P_{trg}, the maximum photon flux measured with BATSE on the *trg* time scale (*trg* = 64, 256, or 1024 ms) in the 50 – 300 keV energy range. These fluxes are taken from the BATSE 3B Catalog. We assume a narrow distribution of burst peak luminosities (i.e. standard candles) on the *trg* time scale.

Because each of the 3 BATSE trigger time scales examine different ranges of redshift, and because the redshift range also depends on the burst pulse shape during the trigger time interval, we consider each trigger time scale separately. We accomplish this by requiring that $(C_{max}/C_{min})_{trg} > 1$, where C_{max} and C_{min} are the maximum burst count rate and threshold count rate for triggering, respectively, measured on the *trg* trigger time scale.

A second selection criterion is motivated by the prediction made by standard cosmological models that burst intensity should be correlated with E_p and anticorrelated with T_{90}. Because of this, the trigger efficiency must be appreciable over the range of fluxes used in the study. Therefore, we will only select those bursts with peak fluxes P_{trg} which are above the value $(P_{trg})_{crit}$ at which the BATSE trigger efficiency is 75%. These critical values for the 3 trigger time scales are given by $(P_{64})_{crit} = 1.33$ phot cm^{-2} s^{-1}, $(P_{256})_{crit} = 0.66$ phot cm^{-2} s^{-1}, and $(P_{1024})_{crit} = 0.34$ phot cm^{-2} s^{-1}.

To avoid the introduction of a correlation between peak flux P_{trg} (our distance indicator) and T_{90}, we require that $T_{90} > trg$ ms. Otherwise, bursts with $T_{90} < trg$ ms would be considered weaker (and thus further away) than they would had their peak flux been measured on a time scale that was more characteristic of the burst temporal variability.

Based on their values of P_{trg}, we bin the bursts into N bins, with an equal number of bursts per bin. The number of bins must be large enough such that the weakest burst in the most intense bin has a peak flux P_{trg} that is at or above the turnover in the integral $\log N(>P) - \log P$ curve. We then calculate $\langle T_{90}\rangle_j$, $\langle T_{50}\rangle_j$, and $\langle E_p\rangle_j$ for each jth bin, where $j = 1$ (weak), ...N (strong).

If we let T represent either T_{90} or T_{50}, then in a standard cosmology the observed values T_{obs} will be related to the intrinsic value T_{intr} through

$$T_{obs} = (1+z)\,T_{intr}. \tag{1}$$

The time dilation factor (TDF) is defined as

$$TDF \equiv \frac{1+z_{dim}}{1+z_{bright}}. \tag{2}$$

Using equation 1, and assuming no evolution in T_{intr} as a function of redshift z, the time dilation factor can then be found for the jth bin from

$$TDF_j = \frac{\langle T \rangle_j}{\langle T \rangle_N}. \tag{3}$$

Similarly, we define the energy shift factor (ESF) as

$$ESF \equiv \frac{1 + z_{\text{dim}}}{1 + z_{\text{bright}}}. \tag{4}$$

Assuming no evolution in the intrinsic value of E_p as a function of redshift z, the ESF for the jth bin is given by

$$ESF_j = \frac{\langle E_p \rangle_N}{\langle E_p \rangle_j}. \tag{5}$$

Time dilation factors and energy shift factors greater than unity would indicate observations consistent with the presence of cosmological effects. We would expect these effects to be strongest for the furthest bursts; this would manifest itself as an inverse correlation between the intensity of each bin and the TDF or ESF. We expect that the ESF and the TDF should be equal for any individual intensity bin.

RESULTS

For this analysis we binned the bursts into 6 bins, using the peak flux P_{64} as our measure of burst intensity. Figure 1a shows the distributions of E_p for the weakest (dotted) and the brightest (dashed) bursts. The logarithmic mean values are given by $\langle E_p \rangle_1 = 140 \pm 35$ keV and $\langle E_p \rangle_6 = 355 \pm 42$ keV. The uncertainties in this work include contributions from both the instrumental uncertatinties as well as those due to finite sampling of the parent distribution. In addition, the uncertainties in logarithmic means are nominally quoted in linear space on both the upper and lower side of the measurement, but as a simplification, we will conservatively adopt the larger of the two and apply it in both directions. We note that the KS test rejects the hypothesis that the two distributions shown in Figure 1a are sampled from the same parent distribution with a KS probability of $P(d, N) = 7. \times 10^{-6}$.

Figure 1b shows the distributions of T_{90} for the weakest (dotted) and the brightest (dashed) bursts. The logarithmic mean values are given by $\langle T_{90} \rangle_1 = 18 \pm 5$ s and $\langle T_{90} \rangle_6 = 8 \pm 3$ s. However, with a KS probability of $P(d, N) = 0.14$ the KS test can not reject the hypothesis that the two distributions are drawn from the same parent population.

The time dilation factors and energy shift factors between all bins and the bin containing the strongest bursts were calculated. The TDFs calculated using the logarithmic means of T_{90} and T_{50} are shown in Figures 2a and 2b, respectively. The ESFs for the same binning are shown in Figure 2c. The

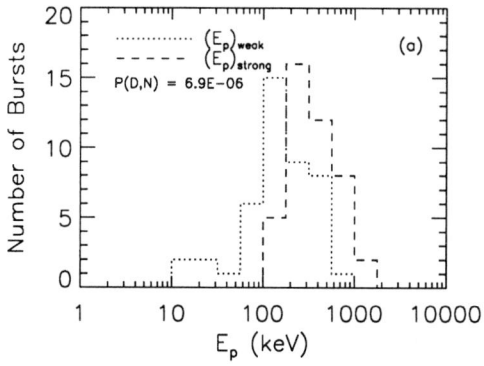

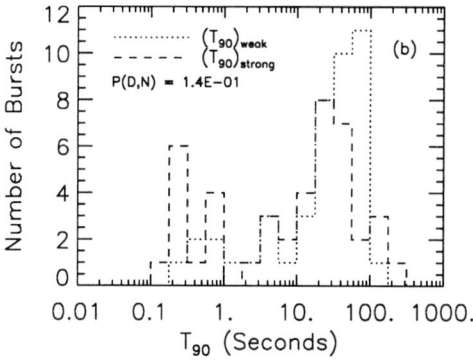

FIG. 1. Distributions of (a) E_p and (b) T_{90} for the weakest and strongest bursts.

TDFs are consistent with the hypothesis that time dilation effects are present in the BATSE T_{90} and T_{50} data. We also find that the ESFs are consistent with the observation of redshift effects previously reported (4). In addition, the TDFs and the ESFs are consistent with each other for each of the bins. However, the TDFs are much less restrictive; they are also consistent with the hypothesis that time dilation is not present in the BATSE data.

CONCLUSIONS

We analyzed the BATSE data set, being careful to avoid the introduction of selection effects, for consistency with the hypothesis that cosmological effects are present in the data. Using T_{90} and T_{50}, we find time dilation factors that are consistent with the presence of cosmological effects and that are consistent with those previously reported in other investigations. We also find energy shift factors that are consistent with the time dilation factors found in this

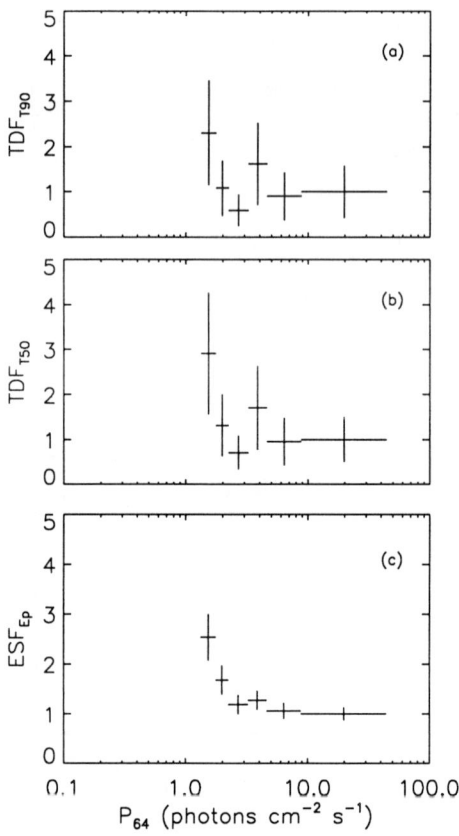

FIG. 2. Measured TDFs and ESFs as a function of intensity.

study, as well as those found in other (though not completely independent) investigations. The time dilation factors found are also consistent with the hypothesis that time dilation is not present in the BATSE data.

REFERENCES

1. J.P. Norris, et al., ApJ **439**, 542 (1995).
2. R. Wijers & B. Paczynski, ApJ **437**, L107 (1995).
3. C. Kouveliotou, et al., ApJ **413**, L101 (1993).
4. R. Mallozzi, et al., ApJ **454**, 597 (1995).
5. C.A. Meegan, et al., ApJ Suppl., in press (1996).

Clusters of cosmic Gamma-Ray Bursts as a manifestation of the local sources

A.V. Kuznetsov

Space Research Institute, Moscow

We have searched clusters of cosmic gamma-ray bursts (GRBs) in the 2B(3B) catalog and in pre-BATSE information. Our search is based on a selection of events coinciding in space and time. A study of the clusters found this way, shows that their angular size is $\sim 30°$ and their time of activity (lifetime) is about 6 months. The best interpretation of these results is that GRB sources are of a local (heliospheric) origin. The main characteristics of some clusters (sources) are collected in a table, which can serve as a basis for creating a catalog of GRB sources. The heliospheric model of GRBs (3) predicts, that the local sources should exist a limited time because of their movement away from the Sun. In this case reduction in time of the GRB fluence from a given source should be observed. Signatures of intensity reduction from a source have in fact been observed. These effects in the framework of the heliospheric model enable us to estimate the burst energies and the distance to the burster sources.

INTRODUCTION

The information provided from the BATSE GRB database is unique in its volume and its continuity and has in many respects supplied a new view on the nature of the cosmic gamma-ray bursts (2). This has resulted in extremely high interest to this phenomenon at present, enhanced by the understanding that we have sufficient data to explain the main characteristics and maybe solve the enigma of the GRB origin. The isotropic and inhomogeneous GRB distribution in space (4,1) leads to the conclusion, that the cosmological models of GRB or the models associated with the heliosphere have a priority. Several authors have researched the small-scale clustering to find gamma-ray burst repeating sources. Quashnock & Lamb (5) in the 1B catalog find that classical bursts repeat on timescales of the order of months. Wang & Lingenfelter (6) have discovered five classical bursts, which, they believe, are emitted from one source GBS 0855−00. The analysis of the repeating gamma-ray bursts in the 1B catalog (3,7) has shown an excess of GRB pairs which are associated in both space and time. The time interval between the bursts of each pair is ~ 4 days or less.

In this paper we consider the suggestion (3) that the GRB clusters can point

© 1996 American Institute of Physics

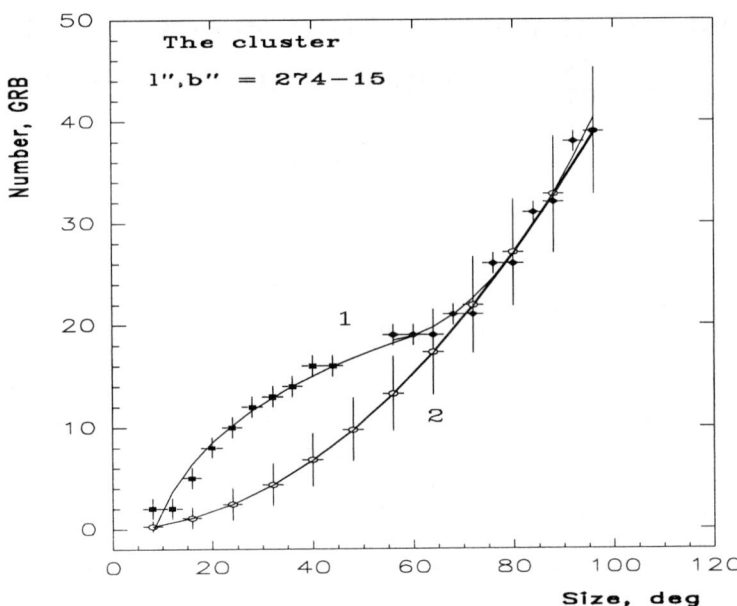

FIG. 1. Dependence of the number of GRBs in a typical cluster on the cluster area.

to local (heliospheric) sources. This work was carried out on the 2B catalog data; the results are confirmed by preliminary analysis on the 3B catalog.

METHOD OF THE SEARCH AND ANALYSIS OF THE RESULTS

Only bursts with a statistical error $\leq 10°$ in their locations are selected for this analysis. There are 422 such events in 2B. First, events which coincide in position within the limits of 2σ of their error boxes (criterion 2σ), are picked out. Then the group of gamma-ray bursts selected on this space signature, is limited in time and a geometrical centre of possible cluster is determined. The significance of a given cluster (source) was confirmed by a statistical estimate. A typical picture of a GRB cluster is presented in Fig. 1.

We have constructed the experimental (curve 1) and calculated (curve 2) dependencies from the centre of a possible GRB cluster on the area (Fig. 1). The calculated curve is constructed by assuming a uniform distribution of the total number of GRBs in the area, and assuming an absence of any clustering. After correction on the GRB cluster number, the calculated curve can serve as a background for a given cluster. By this method the areas containing 4 or more GRBs stood out against the background with good reliability. The detected sources were checked using other coordinate systems.

For the statistical estimation of the GRB excess we used the mean value of the background. The background is well fitted by the dependence of type $Y = A_1 + B_1 * x + C_1 * x^2$ with typical $\chi^2_{10} < 10^{-3}$ and a cluster is approximated

TABLE 1. The local GRB sources (LGBS) in the 2B & 3B catalogs.

N	LGBS(l",b") center, deg	Time of activity	Day number,d	Size, deg	GRB number	The level of significance
1.	2B: 245 - 1	910425-920218	299	30 ± 2	8(10)	5.5 σ
	3B: 245 - 1	910827-920501	248	34 ± 2	12(13)	
2.	2B: 91 - 30	910425-920302	312	34 ± 2	12(14)	6.6 σ
	3B: 89 - 30	- " -	- " -	30 ± 2	9(10)	
3.	2B: 222 + 20	910430-910827	119	30 ± 2	6(7)	6.2 σ
	3B: 223 + 26	910430-910722	83	12 ± 2	4(4)	
4.	2B: 141 + 12	911119-920617	211	26 ± 2	8(9)	7.3 σ
	3B: 136 + 15	- " -	- " -	36 ± 2	8(10)	
5.	2B: 355 - 30	920222-920718	147	34 ± 2	8(9)	7.0 σ
	3B: 349 - 28	920325-930310	350	30 ± 2	12(16)	
6.	2B: 274 - 15	920302-930204	339	42 ± 2	12(16)	5.6 σ
	3B: 273 - 14	920302-930125	329	38 ± 2	12(16)	
7.	2B: 3 + 22	920303-930127	330	34 ± 2	9(11)	4.0 σ
	3B: 3 + 25	920314-921022	222	34 ± 2	9(11)	
8.	2B: 118 - 9	920913-930219	159	26 ± 2	7(8)	8.0 σ
	3B: - " -	- " -	- " -	- " -	- " -	
9.	2B: 309 - 0	930124-930220	28	18 ± 2	4(4)	18.0 σ
	3B: 308 - 0	930206-930503	88	30 ± 2	5(5)	
b) on the preBATSE data						
1.	308 - 0	820825-821124	92	32 ± 2	5(6)	4.9 σ

by a curve of type $Y = A_2 + B_1 x + C_1 x^2$ with typical $\chi_6^2 < 1$. The size of a cluster is determined by comparing the differential curves, the theoretical and real, and the level of significance for a source is determined from the integral curves.

A list of the most interesting sources is presented in Table 1. The main characteristics of the sources are: the coordinates of centre, the time of activity (lifetime), the angular size and the burst number observed from the source. This table can serve as a basis for creating a the catalog of gamma-ray burst sources. Evidently it is possible to assume that the source can consist of 3 or 2 bursts. So the existent catalogs are practically the catalogs of GRBs. For the sensitivity of BATSE instrument the maximum number of the bursts in the source is about 15 and the time of activity is less than 1 year.

The heliospheric model of gamma-ray bursts based on the real phenomenon – the magnetic field reconnection in the magnetosphere of a solar coronal mass ejection under propagating in the heliosphere (3) – predicts a limited lifetime and the decreasing of the radiation flux from the source because of its moving away from the Sun. The time of activity (limited lifetime) is shown in Table 1. The more interesting effect is the fluence decrease in time from the bursts of one source because it gives us the possibility to estimate the burst energy and the distance to the burst. We will present 2 examples on the 3B catalog. We observe this effect in the source, which was first found by Wang & Lingenfelter (6). The first source is number 3 from Table 1, which includes

4 bursts inside a 12 degree box: 910430, 910501, 910626 and 910722 with the corresponding fluences 3.87^{-5}, 5.75^{-6}, 1.5^{-6}, 8.77^{-7} erg/cm^2. Second is source 6 from table 1, which includes 5 bursts in the limits of a 36 degree box: 920511.1, 920617.1, 920619.1, 920802 and 920924 with corresponding fluences 2.08^{-5}, 5.36^{-6}, 4.16^{-6}, 1.34^{-6} and 7.54^{-7} erg/cm^2. The random probability for a sequence of the five events is $8*10^{-3}$.

With reasonable assumptions we can calculate the distance to the bursts and its energy. The observed damping of the source in terms of the heliospheric model can be explained if the burst energy will practically have the constant value for all bursts of given source. Equating the energies of two bursts we will get: $F_1 * R_1^2 = F_2 * R_2^2$, where F_1 and F_2 are the measured fluences from the bursts, R_1, R_2 are the distances from the observer to a burst. Then $R_2 = R_1 + S$, where S is the distance between these bursts. $S = V_{sw} * (T_2 - T_1)$, where V_{sw} is the solar wind velocity, which we take to be 500 km/s for the calculations. So the calculated distances for the 920511.1 and 930924 GRBs of the source $l^{II} = 273°$, $b^{II} = -14°$ are 9 and 48 AU. The energy of these bursts is $4*10^{22}$ ergs.

SUMMARY

In this work it is shown that gamma-ray bursts clustered both in time and space which point to, as we believe, the heliospheric origin of this phenomenon. For the experiments with better sensitivity the maximum number of GRBs in the source and activity time (lifetime) must be increased. The detected sources are the middle-scale clustering of GRBs with a size which exceeds significantly the burst location error. The presence of GRB sources with large sizes by itself speaks in favor of the heliospheric origin. For the galactic extended halo, for example, the linear size of the object, which is the GRB source, will be of order of 100 kpc. The most important effect which can confirm the nearest origin of gamma-ray bursts and provide the practical calculations is the regular decrease of fluence from the GRBs of one source.

On the basis of estimations we can assume that the sources begin to radiate gamma-ray bursts on distances from the Sun of about 5 AU and the boundary of sources observation is near 100 AU. The sensitivity of BATSE permits to observe the GRBs of a local origin with energy 10^{25} ergs at a distance from the Sun of order of 1000 AU.

REFERENCES

1. E.E. Fenimore, et al., Nature **366**, 40 (1993).
2. C. Kouveliotou, ApJ Suppl. 92, **637** (1994).
3. A.V. Kuznetsov, Preprint 1913, Space Research Institute, Moscow (1995).
4. C.A. Meegan, et al., Nature **355**, 143 (1992).
5. J.M. Quashnock, & D.Q. Lamb, MNRAS **265**, L59 (1993).
6. V.C. Wang & R.E. Lingenfelter, ApJ **416**, L13 (1993).
7. V.C. Wang & R.E. Lingenfelter, ApJ **441**, 747 (1995).

GRB Brightness Ratio Distribution Analysis

J. G. Laros*

University of Arizona, Tucson, AZ 85721

> The objective of this analysis is to obtain insight into whether positionally close pairs of GRBs are due to repetitions, clustering, or random chance. We consider the Brightness *Ratio* Distribution (BRD) of pairs of events. Here, brightness is used as a generic term for any quantity related to the observed intensity of an event. The BRD has the interesting property that if one can select pairs whose components are at the same distance—such as, by considering only close-together pairs—then the distance dependence "drops out" of each brightness ratio and the BRD becomes narrower because its width no longer has a component caused by the sources' differing distances. We have begun to apply this analysis to the BATSE events for which location and brightness data are available, comparing the BRD for close-together event pairs to the BRDs for the other (presumedly unrelated) pairs. Preliminary results do not show any clear indication that close-together pairs are related. However, this work is at a very early stage with regard to optimizing the method and understanding its properties.

INTRODUCTION

The question of whether or not positionally close-together pairs in BATSE catalogs are physically associated has attracted attention and controversy since the 1B Catalog became available. The central problem is that the BATSE locations are not precise. Even with the improved locations of the 3B Catalog, one must deal with a background of thousands of random-chance close pairings, where "close" means that the members of the pair have a reasonable probability of actually being connected. It is logical to try to reduce this background by incorporating as much as possible of the available information on each GRB. In a previous analysis (1) the time of occurrence of each burst was used along with the location information to search for spatial-temporal correlations against an expected background of only a few pairs. Unfortunately, the initially tantalizing results using 1B data were not borne out as more bursts were detected. In this paper we present a new analysis that uses primarily the intensity data to augment the location information. The method also can be extended to incorporate temporal and other data. We will first describe the method, and then present preliminary results obtained with the 3B Catalog.

© 1996 American Institute of Physics

THE METHOD

Consider the Brightness *Ratio* Distribution (BRD) of pairs of events.

$$R_{ij} \equiv \frac{B_i}{B_j} = \frac{\mathcal{B}_i f(D_i)}{\mathcal{B}_j f(D_j)} \qquad (1)$$

Here, brightness B is used as a generic term for any quantity related to the observed intensity of an event. $\mathcal{B}$ is the intrinsic source brightness and $f(D)$ is some function of distance, such as $1/D^2$ or (for cosmological models) something more complicated. Like $\log N$-$\log P$ or $V/V_{\max}$, the BRD contains convolved distance and intrinsic luminosity information plus experimental effects. If the ratios are all calculated with the same time sense–e.g., with the later brightness always in the numerator–then the BRD contains additional information. Moreover, if one is somehow able to select pairs whose components are at the same distance–such as, by considering only close-together pairs–then the distance functions in each ratio R cancel (regardless of the specific function and even though the distances to each pair are different) and the BRD becomes narrower than for unrelated pairs because its width no longer has a component caused by the sources' differing distances. The extreme case is "standard candle" pairs, whose BRD is a delta-function. If GRB sources tend to repeat with a bright/dim sequence, then the BRD will be skewed or offset. If close pairs are from different sources within the same distant cluster of galaxies, then the BRD will likewise have no distance spreading, but will be symmetrical. In practice, one would hope to see a related-event BRD superimposed on noise from the random alignments.

Clearly, this technique has the potential to return a great deal of information on GRB sources if a positive signal is clearly seen. However, the analysis has two undesireable features that are most prominent when there is no apparent signal, making it difficult to obtain quantitative upper limits. First, the limits will be strongly model-dependent. For example, the technique is much more sensitive to standard-candle repetitions (delta-function BRD) than it is to repetitions characterized by a broad luminosity function. Second, the BRD analysis is intrinsically more complicated than the more familiar distribution analyses it is intended to transcend. The statistics of non-independent pairs is tricky, and the shape of the distribution curve does not readily reveal its information content. The latter feature is mitigated considerably by the fact that the BATSE data set contains a vast reservoir of far apart (thus, presumedly unrelated) event pairs for tests of various types and to provide a control sample.

RESULTS

We have begun to apply this analysis to the BATSE events for which location and brightness data are available, comparing the BRD for close-together event pairs to the BRDs for the other pairs. The data set used was an 807-event subset of the 3B Catalog. We considered only events with cataloged intensities, and we used a random number generator to discard events in such a way as to produce uniform effective sky coverage. We ran several different trials, varying certain analytical details. "Close together" was defined in two different ways: Using the same fixed angle as the criterion for every event pair, and using a variable angle determined from the published location uncertainty of each GRB. We tried both the 256-ms peak flux and the 50-300 keV fluence as brightness measures. Finally, we carried out the analysis for all $n(n-1)/2$ possible pairs and with the data set divided into 10-event blocks to approximate the Wang and Lingenfelter (1) analysis augmented with intensity information. Figure 1 shows the results of one particular trial. The two curves are normalized by the ratio of the solid angles within and outside the 13.729 degree angle. (We determined that 13.729 degrees yields nearly the optimum signal-to-noise ratio for the fixed-angle case, given the published BATSE position uncertainties.)

Brightness Ratio Distribution

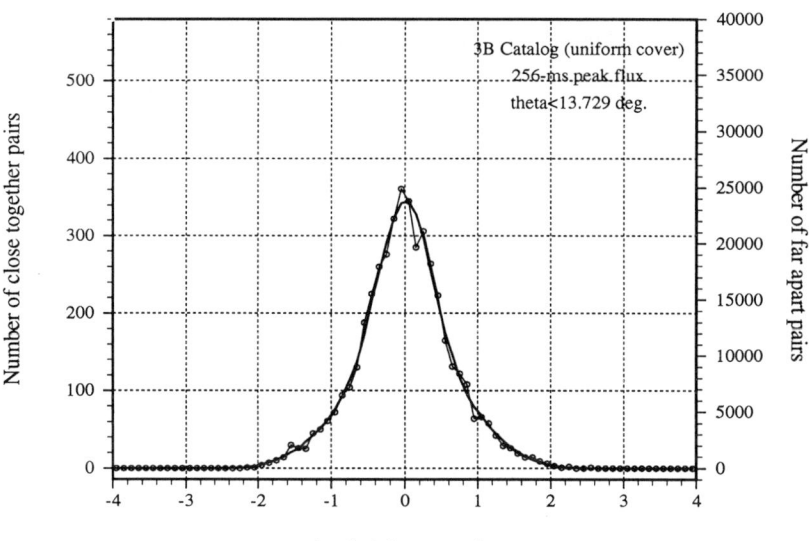

FIG. 1. BRDs for 3B Catalog pairs closer together than and farther apart than a fixed angle of 13.73 degrees (fine line with circles and heavy line, respectively). 256-ms peak flux is the intensity measure.

It can be seen that the curves coincide to within the statistical errors, which are given approximately by the square roots of the numbers of counts in each data point. We chose to display one of the "fixed angle criterion" trials because of a complication associated with the variable criterion. Namely, weak GRBs (with typically larger position uncertainties) are over-represented in the "close-together" curve, thus changing its shape in a way that we do not yet quantitatively understand. In none of our trials did we see anything that could be construed as a significant positive signal. We derive uncorrected model-dependent upper limits of roughly 10 to 100 related pairs for the trials that we ran. These limits will not become much lower as time goes on because of the way the number of background pairs increases with time. On the other hand, the number of physically related pairs could begin increasing more rapidly if the typical repetition time for a burster is a few years. Of course, there exist many other possible trial variations that might reveal a connection between close-together pairs. One could even use a quantity such as hardness or peak-width in place of brightness. Unfortunately, the required significance for a positive result increases with the number of trials. A more fruitful approach for the future would probably be to first run controlled tests designed to produce a better understanding of the strengths and weaknesses of this analysis, and then to apply the analysis selectively in a few trials guided by the burster models.

REFERENCES

1. V.C. Wang & R.E. Lingenfelter, ApJ **441**, 747 (1995).

GRB Localizations from BATSE, Mars Observer, and Ulysses Observations

J. G. Laros[*], W. V. Boynton[*], K. C. Hurley[†], C. Kouveliotou[‡],
G.J. Fishman[‡], C. A. Meegan[‡], T. L. Cline[§], D. M. Palmer[§], R.
D. Starr[§], J. I. Trombka[§], M. Boer[**], M. Niel[**], M. Sommer[††],
and A. E. Metzger[‡‡]

[*] *U of Arizona, Tucson, AZ 85721*
[†] *UC Berkeley, Berkeley, CA 94720*
[‡] *NASA/MSFC, Huntsville, AL 35812*
[§] *Nasa/GSFC, Greenbelt, MD 20771*
[**] *CESR, F-31029 Toulouse Cedex, France*
[††] *MPI, 8046 Garching, Germany*
[‡‡] *JPL, Pasadena, CA 91103*

Between 1992 October 4 and 1993 August 1, 11 GRBs were concurrently observed by CGRO, Mars Observer (MO), and Ulysses. Unfortunately, for all but one of the events MO was in a low-time-resolution mode. Arrival time analysis was used to compute error boxes whose larger dimensions are typically of the same order as the diameters of the BATSE-only boxes and whose smaller dimensions are in the arc minute range. By and large, our arrival-time results are consistent with the BATSE RMS errors stated in the 3B Catalog. That is, the mean uncertainty in the locations of the stronger BATSE events appears to be about 2 degrees.

INTRODUCTION

Accurate GRB locations are obviously valuable for such purposes as counterpart searches and to gain insight on whether or not the sources repeat. They are also needed for confirmation of the BATSE localization algorithms and claimed accuracy, which is the main subject of this paper. In this context we will discuss the 11 IPN localizations that use data from CGRO, MO, and Ulysses. Quantitative information on the error boxes will appear in a future journal publication.

OBSERVATIONS

Between 1992 October 4 and 1993 August 1, concurrent coverage by CGRO, Mars Observer (MO), and Ulysses, was obtained for 78 GRBs. Although most were below the MO and Ulysses thresholds, eleven (Table 1) were positively

© 1996 American Institute of Physics

TABLE 1. Bursts observed by CGRO, MO, and Ulysses

Date	Time @ CGRO	BATSE#	Fluence[a]
93/ 1/20	23:31:40	2138	3.1E-05
93/ 1/27	16:14:25	2149	7.1E-06
93/ 6/12	00:44:17	2387	1.7E-05
93/ 6/12	08:53:57	2389	1.8E-05
93/ 7/ 4	16:49:05	2428	1.2E-05
93/ 7/ 6	05:13:30	2431	1.5E-05
93/ 7/20	14:35:09	2450	2.1E-05
93/ 7/24	16:46:27	2461	[b]
93/ 7/31	03:16:31	2475	[b]
93/ 7/31	06:23:56	2476	4.3E-06
93/ 8/ 1	02:12:13	2477	1.6E-06[c]

[a]From 3B Catalog, $\geq$ 100 keV.
[b]Could not be determined because of data gaps.
[c]Affected by occultation.

detected by all three spacecraft. For all but one of the events MO was in its "mapping" mode, which had a typical time resolution of only 25 s. The resulting error boxes are long and narrow, with larger dimensions that are generally of the same order as the diameters of the BATSE-only boxes and smaller dimensions in the arcmin range. The exception, GB930706 (BATSE# 2431), had an arcmin error box that was observed by Barthelmy, Palmer, and Schaefer (1) and Palmer et al. (2) within 8 days of the event.

RESULTS

Table 2 compares the angular difference between the BATSE and IPN locations (column 3) to the stated BATSE errors (column 4). The average values of those two quantities (not including event #2475) are 2.0 deg and 1.9 deg, respectively. Thus, our arrival-time results appear to be consistent with the BATSE RMS errors stated in the 3B Catalog. As can be seen from the table, we did encounter a serious difficulty when attempting to incorporate MO data for event #2475. We believe that the apparent MO response might be spurious or that a BATSE data gap is somehow involved, but the matter is not yet resolved. Initially, we had a similar problem with BATSE #2477, but we are now reasonably certain that this was caused by the source being occulted mid-burst. In neither case is there evidence that the BATSE-only location should be questioned. Instead, we conclude that arrival time analysis using BATSE data must be undertaken with an eye on the 3B "comments" file and on the geocenter angle.

TABLE 2. BATSE/IPN Comparison

BATSE#	Date	Delta[a]	Sigma[b]	BATSE Comment
2138	93/1/20	6.4 deg.	2.6 deg.	Solar flare during the event at SPEC detector 2[c]
2149	93/ 1/27	1.0	2.0	
2387	93/ 6/12	0.0	1.6	
2389	93/ 6/12	0.1	1.6	
2428	93/ 7/ 4	2.5	1.7	
2431	93/ 7/ 6	2.7	1.7	
2450	93/ 7/20	2.4	1.7	
2461	93/ 7/24	3.2	2.0	Extensive data gaps. Duration cannot be determined.
2475	93/ 7/31	21 or 0[d]	2.7	Extensive data gaps. Duration cannot be determined.
2476	93/ 7/31	1.0	1.7	
2477	93/ 8/ 1	0.3	2.3	[e]

[a] Delta is the angular difference between the IPN and BATSE locations. It is measured exactly only for BATSE# 2431. For the other IPN boxes, which are long and narrow, it has been estimated as 'closest approach x sqrt(2)'.

[b] Sigma is the 1-sigma uncertainty, defined as sqrt(1.6**2+stat**2), where 'stat' is the statistical error listed in the 3B Catalog.

[c] The solar flare was not visible in the BATSE LADs or at the other spacecraft, so it should not have affected the localizations.

[d] DELTA for this event is near zero only if one ignores MO. The MO response does not look like the GRB seen by BATSE, and might possibly be spurious. Also, there are indications that BATSE could have missed the early part of the event, which would affect the IPN localization. This is under investigation.

[e] Geocenter angle is listed as 72.51 deg. BATSE/IPN consistency requires the source to have been occulted by the Earth during outburst. It may be possible to use this information to refine the error box at a later date.

REFERENCES

1. S. Barthelmy, D. Palmer, and B. Schaefer, in Gamma Ray Bursts, eds. G. Fishman, J. Brainerd, and K. Hurley, AIP Conf. Proc. **307**, 392 (AIP, New York, 1994).
2. D.M. Palmer, T.L. Cline, J.G. Laros, K. Hurley, G.J. Fishman, and C. Kouveliotou, Astrophys. & Sp. Sci. **231(1-2)**, 315 (1995).

Threshold Effects on Gamma-Ray Burst Distributions

Theodore T. Lee[†] and Vahé Petrosian[†]

[†] *Center for Space Science and Astrophysics*
Stanford, California 94305

Many of the important conclusions about Gamma-Ray Bursts follow from the distributions of quantities such as the peak flux. Because of the highly transient nature of the bursts, multiple selection thresholds can lead to various forms of data truncation, which can strongly affect the distributions obtained from the data if not accounted for properly. Therefore, a full multivariate analysis is required. Properly accounting for the effects of truncations and correlations, we extract the distributions of flux and fluence from the BATSE 3B Gamma-Ray Burst catalog. We find that the slope of the $\log N$–$\log S$ relation at low values of S is increased, and that the fluence distribution has a sharper break in it than the peak flux distribution.

INTRODUCTION

BATSE 3B GRBs are subject to two threshold effects due to their extreme variability and short durations. The first effect decreases the trigger sensitivity at low fluxes due to variations in the background count rate. The second effect decreases the trigger sensitivity at short durations, because the integration time of the trigger can exceed the duration of the burst (1). Proper treatment of the data is in terms of a multivariate distribution (2). Many of the important conclusions about GRBs follow from single variable distributions (such as $\log N$–$\log S$) obtained from the multivariate distribution. For accurate distributions it is imperative that the effects of data *truncation* and *correlations* between variables are handled properly. We use methods to handle complex truncations in order to utilize the largest possible data set. In this paper we describe the basic method and its application to obtaining the $\log N$–$\log S$ distributions of peak flux and fluence. The correlations and comparison with cosmological models are described in two accompanying papers (3,4).

THE ANALYSIS METHOD

The steps below describe a general procedure to obtain single variable distributions from the multivariate data. First, we clearly define the threshold effects and the resulting data truncation boundaries. Then, we define the

© 1996 American Institute of Physics

associated sets for the data points. The associated set for data point i is the largest untruncated data set associated with point i (see reference (2) and references therein for more details). Next, we test for correlations. For determining correlations, each point is only compared to the points in its associated set in order to avoid obtaining false correlations due to data truncations (5). In the event the test shows a correlation, the resulting single variable distribution may be biased due to the combination of the truncation and the dispersion in the correlation. We transform the data via a simple power-law parametrization in such a way as to remove the correlation (6,7). The data truncation boundaries are transformed accordingly, which of course changes the associated sets. Finally, using these new associated sets, we obtain the single variable distributions. Each data point forming the distribution is weighed by a function of the number of points in its associated set (8).

DEFINING THE THRESHOLD EFFECTS

We first define the threshold effects appropriate for the BATSE 3B data. The first important effect, arising from a variable threshold count rate $\overline{C}_{\text{lim}}$ averaged over a trigger integration time Δt, leads to a truncation of the form

$$\overline{C}_P > \overline{C}_{\text{lim}}, \tag{1}$$

where $\overline{C}_P$ is the peak flux of the burst averaged over the trigger integration time. However, for comparison with models we need the distribution of physically more meaningful quantities such as the peak photon flux or the photon (or energy) fluence. The BATSE catalog, in addition to providing $\overline{C}_P$ and the ratio $c \equiv \overline{C}_P/\overline{C}_{\text{lim}}$, provides the average peak photon flux $\overline{f}_P$ (in cm^{-2} s^{-1}) and the energy fluence $\mathcal{F}$ (in erg cm^{-2}). These quantities depend on the angular orientation of the spacecraft with respect to the burst and on the burst spectrum. For a given burst i with measured $\overline{f}_{P,i}$ and c_i, it is clear that the threshold flux (averaged over Δt) would be $\overline{f}_{\text{lim},i} = \overline{f}_{P,i}/c_i$. If the flux was lower than this limit, the peak photon count rate would have been smaller than $\overline{C}_{\text{lim},i}$ and the burst would not have triggered. Thus we may transform the data and analyze the distributions of $\overline{f}_P$ subject to the threshold $\overline{f}_P > \overline{f}_{\text{lim}}$.

Similar thresholds can be defined for the energy fluence $\mathcal{F}$, except now the pulse profile and spectrum also come into play. For a burst of measured fluence $\mathcal{F}_i$ with its unique spectrum, light curve, and angular position, the limiting fluence will be $\mathcal{F}_{\text{lim},i} = \mathcal{F}_i/c_i$. Bursts with similar spectral, temporal, and spatial properties but with $\mathcal{F} < \mathcal{F}_{\text{lim},i}$ would have a peak count rate below the threshold $\overline{C}_{\text{lim}}$ and would not have been observed. Thus we can obtain the distribution of $\mathcal{F}$.

A second important selection effect on the peak flux occurs because the integration time of the trigger is not infinitely short. The "peak flux" $\overline{f}_P$ is averaged over the trigger time Δt, meaning that for bursts with durations $T \lesssim \Delta t$, $\overline{f}_P$ represents the photon fluence rather than the peak flux. To estimate the true peak flux f_P we use the approximation

$$f_{\rm P} = \overline{f}_{\rm P}\left(\frac{T+\Delta t}{T}\right). \tag{2}$$

As indicated by the ratio of average peak count rates at two different values of Δt (1,2), this seems to work for most burst pulse shapes. This correction raises both the peak and limiting flux for short duration bursts, especially for the $\Delta t = 1024$ ms triggered data. Following the same argument outlined above it is clear that $f_{\rm lim} = f_{\rm P}/c$ so that

$$f_{\rm lim}(T) = \overline{f}_{\rm lim}\left(\frac{T+\Delta t}{T}\right). \tag{3}$$

This transformation complicates the issue because now the duration distribution may affect the flux distribution. Even if as expected $\overline{f}_{\rm P}$ and $\overline{f}_{\rm lim}$ are independent of each other, $f_{\rm P}$ and $f_{\rm lim}$ might not be due to a possible correlation between $f_{\rm P}$ and duration. Using our methods we can test and correct for such correlations and thus obtain unbiased univariate distributions (3).

RESULTS

The resultant $\log N$–$\log S$ distributions are represented in three different ways ($N(S)$, $S^{5/2}dN/dS$, and $d\log N/d\log S$) in Figure 1 for peak flux and in Figure 2 for fluence. The peak flux distribution steepens when the selection bias due to variable threshold rate is accounted for. We also note that this distribution differs from the distribution of average peak flux $\overline{f}_{\rm P}$, due to the duration bias. We find that using different measures of duration (T_{90}, T_{50}, or $T_{\rm eff} = (f_{\rm P}\langle h\nu\rangle)^{-1}\mathcal{F}$, where $\langle h\nu\rangle$ is the average energy per photon) in equations (2) and (3) does not significantly change the results.

We find a positive correlation between fluence and duration. Thus Figure 2 shows not only the distribution corrected for the threshold effects, but also the distribution additionally corrected for the correlation. The fluence distribution has a sharper break in it than the peak flux distribution, with logarithmic slope changing rapidly from $-3/2$ to a relatively constant $-1/2$ (Figure 2). By dividing the corrected fluence distribution by the raw distribution, we may obtain the BATSE trigger efficiency as a function of fluence (Figure 3). This graph should be compared to the results of Fenimore and collaborators (9,10), who model the trigger efficiency through extensive simulations.

CONCLUSIONS

Threshold effects and the multivariate nature of the data can typically result in GRB distributions that are too shallow at small fluences or peak fluxes. We have demonstrated how to obtain the distribution of true peak fluxes and fluences from the BATSE data even though the bursts are triggered based on the peak photon count rates averaged over finite trigger intervals. These distributions are steeper than the uncorrected distributions. The observed

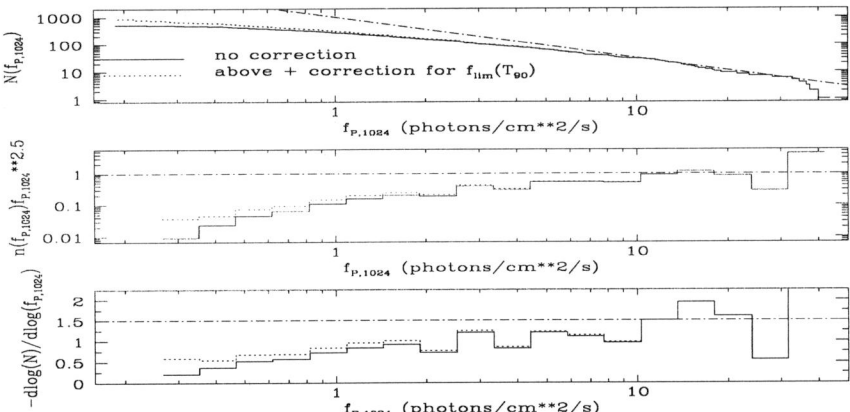

FIG. 1. True peak flux (f_P) distributions. From top to bottom: the cumulative distribution, the differential distribution multiplied by $f_P^{5/2}$, and the logarithmic slope. The solid histograms are what would be found without consideration of the truncation. The dotted histograms are corrected for varying $f_{\lim}(T_{90})$. The dot-dashed lines indicate the homogeneous isotropic Euclidean (HISE) predictions. For the middle graph, the vertical scale is arbitrary.

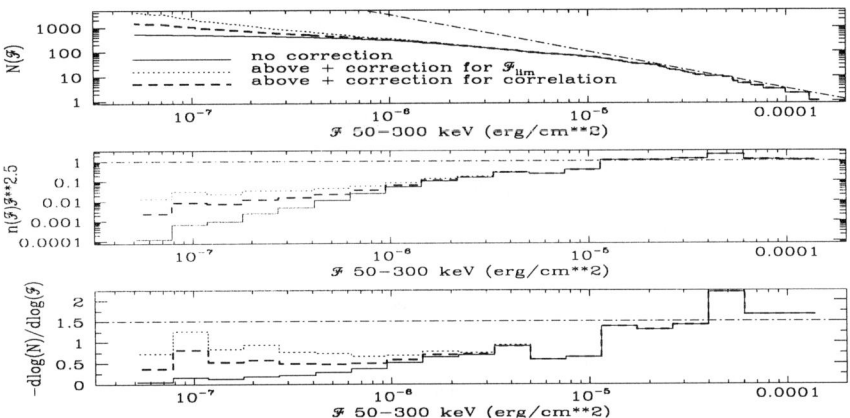

FIG. 2. Fluence ($\mathcal{F}$) distributions. From top to bottom: the cumulative distribution, the differential distribution multiplied by $\mathcal{F}^{5/2}$, and the logarithmic slope. The solid histograms represent no correction, the dotted histograms are corrected for varying $\mathcal{F}_{\lim}$, and the dashed histograms are further corrected for the significant correlation between $\mathcal{F}$ and $\mathcal{F}_{\lim}$. The dot-dashed lines indicate the HISE predictions. For the middle graph, the vertical scale is arbitrary.

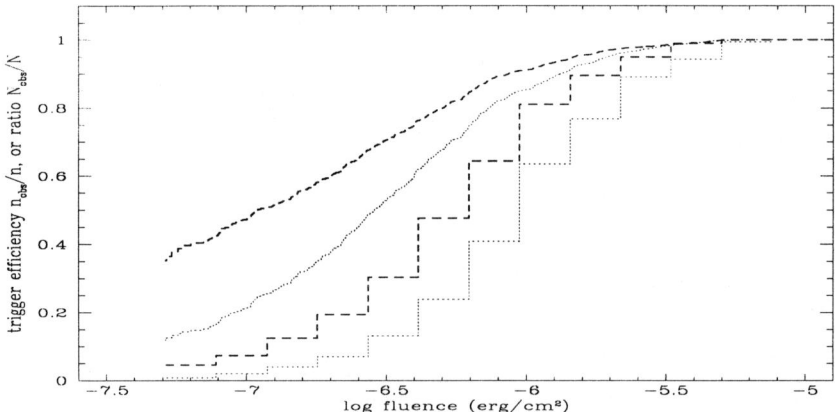

FIG. 3. The top two curves are the trigger efficiency as a function of fluence, while the bottom two curves are the ratio of the total observed number of bursts to the total corrected number of bursts. The heavy dashed lines represent ratios or efficiencies corrected for correlation, while the dotted lines are what would be obtained neglecting the effects of correlation.

sharper break in the fluence distribution as compared to the peak flux distribution may indicate that the total luminosity or energy might be a better standard candle than the peak luminosity (10,11). Analysis and comparison of the data with different trigger intervals, duration ranges, etc., will be described in future publications.

REFERENCES

1. V. Petrosian, T. Lee, and W. Azzam, in Gamma-Ray Bursts, eds. G. J. Fishman, J. J. Brainerd, & K. Hurley, AIP Conf. Proc. **307**, 93 (AIP, New York, 1994).
2. T. Lee and V. Petrosian, ApJ, **470**, in press (1996).
3. V. Petrosian and T. Lee, these proceedings (1996).
4. W. Azzam and V. Petrosian, these proceedings (1996).
5. B. Efron and V. Petrosian, ApJ **399**, 345 (1992).
6. T. Lee, V. Petrosian, and J. McTiernan, ApJ **412**, 401 (1993).
7. T. Lee, V. Petrosian, and J. McTiernan, ApJ **448**, 915 (1995).
8. V. Petrosian, in Proceedings of the Conference on Statistical Challenges in Modern Astronomy, eds. E. Feigelson & G. J. Babu, New York: Springer-Verlag, p. 173 (1992).
9. J. in't Zand and E. Fenimore, these proceedings (1996).
10. D. Bloom and E. Fenimore, these proceedings (1996).
11. C. Madras and E. Fenimore, these proceedings (1996).

Expected Gamma-ray Burst Excess from M31 Based on Halo Models

Hui Li* and Edison Liang[†]

*NIS-2, MS D436, Los Alamos National Lab, Los Alamos, NM 87545
[†]Dept. of Space Physics and Astronomy, Rice University, Houston, TX 77251

We calculate the expected excess of gamma-ray bursts (GRBs) from the direction of M31 assuming that GRBs come from an extended halo around both M31 and our own galaxy. Specifically, we consider the "delayed turn-on" model and the halo beaming model for GRBs. We express the GRB rate from M31 as functions of instrument sensitivities, and the observing solid angle towards M31. We conclude that, with a gamma-ray instrument that is ten times more sensitive than BATSE, delayed turn-on model gives a strong excess of GRBs from M31 but halo beaming model gives only a possible marginal detection in over a year of observations.

INTRODUCTION

A potential test of the galactic halo models for Gamma-ray Bursts (GRBs) is to observe the nearby Andromeda galaxy (M31) since it is also a spiral galaxy with similar mass to our own Milky-Way (MW). Presently BATSE, however, is not sensitive enough to reveal the signal from M31 (1,2). It is thus important to find out under what conditions the observations towards M31 can put strong limits on the existing halo models.

HALO MODELS

By assuming that M31 has the same high velocity neutron star (HVNS) population and the same bursting rate as our own galaxy, we calculate the expected rates of GRBs from M31. We will concentrate on two currently viable versions of halo models, (i.e. the "delayed turn-on" (DTO) model (3,4) and the "halo beaming model" (5,1), see (2)). We also discuss the implications of our results for designing future gamma-ray instruments, especially for those aiming at M31.

The Delayed Turn-on Model

Li & Dermer (3) have shown that the spatial distribution of GRBs observed by BATSE, combined with PVO (6), can be fitted by a HVNS population with

© 1996 American Institute of Physics

a gradual turn-on of the burst rate, which approaches constant after tens of million years (see also (4)). This gives an effective GRB spatial density, produced by HVNSs from our galaxy, as follows:

$$n(r) \propto \begin{cases} \text{constant} & \text{for } r \leq R_c \\ (R_c/r)^2 & \text{for } r > R_c \end{cases} \quad (1)$$

where r is the distance to us and R_c is the so-called "core-radius" (30 − 40 kpc and is fixed as 35 kpc in this study). Meanwhile, the spatial density of GRBs from M31 will have the same profile as Eq.(1) with r being the distance of GRB to M31. Note that the M31 core region extends $\sim 12°$ on the sky ($D_{\text{M31}} = 670$ kpc).

Assume that the GRBs from our galaxy and M31 have the same single luminosity L_0, the ratio R of GRB rate from M31 to our galaxy can be numerically solved using the density profile described in Eq.(1). Here, we only give an *approximate* expression:

$$R \approx \frac{\left(\frac{R_c^2}{2D_{\text{M31}}}\right) \int_0^{D_s} dr\, r \ln[1 + \frac{2D_{\text{M31}} r}{(D_{\text{M31}} - r)^2}(1 - \mu_{\text{obs}})]}{\{\frac{1}{3}\min[D_s, R_c]^3 + H[D_s - R_c]R_c^2(D_s - R_c)\}(1 - \mu_{\text{obs}})} \quad (2)$$

where μ_{obs}, D_s are the cosine of the half opening angle and the sampling depth (given luminosity L_0) of the detectors, respectively. Again, r is the distance to us. Eq.(2) gives $\frac{1}{3}(D_s/D_{\text{M31}})^2$ when $D_s/D_{\text{M31}} < 0.1$ but approaches $(R_c/D_{\text{M31}})^2$ when $D_s < R_c$. H is the Heaviside unit function.

In the left panel of Figure 1, we plot the ratio of expected GRB rates from M31 and MW from the *exact* calculations (solid curves) and Eq.(2) (dashed curves). Note that Eq.(2) fails *only* at $D_s = D_{\text{M31}}$ but fits the exact solution rather well elsewhere. Different curves correspond to different detector half opening angles. BATSE has a sampling depth of 160 − 200 kpc (2), so a very small signal is expected from M31.

The Halo Beaming Model

Another viable halo model for GRBs is the so-called "Halo Beaming Model" (HBM), which was first proposed by Duncan, Li & Thompson (5). Subsequently, it has been shown that HBM can fit the data from the first, second and third BATSE catalog (2). In this model, gamma-rays are assumed to be beamed in a cone with half opening angle ϕ_b centered along the stellar velocity vector. $\phi_b = 20°$ is favored by fitting the BATSE observations though $10° < \phi_b < 30°$ is allowed. Consequently, due to the offset of us from the Galactic center (8.5 kpc), a large fraction of the HVNSs is not visible *initially*, until they have gone beyond $\sim 8.5/\sin(\phi_b) \approx 25$ kpc, when most of them become visible to us. Note that for simplicity, all HVNSs are assumed to have velocity ≥ 800 km/s, so they all essentially move in straight lines.

Again, the observable GRBs from *our galaxy* have a spatial density similar to Eq.(1), but the *observable* GRBs from M31 form a peculiar pattern on

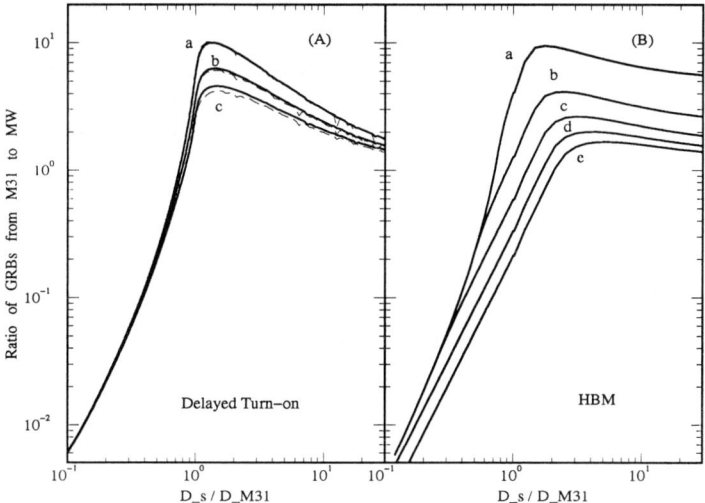

FIG. 1. The ratio of GRBs from M31 to MW as a function of the instrument sampling depth D_s. A single luminosity is assumed. *left*: The Delayed turn-on model. Solid curves (a,b,c) correspond to $\theta_{obs} = 20°$, $30°$, $40°$, respectively. The dashed curves are approximate solutions. *right*: The Halo Beaming Model. Curves (a,b,c,d,e) correspond to $\theta_{obs} = 5°$, $10°$, $15°$, $20°$, $25°$, respectively.

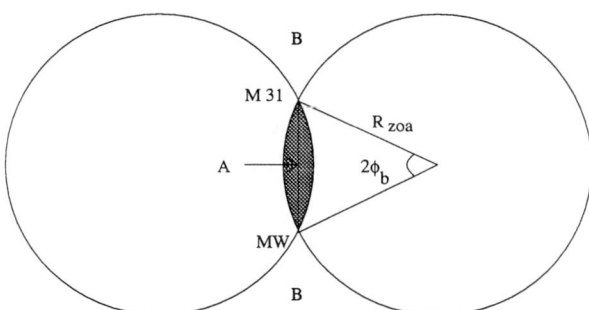

FIG. 2. A cross section of the geometry formed by the observable GRBs from M31 in HBM. Regions "A" and "B" are observable but the "torus" formed by two big circles revolving around the axis is the "Zone of Avoidance". $R_{zoa} \approx 1$ Mpc, and $D_{M31} = 670$ kpc.

the sky assuming both forward and backward beaming. Figure 2 depicts the cross section of this geometry. If we assume for a moment that all HVNSs originate from the center of M31, then only those moving towards us within ϕ_b of the axis connecting M31 and MW have a chance to be seen, we call this region "A". There is also another region "B" beyond M31 that is visible. A prominent feature is the large unobservable region, the "zone of avoidance", which is formed by two big circles (with a small overlap, region "A") revolving around the axis. The radius of the circle is $\frac{1}{2}D_{\rm M31}/\sin(\phi_b) \sim 1$ Mpc. Hence in the HBM picture, region "A" is the prime candidate for observing GRBs from M31 using the gamma-ray telescope.

Following the previous steps, we can again evaluate the ratio of GRB rate from M31 to MW. From region "A" only (i.e. $D_s \leq D_{\rm M31}$) for an instrument with half opening angle $\theta_{\rm obs}$, the ratio R_A is approximately

$$R_A = \begin{cases} \frac{1}{3}\eta^2/\kappa^2 & \text{for } \eta \leq 1, \ \kappa \geq 1 \\ 1 + \frac{1}{1-\eta} + \frac{2}{\eta}\ln[1-\eta] & \text{for } \eta \leq 1-\kappa, \ \kappa < 1 \\ \frac{1}{\eta}\{\frac{1}{\kappa} - \kappa + 2\ln\kappa + \frac{1}{3\kappa^2}[\eta^3 - (1-\kappa)^3]\} & \text{for } 1-\kappa < \eta \leq 1, \ \kappa < 1 \end{cases}$$

(3)

where $\eta = D_s/D_{\rm M31}$ and $\kappa = \theta_{\rm obs}/\phi_b$. Contributions from region "B" can also be calculated, but we omit the tedious derivations and expressions here (but see (8)).

The results of HBM are shown in the right panel of Figure 1. Note that in order to maximize the ratio of M31 to MW from region "A", the half opening angle of the instrument should be less than $10° - 15°$.

CONCLUSIONS AND DISCUSSION

It is evident from Figure 1 that, when the instrument sampling depth reaches close to M31, the ratio $R \gtrsim 1$ for most of the observing angles in DTO and for $\theta_{\rm obs} \lesssim 10°$ in HBM.

If the Log N – Log P distribution of GRBs continues to lower fluxes with a slope of $\beta = -0.7$, then ~ 4000 bursts are expected per year per 4π at the threshold flux of $\sim 10^{-8}$ ergs cm^{-2} s^{-1}. Taking the sampling depth of BATSE as 180 kpc, then an instrument which is ten times more sensitive than BATSE will have $D_s/D_{\rm M31} \approx 0.85$. In Figure 3, we summarize our results by plotting the expected number of bursts per year for a number of solid angles.

The effects of luminosity function can also be included, but we do not show any figures here due to space limit. We can readily find out the new ratio R, by adding an integral over luminosity to equation (2) and replacing D_s with $(L/4\pi f_{\rm thre})^{1/2}$, where $f_{\rm thre}$ is the threshold flux of the instrument, (i.e. including all the bursts with flux higher than $f_{\rm thre}$). The conclusions remain unchanged with the luminosity function when compared to the mono-luminosity cases presented above. But, when the luminosity function has a positive slope, (i.e. more bursts with larger luminosities), the peak of the ratio is brought to higher fluxes (see (8)).

In conclusion, for the delayed turn-on model, a strong signal ($\sim 20\sigma$) is expected towards M31 with a gamma-ray (or hard X-ray) instrument ten times more sensitive than BATSE, and a null result from M31 should be able to rule it out. But for the halo beaming model, a marginal signal is expected

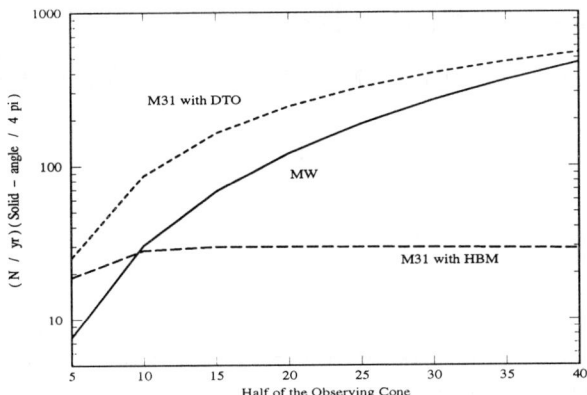

FIG. 3. Number of bursts per year in the observing solid angle with fluxes higher than 10^{-8} ergs cm^{-2} s^{-1}, as a function of the instrument half opening angle $\theta_{\rm obs}$. A slope of -0.7 is assumed for the Log N – Log P distribution extending to 10^{-8} ergs cm^{-2} s^{-1}. Solid, dashed and long-dashed curves are for MW, M31 assuming DTO, and M31 assuming HBM, respectively.

$(2-3\sigma)$. Only if $\theta_{\rm obs}$ of the instrument is $< 10°$, the excess towards M31 may be higher than 5σ. A small $\theta_{\rm obs}$ will always give a higher ratio of $\dot{N}_{\rm M31}/\dot{N}_{\rm MW}$, but the total number of observed bursts goes down dramatically, hence smaller $\theta_{\rm obs}$ is probably not useful. Note that Figure 3 gives the expected number of bursts in a *year* when aimed at M31. In practice, another year is needed to determine the blank field rate for the experiment.

Acknowledgments. Some of the initial calculations on HBM were done together with R. Duncan. H.L. gratefully acknowledges the support by the Director's Postdoc Fellowship at Los Alamos National Lab. At Rice this work was partially supported by NASA NAG5-1515.

REFERENCES

1. H. Li, R. C. Duncan, & C. Thompson, in AIP Conference Proceedings, 307: Gamma-Ray Bursts, eds. G. J. Fishman et al. (NY: AIP), 600 (1994)
2. H. Li, & R. C. Duncan, ApJL, submitted (1995a)
3. H. Li, & C. D. Dermer, Nature, **359**, 514 (1992)
4. T. Bulik, & D. Lamb, Astrophys & Space Sci., in press (1995)
5. R. C. Duncan, H. Li, & C. Thompson, in AIP Conference Proceedings, 280: Compton Gamma Ray Observatory, eds. M. Friedlander et al. (NY: AIP), 1074 (1993)
6. E. E. Fenimore, et al. Nature, **366**, 40 (1993)
7. H. Li, & R. C. Duncan, this volume (1995b)
8. H. Li, & E. P. Liang, ApJL, submitted (1995)

Constraints from 3B & PVO on the Halo Beaming Model

Hui Li* and Rob Duncan[†]

*NIS-2, MS D436, Los Alamos National Lab, Los Alamos, NM 87545
[†]Dept. of Astronomy, University of Texas, Austin, TX 78712

> We make detailed comparisons between the Halo Beaming Model (HBM) and data from the 3rd BATSE catalog, considering both the spatial and intensity distributions. The effects of a luminosity function are also discussed. In the context of this model, we find that the sampling depth of BATSE is ~ 160 kpc and the burst peak luminosity is $\sim 10^{40}$ erg/s. The distinctive anisotropies predicted by HBM are inconsistent with the 3B data at the 2σ level, (i.e., HBM can not be ruled out at the 2σ level). Future observations will be capable of placing more stringent limits. We emphasize that GRB sources born in M31 are almost entirely undetectable by BATSE in this model. We briefly discuss the energy budgets for the bursters and the mechanism of spindown alignment.

I. INTRODUCTION

The third BATSE catalog (3B), with more than 1000 gamma-ray bursts (GRBs), has revealed a nearly isotropic but inhomogeneous sky distribution of bursts (1). In addition, the Pioneer Venus Orbiter (PVO) experiment demonstrated that the intensity distribution of bright bursts is consistent with a locally-uniform density of sources (2). These facts favor cosmological and extended galactic halo models of GRBs, because the distribution of known matter about the Earth is not strongly anisotropic at these distances. Here we consider the galactic "Halo Beaming Model" proposed earlier (3–5).

II. THE HALO BEAMING MODEL

In the HBM, GRBs originate from high velocity neutron stars which emanate from the galactic disk like a "wind" extending into the galactic corona. The HBM *does not require that the bursters have any special orientation in a galactic coordinate frame*, but it does invoke the physically-reasonable condition that gamma-rays are produced only within a cone of angular radius ϕ about the star's magnetic axis $\pm\overline{\mu}$, which is assumed to be roughly aligned (within $\sim 20°$) with the randomly-oriented recoil velocity $\mathbf{V_r}$. References (3,4) discuss the physics of such beaming and alignment. As the stars sail into the halo, the fraction that are potentially detectable at Earth increases

© 1996 American Institute of Physics

in proportion to the transverse area of the beaming cone, or $\sim r^2$. This counter-balances the free-streaming trend in the density of bursters, $n \propto r^{-2}$, within a "core radius" $R_c \sim R_o/\phi$, where $R_o \sim 8.5$ kpc is a galactic disk dimension. This accounts for the "homogeneous" nearby distribution implied by PVO, and it greatly reduces the dipole anisotropy of observable bursts in the direction of the galactic center. At distances larger than R_c, *all* bursters are detectable at Earth, and the $n \propto r^{-2}$ trend prevails, accounting for the "boundedness" found by BATSE.

The particular version of HBM which we analyze here has the following simple properties: [1] bursters are born at positions distributed like young Pop. I stars in the galactic disk; [2] with randomly directed recoils $V_r = 1000$ km/s; [3] they emit GRBs at a constant rate, with [4] constant luminosity, and [5] the gamma ray emission is beamed parallel and anti-parallel to $\mathbf{V_r}$, within an angular radius $\phi = 20°$.

III. MODEL COMPARISONS WITH THE 3B CATALOG

By following a large number of neutron star trajectories in the galactic potential and including the effects of beaming, we get the sky distribution of bursts in the HBM. In reference (4), we carried out a detailed comparison of HBM with the first BATSE catalog. Here, we will follow the same steps but using the 3rd BATSE catalog (1).

A. The Log $\mathcal{N}$–Log P Distribution: There are two free parameters in fitting the Log $\mathcal{N}$–Log P distribution of GRBs using HBM: the absolute normalization (number of GRBs per year) and the peak luminosity, assuming that all GRBs are standard candles. In addition, we filtered the HBM results by the BATSE sky coverage and detection incompleteness, using tables provided by BATSE (6).

In Figure 1, we show the cumulative Log $\mathcal{N}$–Log P data from the combined 3B catalog (solid) and PVO (dashed), compared with the best-fit HBM result (long-dashed curve). Only the long-duration ($T_{90} > 2$ s) bursts are plotted, with P_{1024}, the peak photon flux in a 1024-ms time-bin, used as a brightness measure. We also tried other ways to select bursts, such as $C/C_{\min 256} \geq 1$ or $C/C_{\min 1024} \geq 1$; the resultant GRB brightness distributions are almost identical to that shown in Figure 1. For short ($T_{90} < 2$ s) bursts, significant breaks in the Log $\mathcal{N}$–Log P curve are apparent in the BATSE data. *If these breaks are true features in the GRB population, they could be an important clue to the burster physics, however, we suspect that they are artifacts of the detection process.* In any case, these breaks make fits to the short-burst intensity distribution difficult in the simple HBM, thus we avoid any analysis of the short burst data here.

The peak photon emission rate $\dot{N}$ of the HBM was adjusted to make the fit of Figure 1, giving a best-fit "standard candle" value $\dot{N}_{\text{long}} \simeq 5.4 \times 10^{46}[(1-\cos\phi)/0.06]$ photons s^{-1}. This fit has a $\chi^2 \simeq 10$ with 11 degrees of freedom. This implies a BATSE sampling depth $D \approx 160$ kpc and a peak luminosity $\dot{E} \approx 9 \times 10^{39}$ erg s^{-1} if the average photon energy is ~ 100 keV.

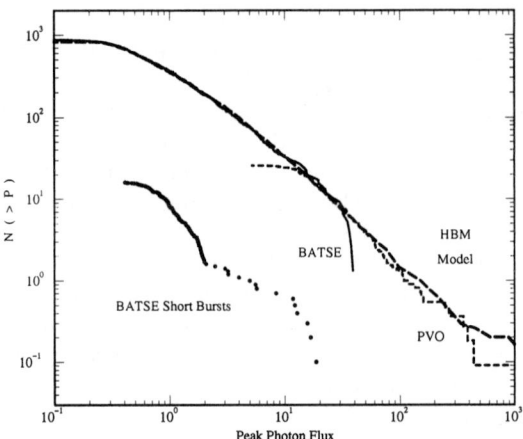

FIG. 1. Number of bursts per year, plotted versus peak photon flux P_{1024} from the 3B catalog (solid) and PVO (dashed), are compared with best-fit HBM curves (long-dashed curve). Here, we only used the *long-duration* bursts from 3B. The distribution of the short bursts alone ($T_{90} < 2$ s) is plotted in the lower left corner with filled dots. Breaks are visible. We have scaled down its normalization for clarity.

B. The Angular Distributions: In Figure 2, we plot measures of angular (an)isotropy in galactic-based coordinates (vertical axis) for the long bursts ($T_{90} > 2$ s) from 3B catalog (567 in total) in which the peak photon flux (P_{1024}) is greater than or equal to a given value, shown on the horizontal axis.

The upper graph in Figure 2, gives the dipole anisotropy toward the galactic center, $\langle \cos \Theta \rangle$. The middle graph shows the galactic disklike quadrupole, $\langle \sin^2 b \rangle$, and the bottom graph gives the galactocentric quadrupole, $\langle \cos^2 \Theta \rangle$. Large error bars on the BATSE data (filled dots) are the one σ deviations evaluated according to $\sigma(\langle X \rangle) = N^{-1/2}[\langle X^2 \rangle - \langle X \rangle^2]^{1/2}$ using the HBM, where $X \equiv \cos \Theta$, $\sin^2 b$ or $\cos^2 \Theta$. These error bars, appropriate for comparing observations with the HBM (solid lines), are slightly larger than the error bars that would be appropriate for comparing with the isotropic model (dashed lines).

The HBM curves (solid lines) make the distinctive prediction that $\langle \cos^2 \Theta \rangle > 1/3$, as explained in reference (3). Dashed lines show model predictions if gamma-ray bursts are isotropic on the sky at *all intensity levels*, as in the cosmological burster hypothesis. Note that both model curves (HBM and isotropy) are corrected for the incomplete BATSE sky coverage. The 1σ numerical (Monte Carlo) uncertainties on the solid lines are very small on the scale of this graph [see (4) for a plot of these uncertainties].

C. Results and Effects of Luminosity Functions: Figure 2 shows that the 3B catalog is statistically consistent with both isotropy and HBM, especially at the faint end, although the HBM can not be ruled out at $\sim 2\sigma$ level. Isotropy fits the observations better than HBM in every graph, unlike in our earlier study based on 1B and 2B (4,5). At the bright end, the HBM

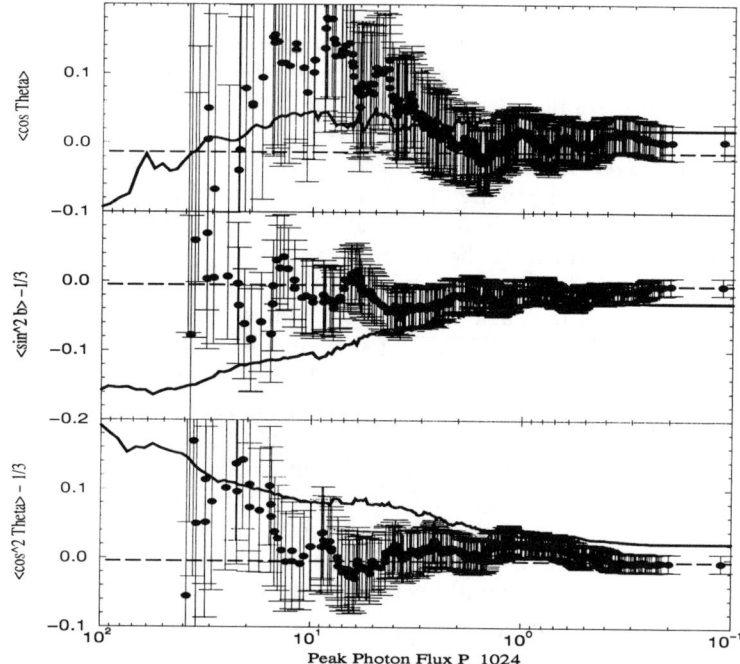

FIG. 2. Cumulative plots of angular (an)isotropy measures versus peak photon flux, for long bursts only in the 3B catalog. Solid curve is from HBM, dots are from BATSE and dashed line is the expected isotropic values. HBM is consistent with 3B at 2σ level. The errors are estimated using HBM (see text **III.B**). Both models are corrected for the incomplete BATSE sky coverage.

offers several clear predictions which may be tested as BATSE accumulates more data.

Luminosity functions could wash out some of the strong anisotropies predicted by HBM at the bright end. For example, if the luminosity function of bursts is a positive power-law, then at the same observed flux level, a greater number of more distant bursters are counted than in the mono-luminosity case. They will reduce the anisotropy signals since the high velocity neutron star distribution appears more and more isotropic at greater distances from the Earth.

D. Bursters from M31 are INVISIBLE to BATSE: Note that in the HBM, almost all bursters born in M31 are unobservable at Earth because of misdirected beaming (4). Only when the sampling depth of the instrument reaches M31 can the effects of M31 become visible (7).

IV. CONCLUSIONS AND DISCUSSIONS

- *GRBs, SGRs and Repetition Rate.* We have suggested that classic GRBs could be emitted by aged soft gamma repeaters (SGRs) that have escaped into the galactic halo (8,9). Since the age of N49 is $\sim 10^4$ yrs and the galactic SN rate is $\sim 10^{-2}$ yr^{-1}, the galactic birth rate of these bursters is roughly $10^{-4} < \Gamma < 10^{-2}$ yr^{-1}. This implies a mean GRB repetition rate of once per 10 to 1000 years in order to produce the BATSE burst rate (~ 800/yr if extrapolated to full-time, full-sky coverage).

• *Energy Budget.* Assuming each burst lasts, on average, 10 seconds, our fitted parameters imply a required total energy budget of 10^{46-48} ergs in 10^8 yrs for each burster, where we have included in the relevant beaming factor. This could easily be provided by the free magnetic energy in "magnetars" (8,10).

• *Near-Alignment.* We have argued (3,4) that *initially* Ω, $\pm \mathbf{V_r}$ and $\pm \overline{\mu}$ are roughly aligned in magnetars. As the star spins down, magnetospheric currents might drive some degree of misalignment; however, once the star spins down past the "death line" at time $\sim 10^5 \, B_{14}^{-1}$ yrs (11), further spindown certainly enforces alignment (12), since the magnetostatic stellar distortion in magnetars is large enough to damp nutations of Ω about μ (13,3).

The HBM shown here gives a acceptable fit to observations, but it requires some special conditions (e.g., uniform bursting rate, $10° < \phi < 30°$) which, although possible, have no clear and compelling theoretical justification. The cosmological hypothesis is more generic and thus more attractive in this sense. Moreover the data are generally better fitted by the cosmological models.

Acknowledgements: We thank Chris Thompson for many contributions and discussions. We also thank Ed Fenimore for providing the combined BATSE and PVO data. HL gratefully acknowledge the support of the Director's Postdoc Fellowship at LANL.

REFERENCES

1. C. A. Meegan, et al., ApJ, submitted, (1995).
2. E. E. Fenimore, et al., Nature **366**, 40 (1993).
3. R. C. Duncan, H. Li, and C. Thompson, in CGRO, eds. M. Friedlander, N. Gehrels & D. Macomb, AIP Conf. Proc. **280**, 1074 (AIP, New York, 1993).
4. H. Li, R. C. Duncan and C. Thompson, in Gamma-Ray Bursts, eds. G. Fishman, J.J. Brainerd & K. Hurley, AIP Conf. Proc. **307**, 600 (AIP, New York, 1994).
5. H. Li and R. Duncan, in High Velocity Neutron Stars and GRBs, eds. R. Rothschild & R. Lingenfelter, AIP Conf. Proc. **366**, 244 (AIP, New York, 1996).
6. The Third BATSE Catalog, GRONEWS (1995).
7. H. Li and E. P. Liang, ApJ Letters, submitted (1995).
8. R. C. Duncan and C. Thompson, ApJ **392**, L9 (1992).
9. E. E. Fenimore, R. W. Klebesadel, and J. G. Laros, ApJ, in press (1996).
10. P. Posiadlowski, M. Rees and M. Ruderman, MNRAS **273**, 755 (1995).
11. K. Chen and M. Ruderman, ApJ **402**, 264 (1993) .
12. F. C. Michel and H. C. Goldwire, Astrophys. Lett. **51**, 21 (1970).
13. P. Goldreich, ApJ **160**, L11 (1970).

Likelihood Analysis of GRB Repetition

Shan Luo, Tom Loredo and Ira Wasserman

Astronomy Department
Cornell University
Ithaca, NY 14853

We develop a framework for computing the likelihoods for GRB repetition models. Our approach is capable of fully and consistently accounting for positional uncertainty, which is crucial to resolving the issue of GRB repetition. We apply the simplest model, which assumes that each site produces at most two bursts, to various burst samples constructed from the BATSE catalogue. The uneven exposure in the BATSE catalogue is taken into account. We find no conclusive evidence of repetition in the spatial distribution of bursts over the entire observation time span. If we divide the catalogue into two-week time intervals, we find one interval with 14 bursts yielding odds of 356 to 1 on 5 repeating sites over no repetition. However, the probability of finding one such interval in the entire catalogue is not small, $\sim 4.6\%$. Nevertheless, this supports the idea that the timescale for repetition may be significantly smaller than the entire duration of the data set.

INTRODUCTION

If the directions to GRB bursts were known with negligible uncertainty, deciding whether bursters repeat would be trivial: one would merely determine whether two bursts had ever been seen from the same direction. But there is substantial uncertainty in the inferred positions for bursts, and this uncertainty is what makes the repetition issue vexing. Since directional uncertainty is the source of the difficulty, resolving the repetition issue demands a methodology that can fully and consistently take this uncertainty into account. This requires more than a simple "smearing" of the position data, as has been done by investigators using correlation functions and other statistics.

We develop a framework to compute the likelihood function for a specific model for repetition and compare it with the likelihood for a model with no repetition. We assume a Fisher distribution for the position uncertainty associated with each burst. The position uncertainty enters the computation in a natural way, and the odds in favor of a repetition model depend not only on the separation between two bursts but also on the position errors of the bursts. Because of the tremendous amount of time needed to compute the likelihood function for a modest size data sample, we limit our analysis to small subcatalogues constructed from the BATSE catalogue. Details of our method and results will appear in a forthcoming publication; we here

© 1996 American Institute of Physics

summarize our main results.

THE FORMALISM

We focus our analysis on the BATSE position catalogues. Three catalogues have been released by the BATSE team, the 1B, 2B, and 3B catalogues. The positions in the 1B and 2B catalogues are computed by the same algorithm with a systematic error of 4°. The positions in the 3B catalogue are computed by a new algorithm with systematic error reported as 1.6°. The 3B positions for the 2B bursts differ sometimes significantly from the values reported earlier. A detailed analysis by Graziani and Lamb (1) indicates that the systematic error may be closer to 4° than to 1.6°, and probably depends on burst intensity. We use a 4° systematic error for our analysis.

We distinguish gamma ray *bursts* from gamma ray *sites*. A site is the true position of a gamma ray source, which we do not know. A burst is the inferred position from a gamma ray detector, which is usually different from the true site position because of statistical and systematical errors. A burst and the corresponding site are linked by the likelihood for the position of the burst, $\mathcal{L}(\hat{n}_s|\hat{n}_i, \kappa_i)$, the probability for seeing the burst at $\hat{n}_i$ if the source site is in direction $\hat{n}_s$. Ideally this function would come from rigorous fitting of the burst data. In practice, the BATSE team provides simplified summaries of each likelihood function in the form of a best-fit direction (the inferred direction) and the angular radius of an azimuthally symmetric 68.3% confidence region (which must be enlarged to include the systematic error). In our calculations, we use these quantities to approximate the likelihood function with a Fisher distribution, namely, $\mathcal{L}(\hat{n}_s|\hat{n}_i, \kappa_i) \propto \exp(\kappa_i \hat{n}_i \cdot \hat{n}_s)$, where κ_i is determined by the angular radius of the confidence region.

For a catalogue of N bursts, a general model for repetition should consider the possibility that there are N_1 single sites each producing only one burst, N_2 double sites each producing two bursts, and so on. Under this assumption, the odds in favor of a particular model of repetition, $\{N_k\}$, over no repetition can be shown to be

$$\mathcal{O}(N_1, \cdots, N_k) = \frac{1}{\mathcal{N}} \sum_{\text{partitions}} \prod_{j=1}^{N_2} \mathcal{O}_2(\hat{n}_{1j}, \hat{n}_{2j}) \cdots \prod_{l=1}^{N_k} \mathcal{O}_k(\hat{n}_{1l}, \cdots, \hat{n}_{kl}), \quad (1)$$

where the sum is over all partitions of the bursts into a set of repetition sites $\{N_k\}$, and $\mathcal{N} = N!/\prod_{i=1}^{k} N_k!(k!)^{N_k}$ is the number of possible partitions. $\mathcal{O}_k(\hat{n}_{1i}, \cdots, \hat{n}_{ki})$ is the odds in favor of the repetition model that the k bursts are all from the same site over the model that they are from different single sites. If we assume a uniform distribution for the burst sites, we have

$$\mathcal{O}_k = \frac{\sinh(R)}{R} \prod_{i=1}^{k} \frac{\kappa_i}{\sinh(\kappa_i)}, \quad (2)$$

with $R^2 = (\sum_{i=1}^{k} \kappa_i \hat{n}_i)^2$. If the burst sites are nonuniform (e.g. uneven exposure over the sky), $\mathcal{O}_k$ can be computed numerically. The effect of position uncertainty has been naturally accounted for through the dependence of $\mathcal{O}$ on κ. In order to assess the plausibility of the repeating hypothesis, we must consider different sets of "occupation numbers" $\{N_k\}$, and calculate the probability of the data for each such choice. The full probability for the data can be shown to be an average of these simpler probabilities over all partitions, and in this way takes into account both our uncertainty regarding the repetition multiplicity, and our uncertainty about which bursts to associate with repeating sites.

SIMPLIFICATION

Clearly it is impossible to explicitly consider all possible multiplicities and site assignments. We restrict ourselves to a very simple model that each source gives at most two bursts in the observation time span, i.e., $N_k = 0$ for $k > 2$.

Even with this simplification, the number of terms to be summed over for a BATSE catalogue can easily be too large to be handled by current computers. Tests show that N has to be less than 20 in order to compute $\mathcal{O}(N_2)$ for N_2 up to $[N/2]$ on a workstation within a few days. This means we have to construct much smaller burst samples out of the BATSE catalogues to apply this approach.

It is possible that whatever physical mechanism underlies repeating bursts only act over a certain lifetime. Wang and Lingenfelter (2) have claimed to find evidence for repetition within a few days from an analysis of temporal-spatial correlation of the 1B catalogue. Therefore we divide the BATSE catalogues into subcatalogues with a time interval of two weeks. We end up with 83 subcatalogues from the 3B catalogue.

Figure 1a is a scatter plot showing the values of N_2 where $\mathcal{O}(N_2)$ peaks vs. the maximum $\mathcal{O}$ at that N_2 for the 83 subcatalogues. The numbers above each N_2 show the number of subcatalogues with $\mathcal{O}$ peaking at that N_2. For most of the subcatalogues, $\mathcal{O}(N_2)$ peaks at $N_2 = 0$ with $\mathcal{O}(0) = 1$; for a few, $\mathcal{O}(N_2)$ peaks at $N_2 \neq 0$ with $\mathcal{O} > 1$. While the peak value of $\mathcal{O} \leq 20$ typically, one subcatalogue peaks at $N_2 = 5$ with $\mathcal{O}(5) = 356$. This is from 14 bursts in the 3B-2B catalogue. Taken at face value, it means the odds in favor of 5 double repetition sites over no repetition in this sample is 356 to 1, a fairly large value. Figure 1b shows $\mathcal{O}(N_2)$ as a function of N_2 for this subcatalogue.

However, this sample is the 'worst' case among 83 samples. To calibrate the probability of getting $N_2 \geq 5$ and $\mathcal{O} \geq 356$ from an $N_2 = 0$ catalogue, we ran 10^5 simulations with 14 points randomly distributed on the sky according to the BATSE exposure map. We found 56 simulations with the likeliest $N_2 \geq 5$ and $\mathcal{O} > 356$, implying a formal probability of 5.6×10^{-4}. The probability of getting such a signal from 83 realizations is $\sim 83 \times 5.6 \times 10^{-4} = 0.046$, a suggestive, but at best speculative, signal that the repetition is real.

Another way of reducing the number of bursts in a sample is to consider only a portion of the sky. We grid the sky into equal area patches and apply our analysis to each patch separately. We assume that the sites responsible for the bursts in a patch are all within that patch.

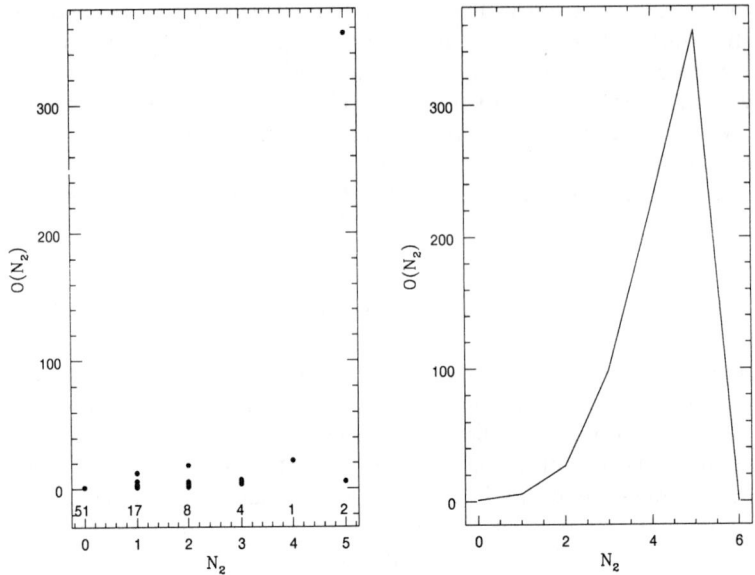

FIG. 1. Results from time slicing of the 3B cataloge. a. left: N_2 and $\mathcal{O}(N_2)$ for the maximum $\mathcal{O}(N_2)$ for each subcatalogue. b. right: $\mathcal{O}(N_2)$ as a function of N_2 for the subcatalogue which has maximum $\mathcal{O}(N_2)$ for $N_2 = 5$.

We divided the 1B, 2B and 3B catalogues into 20, 36, and 90 patches, respectively. For most of patches $\mathcal{O}(N_2)$ peaks at $N_2 = 0$, but some patches peak at $N_2 \neq 0$ with $\mathcal{O} \sim 10$ typically. To see whether these nonzero N_2's can arise from a catalogue with $N_2 = 0$, we ran 10 simulations of 262 bursts each with κ's from the 1B catalogue. No significant difference between the 1B data and the simulated data was found.

We can also get some idea about the important effect of the position uncertainty on the issue of repetition by comparing the simulations with time slicing and spatial patches. In a simulation with 14 bursts in a 2 week interval, the mean separation of bursts is about 46°, much greater than the mean position error of 7.5° of the 1B catalogue. Only rarely will two bursts in such a small sample be close enough by chance to favor a model with $N_2 > 0$. In practice, the probability of getting $N_2 = 7$ is $\sim 3 \times 10^{-5}$. In a spatial patch, the number density of bursts is much higher; the mean separation between two bursts is about 12°, comparable to the 7.5° position error. Pairs of bursts can lie close to each other by chance mimicking repetition. The probability of getting $N_2 = 7$ is now 0.045, increased by a factor of 1500. This suggests that our ability to detect or constrain repetition may not improve with larger catalogues if the position errors are large compared to the mean separation of bursts in the catalogue.

REFERENCES

1. C. Graziani and D.Q. Lamb, this volume (1996).
2. V. Wang and R. Lingenfelter, ApJ **416**, L13 (1994).

Gamma-Ray Burst Redshift Constraints from BATSE Spectral Data

Robert S. Mallozzi, Geoffrey N. Pendleton, William S. Paciesas

University of Alabama in Huntsville
Department of Physics, Huntsville, AL, 35899

> We have used the photon spectra of gamma–ray bursts (GRBs) obtained from data from the Large Area Detectors of the Burst and Transient Source Experiment (BATSE) on the Compton Gamma–Ray Observatory to produce cosmological models of the burst number-intensity distribution. Since it has become common to assume a canonical or average burst photon spectrum in computations of this type, we have examined the consequences of this assumption by using a range of observed burst spectra to create theoretical intensity distributions. We used a conventional Friedmann cosmology to create the models, and assumed that there is no burst source rate density evolution and that the sources are monoluminous (standard candles). This enabled us to focus on the effects of spectral shape on the $\log N$–$\log P$ model parameters (peak luminosity and redshift); the shape of the burst spectrum is found to have an influence on the maximum redshift consistent with the BATSE data. The use of a canonical burst spectrum in modelling the number-intensity distribution yields results that are moderately dependent upon the assumed spectral shape.

INTRODUCTION

Results from BATSE (1) have revealed that the GRB angular distribution, in conjunction with the intensity distribution, is unique among all known astrophysical objects. The angular distribution is consistent with isotropy (1-3), yet the integral intensity distribution ($\log N$–$\log P$) exhibits a slope that flattens from the value of $-(3/2)$ measured for the strong bursts to $\sim (-0.8)$ for the weak bursts (weak/strong denotes low/high peak flux).

A plausible explanation of gamma–ray burst sources is a population at cosmological distances (4,5), a scenario that naturally satisfies the isotropy constraints. This paper supports results (6-12) of analyses of the integral number–intensity distribution that suggest that the sources of the weakest bursts observed with BATSE are consistent with a maximum redshift of $z \sim 1$–2. Some of these analyses utilized a canonical or average burst spectrum to compute the model intensity distributions. We have instead incorporated measured burst spectra into the $\log N$–$\log P$ models to investigate the conse-

© 1996 American Institute of Physics

quences of using different spectral shapes on the derived values of redshift. We employed a direct spectral inversion technique to create the burst photon number spectra, and have chosen to analyze the differential rather than the integral intensity distribution since the data points of the integral distribution are not statistically independent.

Additionally, a common average spectrum to use for gamma–ray bursts is a power law. Since nearly all the bursts in the data set exhibit some curvature in their spectra in the energy range of the BATSE Large Area Detectors (LADs), we have investigated the impact of assuming that burst spectra are described by a power law on the value of the maximum redshift. Power law indices in the range $\alpha \sim (-1.1 \pm 0.3)$ yield maximum model redshifts that are in general agreement with the maximum redshifts derived from models incorporating the curved photon spectra obtained from observed bursts.

MODEL CONSTRUCTION

The expected flux P_i for an event of peak luminosity L_i and intrinsic photon spectrum $\phi(E)$ is given by (9,12)

$$P_i = \frac{L_i}{\int E\,\phi(E)\,dE} \frac{\int \phi[(1+z_i)E]\,dE}{4\pi\,r_i(z)^2}, \tag{1}$$

where $r_i(z)$ is the radial coordinate whose form depends on the parameters of the cosmological model. The range of integration for the numerator of equation (1) is 50–300 keV, corresponding to the energy range used to compute the burst peak flux. The peak luminosity L_i (ergs s^{-1}) is normalized by the peak luminosity L_0 given by the integral in the denominator of equation 1:

$$L_0 = \int_{E_1}^{E_2} E\,\phi(E)\,dE, \tag{2}$$

where E_1 and E_2 represent the observing range of the Large Area Detectors ($\sim$20–2000 keV).

The observed number rate of events with flux in the range $[P_i, P_{i+1}]$ is (e.g. (6))

$$N(P_i, P_{i+1}) = 4\pi \int_{r_i}^{r_{i+1}} \frac{n(r)}{1+z} \frac{r^2}{\sqrt{1-k\,r^2}}\,dr \int \Phi(L)\,dL, \tag{3}$$

where for non–evolving sources the burst rate per unit comoving volume is given by the free parameter $n(r) \equiv n_0 = $ constant, k is the spatial curvature, and the factor of $1/(1+z)$ is included to account for the time dilation of the interval between detected bursts. Assuming the sources are standard candles, the integral over the distribution function of burst luminosity $\Phi(L) = \delta(L-L_0)$ is unity, and each burst source can be associated with a unique value of redshift for a given value of photon flux. With the peak luminosity L_i as a second free parameter, we have inverted equation (1) in an Einstein–deSitter universe ($\Lambda = 0$, $q_0 = 1/2$) to find the values r_i corresponding to the peak flux data recorded by BATSE. The photon spectra used in equation (1) were peak flux spectra of the brightest bursts in the BATSE 3B catalog (13). The brightest 100 bursts were selected from each of the three trigger timescales, where brightness was based on the photon model dependent energy flux (ergs cm^{-2} s^{-1}) in the 50–300 keV band. We then used equation (3) to compare the three sets of independent log$N$–log$P$ models corresponding to the three trigger timescales to the appropriate measured differential peak flux distributions via χ^2 minimization.

RESULTS

Figure 1 shows a sample model fit (histogram) to the BATSE differential 1024 ms peak flux data (crosses). The bins to the left of the dotted line were excluded from the fit to avoid threshold effects. The photon spectrum used in computing this particular model log$N$–log$P$ curve is shown in the inset, and is typical of the 300 log$N$–log$P$ models created from the observed burst spectra. However, the shape of the spectrum used in equation (1) can have a significant impact on the log$N$–log$P$ model parameters. Figure 2a shows the distributions of maximum redshifts derived from models incorporating the 1024 ms peak flux spectra. The dotted histogram is the redshift distribution at the 90% efficiency level. The solid histogram is the distribution using all bursts in the peak flux data set. The distributions of z for the remaining two timescales are similar. A summary of the average values of the model parameters is given in Table 1. Redshift values at the 90% burst efficiency level (dotted line in Figure 1) and the minimum observed flux level are denoted by $\langle z_{\max} \rangle$ (90%) and $\langle z_{\max} \rangle$ (Absolute).

We created intensity distribution models using simulated photon spectra of simple power laws in equation (1) to investigate the impact of the curvature of the observed spectra on the resulting model parameters. We produced number–intensity distribution models using power laws with indices ranging from $\alpha = -1.0$ to $\alpha = -3.0$. The models were compared to the 1024 ms peak flux data. Figure 2b shows the dependency of $z_{0.4}$ on the index of the simulated power law spectrum (circles). Also shown are effective spectral indices in the 50–300 keV band of the 100 models incorporating the 1024 ms peak flux spectra (crosses).

TABLE 1. Summary of Burst Group Properties.

Timescale (ms)	Number of Models	$\langle L_{50} \rangle$ (10^{50} ergs s^{-1})	$\langle z_{\min} \rangle$	$\langle z_{\max} \rangle$ (90%)	$\langle z_{\max} \rangle$ (Absolute)
64	100	31.33 ± 1.81	0.087 ± 0.0007	0.95 ± 0.02	1.60 ± 0.03
256	100	17.20 ± 0.90	0.089 ± 0.0006	1.05 ± 0.02	1.77 ± 0.03
1024	100	13.96 ± 0.73	0.14 ± 0.0008	1.45 ± 0.02	2.37 ± 0.05

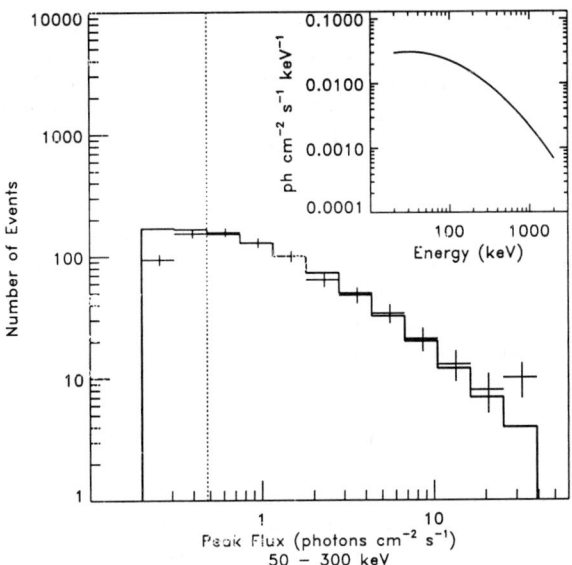

FIG. 1. Model fit to the BATSE $\log N$–$\log P$.

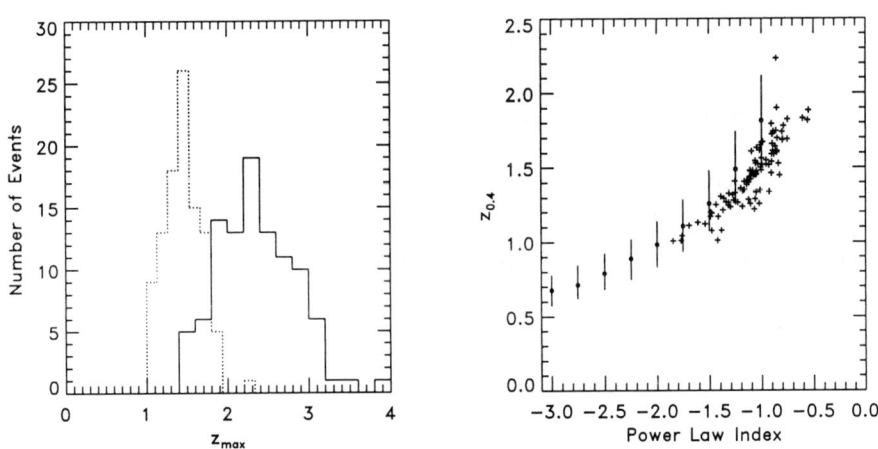

FIG. 2. a) Distributions of z for 100 $\log N$–$\log P$ models incorporating the 1024 ms peak flux spectra. b) Redshift values for models using simulated power law spectra (circles). Also shown are effective spectral indices (50–300 keV) of 1024 ms peak flux spectra (crosses).

CONCLUSIONS

The shape of the burst spectrum used in creating cosmological models of the BATSE number–intensity distribution can have a significant effect on the value of the maximum redshift derived from the models, varying by as much as a factor of 3, with an average variation of ~20%. The average values of the redshift of the faintest detectable bursts for the three peak flux timescales (64 ms, 256 ms, and 1024 ms) derived from models of the BATSE $\log N$–$\log P$ data are self-consistent, and in good agreement with redshifts derived from other analyses (e.g. (10,9,14,15,11).

Simulated spectra of power laws result in model redshifts that vary for different values of the spectral index. Power law indices in the range $\alpha \sim (-1.1 \pm 0.3)$ yield model redshift values that are quantitatively similar to those derived from the observed burst spectra, suggesting that the power law spectrum approximation is acceptable in computations of this type, provided one employs a spectral slope in this range.

REFERENCES

1. C. A. Meegan, et al., Nature **355**, 143 (1992).
2. M. S. Briggs, et al., in Gamma-Ray Bursts, eds. G. J. Fishman, J. J. Brainerd, K. Hurley, AIP Conf. Proc. **307**, 43 (AIP, New York, 1994).
3. M. S. Briggs, et al., ApJ **459**, 40 (1996).
4. V. V. Usov & G. V. Chibisov, Soviet Astron.-AJ **19**, 115 (1975).
5. B. Paczyński, B. ApJ **308**, L51 (1986).
6. S. Mao & B. Paczyński, ApJ Letters **388**, L45 (1992).
7. T. Piran, ApJ **389**, L45 (1992).
8. E. E. Fenimore, et al., Nature **357**, 140 (1992).
9. E. E. Fenimore, et al., Nature **366**, 40 (1993).
10. W. A. D. T. Wickramasinghe, et al., ApJ **411**, L55 (1993).
11. A. G. Emslie & J. M. Horack, ApJ **435**, 16 (1994).
12. E. E. Fenimore & J. Bloom, ApJ **453**, 25 (1995).
13. C. A. Meegan, et al., ApJ Suppl., in press (1996).
14. J. P. Norris, et al., ApJ **439**, 542 (1995).
15. R. S. Mallozzi, et al., ApJ **454**, 597 (1995).

A New Gravitational Lens Search for Gamma Ray Bursts

G. F. Marani[*], R. J. Nemiroff[*,†],
J. P. Norris[†], and J. T. Bonnell[††]

[*]*CSI/George Mason University, Fairfax, VA 22030*
[†]*NASA/Goddard Space Flight Center, Greenbelt, MD 20771*
[††]*USRA/Goddard Space Flight Center, Greenbelt, MD 20771*

In the new 3rd BATSE catalog, many new gamma ray bursts have been added, and all GRB positions and error boxes have been recomputed, yielding many significantly different GRB locations and smaller error boxes. These improvements mandate a new search for GRB gravitational lensing in the new catalog. We are conducting an automated search for echo signals between the 834 triggered bursts with recorded T90s and angular positions within 3σ of their combined positional error. The GRB light curves are compared using different statistical techniques. So far, no convincing gravitational lens pairs have been found.

INTRODUCTION

BATSE results (1) show GRBs are distributed isotropically across the sky, and a log N–log S relationship that yields an inhomogeneous distribution of GRBs in space. These two facts favor a cosmological origin of GRBs. If bursters are at cosmological distances, some of them (0.2% to 10% depending on their distance and the lens properties) will undergo gravitational lensing by foreground galaxies so that a lens echo is produced. This possibility was first suggested by Paczyński (2,3), and later discussed by other authors (4–7).

Time-delayed multiple images of the same burst might be detected. The echoes could appear dimmer or brighter but they would have identical spectra because, by the equivalence principle, gravitational fields bend the light of different energy in the same way. Therefore the images are expected to have identical spectra at all times during the burst. The time delay between the images may vary from seconds (if the lens is a black hole with mass of about $\sim 10^{6.5}$ $M_\odot$) to months (if the lens is a typical galaxy) or years (for the case of a cluster lens), depending on the lens geometry and the position of the source behind the lens.

Different echo searches have been done in the past without any lens detection, although interesting limits on the cosmological abundance of compact dark matter were found (8–10). In the present work, the search is done in an automated procedure and only for those bursts with published T90.

© 1996 American Institute of Physics

SEARCH PROCEDURE

We search the 400 bursts with largest recorded T90s. We use the continuous data (CONT) consisting of count rates in 16 energy channels with time resolution of 2.048 s (11). The data are summed over each of the triggered BATSE detectors. The background is subtracted from each energy channel by fitting user-defined background intervals with a polynomial of degree up to 4. The degree of the polynomial is automatically selected so that it yields the best fit according to the F-test technique (12). This fit is then interpolated across the whole time interval.

Light curves are compared only for those GRB pairs with angular distance less than 3 standard deviations of their combined positional error and, at the same time, less than 30°.

Many statistical comparison methods can be applied (10,13). In our work we proceed as follows. Two different hardness ratio tests are computed. The first one is defined as the ratio of the total counts summed over the energy channels 3, 4, 5, and 6 (energy range $\sim$ 100–300 keV), and the total counts summed over the energy channels 7, 8, 9, and 10 (energy range $\sim$ 40–100 keV), $HR \equiv C_{3,4,5,6}/C_{7,8,9,10}$, both summed over the duration of time profile. The second one is defined as the ratio between the same summed energy channels as mentioned before, but this time over the duration of the brightest peak, $HR^p \equiv C^p_{3,4,5,6}/C^p_{7,8,9,10}$. We define this duration as the time interval around the brightest peak (summed in all the energy channels) in which the counts in each bin are greater than 5% of the peak counts.

We also apply the Kolmogorov-Smirnov test (12) to compare the spectra of the GRB pair. We first obtain the total counts per energy channel by summing the count rates over the whole time profile. Then the cumulative distribution of normalized total counts for the 16 energy channels is computed and tested with the Kolmogorov-Smirnov method. We expect a gravitational lens candidate pair to have the maximum distance between the two distribution functions close to 0 and both bursts be bright to be a convincing candidate pair. Figure 1 shows the distribution of the count rates at the brightest peak (obtained by summing over all the energy channels) of the dimmer burst of the pair as function of the maximum distance. No excess is found for small values of the maximum distance and large values of the count rates (top left corner of Figure 1).

Finally, we test if the time profiles are identical in shape by introducing a modified definition of the cross-correlation. After normalizing the data for each burst of the pair, we find the position of their brightest peaks (summing over all the energy channels), align those peaks, and compute the dot product between the two lightcurves for each energy channel, as if they were vectors in a multidimensional space. If the two time profiles have different number of bins, we fill up the shortest with zeros. If the data are normalized, the dot product returns the cosine of the angle between these two "vectors". If this value is close to 1 (angle close to 0) for each one of the 16 channels, this implies that the two time profiles are similar in shape. We also shift in time

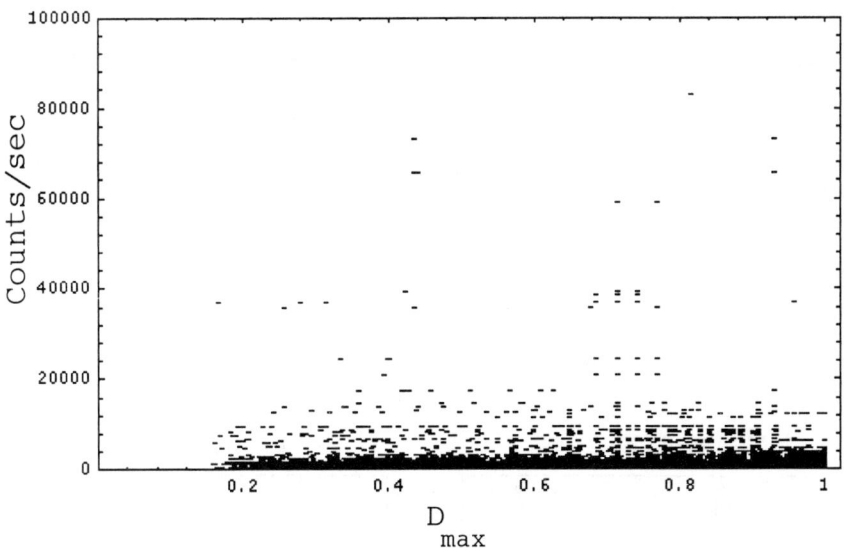

FIG. 1. Maximum distance between the two distribution functions versus the count rates at the brightest peak of the dimmest burst of the pair.

one of the profiles around the brightest peak of the other to avoid problems of different binning.

The criteria for preliminary acceptance or rejection of a GRB pair in all the tests is the following: we perform the tests for those GRBs that we consider too far from each other to be a gravitational lens pairs (for our work, we use angular distances greater than 5 standard deviations of the combined positional error). We search among these pairs, those which yield the smallest differences in T90, hardness ratios (for the two definitions), and the best cross-correlation. These pairs are the ones which have closest statistical properties, but yet are not lenses, thus they define the best "hit by chance". This gives the best limits for our tests in the case of a hit by chance. Finally, we search for lens pairs by requiring that the angular distance be less than 3 standard deviations, and that the statistical properties be better than those of the best hit by chance. If this happens, we can consider the pair as a lens candidate. With these different tests, we take into account the fact that gravitational lensing effects are independent of the energy, and that the time profiles of the images may be time-delayed.

At this moment, no gravitational lens pairs have been found. None of the possible candidate pairs satisfy all the tests at the same time. The best candidate pairs are those shown in Figures 2 and 3, which yielded the best cross-correlations, but both have angular distances greater than 3σ. In the first case, the angular distance between them is 41°, about 8.6σ of their com-

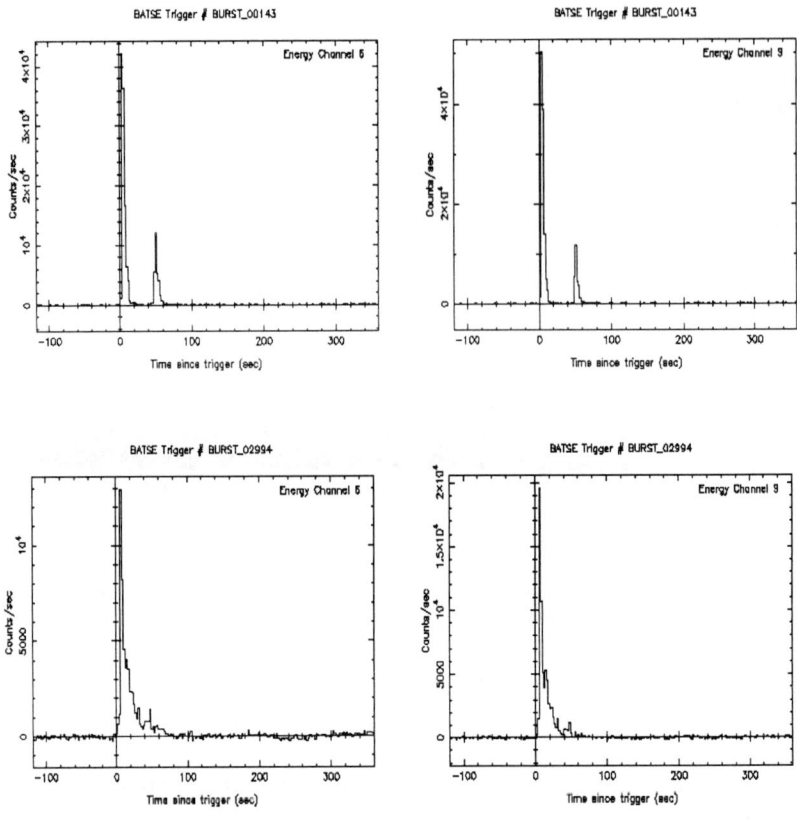

FIG. 2. Time profile for BATSE GRBs with triggered numbers 143 and 2994. Channels 5 and 9 are shown.

bined positional error. In the second case, the angular distance is 21.27°, about 7.8σ. In both cases they also have very different T90's and hardness ratios.

This search is still being carried out and will continue including all the GRBs in the 3rd BATSE catalog and more statistical comparison tests.

REFERENCES

1. C. A. Meegan et al, Nature **355**, 143 (1992).
2. B. Paczyński, Ap. J. **308**, L43 (1986).
3. B. Paczyński, Ap. J. **317**, L51 (1987).
4. S. Mao, Ap. J. **389**, L41 (1992).
5. A. Gould, Ap. J. **386**, L5 (1992).

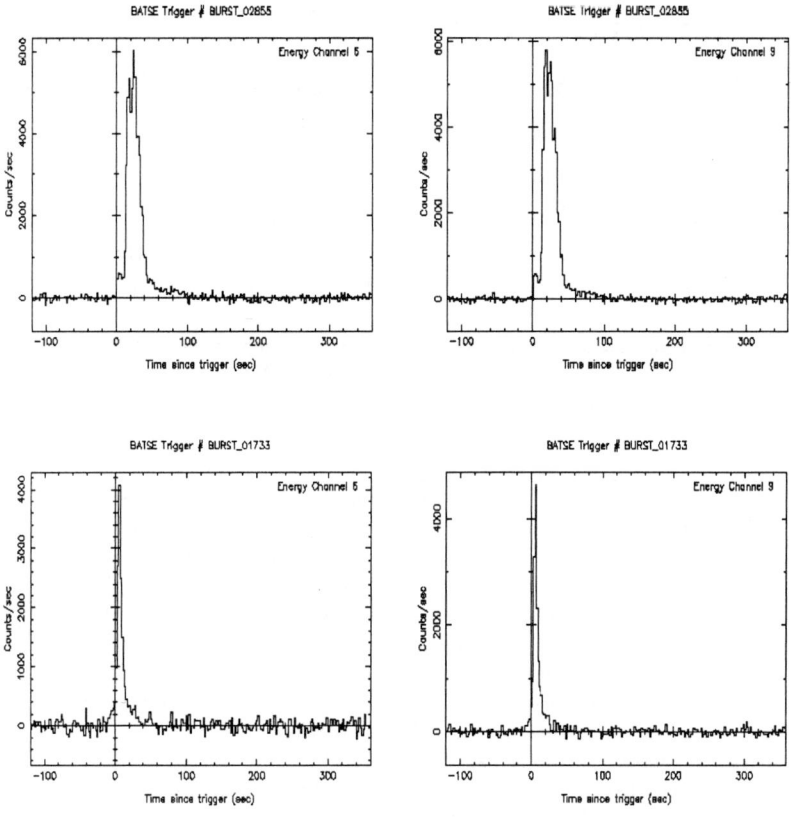

FIG. 3. Time profile for BATSE GRBs with triggered numbers 2855 and 1733. Channels 5 and 9 are shown.

6. O. M. Blaes & R. L. Webster, Ap. J. **391**, L63 (1992).
7. R. Narayan & S. Wallington, Ap. J. **399**, 368 (1992).
8. R. J. Nemiroff et al., in Compton Gamma-Ray Observatory, eds. M. Friedlander, N. Gehrels, and D. J. Macomb, AIP Conf. Proc. **280** 974 (AIP, New York, 1993).
9. R. J. Nemiroff et al., Ap. J. **414**, 36 (1993).
10. R. J. Nemiroff et al., Ap. J. **432**, 478 (1994).
11. G. J. Fishman et al., in Proceedings of the GRO Science Workshop, 2-34 (1989).
12. W. Press, B. Flannery, S. Teukolsky, and W. Vetterling, Numerical Recipes, Cambridge University Press, (1986).
13. J. Wambsganss, Ap. J. **406**, 29 (1993).

Progress with the Konus-W Gamma-Ray Burst Spectrometer on GGS-Wind

E. P. Mazets*, R. L. Aptekar*, D. D. Frederiks*,
S. V. Golenetskii*, V. N. Ilynskii*, M. M. Terekhov*,
T. L. Cline[†], P. S. Butterworth[†‡], and D. E. Stilwell[†]

*Ioffe Physical-Technical Institute, St.Petersburg, Russia 194021
[†]NASA/GSFC, Code 661, Greenbelt, MD 20771
[‡]Hughes STX, 4400 Forbes Boulevard, Lanham, MD 20706

The cosmic gamma-ray burst spectrometer Konus-W has been successfully making observations for nearly one year, since the launch of the GGS-Wind spacecraft. The instrument consists of two large scintillator units of size and shape very nearly the same as the spectroscopy detectors on CGRO BATSE. These face towards the ecliptic poles so as to survey the sky in a moderately uniform fashion. At least 114 gamma ray bursts have triggered the system in the first 330 days of operation, yielding detailed time histories and spectra. A large number of additional events are seen in the background mode at much coarser resolution. These observations can be combined with those of the Interplanetary Network to reduce the total area of the segmented annular source fields derived from several degrees to about one degree in length, although the data cannot obtained from this spacecraft in the rapid turnaround mode needed to benefit the BACODINE system. The Konus spectra can be summarized presently as providing little indication of the frequent occurrence of major spectral features.

INSTRUMENT CAPABILITIES

The Konus-Wind instrument has two sensors, at opposite ends of the GGS-Wind spacecraft, which provide all-sky monitoring. In the default 'background' operating mode, each sensor counts photons in three energy ranges (12–45 keV, 45–190 keV and 190–760 keV) every 1.472 or 2.944 s (depending on the telemetry rate). When 'burst' mode is triggered, the triggering sensor returns 4096 count values for each energy range, including 256 2 ms pre-trigger measurements and post-trigger values integrated in increasing timebins from 2 to 256 ms; giving time-histories of 230.144 s duration. Burst mode also generates sets of 64 energy spectra in each of two energy ranges (12–760 keV and 200–10000 keV) where the first 4 in a set are accumulated in 0.064 s, the next 52 in the set from 0.256 to 8.192 s (adapted to the observed intensities) and the last 8 in the set are 8.192 s each. For more details, see (1).

© 1996 American Institute of Physics

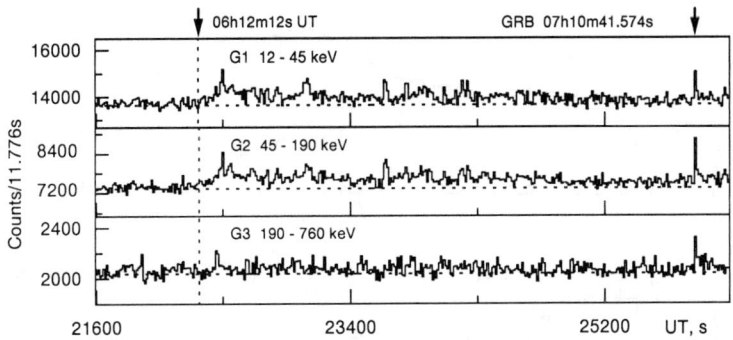

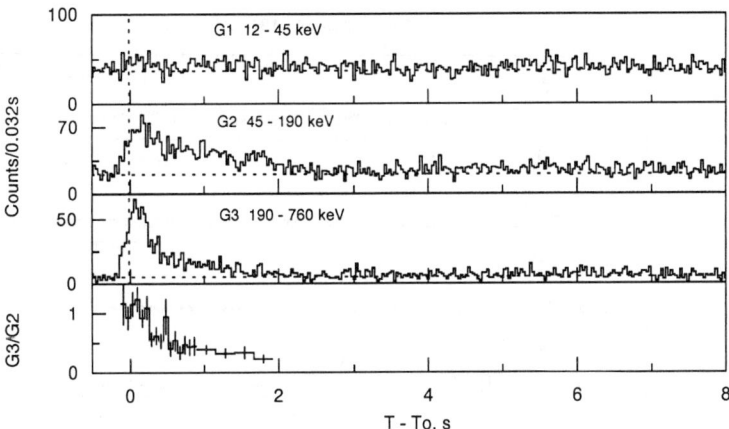

FIG. 1. GRB 950325a, which was seen in background data for an hour (upper frame) before triggering burst mode (lower frame)

DATA OBTAINED

From November 11, 1994 to October 16, 1995 (330 days of observations) 226 GRBs were detected (including 114 in burst mode) together with 73 solar flares (15 in burst mode) and 47 events due to charged particles, trapped radiation and solar cosmic rays. The Transient Gamma-Ray Spectrometer (TGRS) is located next to the Konus detector which faces the Southern ecliptic hemisphere. TGRS has better spectral resolution than Konus, but lower sensitivity. See (2) for some comparisons of Konus and TGRS results.

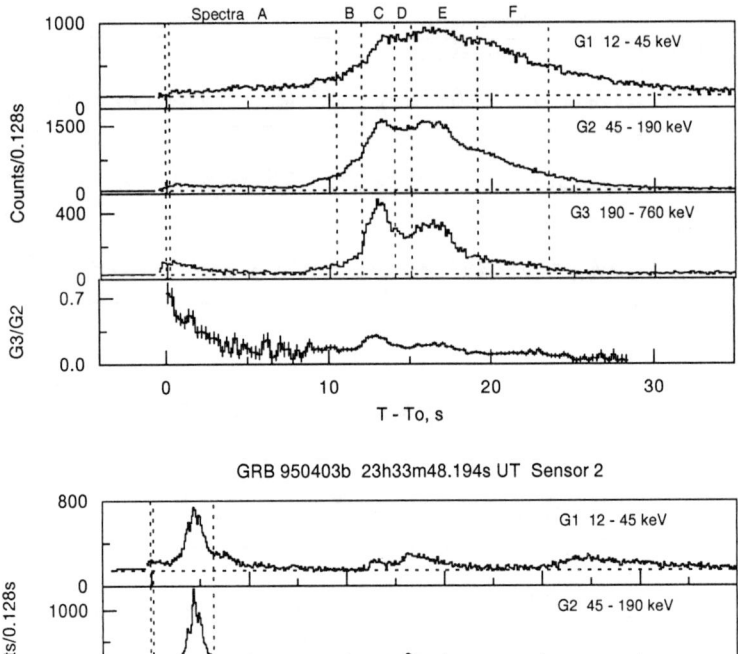

FIG. 2. Burst mode time histories of the two strongest events seen so far, GRB 950822 (upper frame) and GRB 950403b (lower frame)

EXAMPLES

The examples of Konus data presented here are for one of the longest events observed in background mode(950325a), the two strongest events detected so far (950403b and 950822) and the most recent trigger examined (951016).

Background Mode A long weak event with intensity below trigger threshold was detected on 25 March, 1995 at 6h 12m 12s UT (Fig. 1). It lasted more than an hour. An intense narrow pulse at 7h 12m 41.574s triggered a burst mode record in sensor 2 (the sensor facing the Northern ecliptic hemisphere). Note that the ratio of count rates in sensor 2 to sensor 1, which is a rough

Spectra of GRB 950822

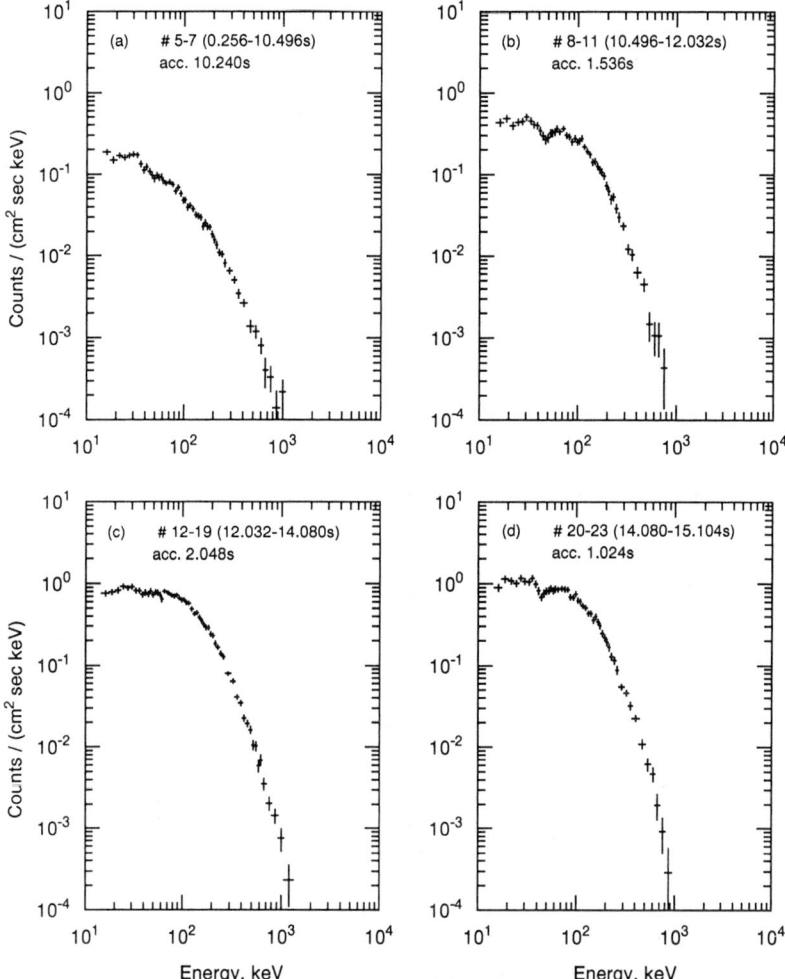

FIG. 3. Spectra of the strongest event seen so far, GRB 950822, accumulated during four different time intervals. The intervals, accumulation times and individual spectra used are indicated on each frame

measure of angle of incidence of radiation, is approximately the same for both records, N(S2)/N(S1) = 6.

Burst Mode Only a very small fraction of our burst mode data can be presented here. We display time history data on two of the brightest GRBs detected (Fig. 2). We show some integrated count rate spectra for the same events, and for the most recent bright event (Fig. 3 and Fig. 4). Our work to obtain incident photon spectra is still in progress. What is more, information

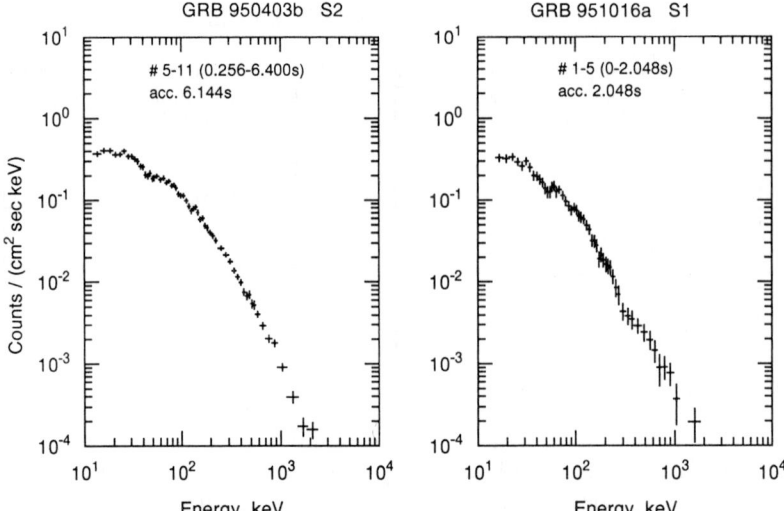

FIG. 4. Spectra of the second strongest event seen (GRB 950403b, left frame) and one of the most recent (GRB 951016a, right frame)

on burst location and hence on angles of incidence of radiation is absent in many cases. The figures presented therefore represent the instrumental energy loss spectra only.

CONCLUSIONS

These are preliminary results. No deconvolution procedures have been used here. Some of the weaker spectral 'features' are expected to be removed when the count rate distributions are converted into photon spectra. (See, for example, 950822 (c)). The deconvolution procedures will take into account such elements as non-linearity of the scintillator output. Some stronger 'features' may remain. (See, for example, 950822 (b) and (d)). Such 'features' are relatively rare. So far, no BATSE observations of the same events allow for direct comparison. The accumulation of more events, and especially of more events in common with BATSE, may help to clear up this mystery. Our position at the present is not to claim the confirmation of spectral features until more data have been accumulated and analyzed.

REFERENCES

1. R. Aptekar, *et al.*, Space Sci. Rev. **71** 265 (1995).
2. D. Palmer, *et al.*, these proceedings.

Geocenter Angle Distribution of the 3B Catalog

M. L. McCollough[*], C. A. Meegan[†], and G. N. Pendleton[‡]

[*]*MSFC/Hughes STX, Huntsville, AL 35812*
[†]*MSFC, Huntsville, AL 35812*
[‡]*MSFC/University of Alabama in Huntsville, Huntsville, AL 35812*

We examine the geocenter angle distribution of the 1122 bursts contained within the 3B catalog. The distribution is modeled using an isotropic distribution of gamma-ray bursts and a sharp cutoff due to the presence of the earth. An excellent agreement between the distribution and the model is found, which provides an independent assessment of the BATSE location errors. Comparisons between the geocenter angles determined from different data types are made. Examination of the error distribution near the earth cutoff angle is presented.

INTRODUCTION

The 3B catalog of BATSE observations of gamma-ray bursts (GRBs) has confirmed with greater statistical significance that GRB's angular distribution is consistent with isotropy (5,3). The angular distribution of bursts has also been used to set limits on burst repetition (4,1). Thus it is important to check the accuracy of the locations in as many ways as possible. Here we will examine location errors by looking at the geocenter angle distribution of GRBs. The geocenter angle is the angle between the GRB location and the earth's center as viewed from the spacecraft. For an isotropic distribution of GRBs the geocenter angle distribution is easily modeled and the sharp cutoff of the earth's limb provides a way to estimate the errors associated with the burst locations.

MODEL OF GEOCENTER ANGLE DISTRIBUTION

Assuming the GRB distribution is isotropic, as is indicated by the observations, the number of bursts per unit solid angle (N_0) will be constant and the GRB distribution on the sky is given by:

$$N_0 d\Omega = N_0 \sin\theta d\theta d\phi \qquad (1)$$

Taking θ to be the geocenter angle (θ_g) and using azimuthal symmetry we can write the distribution of the bursts as a function of geocenter angle as,

$$dN(\theta_g) = 2\pi N_0 \sin\theta_g d\theta_g \qquad (2)$$

© 1996 American Institute of Physics

where $d\theta_g$ is the angular binning of the data. The above distribution should give the angle distribution for geocenter angles greater than or equal to the cutoff angle (θ_e) due to the earth's limb. Below θ_e the distribution will have a value of zero. The constant N_0 can be determined by integrating the distribution from θ_e to π. This should yield the total number of bursts (N_T). The distribution is thus given by:

$$dN(\theta_g) = \begin{array}{ll} \frac{N_T}{1+\cos\theta_e} \sin\theta_g d\theta_g, & \theta_g \geq \theta_e \\ 0 & , \theta_g < \theta_e \end{array}$$

θ_e is given by

$$\theta_e = \arcsin(\frac{r_{\text{limb}}}{r_{\text{sp}}}) \qquad (3)$$

where
r_{sp} = distance from the center of the earth to the spacecraft
$r_{\text{limb}} = r_e + r_{\text{atm}}$
$r_e = 6378.164$ km (earth's equatorial radius)
$r_{\text{atm}} = 70$ km (atmospheric cutoff height)

The atmospheric cutoff height (r_{atm}) is taken from the value found from occultation analysis of BATSE data (2). The flux below 100 keV is attenuated by a factor of 2 at this height and the occultation step is quite sharp. The estimated error in θ_e is no larger than $\sim 0.5°$. θ_e is calculated for each burst and a median value is used in the model; it has a range of 70.9° to 73.9° with a median value of 71.9°.

GEOCENTER ANGLE DISTRIBUTION

In Figure 1 we plot the distribution of GRBs as a function of geocenter angle (in 3.1° bins). Also plotted is the model curve, without location errors, for $N_T = 1122$ bursts and a θ_e of 71.9° (the median value of the 1122 bursts). It is apparent from Figure 1 that the observed distribution is very well described by the model. One can also see the sharp drop off in bursts at θ_e.

The most notable variation from the model occurs near the cutoff angle. It is apparent there are GRBs below θ_e. The explanation for this lies in the location uncertainties associated with the GRBs. By examining the bursts that appear to lie below the horizon, estimates of the location errors can be estimated and checked against the errors found by other means (IPN locations, solar flares, or Cygnus X-1 events). It should be noted that the number of bursts below θ_e rapidly falls off as one goes to lower θ_e. Also there is a deficit in bursts for the first few bins just larger than θ_e. This deficit corresponds to the number of bursts that are found below the horizon.

An examination of the geocenter angle distribution of the bursts by BATSE data type show that the geocenter angle distributions and location uncertainties are consistent between data types.

BURSTS LOCATED BELOW THE CUTOFF ANGLE

As noted the bursts with geocenter angles less than θ_e can be used to examine the location errors. There are 26 bursts that locate below the horizon. In Figure 2 we plot the angle by which a burst falls below the horizon *vs* the total error for each of the 26 bursts. The total error is the statistical error plus a systematic error of 1.6° added in quadrature. The systematic error was determined by looking at location errors from IPN, Comptel, and WATCH locations (6). In Figure 2 it is apparent that none of the bursts represent a 3σ or greater deviation from being above the horizon. In fact, 20 bursts (77%) are less than 1 σ away from the horizon, 24 bursts (92%) are less than 2σ away from the horizon, and 26 bursts (100%) are less than 3σ away from the horizon.

MODELING OF THE BURSTS LOCATED BELOW THE HORIZON

To compare the distribution of bursts locating below the horizon with the expected distribution, a Monte Carlo simulation was performed treating the sky as being flat (constant number of bursts with geocenter angle and no variation due to changing solid angle) with a sharp cutoff at θ_e. Events were chosen to occur at random locations near, but above, the horizon and then given random statistical errors in location (the errors were multiplied by $\frac{1}{\sqrt{2}}$ since the errors were for a circular error box and the geocenter angle is measured only along one dimension). This was performed for 2 million events. The events found below the horizon from the simulation represent the distribution of events one would expect to find below the horizon as a result of location uncertainty. To compare with the observations the number of events below the horizon in the simulation are normalized to the number found in the observations below the horizon (26 bursts). In Figure 3a we plot the number of bursts below the horizon verses the number of sigma below the horizon. The points marked with + are the result of the simulation (in .2σ bins) and the the observations are represented by the histogram. The data are fit (through the total error) using a systematic error of 1.6°. A comparison of the data with a larger systematic error (10°), shown in Figure 3b, does not fit the data as well.

CONCLUSIONS

- The geocenter angle distribution for the 3B catalog is well fit by the model presented above.

- The geocenter angle distribution appears to be the same for the different BATSE data types used to determine locations.

- The bursts which locate below the horizon are consistent, within their statistical and systematic error (1.6°), with being above the horizon.

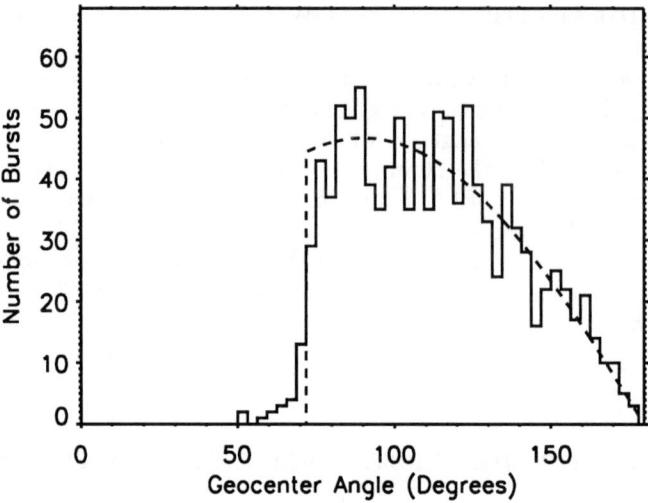

FIG. 1. Geocenter angle distribution. The dashed line represents the expected distribution assuming no location errors. The horizon location θ_e (71.9°) includes the effect of atmospheric absorption. Location errors produce a small deficit in the number of bursts just above the horizon and a small excess of bursts just below the horizon.

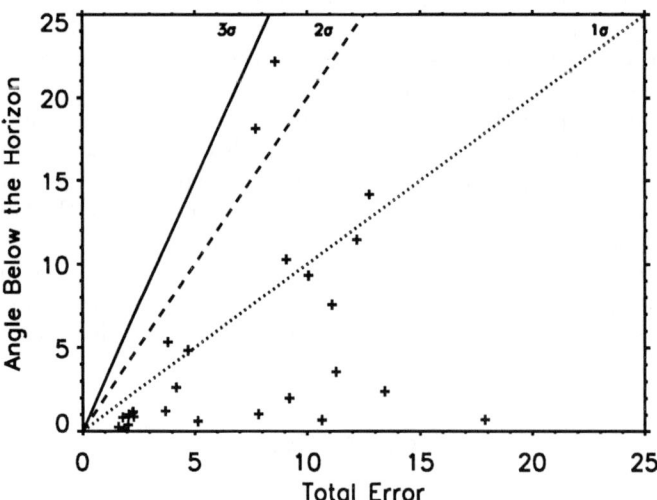

FIG. 2. For the bursts which appear below the horizon are plotted angle below the horizon verses the total error. The dotted line corresponds to a 1σ deviation from the horizon. The dashed line corresponds to a 2σ deviation from the horizon. the solid line corresponds to a 3σ deviation from the horizon. Note that none of bursts below the horizon are 3σ or greater from the horizon.

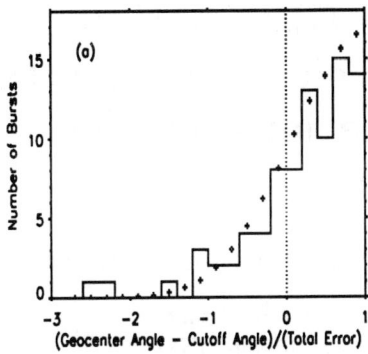
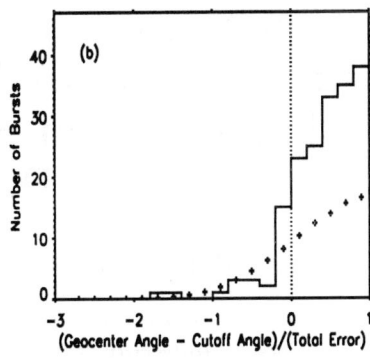

FIG. 3. (a) A comparison between a model for the bursts below the horizon to the observed distribution is shown in this figure. The +s represent the results of a Monte Carlo simulation for a flat sky with a sharp cutoff at θ_e. The binned data are the geocenter angle of the burst minus its θ_e divided by its total error. The total error is the statistical error for each burst plus a systematic error of 1.6° added in quadrature. (b) A comparison between a model for the bursts below the horizon to the observed distribution with a systematic error of 10° added in quadrature. Note the difference in the shape of the simulation and the observation.

- The distribution of bursts below the horizon is well fit by a Monte Carlo simulation for the bursts having a systematic error of order 1.6°. A significantly larger systematic error appears to be excluded by these data.

REFERENCES

1. J.J. Brainerd, et al., ApJ Letters **441**, L39 (1995).
2. B.A. Harmon, et al., in Compton Observatory Science Workshop, eds. C.R. Shrader, N. Gehrels, & B. Dennis, 69 (NASA CP 3137, 1992).
3. C.A. Meegan, et al., Nature **355**, 143 (1992).
4. C.A. Meegan, et al., ApJ **446**, L15 (1995a).
5. C.A. Meegan, et al., ApJ submitted (1995b).
6. G. Pendleton, et al., in preparation (1995).

Constraints on the Distribution of Neutron Star Birth Velocities from the Properties of Rotation-Powered Pulsars

Lucia Munoz-Franco, D. Q. Lamb and Tomasz Bulik

Department of Astronomy and Astrophysics
University of Chicago, Chicago, Illinois 60637

Knowledge of the distribution of neutron star birth velocities in the Milky Way is essential to Galactic corona models of gamma-ray bursts, and to understanding the intrinsic properties of neutron stars. The recent discovery of very high velocity rotation-powered pulsars shows that we are far from understanding the shape of this velocity distribution. We investigate this distribution using a maximum likelihood approach. We carry out numerical simulations of the evolution of a population of pulsars in the Galaxy, assuming various models for the distribution of birth velocities and for the initial values and temporal evolution of the other intrinsic properties of the pulsars. We use these simulations and a characterization of the properties of radio instruments to calculate the expected distribution of observed properties of the pulsars. In this paper, we derive the resulting likelihood function for a nearly ideal instrument. As a simple illustrative example, we consider the expected and observed proper motions in the second Molonglo survey.

INTRODUCTION

A number of important questions about rotation-powered pulsars remain unanswered (1). What is the dependency of pulsar radio luminosity L_{radio} on spindown luminosity $L_{spindown}$, rotation period P and the derivative of the rotation period $\dot{P}$? Do pulsar magnetic fields evolve or not? Does a "death line" exist in the $(\log P, \log \dot{P})$-plane? Is there evidence for more than one population of pulsars? Finally, what is the distribution of pulsar birth velocities? Knowledge of this velocity distribution is essential to Galactic corona models of gamma-ray bursts.

There are more than 630 rotation-powered pulsars known, a number large enough to admit a statistical analysis of the properties of the pulsar population in our galaxy. We carry out such an analysis using a maximum likelihood approach, and use it to address these questions.

© 1996 American Institute of Physics

NUMERICAL SIMULATIONS

We simulate a population of rotation-powered pulsars in the galaxy assuming various models for the distribution of initial values of the intrinsic properties of pulsars: magnetic field B_0, period P_0, period derivative $\dot{P}_0$, angle of magnetic dipole to rotation axis ξ_0, radio luminosity $L_{\mathrm{radio},0}$, pulse width w_0, position $\vec{x}_0$, and velocity $\vec{v}_0$. We then assume various models for the temporal evolution of these quantities.

We characterize the radio instrument by considering its properties: antenna resolution BWHM (beam width at half maximum), bandwidth B, receiver noise temperature T_r, sky temperature in the direction searched $T_{\mathrm{sky}}(\hat{n})$, and integration time T. These properties are summarized by a limiting radio flux density $S_{\min}$.

We then use $S_{\min}$ to calculate, from our dynamical calculations and calculations of evolution of intrinsic properties, the observed pulsar quantities, $\{\mathcal{P}\}$: period P, period derivative $\dot{P}$, observed radio flux density S_{obs}, proper motion $\vec{\mu}$, angular position $\hat{n}$, dispersion measure DM, and observed width of pulse w.

We then use a likelihood approach to compare the observed data with the expected data for a given model.

SIMULATED DATA

As an illustrative example, we show the results of simulations of the magnitude of the proper motion μ for a population of rotation-powered pulsars. We evolve 1.5×10^6 neutron stars for 50 Myrs and we calculate the resulting angular distribution and the distributions of magnitude of proper motion μ. We show those distributions in figures 1 and 2 respectively. We assume the sources to be standard candles visible to a minimum flux density $S_{\min} = 15$ mJy, the limiting flux density of the second Molonglo survey (3).

DERIVATION OF THE LIKELIHOOD FUNCTION

In our derivation of the likelihood function we follow closely the treatment presented by Loredo and Wasserman (2). We assume that we have a model that produces a probability function of pulsars as a function of each of the observable quantities,

$$\frac{dR(\hat{n},\ldots,w)}{dPd\dot{P}dS_{obs}d\vec{\mu}d\hat{n}dDMdw}. \tag{1}$$

We separate the data into two groups, *detection data*, d_i, and *non-detection data*, $\bar{d}_j$, which we define as follows:

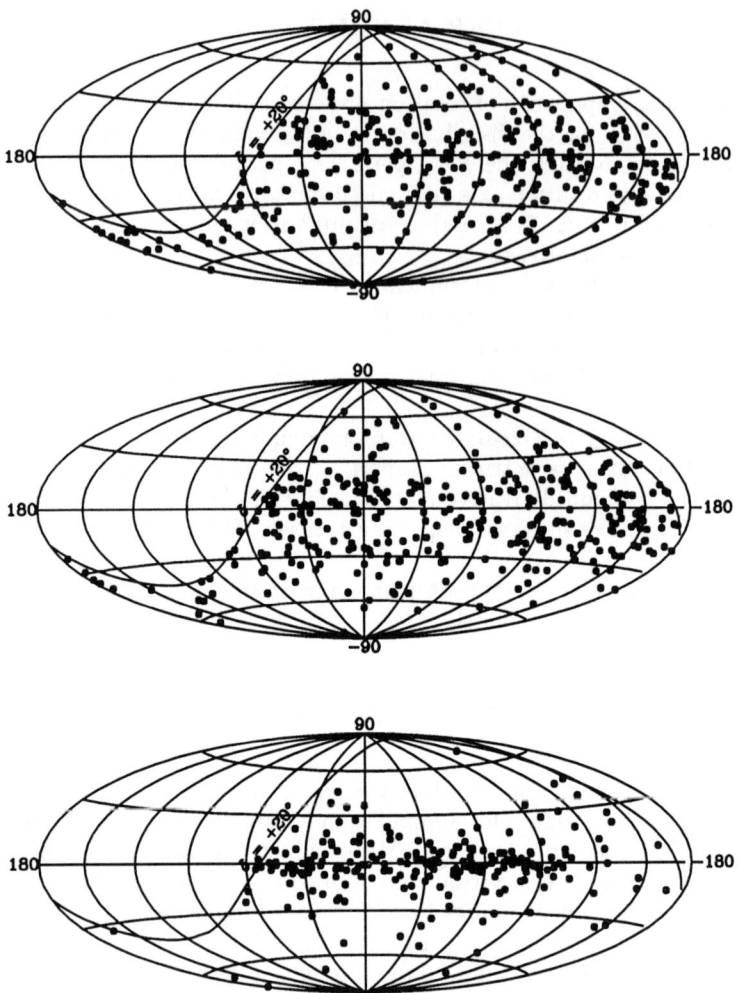

FIG. 1. Distribution of sources on the sky for simulations in which the pulsar radio luminosity is assumed to be a standard candle and the pulsar birth velocities are $v_0 = 400$ km s^{-1} (top), and $v_0 = 800$ km s^{-1} (middle); and for the second Molonglo survey (bottom) (4).

$d_i \equiv$ A detected pulsar has direction $\hat{\mathbf{n}} \in [\hat{\mathbf{n}}_i, \hat{\mathbf{n}}_i + \delta\hat{\mathbf{n}}]$ period $P \in [P_i, P_i + \delta P], \ldots$, pulse width $w \in [w_i, w_i + \delta w]$.

$\bar{d}_j \equiv$ No detected pulsars have direction $\hat{\mathbf{n}} \in [\hat{\mathbf{n}}_j, \hat{\mathbf{n}}_j + \delta\hat{\mathbf{n}}]$ period $P \in [P_j, P_j + \delta P], \ldots$, pulse width $w \in [w_j, w_j + \delta w]$.

We assume that the volume in the space of observable parameters defined by $\delta\hat{\mathbf{n}}, \delta P, \ldots, \delta w$ contains at most one pulsar.

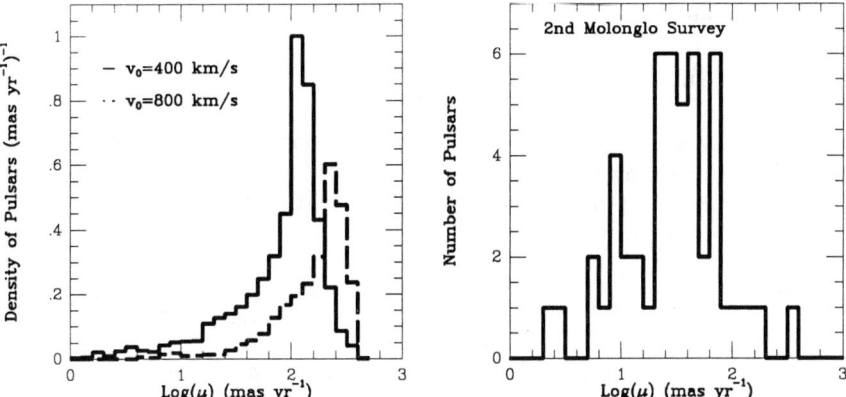

FIG. 2. Left panel: Distribution of μ for simulations in which the pulsar radio luminosity is assumed to be a standard candle and the pulsar birth velocities are $v_0 = 400$ km s^{-1} (solid line) and $v_0 = 800$ km s^{-1} (dashed line). Right panel: Distribution of μ in the second Molonglo survey (4).

Since the probability of detecting a pulsar in a certain volume of our observed quantities space is independent of whether or not another pulsar was detected in another volume, the likelihood function is the product of the independent probabilites for detections and non-detections,

$$\mathcal{L}(\mathcal{P}) = \prod_{i=1}^{N_d} p(d_i|\mathcal{P}) \prod_{j=1}^{N_{nd}} p(\bar{d}_j|\mathcal{P}) , \qquad (2)$$

where N_d is the total number of pulsars detected and N_{nd} is the number of volumes in parameter space with no detected pulsars.

Furthermore, the probability of detecting a pulsar in a given volume of our observable parameter space is given by the Poisson distribution:

$$p_m = \frac{\langle n \rangle^m}{m!} e^{-\langle n \rangle}, \qquad (3)$$

where $\langle n \rangle$ is the expected number of pulsars.

Detection Probability. The detection probability, $p(d_i|\mathcal{P})$, can be calculated using the following complete set of propositions:

$(\hat{n}_i, P_i, \ldots, w_i) \equiv$ The pulsar has position $\hat{n} \in [\hat{n}_i, \hat{n}_i + \delta \hat{n}]$, ..., pulse width $w \in [w_i, w_i + \delta w]$.

So that,

$$p(d_i|\mathcal{P}) = \int d\vec{n} \int d P \ldots \int dw \, p(d_i|\hat{n}_i, \ldots, w_i, \mathcal{P}) p(\hat{n}_i, \ldots, w_i|\mathcal{P}) \qquad (4)$$

The first term in the integrand is the individual pulsar likelihood, $\mathcal{L}_i(\hat{n}_i, \ldots, w_i)$. We will consider an almost ideal detector so that $\mathcal{L}_i$ is equal to a constant if $\hat{n} \in [\hat{n}_i, \hat{n}_i + \delta\hat{n}]$, ... etc. and zero otherwise,

$$\mathcal{L}_i(\hat{\mathbf{n}}_i, \ldots, w_i) = \begin{cases} \frac{1}{\delta\hat{n}\ldots\delta w} & \text{if } \hat{n} \in [\hat{n}_i, \hat{n}_i + \delta\hat{n}], \text{etc} \\ 0 & \text{otherwise.} \end{cases} \quad (5)$$

We find the second term by breaking the proposition into two constituent propositions asserting (1) occurrence of a pulsar within $\delta\hat{n}, \ldots, \delta w$; and (2) non-occurrence of a pulsar within $\delta\hat{n}$, say, but with different $P, \ldots, w$.

So we have the probability,

$$p(d_i|\mathcal{P})d\hat{\mathbf{n}}\ldots dw = d\hat{\mathbf{n}}\ldots dw \frac{dR(\hat{\mathbf{n}}_i, \ldots, w_i)}{d\hat{\mathbf{n}} dP \ldots dw} \exp\left[-\frac{dR(\hat{\mathbf{n}}_i, \ldots, w_i)}{d\hat{\mathbf{n}} dP \ldots dw} d\hat{\mathbf{n}}\ldots dw\right]$$

$$\times \exp\left[-\left(\frac{dR}{d\hat{\mathbf{n}}} - \frac{dR}{dP \ldots dw} dP \ldots dw\right) d\hat{\mathbf{n}}\right]. \quad (6)$$

Finally, from equations 5, 6 and 4 we obtain the detection probability,

$$p(d_i|\mathcal{P}) = \frac{dR(\hat{\mathbf{n}}_i, \ldots, w_i)}{d\hat{\mathbf{n}} dP \ldots dw}. \quad (7)$$

Non-detection Probability. The non-detection probability, $p(\bar{d}_j|\mathcal{P})$, is obtained simply from the Poisson distribution with $m = 0$ and

$$\langle n_j \rangle = \int_{\delta\hat{\mathbf{n}}_j} d\hat{\mathbf{n}} \ldots \int_{\delta w_j} dw \frac{dR(\hat{\mathbf{n}}, \ldots, w)}{d\hat{\mathbf{n}} dP \ldots dw} \eta(P, \ldots, w) \equiv R_{\text{eff}}(\hat{\mathbf{n}}_j, \ldots, w_j) \quad (8)$$

where $\eta(P, \ldots, w)$ is the efficiency of detection.

Then, we have the non-detection probability,

$$p(\bar{d}_j|\mathcal{P}) = e^{-R_{\text{eff}}(\hat{\mathbf{n}}_j, \ldots, w_j)}. \quad (9)$$

This expression shows that non-detections provide information about the total, not the differential, pulsar probability, and only about that part of the total probability that is detectable.

Likelihood Function. Finally, we obtain for a nearly ideal instrument,

$$\mathcal{L}(\mathcal{P}) = \prod_{i=1}^{N_d} \left[\frac{dR(\hat{\mathbf{n}}_i, \ldots, w_i)}{d\hat{\mathbf{n}} dP \ldots dw}\right] \prod_{j=1}^{N_{\text{nd}}} \left[e^{-R_{\text{eff}}(\hat{\mathbf{n}}_j, \ldots, w_j)}\right]. \quad (10)$$

The likelihood function depends on both detections and non-detections. The efficiency of the detector appears only through the non-detection events.

REFERENCES

1. D. Q. Lamb, in *Frontiers in X-Ray Astronomy*, eds. Y. Tanaka and K. Koyama, (Japan: Yamada Science Foundation and Universal Academy Press, 1992).
2. T. J. Loredo and I. M. Wasserman, Ap. J. Suppl. **96**, 261 (1995).
3. R. N. Manchester, et. al., M.N.R.A.S., **185**, 409 (1978).
4. J. H. Taylor, R. N. Manchester and A. G. Lyne, Ap. J. S. **88**, 529 (1993).

Time-Dilation, Log N - Log P, and Cosmology

R.J. Nemiroff,[1,2] J.P. Norris,[1] J.T. Bonnell,[3] and J.D. Scargle[4]

[1] *NASA/Goddard Space Flight Center, Greenbelt, MD 20771*
[2] *George Mason University, Fairfax, VA 22030*
[3] *Universities Space Research Association*
[4] *NASA/Ames Research Center, Moffett Field, CA 94035*

We investigate whether a simple cosmology can fit GRB results in both time dilation and Log$N$ - Log$P$ simultaneously. Simplifying assumptions include: all GRBs are spectrally identical to BATSE trigger 143, $\Omega = 1$ universe, and no luminosity and number density evolution. Observational data used include: the BATSE 3B peak brightness distribution (64-ms time scale), the Pioneer Venus Orbiter (PVO) brightness distribution, and the Norris et al. (8) time dilation results for peak aligned profiles presented at this meeting. We find acceptable cosmological fits to the brightness distributions when placing BATSE trigger 143 at a redshift of 0.15 ± 0.10. This translates into a $(1 + z_{\rm dim})/(1 + z_{\rm bright})$ factor of about 1.50 ± 0.50 between selected brightness extremes of the Norris et al. sample. Norris et al. (8,10,13) estimate, however, that $(1+z_{\rm dim})/(1+z_{\rm bright}) \approx 2.0 \pm 0.5$ when considering duration tests. The difference is marginal and could be accounted for by evolution. We therefore find that evolution of GRBs is preferred but not demanded.

INTRODUCTION

There are two main results evident with BATSE data which, when taken together, appear consistent with a cosmological origin (1,2). The first is the isotropic nature of the BATSE GRB distribution, and the second is the apparently confined nature of the peak brightness distribution (3,4). Cosmological fits to the brightness distribution have been made by several research teams (5-7). An independent third result is now claimed to be consistent with a cosmological origin of GRBs: time dilation (8-12). In this paper we investigate whether the most recent time dilation results of Norris et al. (13) and Bonnell et al. (14) are consistent with the 3B peak brightness distribution. Fenimore & Bloom (15) imply that the two might not be consistent with a single cosmological setting, as the redshifts they derived from the Norris et al. (8) time dilations are much greater than those implied by the brightness distribution and uniformity assumptions.

Data from the BATSE 3B catalog were accessed from the Compton GRO Science Support Center. BATSE bursts with T90 durations greater than 2

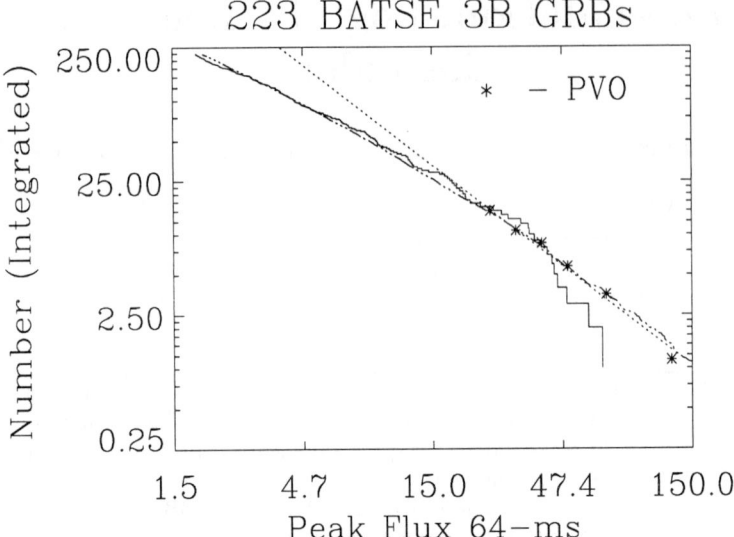

FIG. 1. The combined BATSE - PVO brightness distribution fit to an $\Omega = 1$ cosmology. The solid histogram steps show BATSE 3B data. The asterisks show PVO data normalized to the BATSE rate. The dotted line represents a canonical uniform distribution with -1.5 slope line. The dot-dashed line shows a Monte Carlo fit to an $\Omega = 1$ cosmology of standard candle GRBs with no evolution. The best fit had BATSE trigger 143 placed at a redshift of 0.15, and statistically acceptable fits placed 143 within a redshift of 0.1 of this value.

seconds and peak fluxes greater than 1.827 photons cm^{-2} sec^{-1} were included. BATSE is at least 95 % complete to this peak flux, according to the published BATSE 2B trigger efficiency table.

PVO brightness distribution data were taken from Table 1 of Fenimore and Bloom (15). The conversion from PVO rate to equivalent BATSE rate was found by demanding that the rates be equivalent at the dimmest complete PVO peak flux bin. The conversion factor of 1.25 between peak flux in the PVO energy band 100 - 500 keV and peak flux in the BATSE energy band 50 - 300 keV was given by Fenimore in a private communication.

To test for consistency we created Monte Carlo simulations. We first generated theoretical $\log N$ - $\log P$ (here P stands for peak flux in photons cm^{-2} sec^{-1} on the 64-ms time scale) by considering BATSE trigger 143 a burst with a canonical spectrum. BATSE trigger 143 is one of the brightest and best studied GRBs: its spectrum is well known because of good counting statistics and because it was also seen by the EGRET and COMPTEL instruments on board CGRO; its time series is well studied because of good counting statistics and because it was only the 10th cosmic GRB detected by BATSE. We use only the spectrum at the peak - more specifically for 1 second centered on the bin of peak counts.

We quantified 143's photon spectrum and generated a theoretical cosmology

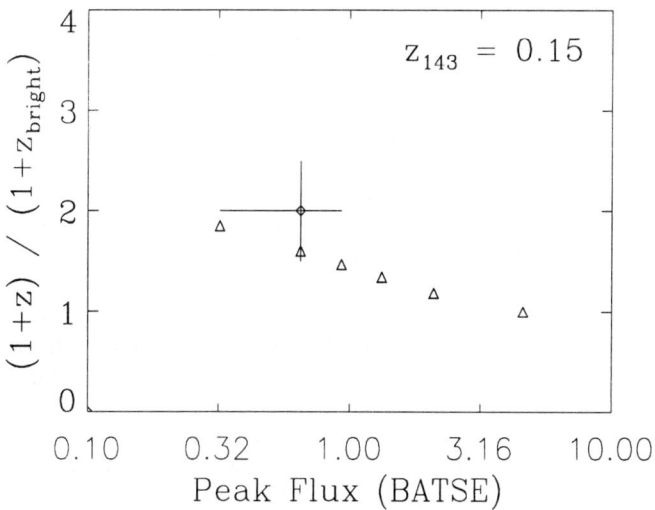

FIG. 2. Comparison of the time-dilation results implied by the brightness distribution plots with the time-dilation results measured by Norris et al. (8) for a simple cosmological model. The triangles represent the expected time-dilation from fits to the Log$N$ - Log$P$ for the simple cosmology described in the text and for BATSE trigger 143 placed at a redshift of 0.15. The diamond with the error bars is the time-dilation result extracted from Norris et al. measured by the peak alignment technique (13).

by throwing it randomly in an $\Omega = 1$ universe. In the subsequent Monte-Carlo simulation, we then re-measured each new "143" burst being careful to consider all cosmological dimming, time-dilating, and reddening factors on the actual 143 spectrum and burst rate. No evolution was considered, however.

We then numerically compared the theoretical brightness distributions generated with a combined BATSE - PVO brightness distribution. The comparison was done using the two-distribution KS test. The test is sensitive to not just the shape of the two distributions but their dynamic range in peak flux. The only parameter varied to maximize the goodness of fit was the actual redshift of the measured 143 spectrum itself.

RESULTS

First, ignoring time dilation, we found it quite possible to fit the combined BATSE - PVO brightness distribution with this simple cosmological paradigm. Such a fit is shown in Figure 1. In this fit BATSE trigger 143 was placed at a redshift of 0.15. Acceptable fits at the 1σ level were found for redshifts between 0.05 and 0.25.

Next, we wished to test whether these fits were consistent with the most recent Norris et al. (13) measurement of time-dilation. Figure 2 shows effective time dilations from the best fit to the cosmology implied by Figure 1, along with the recent time dilation result for the peak alignment method (13). Only one redshift-corrected time-dilation point can be used to date.

Figure 2 shows that the time-dilation factor between the brightness groups in question is about 2.0, while the time-dilation factor implied by the brightness distribution fit in the above cosmology is about 1.5. There is some uncertainty in the Norris et al. (13) factor however, which is about 0.50 in redshift based on the quoted uncertainties and the jitter of the time dilation points. This places the two time-dilation factors about one σ apart. One may note that the time dilation in this bin is only about two σ away from no time-dilation at all - but the inference that time-dilation itself has only marginal statistical significance would be deceiving. This is because neighboring unrelated data points also stray in the same sense from the line, so that the cumulative probability of all the points straying is greater (10,13). As the full time-dilation redshift-decomposition has only been estimated for a single point as yet, we can only estimate that the statistical probability would be at least a factor of a few higher, were points from all brightness classes accounted for.

We also note that were a $1 - \sigma$ higher fit redshift used for 143 in the brightness distribution, a much better agreement would be found between the time-dilation factor implied by the brightness distribution and that measured by the Norris et al. (13) time dilation analysis.

CONCLUSIONS

In general we find the cosmologies implied by the Log$N$ - Log$P$ and the time-dilation are only in good agreement for some methods of time-dilation determination. For peak alignment and auto-correlation measures there is a disagreement in the sense that the burst redshifts implied by the Norris et al. (8,10,13) time dilations are greater than those implied by the combined BATSE - PVO brightness distributions. Since the measured errors are large, however, even this simple non-evolutionary cosmology is not rigorously excluded - to greater than about 2σ - from explaining both results simultaneously. *We therefore conclude that evolution is preferred but not demanded.*

REFERENCES

1. B. Paczynski, Nature **355**, 521 (1992).
2. T. Piran, ApJ **389**, 45 (1992).
3. G.J. Fishman et al., ApJ Suppl. **92**, 229 (1994).
4. C.A. Meegan et al., these proceedings (1996).
5. C.D. Dermer, Phys. Rev. Letters **68**, 1799 (1992).
6. W.A.D.T. Wickramasinghe, et al., ApJ **411**, L55 (1993).
7. A.G. Emslie & H.M. Horack, ApJ **435**, 16 (1994).
8. J.P. Norris, et al., ApJ **424**, 540 (1994).

9. R.J. Nemiroff, et al., ApJ **435**, L133 (1994).
10. J.P. Norris, et al., ApJ **439**, 542 (1995).
11. E. Cohen & T. Piran, ApJ **444**, 25 (1995).
12. R.S. Mallozzi, et al., ApJ **454**, 597 (1995).
13. J.P. Norris, et al., these proceedings (1996).
14. J.T. Bonnell, et al., these proceedings (1996).
15. E.E. Fenimore and J.S. Bloom, ApJ **453**, 25 (1995).

A Constraint on the Distance Scale to Cosmological Gamma–Ray Bursts

Jean M. Quashnock[1]

Department of Astronomy and Astrophysics
University of Chicago, Chicago, Illinois 60637

If γ–ray bursts have a cosmological origin, the sources are expected to trace the large–scale structure of luminous matter in the universe. I use a new likelihood method that compares the counts–in–cells distribution of γ–ray bursts in the BATSE 3B catalog with that expected from the known large–scale structure of the universe, in order to place a constraint on the distance scale to cosmological bursts. I find, at the 95% confidence level, that the comoving distance to the "edge" of the burst distribution is greater than 630 h^{-1} Mpc ($z > 0.25$), and that the nearest burst is farther than 40 h^{-1} Mpc. The median distance to the nearest burst is 170 h^{-1} Mpc, implying that the total energy released in γ–rays during a burst event is of order 3×10^{51} h^{-2} erg. None of the bursts that have been observed by BATSE are in nearby galaxies, nor is a signature from the Coma cluster or the "Great Wall" likely to be seen in the data at present.

INTRODUCTION

The origin of γ–ray bursts is still unknown and is currently the subject of a "great debate" in the astronomical community. Do the bursts have a galactic origin (1) or are they cosmological (2) ? And what is their distance scale?

In this paper, I do not attempt to answer the first question, but rather, I show that *if* one assumes that γ–ray bursts are cosmological in origin, one can begin to answer the second question and place a constraint on the distance scale to the bursts. This is because cosmological bursts are expected to trace the large–scale structure of luminous matter in the universe (3) . The constraint comes from comparing the *expected* clustering pattern of bursts on the sky — which will depend on their distance scale because of projection effects — with that *actually observed*. The observed angular distribution is in fact quite isotropic (4) ; hence, only a lower limit to the distance scale can be placed because a sufficiently large distance will always lead to a sufficiently isotropic distribution on the sky.

Here I use a powerful new likelihood method (5) , which I had previously developed to analyze repeating of γ–ray bursts in the BATSE 1B and 2B

---

[1] *Compton GRO Fellow*

catalogs (6) , to compare the observed counts–in–cells distribution in the new BATSE 3B catalog (7) with that expected for bursts at cosmological distances.

Here I will assume for simplicity that $\Omega_0 = 1$ and $\Lambda = 0$, and that the large–scale structure clustering pattern is constant in comoving coordinates. The results are in fact insensitive to these assumptions because of the small redshifts that are involved. I follow the usual convention and take h to be the Hubble constant in units of 100 km s^{-1} Mpc^{-1}.

LIKELIHOOD METHOD

Let N_{cell} be a large number of circular cells, each centered on a random position on the sky. Each cell is of fixed solid angle size $\Omega = 2\pi(1 - \cos\theta_{\text{rad}})$, where θ_{rad} is the angular radius of the cell. I set the number of cells to be such that any part of the sky is covered, on average, by one cell; hence, $N_{\text{cell}} = 4\pi/\Omega$. Let C_N to be the number of these cells having N γ–ray bursts in them, out of the $N_{\text{tot}} = 1122$ in the BATSE 3B catalog, where $N = 0, 1, 2, ...$ I then define the observed counts–in–cells distribution, $P_N \equiv C_N/N_{\text{cell}}$, as the probability that a randomly chosen cell of size Ω has N bursts in it. The counts–in–cells distribution contains information about clustering of γ–ray bursts on scales comparable to the angular size θ_{rad} of the cell.

I now define Q_N to be the counts–in–cells distribution that is expected if γ–ray bursts are cosmological in origin and trace the large–scale structure of luminous matter in the universe. This expected distribution depends on only one unknown parameter, the effective distance D to γ–ray bursts (which I define below), because the angular clustering pattern of bursts on the sky will depend by projection on this distance.

The *likelihood* $\mathcal{L}$ measures how likely it is that the observed counts–in–cells distribution P_N is drawn from the expected distribution Q_N. Since Q_N depends on the unknown effective distance D to γ–ray bursts, the likelihood is really a measure of how likely a given value of D is. I find that (5) :

$$\log \mathcal{L} = N_{\text{cell}} \sum_N P_N \log Q_N + \text{constant} . \qquad (1)$$

Now the cumulative $C_{\max}/C_{\min}$ distribution of γ–ray bursts seen by BATSE begins to roll over from a $-3/2$ power–law for bursts fainter than $C_{\max}/C_{\min} \sim 10$. Since this is many times above threshold, it suggests that BATSE sees most of the source distribution and that this distribution is not spatially homogeneous (8) . I define D as the comoving distance beyond which the source density drops appreciably. It is not the distance to the very dimmest burst in the BATSE catalog, but rather the typical distance to most of the dim bursts in the sample; thus, D is the *effective distance* to the "edge" of the source distribution in the BATSE catalog.

I take the power spectrum which characterizes the large–scale clustering of γ–ray burst sources to be the same as that determined from a redshift survey of radio galaxies (9) . This power spectrum is characteristic of moderately rich environments, and is intermediate between that of ordinary galaxies and clusters. Because the exact bias factor relating the clustering of γ-ray burst sources to that of luminous matter is unknown, such an intermediate ansatz is reasonable. In any case, the resultant distance limit depends only weakly on the bias factor (roughly as the square root).

Knowledge of the power spectrum permits a calculation of the expected angular clustering pattern, the expected counts–in–cells distribution Q_N, and finally the likelihood $\mathcal{L}$ [from equation (1)], all as a function of the effective distance D to γ-ray bursts (5). I have included the smearing due to finite positional errors on the clustering on small scales (10). Indeed, each burst in the BATSE catalog is assigned a positional uncertainty $\theta_{\rm err}$ corresponding to a 68% confidence that the true burst position is within an angle $\theta_{\rm err}$ to the position listed in the catalog.

I have chosen the cell size $\theta_{\rm rad}$ in order to maximize the sensitivity of detection, or signal–to–noise, given the strength of the signal expected. For a sample of 1122 bursts (the total number of bursts in the BATSE 3B catalog) with positional smearing of $\theta_{\rm err} = 3.8°$ (the median value in the 3B catalog), the signal–to–noise is maximized when cells of $\theta_{\rm rad} = 5°$ are used (5).

RESULTS

Figure 1 shows the likelihood of the BATSE 3B catalog data as a function of the effective comoving distance D, calculated using cells of size $\theta_{\rm rad} = 5°$. The likelihood is normalized to that expected for an isotropic distribution on the sky. At large values of D (the maximum value allowed is $D = R_{\rm H} = 6000\ h^{-1}$ Mpc, the size of the horizon in a closed universe), the likelihood goes to unity, because by projection a sufficiently large distance will always lead to an isotropic distribution on the sky. Note also that there is no value of D for which the likelihood is greater than 1; thus, the maximum likelihood value for D is $R_{\rm H}$ and the 3B data are consistent with isotropy.

The solid line in Fig. 1 shows the likelihood for a positional smearing of $\theta_{\rm err} = 3.8°$, corresponding to the median value in the 3B catalog. To illustrate the dependence of these results on positional errors, I also show (dashed line) the results for a larger positional smearing[2] of $\theta_{\rm err} = 6.6°$ (with cells of size $\theta_{\rm rad} = 9°$ to maximize signal–to–noise).

Small values of the effective comoving distance to γ-ray bursts are unlikely, according to Fig. 1: I find, at the 95% confidence level, that for the 3B median positional error of $\theta_{\rm err} = 3.8°$, D must be greater than 630 h^{-1} Mpc, corresponding to a redshift $z > 0.25$. If the positional errors are larger than quoted and are better characterized by $\theta_{\rm err} = 6.6°$, these results are only slightly weakened: At the 95% confidence level, D must be greater than 500 h^{-1} Mpc, corresponding to a redshift $z > 0.19$.

---

[2] Graziani & Lamb (11) compare the 3B positions with those from the IPN network, and conclude that the systematic errors are larger than the 1.6° value quoted in the 3B catalog. Their best-fit model gives a median positional error of 6.6°.

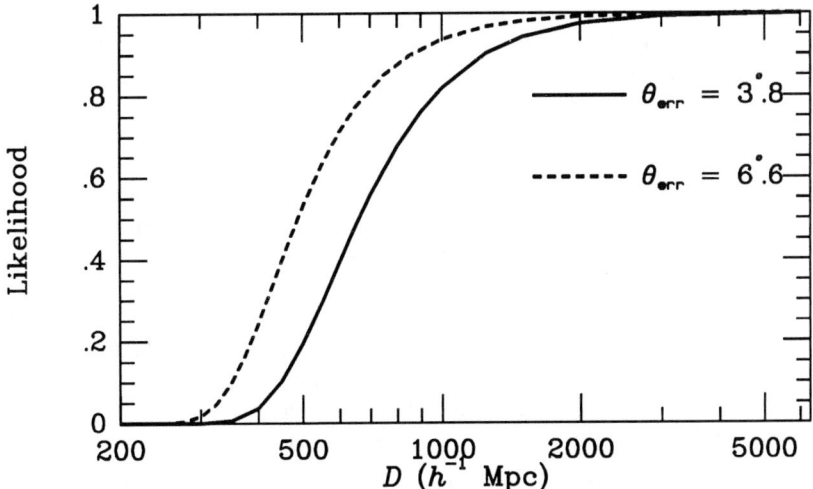

FIG. 1. The likelihood of the BATSE 3B catalog data as a function of the effective comoving distance D to γ-ray bursts, shown with a smearing of $\theta_{\rm err} = 3.8°$ and $6.6°$. See ref. (5), ©1996 by The American Astronomical Society.

These limits are not sensitive to earlier assumptions on cosmology and clustering evolution since these only become important at higher redshifts. They are also conservative limits in that a constant median value for the positional errors was used, rather than the entire distribution of errors. This is because the bright bursts, which ostensibly are nearer to us, are more clustered and are responsible for the bulk of the expected signal, but in fact have smaller errors than the median value. The faint bursts, which are far away, are hardly clustered to begin with (even before smearing), but have errors larger than the median value. Hence the expected clustering pattern has been smeared more by using a constant median value (this permits a simpler calculation) than by smearing using the entire distribution of errors. So the counts–in–cells statistic has been weakened somewhat and thus the quoted lower limits are in fact conservative.

CONCLUSIONS

If γ-ray bursts are cosmological and trace the large–scale structure of luminous matter in the universe, and their positional errors are as quoted in the 3B catalog, then the lack of any angular clustering in the data implies that the observed distance to the "edge" of the burst distribution must be farther than 630 h^{-1} Mpc. Since there are 1122 bursts in the catalog, an effective limit on the *nearest* burst to us can be placed by convoluting the likelihood as a function of D (Fig. 1) with the nearest neighbor distribution of 1122 bursts inside a sphere of radius D. I find that the nearest burst must be farther than 40 h^{-1} Mpc at the 95% confidence level, and farther than 10 h^{-1} Mpc at the 99.9% level. At this level of confidence, then, none of the bursts that have been observed by BATSE are in nearby galaxies. A signature from the Coma cluster or the "Great Wall" ($\sim 70\ h^{-1}$ Mpc) is not likely to be seen in the data at present, since only a few bursts could have originated from these distances.

The median distance to the nearest burst is 170 h^{-1} Mpc. Since the brightest burst in the 3B catalog has a fluence of 7.8×10^{-4} erg cm^{-2} in γ–rays, this implies that the total energy released in γ–rays during a burst event is of order $3 \times 10^{51}\ h^{-2}$ erg.

As the number of observed γ–ray bursts keeps increasing, the distance limit will improve. In fact, with 3000 burst locations, the clustering of bursts might just be detectable (3) and would provide compelling evidence for a cosmological origin. If it is not detected, the redshift to the "edge" of the burst distribution would be put at $z \sim 1$ or beyond.

Acknowledgments. I would like to acknowledge useful discussions with Carlo Graziani, Don Lamb, Cole Miller and Bob Nichol. This research was supported in part by NASA through the *Compton* Fellowship Program — grant NAG 5-2660, grant NAG 5-2868, and contract NASW-4690.

REFERENCES

1. D. Q. Lamb, Proc. Astr. Soc. Pac., in press (1995).
2. B. Paczyński, Proc. Astr. Soc. Pac., in press (1995).
3. D. Q. Lamb and J. M. Quashnock, Astrophys. J. Lett. **415**, L1 (1993).
4. M. S. Briggs et al., Astrophys. J. **459**, 40 (1996).
5. J. M. Quashnock, Astrophys. J. Lett. **461**, L69 (1996).
6. J. M. Quashnock, Astrophys. Sp. Sci. **231**, 35 (1995).
7. C. A. Meegan et al., Astrophys. J., submitted (1996).
8. C. A. Meegan et al., Nature **355**, 143 (1992).
9. J. A. Peacock and D. Nicholson, Mon. Not. Roy. Astr. Soc. **253**, 307 (1991).
10. D. H. Hartmann, E. V. Linder and G. R. Blumenthal., Astrophys. J. Lett. **367**, 186 (1991).
11. C. Graziani and D. Q. Lamb, these proceedings

Constraints on the Gamma-ray Burst Luminosity-Duration Relationship in the Galactic Scenario

R.E. Rutledge[a], W.H.G. Lewin[a], J. Hakkila[b], G. Pendleton [c], J.P. Lestrade[d], C. Kouveliotou[e], C. Meegan [f]

[a] 37-624B, MIT, Cambridge MA 02139
[b] Dept. Physics and Astronomy, Mankato State University
[c] University of Alabama in Huntsville
[d] Mississippi State University
[e] Universities Space Research Association
[f] NASA/Marshall Space Flight Center

> We present a method to investigate constraints on the luminosity-duration relationship in the context of galactic models. The observed duration distribution cannot be explained by a one-to-one correspondence between luminosity and duration; a distribution of durations as a function of luminosity is required.
>
> We also comment on the preliminary report of a correlation between burst spectrum and spatial distribution. If confirmed, this correlation may provide the first evidence for a finite-width luminosity function in GRBs. Observational parameters derived from this result permit a measurement of a power-law luminosity function slope, as well the relative frequency density of GRBs of the two different spectral types in the analyses.

Several authors have reported a correlation between cosmic GRB duration and peak-fluxes (1–4). These have often been interpreted as being due to cosmological redshift effects; however, blast-wave models predict a luminosity-correlated duration which may also be responsible for all or part of this effect (5). As such, it is useful to develop a formalism to investigate the possible contributions of luminosity-duration correlations to the observed effects.

Also, we comment on the preliminary report by C. Kouveliotou (7) that a sample of spectrally hard bursts have a different peak flux distribution from a sample of soft bursts.

First, we describe a simple way to understand how peak-flux (F_p) correlated durations can result from a Euclidean scenario. We assume a frequency density distribution n(r) (see Fig. 1a), which can be considered constant out to some radius r_b, and a single population of GRB sources, comprised of a fraction (f) with intrinsic luminosity L_1 and duration t_1 and the remaining fraction (1-f) with intrinsic luminosity L_2 and duration t_2. The parameters of the observed duration distribution (or, the distribution of any observational

© 1996 American Institute of Physics

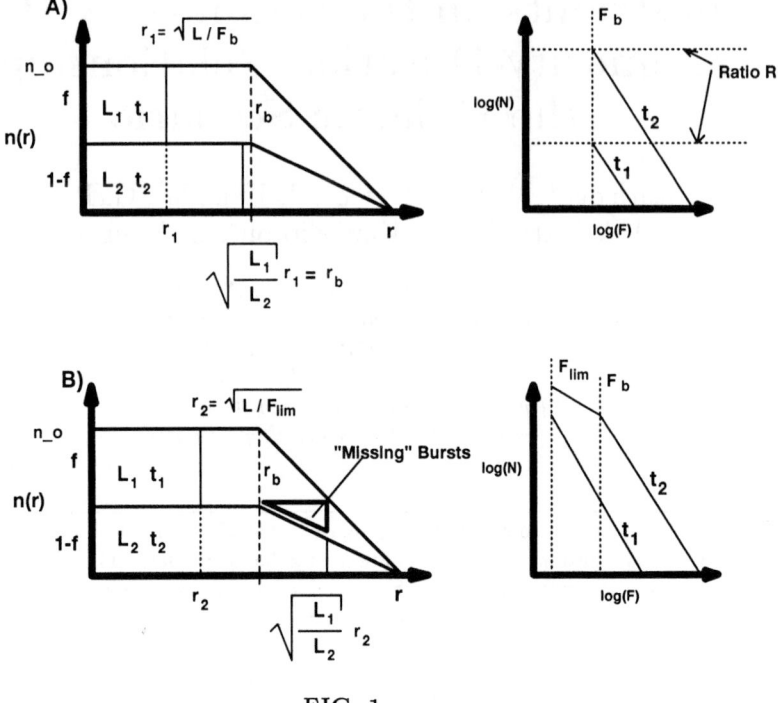

FIG. 1.

property which is not mathematically related to F_p) must remain constant as a function of F_p for $F_p > F_b$ as the proportion of sources of different luminosities (and therefore, different durations) remains the same. Only at $F_p < F_b$ will the duration distribution change, as fewer high luminosity bursts are observed relative to low-luminosity bursts ("missing bursts", Fig. 1b).

In a like fashion, if one divides GRBs by any observational property which is correlated to their intrinsic luminosity, one should observe evidence for deviation from homogeneity in the integrated number vs. peak flux distribution at different F_p values; the sub-sample with higher intrinsic luminosity will show evidence for deviation from homogeneity at a higher F_p value.

MODEL AND ANALYSIS

We assume a power-law intrinsic luminosity distribution:

$$\phi(L) = AL^{-\beta} \quad 1 < L < \chi \tag{1}$$

where A, β, and χ are constants. A is a normalization constant; β is the assumed power law slope of the luminosity function; and χ is the observed full width of the power-law luminosity function.

We assume a power-law relationship between the burst time-scale (t_{dur}) and the burst intrinsic luminosity:

$$t_{\text{dur}}(L) = t_0 L^{-\lambda} \tag{2}$$

where t_0 and λ are constant. We assume a frequency density distribution:

$$n(r) \propto \frac{1}{1+r^\alpha}, r < r_{\text{tidal}} \tag{3}$$
$$= 0, r > r_{\text{tidal}} \tag{4}$$

We define r_{max}:

$$r_{\text{max}} = \sqrt{\frac{\chi}{F_{\text{lim}}}} \tag{5}$$

where χ is the observed full width of the luminosity function and F_{lim} is the minimum observable flux. Thus, a burst of luminosity χ at a distance r_{max} will produce a measured peak flux $F_p = F_{\text{lim}}$.

We assume a frequency density distribution function (α), an intrinsic luminosity function power-law (β) and a width to the luminosity function (χ). The observational parameter we test against is the integrated number distribution as a function of duration, $N(T > T')$:

$$N(T > T') = \int_1^\chi \int_0^{\left(\frac{L}{\chi}\right)^{1/2} r_{\text{max}}} \frac{r^2}{1+r^\alpha} \, dr \, \phi(L) \, dL \tag{6}$$

where the overall constant has been dropped. If we rewrite this using Eq. 2:

$$N(T > T') = \int_{\frac{T'}{T_{\text{min}}}\chi^{-\lambda}}^{1} \int_0^{\left(\frac{T^{-1/\lambda}}{\chi}\right)^{1/2} r_{\text{max}}} \frac{r^2}{1+r^\alpha} \, dr \, T^{1-\frac{1}{\lambda}(1+\beta)} dT \tag{7}$$

where, again we drop the overall constant of normalization. Due to the one-to-one nature of the assumed duration-luminosity relationship (Eq. 2), there is a unique relationship between the duration – luminosity parameter (λ), the observed minimum and maximum burst durations, and the assumed luminosity function width (χ):

$$\lambda = \frac{\ln(t_{\text{dur,max}}/t_{\text{dur,min}})}{\ln(\chi)} \tag{8}$$

Then, we assume values for r_{max}, r_{tidal}, χ, β, α, measure the value of λ from the data sample, produce the integrated duration distribution (Eq. 7) and compare it using a KS test to the observed distribution.

We have done this for the following range of values: ($\alpha = 2.0$; $2 \leq r_{\text{tidal}} \leq 20$, $\delta r_{\text{tidal}} = 20\%$; $3 \leq r_{\text{max}} \leq 100$, $\delta r_{\text{max}} = 20\%$; $2 \leq \chi \leq 100$, $\delta \chi = 25\%$; $-3 \leq \beta \leq 3$, $\delta \beta = 0.25$), where the values indicated by "δ" indicate the increment between parameter values which were tested. For all combinations of these parameters, no acceptable distributions (KS probability > 0.001) were found.

COMMENT ON THE HARDNESS/SPATIAL DISTRIBUTION CORRELATION

Kouveliotou (7) shows preliminary evidence that, when the GRB populations are divided into three spectral hardness groups (hard, mid, and soft), the hard and soft bursts have different peak-flux distributions. Here, we comment on the possible significance of this observational result, if it is confirmed.

For reasons we discuss above (in the context of burst durations, but the discussion holds if we substitute burst spectral hardness), the fact that the integrated-number vs. peak-flux distribution of the hard bursts "rolls over" at a higher peak-flux than the soft-bursts can be interpreted, if we assume all GRBs come from a single population, as meaning that hard bursts have higher intrinsic luminosity than the soft bursts. Using a power-law luminosity function (Eq. 1) and estimating the luminosities of the hard bursts and soft bursts as single luminosities (L_{hard}, L_{soft}), then if soft bursts make up a fraction f of the total population and hard bursts make up the remaining $1 - f$ (Fig. 1), we find the ratio:

$$\frac{f}{1-f} = \left(\frac{L_{\text{hard}}}{L_{\text{soft}}}\right)^{\beta} \qquad (9)$$

From Fig. 1, we define R as the ratio of hard bursts observed with $F > F_b$ to soft bursts observed with $F > F_b$, which can be measured directly to observational accuracy, and is equivalent to:

$$R = \frac{(1-f)\, n_0 \, \frac{L_{\text{hard}}}{L_{\text{soft}}}^{3/2} r_1^3}{f \, n_0 \, r_1^3} \qquad (10)$$

Combining Eqns. 9 & 10 produces an equation for the power-law luminosity slope:

$$\beta = \frac{3}{2} - \frac{\ln(R)}{\frac{L_{\text{hard}}}{L_{\text{soft}}}} \qquad (11)$$

If the value at which the peak flux distribution of hard bursts deviates from homogeneity is observable, while that of the soft bursts is not, then the a lower limit may be placed on the ratio of luminosities (i.e., the width of the luminosity function):

$$\frac{L_{\text{hard}}}{L_{\text{soft}}} \geq \frac{F_{b,\text{hard}}}{F_{\text{lim}}} \qquad (12)$$

Thus, with an observed value of R, and a lower limit on $L_{\text{hard}}/L_{\text{soft}}$, the value of the slope of the luminosity function β (Eq. 11) can be determined to observational certainty – with both upper and lower limits.

Also, one can find a lower limit on the ratio of the frequency density of hard bursts to that of soft bursts:

$$\frac{f}{1-f} = R \left(\frac{L_{\text{hard}}}{L_{\text{soft}}}\right)^{\frac{3}{2}} \tag{13}$$

while an upper limit on the width of the luminosity function (which has been found using other methods – e.g. (6)) permits an upper-limit to be placed on this value.

Importantly, definitive demonstration of different peak flux distributions of bursts separated by spectral hardness can be interpreted as evidence for the correlation between spectral hardness and burst luminosity, and thus for a finite width luminosity function.

DISCUSSION AND CONCLUSIONS

We described our method to investigate the luminosity-duration relationship in galactic source population scenarios. For the range of parameters investigated, no satisfactory combination of parameters exists which provides agreement with the observed duration distribution in GRBs; this may be overcome by discarding the one-to-one duration-luminosity relation and assuming a distribution of durations as a function of luminosity.

We have also commented on the possible significance of the findings of Kouveliotou (7). One interpretation is that the observed difference in the peak-flux distributions of hard and soft bursts is due to hard bursts having higher luminosity than soft bursts. In such a case, using simplified assumptions, one can measure observational parameters which constrain the slope of the observed luminosity function, as well as the relative frequency per unit volume of hard bursts vs. soft bursts.

Acknowledgements: This work is supported by the NASA GSRP under grant NGT-51369 (RR) and the CGRO Guest Investigator Program under NAG5-2046 (WHGL).

REFERENCES

1. J.P. Norris, et al., ApJ **424**, 540 (1994).
2. J.P. Norris, et al., ApJ **439**, 542 (1995).
3. S. P. Davis et al., in Gamma-Ray Bursts, eds. G.J. Fishman, J.J. Brainerd, & K.C. Hurley, AIP Conf. Proc. **307**, 182 (AIP, New York, 1994).
4. R.A. Wijers, & B. Paczyński, ApJ Lett. **437**, 107L (1994).
5. J. Brainerd, ApJ Lett. **428**, L1 (1994).
6. J. Hakkila, et al., ApJ **454**, 134 (1995).
7. C. Kouveliotou, et al., these proceedings (1996).

Search For Constraints on the Gamma-ray Burst Peak Flux-Distance Relation in the Cosmological Scenario

R.E. Rutledge[a], W.H.G. Lewin[a], J. Hakkila[b], T. Koshut[b],
G. Pendleton[b], J.P. Lestrade[d], C. Kouveliotou[e], J. Horack[f],
C. Meegan[f]

[a] 37-624B, MIT, Cambridge MA 02139
[b] Dept. Physics and Astronomy, Mankato State University
[c] University of Alabama in Huntsville
[d] Mississippi State University
[e] Universities Space Research Association
[f] NASA/Marshall Space Flight Center

We describe how we investigate the constraints on the gamma-ray burst (GRB) flux–red shift relationship in the cosmological scenario. Using the burst duration as a standard time-scale, and the burst peak flux as a standard candle, the absolute values of the red shifts of any two bursts can be determined. Using multiple pairings of bursts, a statistical distribution of the red shift of any individual burst can be produced, the properties of which are determined entirely by the properties of the intrinsic duration distribution and luminosity distribution. We apply the method to bursts in the 3B catalog, and find the non-physical result of the brightest bursts having medians near (1+z)=1.0, and 1+z *decreasing* with decreasing burst flux, which appears to be due to the broad log-normal distribution of the observed duration distribution.

INTRODUCTION

A leading paradigm for the spatial distribution of GRBs places the progenitors at significant red shifts. As the progenitor population has not been identified, there exists no *a priori* information on the expected intrinsic characteristics (*i.e.* intrinsic duration distribution, luminosity function, time-dependent spectrum). It cannot be ruled out that GRBs have more than one parent population.

Most previous work has sought to find the *relative* amount of duration stretching between bursts of different peak intensities, to find the time-dilation expected from a cosmological origin. Duration measures, such as t_{50} and t_{90}, are used in spite of large observed dispersion, because there exists few time-scale measurements available to all GRBs which might otherwise be used.

© 1996 American Institute of Physics

One such measurement is the peak in the νF_ν photon energy spectrum (1).

There does exist a method to find the *absolute* red shifts of GRBs. This has been applied recently (2) to burst ensembles, to check for consistency between reported relative amounts of time-dilation in GRBs and the observed different peak fluxes.

Here we describe how to use this method applied to individual bursts, to find a most likely *absolute* red shift for that particular burst.

METHOD

If one has two GRBs which are known *a priori* to have identical co-moving durations and peak fluxes, one may use the values of these properties to find their *absolute* red shift of origin.

To simplify this analysis we make the following assumptions: 1) GRBs durations are not a function of the co-moving energy (which is not necessarily the case (3); 2) the GRB duration distribution does not evolve with red shift; 3) the GRB luminosity function does not evolve with red shift; 4) $\Omega_0 = 1$; 5) the GRB spectrum can be represented as a power-law.

To simplify discussion, consider (first) two GRBs of identical duration, of identical peak flux, at two different red shifts. The ratio of their durations will be:

$$r = \frac{t_1}{t_2} = \frac{(1+z_1)}{(1+z_2)} \frac{(1+z_1)^\tau}{(1+z_2)^\tau} \tag{1}$$

where we have included an extra power of $(1+z)^\tau$ to account for differences in burst duration due to co-moving energy. If $\tau = -1$ then r=1 and no cosmological time-dilation effect would be observed. We will assume throughout a value $\tau = 0$.

The ratio of the burst peak fluxes will be:

$$\chi = \frac{P_2}{P_1} = \left(\frac{1+z_1}{1+z_2}\right)^\alpha \left(\frac{1 - \frac{1}{\sqrt{1+z_1}}}{1 - \frac{1}{\sqrt{1+z_2}}}\right)^2 \tag{2}$$

Inverting these two equations for the red shifts gives:

$$1 + z_1 = \left(\frac{\frac{\sqrt{\chi}}{r^{(\alpha-1)/2}} - 1}{\frac{\sqrt{\chi}}{r^{\alpha/2}} - 1}\right)^2 \tag{3}$$

$$1 + z_2 = \left(\frac{\frac{\sqrt{\chi}}{r^{\alpha/2}} - \frac{1}{\sqrt{r}}}{\frac{\sqrt{\chi}}{r^{\alpha/2}} - 1}\right)^2 \tag{4}$$

These two equations are exactly solvable, given the ratio of durations r and peak fluxes χ.

Now, we take a sample of N bursts, and number them i=1, N. For each burst in turn, beginning at j=1, we calculate $(1 + z_{j,i})$, using bursts with odd values of i for $i < j$ and for even values of i for $i > j$. This way, we compare each burst with the maximum range of peak flux values, and each resulting calculation of $(1 + z_{j,i})$ is statistically independent of all other values of $(1 + z_{j,i})$ produced.

If all GRBs follow the above assumptions, we can expect a mode at the most-likely value of $(1 + z)$; the statistics of the distribution of $(1 + z)$ will be dependent on the GRB luminosity function, duration distribution function, and the accuracy of the measurements of both.

DATA SELECTION

We use data from the BATSE 3B catalog (4). We use bursts with values of t_{90} and P_{1024} listed with relative accuracy of better than 20%, with values of $t_{90} > 2.0$ seconds and of $P_{1024} > 0.5$ phot cm^{-2}sec^{-1}. This gives us a sample of 424 bursts.

We note that applying Spearman rank-order correlation between the duration and peak fluxes gives the probability of producing the observed correlation in this sample as 55%.

SOME RESULTS

We produced $(1 + z)$ values for the 424 bursts in the present dataset. We separated the 424 bursts into 6 groups of approximately 80 bursts each, based on their peak flux values (on the 1024 ms time scale). The $(1 + z)$ values for all 80 bursts in each of the six peak-flux divisions were concatenated into six $(1 + z)$ distributions; these are shown in the figure (panel *a* corresponds to the lowest peak fluxes, panel *f* to the highest peak fluxes). The broken line in each panel is drawn at the median value of the distribution.

Inter-comparing the distributions of different peak flux ranges, there is clear difference in the qualitative shapes of the distributions, and in the median of the distributions, which increases with increasing peak-flux range from an non-physical value of ∼0.3 to 0.9.

To aid in interpreting these results, we compared the observed distributions with three simulated distributions.

1. First, we used a boot-strap method to produce simulated distributions. For each observed burst (with values of peak flux and duration), we drew, with replacement, from the observed peak-flux distribution and duration distribution, to produce 10,000 $(1 + z)$ values. Thus, the resulting distribution was produced with the assumption that there exists no correlation between the duration and peak fluxes of bursts. These 10,000 values for each of the ∼80 bursts in each distribution, were used to produce a simulated distribution, which is included in the figure; the simulated distributions are not visually discernible from the observed distributions.

2. We simulated a data-set of bursts which have the observed peak-flux distribution in a non-evolving standard-candle cosmological scenario with the

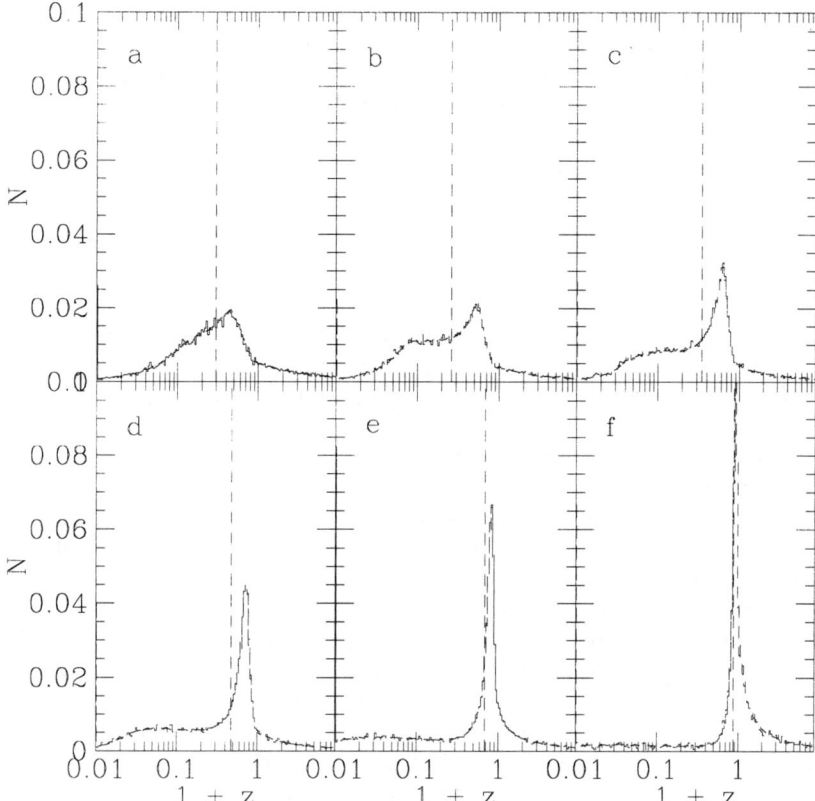

FIG. 1. The $(1+z)$ distributions of 6 groups of ~ 80 bursts, sorted by peak-flux, in order of increasing peak-flux range; the distribution in panel *a* represents GRBs from the lowest 80 peak-flux values, and panel *f* represents the highest 80 peak-flux values. The vertical broken line is at the median value of the distribution in each panel

faintest burst in the present data-set coming from $(1+z)= 2.0$, and a power-law photon spectral slope of 2.0 (cf. (5)) and a single-valued (roughly delta-function) co-moving duration distribution. This simulated data-set was analyzed as the observed data-set was (described above), and produced highly peaked (half-width of $\sim 20\%$) medians at $(1+z)$ values >1, which decrease with increasing peak flux range, as expected, to $(1+z)\sim 1.2$. Visual inspection easily discerns between these simulated distributions and the observed distributions.

3. We simulated a data-set of bursts which have the observed peak flux distribution in a non-evolving standard-candle cosmological scenario (assuming the faintest burst in the present data-set coming from $(1+z)=2.0$, and power-law photon spectral slope of 2.0), but with a co-moving log-normal duration

distribution of width $\sigma = 0.5$ and mean $\log(t_{90})=1.17$. For each peak-flux value, a random co-moving duration was generated and increased by a factor of $(1+z)$ to produce the expected observed duration. The simulated data-set was analyzed as the observed data-set (described above), producing six $(1+z)$ distributions. The simulated $(1+z)$ distributions were qualitatively similar to the observed distributions.

DISCUSSION AND CONCLUSIONS

There is a strong difference between $(1+z)$ distributions of the simulated data-sets, depending on whether a delta-function duration distribution or a log-normal duration distribution is assumed. In the former, the $(1+z)$ distributions are highly peaked with median values >1, which decrease with increasing peak-flux. In the latter, the $(1+z)$ distributions show the complex behavior seen in the observed data, with median values of $(1+z)$ <1, which increase with increasing peak-flux.

We note that the qualitative behavior of the $(1+z)$ distributions as a function of peak-flux range is not due to a correlation between peak-flux and duration, as the behavior is also found in simulated data-sets where there is no (i.e. random) correlation between peak-flux and duration.

We have shown a method which can, in principle, produce the peak-flux–absolute red shift relationship of GRBs in a cosmological scenario. We have found that the $(1+z)$ distributions of bursts found with this method are qualitatively similar to simulated data-sets where there is no intrinsic relationship between burst durations (drawn from a log-normal distribution) and peak-fluxes, as well as to simulated data-sets where there is the expected correlation between peak-flux and parameters of a log-normal duration distribution. This work is on-going, and we will investigate the quantitative limits on the possible correlation between the peak-flux and log-normal duration distribution parameters.

We find that the medians for most of the bursts' $(1+z)$ distributions in the present sample occur at the non-physical values of $(1+z) < 1$; the value of the median increases with decreasing peak flux, until for the brightest bursts $(1+z) \sim 1$–having the opposite sense from what is expected from the simple model above. Based on comparison with our simulated data-sets, this appears to be due to the broad dispersion in the observed log-normal duration distribution.

REFERENCES

1. R. Mallozzi, et al., ApJ, submitted (1995).
2. J. Horack, et al., ApJ, submitted (1996).
3. E. E. Fenimore, et al., ApJ **448**, L101 (1995).
4. C. A. Meegan, et al., 1995, "The Third BATSE Gamma-Ray Burst Catalog", available in electronic form from GROSSC, NASA/GSFC (1995).
5. R. E. Rutledge, L. Hui, & W. H. G. Lewin, MNRAS **275**, 753 (1995).

The Correlation Between Gamma-Ray Burst Duration and Peak Flux

R.E. Rutledge[a], W.H.G. Lewin[a], G. Pendleton[b], J.P. Lestrade[c], C. Kouveliotou[d], C. Meegan[e]

[a] 37-624B, MIT, Cambridge MA 02139
[b] University of Alabama in Huntsville
[c] Mississippi State University
[d] Universities Space Research Association
[e] NASA/Marshall Space Flight Center

> Using non-parametric tests, we examine data in the 3B BATSE catalog for correlations between burst duration (t_{50}, t_{90}) and peak flux (P_{64}, P_{256}, P_{1024}). We find no evidence for such a correlation in the present samples. Comparing duration distributions of bright and faint bursts, we find that the amount of duration-stretching is limited (4σ) to between 0.7 and 1.3. This result is in conflict with previously reported results.

INTRODUCTION

Interest in the relationship between burst duration and peak flux of cosmic Gamma Ray Bursts (GRBs) has grown. Investigations were first motivated by cosmological-origin models, which predict a correlation due to time-dilation. Correlation between peak flux and duration may also be due to an intrinsic relationship between burst luminosity and duration (*e.g.* (1)).

A dependence between the comoving photon energy and the burst duration may also produce a measurable effect in a cosmological scenario. If the dependence of duration on photon energy is $t \propto E^{-1}$, then the effect of time-dilation on the observed duration will be exactly cancelled; if the exponent is < -1 then burst durations will be *anti*-correlated with peak flux; if the exponent is > -1, the burst durations will be correlated with peak flux. Fenimore et al. (11) report that the pulse-width of GRBs as a function of energy is consistent with a power-law dependence of between 0.37-0.42 (no quoted confidence).

Using the Spearman and Kendall non-parametric tests (2), we search for correlations between the duration measures (t_{50}, t_{90}) and peak flux measures (P_{64}, P_{256}, P_{1024}) in the BATSE 3B catalog.

DATA SELECTION

We use data from the BATSE 3B catalog (3). We draw samples comparing both the duration measures (t_{50}, t_{90}; (4); (5)) with all three peak flux measures (P_{64}, P_{256}, P_{1024}), producing six different data samples. In addition, we apply three data-selection criteria (see Table 1), which results in a total of 18 different (but not independent) data samples.

TABLE 1. The Data Selection Criteria

Ref.	Relative Uncertainty	Duration Cutoff
$t_{50} > 4\times$	<0.2	$t_{50} > 4\times$ Peak Flux Integration Time
$t_{50} > 2.0$ sec	<0.2	$t_{50} > 2.0$ sec
No Relative Uncertainty	None	$t_{50} > 2.0$ sec

In the "$t_{50} > 4\times$" sample, data for each of the six pairs of (duration, peak flux) were drawn, requiring that the quoted relative uncertainty in both the peak flux and duration measures of a burst is less than 20% (arbitrarily chosen), and that the value of the duration measure t_{50} for each burst be at least four times the integration time scale of the peak flux measure for that sample.

In the "$t_{50} > 2.0$ sec" sample, data were drawn, requiring that the quoted relative uncertainty in both the peak flux and duration measures of a burst is less than 20%, and that the value of the duration measure t_{50} for each burst be greater than 2.0 s.

In the "No Relative Uncertainty" sample, data were drawn requiring only that the value of the duration measure t_{50} for each burst be greater than 2.0 s.

SPEARMAN AND KENDALL'S TAU CORRELATION

We applied the Spearman rank-order correlation test, and the looser but similar Kendall's Tau test (2).

To gauge the sensitivity of these tests to intrinsic duration stretching, we took the data samples above, and ordered them by peak flux, ranked the bursts by peak flux $i = 1$ to N, where N is the number of bursts. We artificially stretched the duration of the burst ranked i to

$$t_{\text{stretch}} = t_{\text{dur}}(1.0 + \frac{i}{N}) \qquad (1)$$

where t_{dur} is either t_{50} or t_{90} (whichever was used for that sample), resulting in a "hand-inserted" factor of two difference between the duration of the faintest and brightest burst in the sample. The resulting samples were also subjected to the Spearman and Kendall's Tau tests.

In Table 2, we show the resulting probability of the data samples being produced from a randomly distributed population, from the Spearman and Kendall's Tau correlation tests (indicated by t_{dur}).

There is no indication of correlation between peak flux and duration in any of the observed 3B data-sets. Many of the data-sets (usually ones using P_{64} or P_{256}) with duration stretching "hand-inserted" show no correlation either, indicating that the dispersion in the actual data-set is too great to permit detection of a correlation resulting in a factor of two difference in duration over the considered flux range.

For all datasets using (P_{1024}, t_{90}), these tests appear to be sensitive to duration-stretching of a factor of two, but no correlation in the data is present.

TABLE 2. % significance from Spearman / Kendall tests

		$t_{50} > 4\times$							
		\multicolumn{2}{c}{t_{50}}			\multicolumn{2}{c}{t_{90}}		
	N	Spearman		Kendall		N	Spearman		Kendall	
	(bursts)	t_{dur}	$t_{stretch}$	t_{dur}	$t_{stretch}$	(bursts)	t_{dur}	$t_{stretch}$	t_{dur}	$t_{stretch}$
P_{64}	352	3	76	3	39	340	6	42	6	46
P_{256}	471	7	14	7	14	452	44	1	43	0.9
P_{1024}	488	4	3	4	3	488	69	$3\cdot10^{-4}$	69	$3\cdot10^{-4}$
			$t_{50} > 2.0$ sec							
		\multicolumn{2}{c}{t_{50}}				\multicolumn{2}{c}{t_{90}}		
	N	Spearman		Kendall		N	Spearman		Kendall	
		t_{dur}	$t_{stretch}$	t_{dur}	$t_{stretch}$		t_{dur}	$t_{stretch}$	t_{dur}	$t_{stretch}$
P_{64}	304	0.4	95	0.4	92	293	9	17	8	19
P_{256}	457	3	25	14	24	435	39	0.8	37	0.8
P_{1024}	526	17	1	17	2	526	34	$9\cdot10^{-5}$	34	$9\cdot10^{-5}$
			No Relative Uncertainty							
		\multicolumn{2}{c}{t_{50}}				\multicolumn{2}{c}{t_{90}}		
	N	Spearman		Kendall		Spearman		Kendall		
		t_{dur}	$t_{stretch}$	t_{dur}	$t_{stretch}$	t_{dur}	$t_{stretch}$	t_{dur}	$t_{stretch}$	
P_{64}	597	90	0.02	91	0.02	7	$1\cdot10^{-6}$	8	$1\cdot10^{-6}$	
P_{256}	597	78	$4\cdot10^{-3}$	76	$4\cdot10^{-3}$	2	$2\cdot10^{-8}$	2	$3\cdot10^{-8}$	
P_{1024}	597	39	$3\cdot10^{-4}$	39	$3\cdot10^{-4}$	0.1	$1\cdot10^{-10}$	0.2	$2\cdot10^{-10}$	

DROPPING THE HIGH-FLUX BURSTS

The correlation of peak flux with duration should be relatively small (if due to redshift) or nothing (if due to luminosity–duration correlation) in the range where the peak fluxes in the $-3/2$ part of the integrated peak flux distribution.

As this is the case, we took only bursts with peak fluxes below the $-3/2$ part of the integrated peak flux distribution, identified by eye as occurring at 10.0, 6.0, and 3.0 phot cm^{-2}sec^{-1} in the P_{64}, P_{256}, and P_{1024} measures. We then re-performed the above analysis for the resulting burst samples.

Except for one case, there is no significant correlation present in the 3B data between t_{50} or t_{90} with any of the three peak flux measures (P_{64}, P_{256}, P_{1024}). As with the data-sets (above), about half of the data sets are such that a factor of two difference put in "by hand" is undetectable by these tests.

The data which does appear to show a correlation is the sample of (P_{1024}, t_{90}) selected using the "No Relative Uncertainty" criteria. This sample has

7 (of the 489) bursts with peak fluxes lower than the lowest peak flux in the "$t_{50}>$ 2.0 sec" sample. If these seven are removed from the sample, the probability increases slightly from 0.007 to 0.01 %, compared to 5% for the "$t_{50}>$ 2.0 sec" sample, which covers the same peak-flux range, but has 15% fewer bursts (419 compared to 482). From this we conclude that the reason the "No Relative Uncertainty" sample shows a more significant correlation than the "$t_{50}>$ 2.0 sec" sample is not the larger sample size, but the inclusion of bursts with higher relative uncertainties in their duration and peak-flux measures.

DISTRIBUTION COMPARISONS

We investigated the amount of relative "shift" permitted by each of these data sets. This was done in the following manner (similar to that used by (6)). To compare distributions of bursts, the samples must be adequately flux-complete, which we take to be (for each of the time-scales involved) the 96% limit.

For each of the samples used, two separate distributions were produced, of the 10% highest and, separately, the 40% lowest peak flux values (above the 96% flux completion limit).

The durations of the "10% highest" group were multiplied by a constant factor (the "Relative Shift") and the resulting distribution compared with the duration distribution of the "40% lowest" group using the KS test.

If we assume the standard-candle GRBs with fluxes at the 96% completion limit to come from $(1+z) = 1.6$ (cf. (7)), the relative amount of time-dilation is ~ 1.35. In all three data selection sets, for all samples, we find the relative shift between the data-sets are between 0.7 and 1.3, at the 4σ level. For all samples tested, the KS probabilities at a relative shift of 1.0 indicate that the distributions are indistinguishable.

CONCLUSIONS

We find no evidence for correlation between any of the three peak flux measures (P_{64}, P_{256}, P_{1024}) with either of the two duration measures (t_{50}, t_{90}) in the 3B catalog. The amount of "relative shifting" between the brightest 10% of GRBs and the faintest 40% (above the 96% flux completion limit) is between $\sim$0.7-1.3 (4σ limits).

This result is in conflict with several past results, which show correlations between various duration measures and peak flux measures (8,9,6,10). Possible sources of this conflict include:

- The 3B durations (or, other durations) may contain systematic correlations between the duration measure and burst brightness (*e.g.* Bonnel et al. , this conference)

- The non-existence of a physical anti-correlation between luminosity and duration.

- The non-existence of relative time-dilation of the magnitude searched for here.

- The compensation of the amount of relative-time dilation due to red shift with a correlation between burst duration and co-moving photon energy, or a correlation between luminosity and these duration measures, or other compensating effects.

- relative uncertainties due to the use of non-physical units (*i.e.* the use of detector response counts, rather than the burst peak fluxes).

The present analysis is not sensitive to non-monotonic variations in the duration distribution as a function of peak flux.

Acknowledgements: This work is supported by the NASA GSRP under grant NGT-51369 (RR) and the CGRO Guest Investigator Program under NAG5-2046 (WHGL).

REFERENCES

1. J.Brainerd, ApJ Letters **428**, L1 (1994).
2. W. Press, et al. , Numerical Recipes, (Cambridge University Press, 1992).
3. C.A. Meegan, et al. , "The Third BATSE Gamma-Ray Burst Catalog", available in electronic form from GROSSC, NASA/GSFC (1995).
4. C. Kouveliotou, et al. , ApJ Letters **413**, L101 (1993).
5. T. Koshut, PhD Thesis, University of Alabama in Huntsville (1996).
6. J.P. Norris, et al. , ApJ **439**, 542 (1995).
7. R.E. Rutledge, et al. , MNRAS **276**, 753 (1995).
8. J.P. Norris, et al. , ApJ **424**, 540 (1994a).
9. S.P. Davis, et al. , in Gamma-Ray Bursts, eds. G.J. Fishman, J.J. Brainerd, & K.C. Hurley, AIP Conf. Proc. **307**, 182 (AIP, New York, 1994).
10. R.A.M.J. Wijers, & B. Paczyński, ApJ Letters **437**, L107 (1994).
11. E.E. Fenimore, et al., ApJ Letters **448**, L101 (1995).

A New Distance Scale for GRBs: The Local Group Halo

Bradley E. Schaefer

Yale University, Physics Department, New Haven CT 06520-8121

I would like to propose a new distance scale to Gamma Ray Bursters, where the characteristic distance is Megaparsecs. This distance scale is intermediate between those of the familiar galactic models (D~100 kpc) and cosmological models (D~1 Gpc). I envision bursters as being neutron stars shot with high velocities (>1000 km/s) out of our own Milky Way galaxy as well as nearby galaxies in the Local Group. The existence of such high velocity neutron stars has been recently discovered, and the inevitable fate of these is to escape the galaxy at high speeds. After a Hubble Time, these escapees can travel over 10 Mpc. I adopt the proposal by Duncan and Thompson that the high velocities are caused by extremely large magnetic fields, which will also beam the gamma rays into small cones directed along the velocity vector. With this preferred beaming, the Solar System will see bursts primarily coming from galaxies near our own. As such, my new distance scale is associated with a halo of neutron stars emitted by our Local Group of galaxies.

INTRODUCTION

The BATSE Dilemma (an isotropic yet inhomogenous population) killed off the favorite galactic disk models. The dilemma may be solved by either a large halo surrounding our own Milky Way or by a cosmological distance scale. (Yes, there is also the heliocentric distance scale, but this is generally disregarded for good reasons, and all my general comments also apply to it anyway.) This duality of solution has been enshrined in 'The Great Debate', wherein the GRB problem is posed as the competition of the galactic versus the cosmological distance scales.

Nevertheless, both distance scales have major problems. A primary difficulty is the lack of a plausible model to place at the suggested distance. That is, to date, every proposed scenario can be 'refuted' on some grounds or another. For example, cosmological models involving neutron star energies cannot emit enough gamma ray energy for the suggested distances, should emit a thermal spectrum, and should appear in host galaxies. Alternatively, galactic models have trouble both being isotropic and avoiding an M31 excess, while many of the same energetics problems are still present. These problems may be soluble, but none of the solutions are pleasant or natural. During 'The Great Debate', it was notable that neither side discussed any particular models, since neither side dared to suggest any scenario without inviting swift and decisive rebuttal. This compartmentalization is possible since many of the distance scale indicators are independent of the 'black box' which produces the bursts.

Another loose thread (no, make that a loose rope) is the unique nature of the 5 March 1979 burst. This very bright event is definitely associated with the SGR phenomenon. Recently Rothschild et al. (1) have tied this to an x-ray plerion in the N49 SNR, just as the other two SGRs are tied to SNRs. Also, Fenimore et al. (2) have used archival data to show that the spectrum of the main peak has a typical hardness found in classical GRBs, thus forcing the realization that the 5 March 1979 event is indistinguishable from classical GRBs. The one jarring discrepancy is that the energy of the event is greatly too large for a galactic halo distance scale and greatly too small for a cosmological distance scale.

LOCAL GROUP HALO

One of the problems with the galactic halo distance scale is that most (~99%) of the neutron stars ejected from our Milky Way do not remain bound. This arises because the high velocity neutron stars have speeds (V>1000 km/s) much greater than the galaxy's escape velocity (V_{esc} = 360 km/s near the Sun). But what happens to these 'lost' bursters? It is simple; they just keep on moving away from the Milky Way. So we know that there must exist a halo of ejected neutron stars that surrounds our Milky Way.

Since our galaxy has likely been creating these neutron stars since its youth, the ejected halo will now extend quite a large distance. In a billion years, a star moving at 1000 km/s will travel a Megaparsec. So of necessity, our halo must surround the entire Local Group of galaxies. Also, the halos of the other Local Group galaxies must intermingle with our halo. I call this ensemble of high velocity neutron stars the 'Local Group Halo'.

I will adopt the Duncan & Thompson (3) explanation for why the neutron stars have their high observed velocity. Briefly, they use a very high magnetic field (B~ 10^{15} G) to provide an asymmetry for a neutrino jet that will accelerate the neutron star during its formation. Such a high magnetic field will also provide a means of collimating the gamma ray emission into a small cone pointing along the velocity vector. This beaming eases the overall energy requirements [without requiring more sources!] and is essential to get a $-3/2$ slope of the $logN-LogP_{max}$ curve for bright bursts.

Why are burst isotropic? Neutron stars are ejected with random (i.e. isotropic) kick velocities at the time of formation. The size of the Local Group Halo is much greater than our distance to the center of the Milky Way, hence there will be no significant dipole or quadrupole moment.

How to explain the $logN-LogP_{max}$ curve? As the burster gets farther from the Milky Way, its cone of beamed emission will illuminate more of our galaxy; naturally producing a $-3/2$ slope! Distant bursters all illuminate all our galaxy, so the D^{-2} density dependence makes the low flux bursts have a slope of $-1/2$ in the asymptotic limit.

Why don't we see concentrations towards nearby galaxies? We are embedded in the stream of LMC bursters near enough to their source so as to see no anisotropies from the LMC. We are embedded in the stream of M31 bursters, most of which will not be beamed in our direction. Bursters from galaxies outside our Local Group are too far away and only a few of their bursters will be beamed at Earth.

Why don't we see quiescent counterparts? Simple, because they are lone neutron stars of order a Mpc away.

A SPECIFIC MODEL

Let me propose a specific model. I will take bursters to be formed from supernova eruptions which give isotropic kick velocities of magnitude 2000 km/s. These high velocity neutron stars will emit bursts with a luminosity of 3×10^{42} erg/s over a cone of 8° half opening angle directed along the kick velocity vector. The burst rate is uniform over the life of each burster at least until it passes outside BATSE detection distance. The burster formation rate is uniform in time, proportional to the galaxy mass, and with a galactic distribution uniform in radius out to some maximum size. The standard observed rotation curve (with V_{rot} = 220 km/s at the Solar circle) for our Milky Way is assumed.

With this model, I have used a Monte Carlo simulation to deduce the global properties of bursts. They are suitably isotropic with $\langle \sin^2(b^{II}) - 1/3 \rangle =$ −0.017, $\langle \cos(\theta) \rangle = 0.019$, and $\langle \cos^2(\theta) - 1/3 \rangle = 0.014$. The fraction of all BATSE bursts from the LMC is 4%, while M31 contributes 0.3%. These extragalactic bursters are loosely concentrated both towards the galaxy of origin as well as towards the antipodes with no significant clustering. The resulting $LogN-LogP_{max}$ curve is indistinguishable from that of the combined BATSE plus PVO observations (4).

The source of the majority of the Local Group Halo bursts are from neutron stars ejected by our Milky Way. There are a limited number of these neutron stars, so to reproduce the observed rate (800/yr for the whole sky to the BATSE threshold) I must have each burster repeat. [All galactic models have this same feature.] The required repetition rate will depend on our galaxy's production rate, the fraction of neutron stars which burst, and the BATSE sampling distance. If all neutron stars make bursts, then typical numbers give a recurrence time for a single burster of 40,000 yr.

The total energy required for a single event will depend on the solid angle of the emission cone ($\omega/4\pi = 0.5\%$), the luminosity in the cone (3×10^{42} erg/s), and the average duration (∼10 s). For my specific model, an energy of 1.5×10^{41} ergs is needed for 10^4 bursts while the neutron star remains in range. Thus a total of 1.5×10^{45} ergs is needed. This amount of energy is a small fraction of the available gravitational energy ($\sim 10^{53}$ erg), although I cannot imagine how this energy could be released in so many small packets. Another reasonable source of energy is the rotational energy (remember glitches?), where the start period only need be faster than a few seconds. Alternatively, the neutron star has $B_{15}^2 \times 2 \times 10^{47}$ erg of magnetic energy which could possibly get released in many small packets (like with Solar flares). This energy source looks favorable since ~100 times the needed energy is available and since I have already invoked extremely high magnetic fields to get the high velocities and beaming. In other words, the needed energy is relatively small and has a natural source within my distance scale.

There are upper limits to the distance scale and these arise primarily from the rotation of our own galaxy. With this rotation and a small cone opening angle, the bursters' space velocity vector will not point along the kick velocity vector, so the cone of illumination will slowly sweep off the Milky Way and distant bursters will not be seen. With this rotation and a large cone opening angle, the break in the $\mathrm{LogN-LogP_{max}}$ curve must be close to the Local Group.

For the Local Group Halo distance scale, the energy of the 5 March 1979 GRB becomes typical. In this case, the only thing unusual about this burst is that it also happens to be the closest classical GRB. But since it is close (and young), we also could see the SGR events from the same source. So in the Local Group Halo idea, the 5 March 1979 event is a normal classical GRB [no need to require a third class of bursts] and the SGRs come from the same source [no need to require a second class of bursters].

CONCLUSIONS

The Local Group Halo idea fits well with current observations. In particular, the BATSE dilemma (isotropic yet inhomogenous) is *naturally* solved. The isotropy arises from the random (i.e., isotropic) kick velocity while concentrations towards other galaxies do not occur because their bursters are only rarely beamed at our Milky Way. The $-3/2$ slope of the $\mathrm{LogN-LogP_{max}}$ curve for bright bursts is a natural consequence of the emission cone illuminating more of our galaxy as it gets further away. The turnover in the $\mathrm{LogN-LogP_{max}}$ curve for faint bursts is a natural consequence of the D^{-2} density fall off at great distances.

The chief disadvantage of the Local Group Halo distance scale is that no specific model is presented to produce the bursts from the extreme magnetic field. But this same disadvantage is also suffered by all other distance scales. The chief advantage of the Local Group Halo distance scale is that the BATSE dilemma has a natural explanation. By Ockham's Razor, the lack of new hypotheses (the high velocity neutron star population is known) and even the unification of phenomena (classical GRBs + 5 March 1979 GRB + SGRs) argue strongly that the Local Group Halo is the best distance scale.

REFERENCES

1. R. E. Rothschild, S. R. Kulkarni, and R. E. Lingenfelter, Nature **368**, 432 (1994).
2. E. E. Fenimore, R. W. Klebesadel, and J. G. Laros, Ap. J., in press (1995).
3. R. C. Duncan and C. Thompson, Ap. J. **392**, L9 (1992).
4. E. E. Fenimore et al., Nature **366**, 40 (1993).

Contribution of Galactic Arm Sources to the 3B Catalog

I. A. Smith

Rice University

By comparing theoretical models to the BATSE observations, it was shown that Galactic spiral arm sources cannot be the sole population of gamma-ray bursters. However, if there are two populations of burst sources, some of them could be in the spiral arms. If the Galactic arm sources are homogeneous and isotropic in space, it is possible that over $\sim 1/2$ of all the bursters could be in the spiral arms. However, it is illustrated here that for inhomogeneous spiral arm distributions, and for observers offset from the axis of the arm, the allowed fraction of the sources that can be in the arm is significantly reduced. Thus the fraction of spiral arm sources is likely to be at most 10% to 20%.

INTRODUCTION

It was shown in (7) that it is not possible for all the gamma-ray burst sources to be in the Galactic spiral arms. However, if there are two populations of burst sources, some of them could be in the spiral arms. One possible Galactic arm contribution might come from flare stars (2,5,6; but see 1). However, the results shown here apply to *any* Galactic arm population. Liang & Li (2) assumed that the Galactic arm sources are homogeneous and isotropic in space, and found that up to $\sim 1/2$ of the bursters could be in the spiral arms. However, it will be illustrated here that for inhomogeneous spiral arm distributions, and for observers offset from the axis of the arm, the allowed fraction of the sources that can be in the arm $N_{\text{arm}}/N_{\text{total}}$ is significantly reduced.

MODEL DETAILS

In Smith & Lamb (8), an exponential disk was combined with an "ideal" halo to calculate the maximum fraction of sources that could be in the disk. A similar calculation is performed here using the spiral arms of Smith (7) combined with an "ideal" halo.

The geometries chosen for the spiral arms are the same as in (7). A cylindrical geometry is used and the sources are assumed to be standard candles that are distributed axially symmetrically. The Sun is taken to be a distance ρ_0 from the axis of the arm, which is small compared to the distance of the Sun from the Galactic center ($R_0 = 8.5$ kpc). The curvature of the arm is

© 1996 American Institute of Physics

ignored. The following number density distributions $n(\rho)$ have been used for the sources:

(1) Gaussian arm; $n(\rho) = n_a e^{(-1/2)[(\rho-\rho_s)/\sigma]^2}$.

(2) Constant density arm with sharp cutoff; $n(\rho) = n_a$ if $\rho \leq \rho_{cd}$, $n(\rho) = 0$ if $\rho > \rho_{cd}$, n_a is a constant.

(3) Exponential arm; $n(\rho) = n_a e^{-\rho/\bar{\rho}}$.

(4) "Alpha" arm; $n(\rho) = n_a/[1 + (\rho/\rho_c)^\alpha]$.

In (7), the values of $<V/V_{\max}>$, $<\cos\psi>$, and $<\sin^2 b>$ were calculated for different distances D to the faintest arm source that could be detected by BATSE. These numbers can then be combined with those of a halo or cosmological population to generate the average values observed by BATSE. To determine the maximum possible fraction of sources in the arm, an "ideal" halo is used: $<V/V_{\max}>_h = 0$, $<\cos\theta>_h = 0$, and $<\sin^2 b>_h = 1/3$ (8). Note that these numbers cannot be obtained for any real halo.

ALLOWED FRACTION OF GALACTIC ARM SOURCES

If the value of $<\sin^2 b>_{\text{BATSE}} > 1/3$, the source number density distribution cannot peak on the axis of the arm. Such is the case for Gaussian arms that are used in Figures 1 and 2, where the 3B values are used for the BATSE observations: $<V/V_{\max}> = 0.33$, $<\cos\theta> = 0.011$, and $<\sin^2 b> = 0.335$ (4).

Figure 1 plots the allowed fraction of arm sources $N_{\text{arm}}/N_{\text{total}}$ as a function of the maximum distance that an arm source can be detected D for Gaussian distributions of different widths. Note that there are no solutions for D larger or smaller than the values shown for each curve. The observer is on the axis of the arm, so there is no constraint from the $<\cos\theta>$ observation. It can be seen from Figure 1 that it is possible to have $N_{\text{arm}}/N_{\text{total}}$ as large as $\sim 2/3$ for all the different spatial distributions of the Galactic arm sources. However, this only occurs for specially chosen values of D. A broader Gaussian is better than a thin shell for giving a larger $N_{\text{arm}}/N_{\text{total}}$ over the allowed values of D. The fact that the thinner shell arms come closer to the BATSE data point in Figure 2b of (7) is only manifested in the limit to $N_{\text{arm}}/N_{\text{total}}$ at large values of D.

It was shown in (7) that the offset of the observer from the axis of the arm is also an important parameter that has to be taken into account, because there is then an extra constraint from $<\cos\theta>$. Figure 2 shows the effect of different offsets ρ_0 from the axis of the Gaussian arm with $\rho_s = 1$, $\sigma = 1$. It can be seen that as ρ_0 increases, the maximum value of $N_{\text{arm}}/N_{\text{total}}$ is reduced, because of the $<\cos\theta>$ constraint. Also, the allowed range of D is reduced because of the $<\cos\theta>$ constraint, and there are values of ρ_0 that have no solutions (for example $\rho_0 \gtrsim 1.5$ in the example shown).

For values of $<\sin^2 b>_{\text{BATSE}} < 1/3$, it is possible to use arm number density distributions that peak on the axis of the arm. Examples of this are shown in Figure 3, where the previous BATSE values of $<V/V_{\max}> = 0.321$, $<\cos\theta> = 0.031$, and $<\sin^2 b> = 0.326$ are used (3). Again, including an

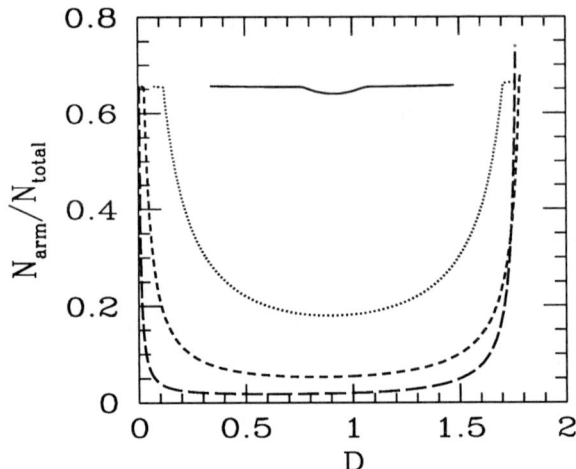

FIG. 1. Fraction of the burst sources that can be in the Galactic arm N_{arm}/N_{total} as a function of the maximum distance that an arm source can be detected D. Arm sources with a Gaussian number density distribution are combined with an "ideal" halo. For all the curves, the Sun is on the axis of the arm, $\rho_0 = 0$, and $\rho_s = 1$. Solid curve: $\sigma = 2$. Dotted curve: $\sigma = 1$. Short dashed curve: $\sigma = 0.5$. Long dashed curve: $\sigma = 0.2$. Distances are dimensionless, because only the ratios of quantities are important, for example σ/D.

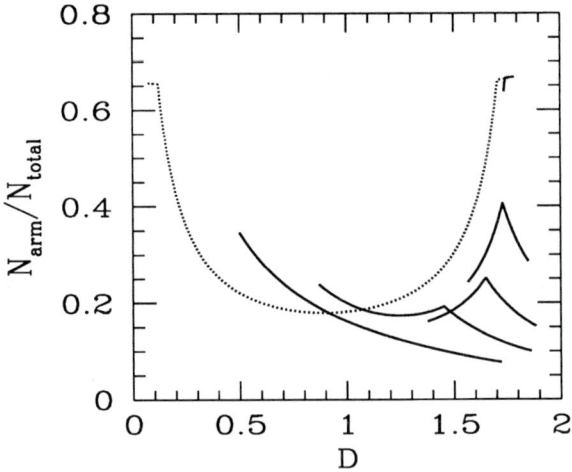

FIG. 2. Fraction of the burst sources that can be in the Galactic arm N_{arm}/N_{total} as a function of the maximum distance that an arm source can be detected D. Arm sources with a Gaussian number density distribution are combined with an "ideal" halo. All the curves use $\rho_s = 1$ and $\sigma = 1$, and different offsets ρ_0 of the Sun from the axis of the arm. Dotted curve: $\rho_0 = 0$ (no $< \cos\theta >$ constraint is applied). Solid curves from top right to bottom left: $\rho_0 = 0.25, 0.5, 0.75, 1.0, 1.25$.

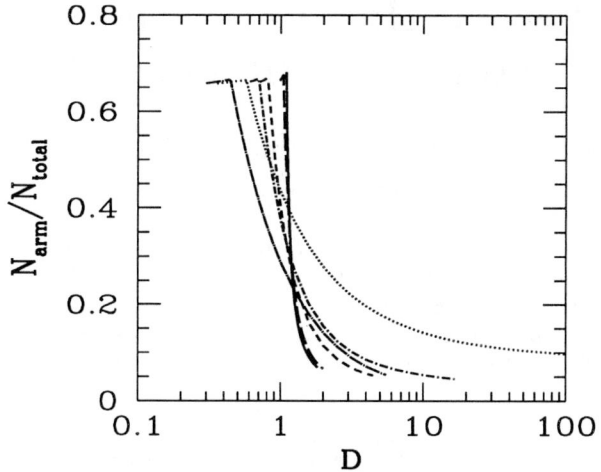

FIG. 3. Fraction of burst sources that can be in the Galactic arm N_{arm}/N_{total} as a function of the maximum distance an arm source can be detected D (log scale, unlike Figures 1 and 2). Arm models with different source number density distributions $n(\rho)$ are combined with an "ideal" halo; $\rho_0 = 0$ for all curves. Solid curve: constant density arm with sharp cutoff, $\rho_{cd} = 1$. Long dash dotted curve: exponential arm, $\bar{\rho} = 1$. Other curves: "alpha" arms, $\rho_c = 1$. Long dashed curve: $\alpha = 10$. Short dashed curve: $\alpha = 3$. Short dash dotted curve: $\alpha = 2$. Dotted curve: $\alpha = 1$.

offset of the Sun from the axis of the arm reduces the allowed fraction of arm sources.

In conclusion, the allowed fraction of spiral arm sources will be at most 10% to 20%, except for particular choices of the arm geometry.

Acknowledgements: This work was supported at Rice University by NASA grant NAG 5-2772.

REFERENCES

1. P. Li, K. Hurley, G.J. Fishman, C. Kouveliotou, & D. Hartmann, ApJ **446**, 267 (1995).
2. E.P. Liang, & H. Li, A & A **273**, L53 (1993).
3. C.A. Meegan, et al., in Gamma-Ray Bursts, eds. G. J. Fishman, J. J. Brainerd, & K. Hurley, AIP Conf. Proc. **307**, 3 (AIP, New York, 1994).
4. C.A. Meegan, et al., ApJ, submitted (1995).
5. A.R. Rao, & M.N. Vahia, A & A **281**, L21 (1994a).
6. A.R. Rao, & M.N. Vahia, A & A **287**, L34 (1994b).
7. I.A. Smith, ApJ **429**, L65 (1994).
8. I.A. Smith, & D.Q. Lamb, ApJ **410**, L23 (1993).

Spherical Harmonic Analysis of the Angular Distribution of GRBs

Max Tegmark[1,2], Dieter H. Hartmann[3], Michael S. Briggs[4], and Charles. A. Meegan[5]

[1] *Max-Planck-Institut für Physik, Föhringer Ring 6, D-80805 München*
[2] *Max-Planck-Institut für Astrophysik, D-85740 Garching*
[3] *Dept. of Physics and Astronomy, Clemson University, Clemson, SC 29634*
[4] *Dept. of Physics, University of Alabama, Huntsville, AL 35899*
[5] *NASA/Marshall Space Flight Center, Huntsville, AL 35812*

We compute the angular power spectrum C_ℓ of the BATSE 3B catalog, and find no evidence for clustering on any scale. These constraints bridge the entire range from small scales, probing source clustering and repetition, to large scales constraining possible Galactic anisotropies, or those from nearby cosmological large scale structures.

INTRODUCTION

The observed angular distribution of γ-ray bursts (GRBs) is isotropic, while their brightness distribution shows a reduced number of faint events. These observations favor a cosmological burst origin. Clustering of bursts could be evidence of actual clustering of sources or of repeated emission. Repetition would call into question the viability of many cosmological burst models. Anisotropies manifest themselves on different angular scales and with different magnitudes. Galactic features cause large-scale distortions, while true repetition would affect small scales. For large-scale signatures, we search for excesses of sources towards some direction or a concentration towards some plane in the sky, i.e., we seek a dipole or quadrupole moment. It is now common practice to apply both coordinate-free and galactic tests (1). Dipole- and quadrupole measures were sufficient when sample sizes were small. Now an extension of moment methods to higher orders is needed. Low order multipoles are not sensitive to instrumental smearing, but higher harmonics are. If associated with galaxies, we expect clustering on very small scales. If bursts repeat, we expect clustering at $\theta=0$. Both effects are diluted by localization uncertainties, and angular power is transferred from small (or zero) angular scales to a scale given by the detector response. One tool for the analysis of source clustering is the two-point correlation function (2), which is related to the power spectrum through a Fourier transform (4).

© 1996 American Institute of Physics

METHOD

We model the GRB distribution as a 2D stochastic point process $n(\hat{\mathbf{r}}) = \sum_i \delta(\hat{\mathbf{r}}, \hat{\mathbf{r}}_i)$ with intensity (average point density per steradian) $\lambda(\hat{\mathbf{r}})$. Here δ denotes the 2D Dirac delta function, and the unit vectors $\hat{\mathbf{r}}_i$ correspond to the various GRB positions. If we had detected a nearly infinite number of bursts, then the function $\lambda(\hat{\mathbf{r}})$ would be known with great accuracy, and the only source of errors when computing its power spectrum would be cosmic variance. Since in practice we have only a finite number of bursts (1122 for 3B), our estimates of λ include shot noise. A Poisson process satisfies the expectation value equations

$$\langle n(\hat{\mathbf{r}}) \rangle = \lambda(\hat{\mathbf{r}}), \tag{1}$$

and

$$\langle n(\hat{\mathbf{r}})n(\hat{\mathbf{r}}') \rangle = \lambda(\hat{\mathbf{r}})\lambda(\hat{\mathbf{r}}') + \delta(\hat{\mathbf{r}}, \hat{\mathbf{r}}')\lambda(\hat{\mathbf{r}}). \tag{2}$$

Here λ is itself a random field, $\lambda(\hat{\mathbf{r}}) = \bar{n}(\hat{\mathbf{r}})[1 + \Delta(\hat{\mathbf{r}})]$, where the underlying density fluctuations Δ are modeled as a Gaussian random field. The function $\bar{n}$, which we will refer to as the *exposure function*, is the number of bursts per steradian expected a priori, not the number density actually observed. In other words, $\bar{n}(\hat{\mathbf{r}})$ is proportional to the exposure time in the sky direction $\hat{\mathbf{r}}$. We assume that $\langle \Delta(\hat{\mathbf{r}}) \rangle = 0$ and that the statistical properties of the field Δ are isotropic, which means that if we expand it in spherical harmonics as

$$\Delta(\hat{\mathbf{r}}) = \sum_{\ell=0}^{\infty} \sum_{m=-\ell}^{\ell} a_{\ell m} Y_{\ell m}(\hat{\mathbf{r}}), \tag{3}$$

then

$$\langle a_{\ell m} a_{\ell' m'} \rangle = \delta_{\ell \ell'} \delta_{mm'} C_\ell, \tag{4}$$

where the coefficients C_ℓ are known as the *angular power spectrum*. There are thus two separate random steps involved in generating n: first the generation of the smooth field Δ, then the Poissonian distribution of points.

Given the field $n(\hat{\mathbf{r}})$, we wish to estimate the coefficients $a_{\ell m}$. We define them as

$$\tilde{a}_{\ell m} \equiv \int Y_{\ell m}(\hat{\mathbf{r}}) \frac{n(\hat{\mathbf{r}})}{\bar{n}(\hat{\mathbf{r}})} d\Omega - \delta_{\ell 0} \delta_{m 0} \sqrt{4\pi}. \tag{5}$$

We now compute the statistical properties of these estimates. By substituton we obtain

$$\langle \tilde{a}_{\ell m} \rangle = \int Y_{\ell m}(\hat{\mathbf{r}}) d\Omega - \delta_{\ell 0} \delta_{m 0} \sqrt{4\pi} = 0, \tag{6}$$

i.e., the expectation values vanish. Since the expectation values of the true coefficients $a_{\ell m}$ vanish as well, this means that our estimates are unbiased.

Using the expressions above, we find that the correlation between two multipole estimates is

$$\langle \tilde{a}_{\ell m} \tilde{a}_{\ell' m'} \rangle = \int \int Y_{\ell m}(\hat{\mathbf{r}}) Y_{\ell' m'}(\hat{\mathbf{r}}') \left[\langle \Delta(\hat{\mathbf{r}}) \Delta(\hat{\mathbf{r}}') \rangle + \frac{1}{\bar{n}(\hat{\mathbf{r}})} \delta(\hat{\mathbf{r}}, \hat{\mathbf{r}}') \right] d\Omega d\Omega', \quad (7)$$

which reduces to

$$\langle \tilde{a}_{\ell m} \tilde{a}_{\ell' m'} \rangle = \delta_{\ell \ell'} \delta_{m m'} C_\ell + \int \frac{Y_{\ell m}(\hat{\mathbf{r}}) Y_{\ell' m'}(\hat{\mathbf{r}})}{\bar{n}(\hat{\mathbf{r}})} d\Omega. \quad (8)$$

Defining the quantities

$$\tilde{C}_{\ell m} \equiv \tilde{a}_{\ell m}^2 - b_{\ell m}, \quad (9)$$

we find that they are unbiased estimates if we choose the *bias correction* as

$$b_{\ell m} \equiv \int \frac{Y_{\ell m}^2(\hat{\mathbf{r}})}{\bar{n}(\hat{\mathbf{r}})} d\Omega. \quad (10)$$

If $\bar{n}$ is constant, then the bias correction becomes simply $b_{\ell m} = 1/\bar{n}$, independent of ℓ and m. The $\tilde{C}_{\ell m}$ are thus good estimates of C_ℓ for each m-value separately. To reduce error bars, we estimate power by averaging the $\tilde{C}_{\ell m}$:

$$\tilde{C}_\ell \equiv \frac{1}{2\ell + 1} \sum_{m=-\ell}^{\ell} \tilde{C}_{\ell m}. \quad (11)$$

Defining b to be the average of the bias corrections $b_{\ell m}$, we find that b is in fact independent of ℓ, and obtain

$$b \equiv \frac{1}{2\ell + 1} \sum_{m=-\ell}^{\ell} b_{\ell m} = \frac{1}{4\pi} \int \frac{d\Omega}{\bar{n}(\hat{\mathbf{r}})}, \quad (12)$$

i.e., b is just the spherical average of $1/\bar{n}$.

It is straightforward to include the effects of position errors in the formalism, which is described in a more detailed ApJ version of this paper (5). We model the BATSE beam function as a Fisher function

$$B(\hat{\mathbf{r}} \cdot \hat{\mathbf{r}}') = \frac{\exp\left[\sigma^{-2} \hat{\mathbf{r}} \cdot \hat{\mathbf{r}}'\right]}{4\pi \sigma^2 \sinh[\sigma^{-2}]}, \quad (13)$$

characterized by a location error σ. This is a spherical version of the Gaussian distribution, and reduces to

$$B(\cos\theta) \approx \frac{\exp\left[-\frac{1}{2} \frac{\theta^2}{\sigma^2}\right]}{2\pi \sigma^2} \quad (14)$$

when $\sigma \ll 1$ radian $\approx 60°$. The Fisher function has the advantage that it is correctly normalized (its integral over the sphere is unity) for arbitrarily large

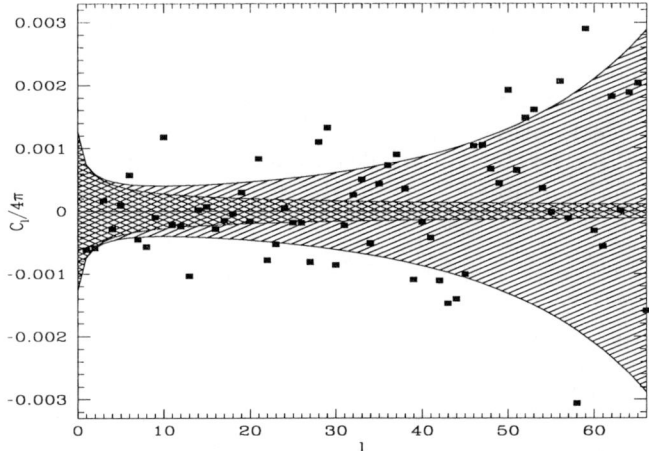

FIG. 1. The shot-noise corrected angular power spectrum of 3B (solid squares). The shaded region shows the 1σ error bars. Any type of clustering would drive the measured points upward. The double-shaded region shows what the errors would be without localization uncertainties.

angles σ, which is not the case for the plane Gaussian. In addition to statistical position errors we include (in quadrature) a $1.6°$ systematic uncertainty. This value is significantly lower than the $4°$ of earlier catalogs, allowing us to extend spherical harmonic analysis to $\ell \sim 60$ before localization uncertainties wash out intrinsic angular power.

RESULTS AND DISCUSSION

The power spectrum $\tilde{C}_\ell$ of the the 3B data (3) is shown in Figure 1. There is no evidence of deviations from isotropy on any angular scale. If the gamma-ray bursts are completely uncorrelated, the points should scatter symmetrically around zero, with about 68% in the shaded region. Since all power is by definition positive, the presence of any type of clustering would shift the distribution upwards, leading to a positive excess. A monopole $C_0/4\pi = 0.0001$ corresponds to a fluctuation of $\sqrt{0.0001} = 1\%$ in the average burst density. Likewise, $[C_\ell/4\pi]^{1/2}$ can be interpreted as the density fluctuation on the angular scale $\theta \approx 60°/\ell$. The size of the error bars (the height of the shaded region) is readily understood. For $\ell = 0$, we have $N_\ell^{\text{eff}} = N = 1122$, so apart from the factor of $\sqrt{2}$, the shot noise gives just the familiar Poisson variance $1/N$. As ℓ increases, the $(2\ell + 1)$-denominator reduces the error bars, since many independent modes are being averaged.

Although the angular power spectrum C_ℓ provides a useful measure of the amount of clustering on different angular scales, it does not contain any information about the relative *phases* of the different multipoles $a_{\ell m}$. The loss of phase-information means that although the power spectrum may tell us that there is extra power on some scale, it does not tell us anything about where in the sky this power is coming from. Fortunately, this type of information is easy to extract through definition of a map $x_\ell(\hat{\mathbf{r}})$, the *multipole map* corresponding to multipole ℓ, as the sky map

$$x_\ell(\hat{\mathbf{r}}) \equiv \sum_{m=-\ell}^{\ell} \tilde{a}_{\ell m} Y_{\ell m}(\hat{\mathbf{r}}), \tag{15}$$

Multipole expansion of the projected distribution of GRBs does not show evidence for clustering on any angular scale. This argues against the recurrence of a large fraction of burst sources (6) and against any source population with strong intrinsic anisotropies. The remarkable degree of isotropy of GRBs severely constrains any burst model that invokes traditional geometric features of the Milky Way (disk, bulge, or halo).

Acknowledgements: This work was supported by NASA under grant NAG 5-1578 and by Deutsche Forschungsgemeinschaft under grant SFB-375.

REFERENCES

1. M. S. Briggs, et al., ApJ **459**, 40 (1996).
2. D. H. Hartmann & G. R. Blumenthal, ApJ **342**, 521 (1989).
3. C. A. Meegan, et al., ApJ, submitted (1995).
4. P. J. E. Peebles, The Large Scale Structure of the Universe, (Princeton University Press, 1980).
5. M. Tegmark, D. H. Hartmann, M. S. Briggs, & C. A. Meegan, ApJ, submitted (1995a).
6. M. Tegmark, D. H. Hartmann, M. S. Briggs, J. Hakkila, & C. A. Meegan, ApJ, submitted (1995b).

DMSP Satellite Detections of Gamma-Ray Bursts

J. Terrell, P. Lee, and R. W. Klebesadel

Los Alamos National Laboratory, Los Alamos, NM 87545

J. W. Griffee

Sandia National Laboratory, Albuquerque, NM 87185

Gamma-ray burst detectors are aboard six U. S. Air Force Defense Meteorological Satellite Program (DMSP) spacecraft, two of which are currently in use. Their 800-km altitude orbits give a field of view to 117° from the zenith. A great many bursts have been detected, usually in coincidence with detections by GRO or other satellites such as PVO or ULYSSES. The directions of the sources can be determined with considerable accuracy from such correlated observations, even when GRO/BATSE with its directional capabilities is not involved. Thus these DMSP data, especially in conjunction with other observations, should be helpful in trying to understand the true nature of gamma-ray bursts.

INTRODUCTION

Six of the DMSP spacecraft launched by the U. S. Air Force have carried gamma-ray burst detectors into orbit at 800 km. Two of these (DMSP 12 and 13) are currently in operation, in near-polar orbits, attaining latitudes of 81° with an inclination of 99°. Each carries two gamma-ray detectors with ~100 cm^2 of NaI each (40 cm^2 of CsI, for DMSP 13) sensitive to gamma-rays of 50 keV or more coming from sources within ~117° of the spacecraft zenith. A fuller description may be found elsewhere (1–3).

RESULTS

Many gamma-ray bursts have been detected by two or more of the DMSP spacecraft. When other spacecraft have also detected these bursts, relative times of detection and fields of view can give considerable information on source location, even when no detection was made by the Compton Observatory (GRO).

Such a burst was detected by DMSP 11, 12, and 13 on 7 April 1995, and also by ULYSSES and WIND. Figure 1 shows the data from DMSP 11 for this hard burst, with gamma rays exceeding 430 keV in both detectors (for DMSP

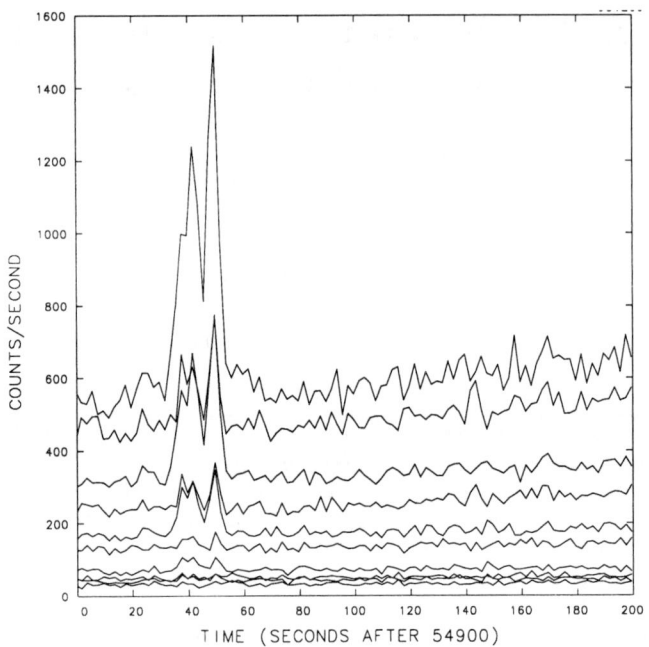

FIG. 1. DMSP 11 data for a hard gamma-ray burst on 7 April 1995.

11 the channel thresholds are at 50, 100, 200, 430, and 550 keV, with cutoff at 1000 keV). The first counting-rate peak in the DMSP data occurred in the 2-second time channel centered on 54932 sec UT. Direction analysis, undertaken in cooperation with other investigators, has not yet been completed for this and other events reported here.

Another hard burst, much more intense, was detected on 22 August 1995. The DMSP 12 data are shown in Figure 2. This was probably the most intense burst ever detected by DMSP satellites, with a counting rate increasing to almost 4000 counts/sec at the peak time of 13767 sec UT. It was also detected by BATSE (GRB No. 3767), KONUS, and ULYSSES.

DMSP 13 detected a softer but interesting burst on 25 August 1995, shown in Figure 3. The first peak count at 15786 sec UT was followed by a brief higher peak at 15864 sec, and then by a long high-energy peak extending from ~15900 to 16950 sec UT. The NaI detectors on DMSP 13 have thresholds at 60, 150, 375 keV. The high-energy peak (lowest counting rates in Figure 2) corresponds to photons of energy >6 MeV.

One of the strongest bursts ever detected by BATSE and COMPTEL (4,5). was GRB No. 2831 ("Olympic"), which triggered BATSE at 82962 sec UT on 17 February 1994. This was also detected by DMSP 10, DMSP 11, and ULYSSES. The data from DMSP 10, shown in Figure 4, indicate a curious pattern of repeated outbursts over ~180 seconds. A Fourier analysis of this data (that corresponding to the highest counting rates) gives some evidence of

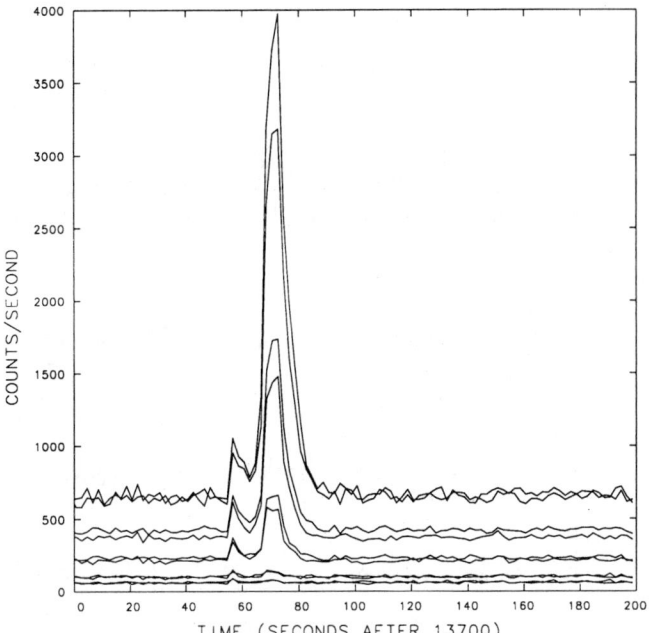

FIG. 2. DMSP 12 data for a hard and very intense gamma-ray burst on 22 August 1995.

a periodicity of ~24 seconds (Figure 5); the same result is given by DMSP 11 data. Thus it is possible that this gamma-ray burst has a pulsating character reminiscent of the 5 March 1979 burst with its 8-sec periodicity (6).

However, the fact that the length of the outburst is only about seven periods necessarily means that the evidence for periodicity is uncertain. Randomly occurring "shot noise" (10-sec pulses) can give a similar Fourier spectrum, with maximum power near zero frequency (7,8). Considering that each point in Figure 5 has a standard deviation of 100%, or 50% even when smoothed over 4 terms, the power spectrum gives no conclusive evidence of periodicity.

CONCLUSIONS

The DMSP spacecraft have produced a large body of data on gamma-ray bursts, representing a considerable resource for the study of these events. The data include time histories and spectral information, and directional information can be greatly enhanced by combining these observations with those of other spacecraft. They may well be of service in obtaining a fuller understanding of the origin of gamma-ray bursts.

Acknowledgments. This work was supported by NASA, by the U. S. Department of Defense, and by the U. S. Department of Energy.

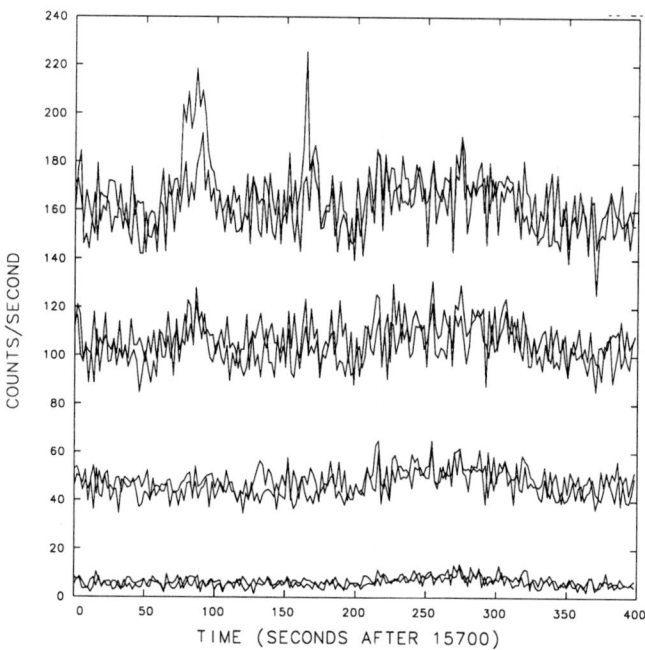

FIG. 3. DMSP 13 data for a soft gamma-ray burst followed by a very long hard afterpulse, on 25 August 1995.

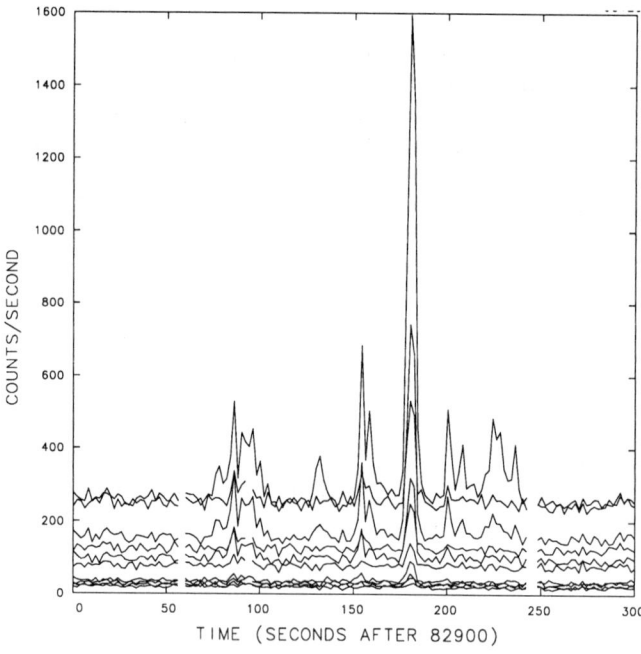

FIG. 4. DMSP 10 data for the intense gamma-ray burst of 17 February 1994.

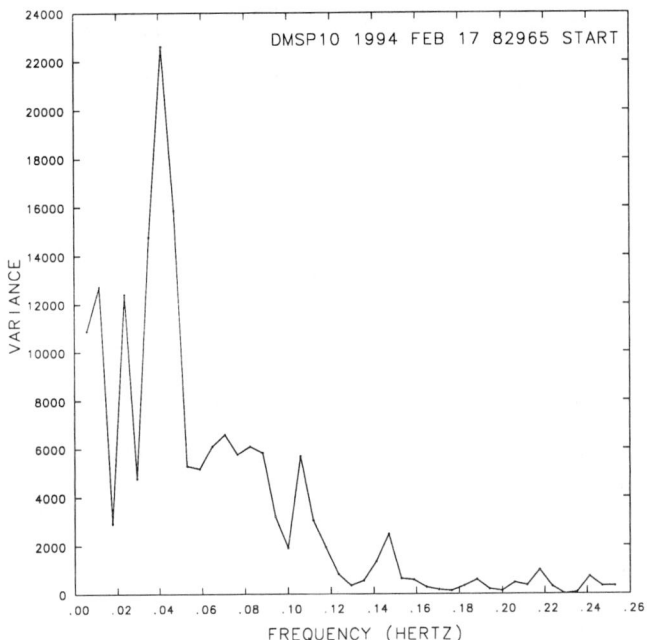

FIG. 5. Fourier power spectrum of the gamma-ray burst data in Figure 4 (DMSP 10; 17 February 1994).

REFERENCES

1. J. Terrell, P. Lee, R. W. Klebesadel, and J. W. Griffee, in Gamma-Ray Bursts, eds. W. S. Paciesas and G. J. Fishman, AIP Conf. Proc. **265**, 48 (AIP, New York, 1992).
2. J. Terrell, P. Lee, R. W. Klebesadel, and J. W. Griffee, in Compton Gamma-Ray Observatory, eds. M. Friedlander, N. Gehrels and D. J. Macomb, AIP Conf. Proc. **280**, 788 (AIP, New York, 1993).
3. J. Terrell, P. Lee, R. W. Klebesadel, and J. W. Griffee, in Gamma-Ray Bursts, eds. G. J. Fishman, J. J. Brainerd and K. Hurley, AIP Conf. Proc. **307**, 34 (AIP, New York, 1994).
4. R. M. Kippen et al., IAU Circular No. 5937 (21 February 1994).
5. K. Hurley et al., Nature **372**, 652 (1994).
6. J. Terrell, W. D. Evans, R. W. Klebesadel, and J. G. Laros, Nature **285**, 383 (1980).
7. J. Terrell and K. H. Olsen, Astrophysical J. **161**, 399 (1970).
8. N. J. Terrell, Astrophysical J. **174**, L35 (1972).

Gamma-Ray Burst Repetition and BATSE 3B Position Uncertainties

V. C. Wang* and R. E. Lingenfelter[†]

*Joint Sciences Department
Claremont-McKenna College, Claremont, California 91711
[†]Center for Astrophysics & Space Sciences
University of California, San Diego, La Jolla, California 92093

> We found eight pairs of candidate repeating bursts in the First BATSE Catalog sample that were so closely clustered in both position and time that the Poisson probability of their random occurrence was only $\sim 2\times 10^{-5}$. But because onboard tape recorder failures seriously reduced the number of accurately positioned bursts in the Second Catalog, it did not allow us to either confirm or refute the predicted repetition. We have now analysed the Third Catalog positions for fast ($<$ few day) repeating burst sources, and we find that neither the revised 1B nor the new 3B burst positions now show any significant evidence for closely clustered bursts in position and time. But the changes in burst positions from the 1B to 3B catalogs show such large systematic errors in position that no significant limits can presently be set on burst repetition in the BATSE data.

INTRODUCTION

The question of whether or not gamma-ray burst sources produce repeated bursts, and, if so, on what time scales, has fundamental implications for the nature and origin of the burst sources. Repeated gamma-ray bursts are expected from a number of suggested burst models, such as episodic accretion, thermonuclear runaway, or starquakes on galactic neutron stars (for reviews see, e.g. (3)). Multiple bursts might also be expected from unrepeatable, cataclysmic events, such as mergers of extragalactic neutron star binaries, which may still appear to repeat on rare occasions through gravitational lensing, but with essentially identical time histories and spectra. Thus, determining whether gamma ray burst sources repeat can place important constraints on burst models and greatly aid in our understanding of the origin of bursts.

We have searched the BATSE gamma-ray burst data for possible repeating sources of classical bursts. Searching the first BATSE catalog for bursts which are closely associated in both position and time, we found (6,7) several candidate repeaters, which appear to be otherwise indistinguishable in spectra, time histories, and durations from other classical bursts. In particular, there is a significant excess in the number of pairs of gamma-ray bursts which have an angular separation between the bursts of less than their estimated

© 1996 American Institute of Physics

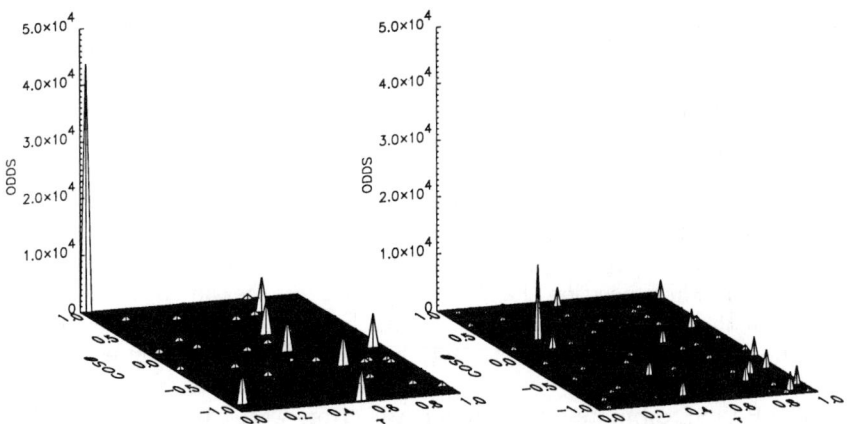

FIG. 1. The Poisson odds $(1/P(>n))$ for the random occurrence of of BATSE burst pairs divided into equal bins of angular and temporal separation in a τ-$\cos\theta$ phase space, showing the significant excess of closely-clustered, candidate repeater pairs with τ close to 0 and $\cos\theta$ close to 1 in the 1B Catalog sample (left) and no significant excess in the new 3B Catalog sample (right).

(1) positional uncertainties and an interval between their occurrence times of less than several days. Optimizing the signal at an angular separation, θ_s, less than the estimated (1) systematic uncertainty of $\sim 4°$ and a temporal separation, t_s, less than ~ 4 days, we found that the probability of observing such a clustered excess from a Poisson ensemble is $\sim 2 \times 10^{-5}$. This can be seen in Figure 1, in τ–$\cos\theta_s$ space, where $\tau \equiv (2t_s/t_{\max}) - (t_s/t_{\max})^2$, suggesting that these bursts arose from repeating sources.

Although the significance of the clustered excess was optimized *a posteriori*, because it depends on temporal and spatial bin sizes that could not be defined *a priori*, we used the optimizations from the first catalog to predict the expected excesses in subsequent BATSE data sets. Unfortunately, during the second BATSE catalog period the failure of the onboard tape recorders seriously reduced the number of accurately positioned bursts so that we could neither confirm, nor refute, the predicted repetition in that sample (7).

ANALYSIS OF THE 3B CATALOG

We now present the results of our analysis of the Third Catalog positions for fast ($<$ few day) repeating burst sources. These positions were calculated with a revised burst location algorithm which included data from detectors where the burst position could be as much as $30°$ behind the plane of the detector. As we discuss below, this may be a source of added error rather than an improvement. We also discuss the burst positional uncertainties and their effects on the limits that can be set on the fraction of the BATSE bursts that could come from sources that repeat in less than a few years.

From our analysis of the new 3B Catalog positions (8), we find that there is no significant evidence (see Figure 1) for closely clustered bursts in position and time. Moreover, we also find that there is now no significant evidence for such closely clustered bursts in the revised positions of the original 1B burst data set.

However, we find that the changes in burst positions from the 1B/2B to 3B Catalogs (see Figure 2) are much greater than expected for the constant, combined systematic errors of 4° and 1.6°, which were estimated by the BATSE team (1,8). These position changes provide a direct measure of the systematic errors in the BATSE location algorithms, because as Lamb (5) has pointed out, both programs use the same data base and thus they are independent of statistical errors.

The BATSE estimates of the systematic errors were based on the brightest bursts for which independent positions could be determined, e.g., by the Interplanetary Network (IPN), or by the imaging COMPTEL detector. It was assumed (8) that the systematic errors were independent of burst intensity and hence of the associated statistical errors. However, as can be clearly seen in Figure 2, the changes in these burst positions are not at all representative of the changes in the overall sample, because the systematic errors are not constant, as was assumed, but show a strong dependence on the estimated statistical error.

In particular, in the 1B/2B sample of 585 bursts for which revised 3B positions were determined, we would have expected only 32%, or 187 bursts, to have position changes greater than the 4.3° combined 1B/2B and 3B 1σ systematic position errors, if the constant values of 4° and 1.6° estimated by the BATSE team (1,8) were correct. Instead, 270 bursts, nearly half the sample, had larger position changes than expected from the BATSE estimated 1σ systematic errors. Moreover, the discrepancy in the nominal 2σ errors is even greater. We would have expected only 5%, or 27 bursts, to have position changes greater than 8.6°, and we find that there are instead 132!

As we see in the lower panel of Figure 2, there is a clear dependence of the systematic error on the estimated statistical error. The systematic error is approximately equal to $4(\sigma_{stat})^{-1/2}$. The correlation of the systematic error with the statistical error also raises the serious question of whether there may be a significant error in the BATSE estimate of the statistical error as well. The total position error is, of course, the sum of the two in quadrature.

Although the 1B/2B to 3B position changes reflect the errors in both location algorithms, other analyses show that the greater part of the error comes from the new 3B algorithm, rather than the old 1B algorithm. This was shown in an extensive analysis of the total errors in BATSE burst position estimates made by Graziani and Lamb (2), who compared both the 1B/2B and 3B catalog positions with nearly 200 IPN position arcs determined by 2 spacecraft. They found that the 1B/2B positions were in fact consistent with a constant systematic error of ~4° as estimated by the BATSE team (1). However, the 3B are not consistent with a constant systematic error, and show instead a

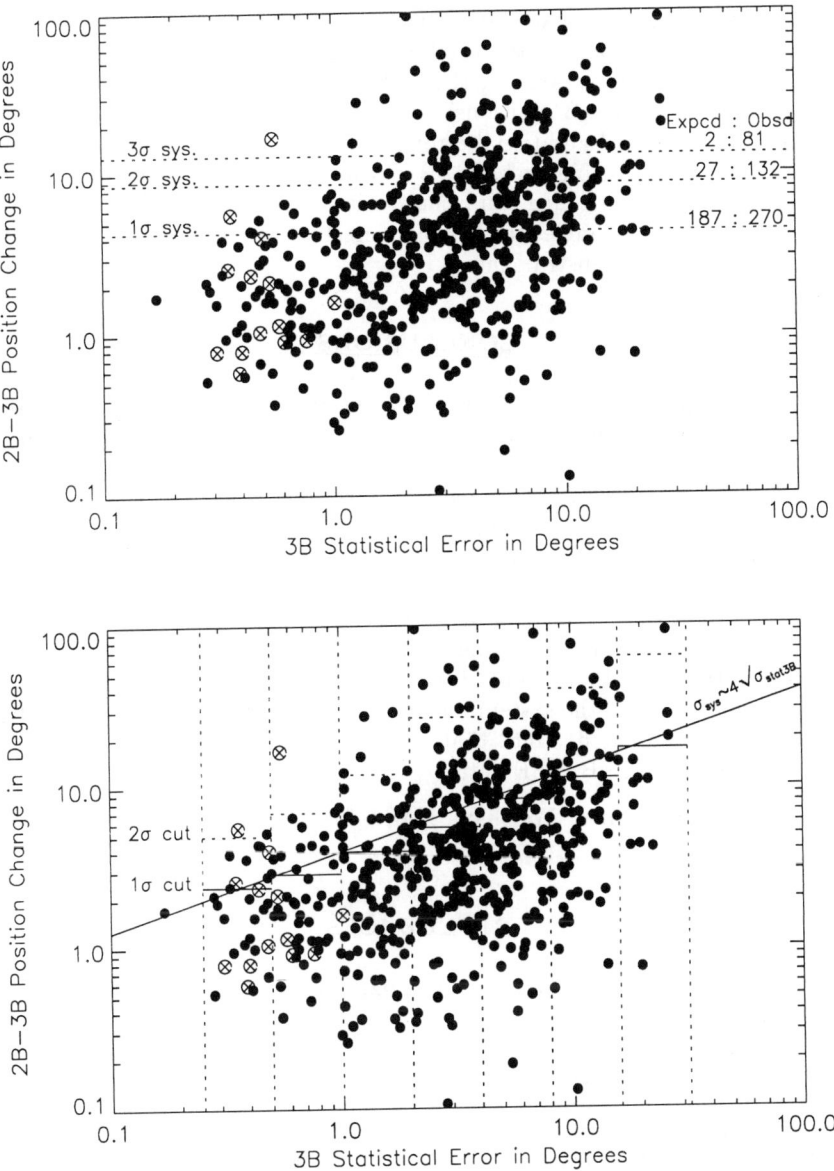

FIG. 2. The burst positions changes from the 1B/2B to 3B Catalogs are shown to be much greater than expected for the constant, combined systematic errors of 4° and 1.6°, estimated by the BATSE team (1,8), and they show a clear dependence of the systematic error on the estimated statistical error, with the systematic error approximately equal to $4(\sigma_{stat})^{-1/2}$. The bursts with IPN locations are shown with a circled X.

correlation with the estimated statistical error. Moreover, they find that although the 3B positions for the brightest bursts are more accurate than the 1B position, the 3B positions for the bulk (~80%) of the bursts are actually less accurate than the 1B positions!

One likely source of additional error in the new 3B position algorithm, as suggested by Hua and Lingenfelter (4), is that when the detector response in the backward 30° was added azimuthal variations were ignored, even though from many azimuthal angles the BATSE detectors are obscured by the spacecraft and neighboring instruments.

CONCLUSIONS

From our analysis of the new 3B Catalog positions, we find that neither the revised 1B nor the new 3B burst positions now show any significant evidence for closely clustered bursts in position and time.

However, we find that the changes in burst positions from the 1B/2B to 3B Catalogs (see Figure 2) are much greater than expected for the constant, combined systematic errors of 4° and 1.6° estimated by the BATSE team (1,8), and other analyses indicate that for the bulk of the bursts the new 3B Catalog positions are in fact less accurate than those of the 1B Catalog.

In order to understand these unexpectedly large 1B-3B position changes, it is essential that both the 1B and 3B BATSE burst location programs be made public. Only then can the systematic and statistical errors be thoroughly and independently evaluated. Until that is done, no significant limits can be set on burst repetition in BATSE data.

Acknowledgements. We thank NASA for support under grant NAG 5-1597.

REFERENCES

1. G. Fishman, et al., Astrophys. J. Supp., **92**, 229 (1994).
2. C. Graziani, & D. Q. Lamb, in these proceedings.
3. J. C. Higdon, & R. E. Lingenfelter, Ann. Rev. Astron. Astrophys, **28**, 401 (1990).
4. X.-M. Hua, & R. E. Lingenfelter, in these proceedings.
5. D. Q. Lamb, in these proceedings.
6. V. C. Wang, & R. E. Lingenfelter, Astrophys. J., **416**, L13 (1993).
7. V. C. Wang, & R. E. Lingenfelter, Astrophys. J., **441**, 747 (1995).
8. C. Meegan, et al., Astrophys. J., submitted (1995).

A Search for Micro Cosmic Gamma-Ray Bursts in BATSE One Second Continuous Data

C.A. Young*, M.B. Arndt*, D.A. Biesecker[†] and J.M. Ryan*

*University of New Hampshire
Center for the Study of Earth, Oceans, and Space
Durham New Hamspire 03824
[†]University of Birmingham
School of Physics and Space Research
Edgbaston, Birmingham, B15 2TT UK

Based on a previous successful search for untriggered solar microflares we have conducted a search for untriggered cosmic gamma-ray bursts (GRBs). The large number of untriggered hard x-ray solar flares suggests that there might exist a class of GRBs with temporal and spectral properties similar to that of hard x-ray solar flares, possibly with similar physics as well. We have scanned a subset of BATSE's continuous 1 second data from four channels independently. This search can identify GRBs with characteristic rise times longer than 1 second, while the electronic identification system only triggers on rise times of 64–1024 ms. We can also search for softer events, not necessarily SGRs, by including the lowest energy channel. A search of continuous unbiased data can lower the search sensitivity to a minimum determined by the intrinsic and orbit-modulated background. We present the results of our progress.

INTRODUCTION

A search for solar flares in the BATSE one second data (1) yields a power law frequency distribution (Fig. 1). This distribution extends that obtained from the flares that electronically triggered BATSE to include flares an order of magnitude smaller in size (Fig. 2), demonstrating that BATSE has difficulty triggering on events with characteristics similar to those of solar flares, i.e., softer spectra and rise times slower than 1024 ms. This suggests that there may exist a population of micro cosmic gamma-ray bursts with softer energies and slower rise times that fail to trigger the BATSE instrument. We have used the experience and techniques of the solar microflare search as a guide in the search for micro GRBs. Here, we outline the techniques of the solar search, how they are modified to search for GRBs, and present our preliminary results.

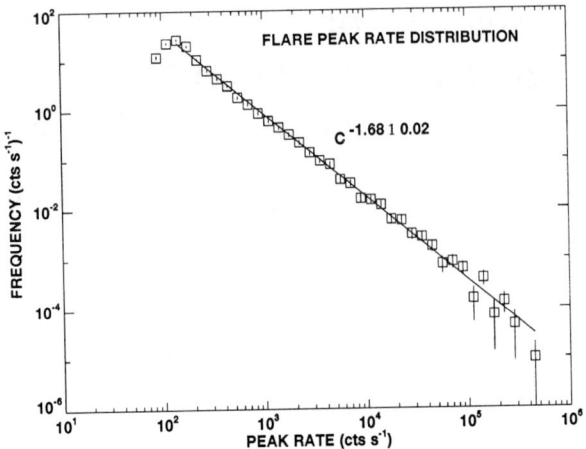

FIG. 1. Full flare distribution

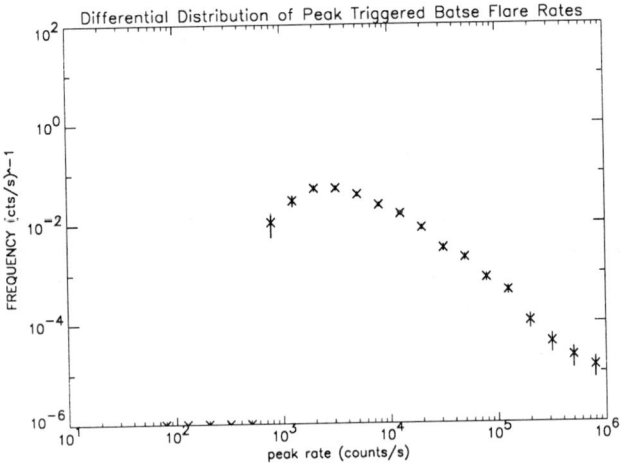

FIG. 2. Distribution of flares that triggered BATSE

SEARCH METHOD

The first task in the search is to locate the events in time. For short periods of time ($\lesssim 30$ s) we assume a linear background count rate in all energy channels and subtract it. A burst reveals itself as a large negative spike in the 2nd derivative with neighboring positive spikes (Fig. 3). However, the discrete second derivative (or second difference) amplifies noise so we must smooth the second difference to reduce this effect, increasing the signal-to-noise ratio. This is done in two steps. First, we average the second difference S_i over an interval of $2M + 1$ centered on the time interval i. The different values of M represent different integration times, i.e., large M corresponds to smoother, longer events and smaller M corresponds to shorter, more variable events. Second, we further improve the signal-to-noise ratio by averaging these smoothed values with their Z nearest neighbors. The signal-to-noise ratio is greatest for $Z = 4$; higher values of Z only add to the computational complexity (2). This study uses $Z = 4$ and M values of $4, 8, 16,$ and 32. Once

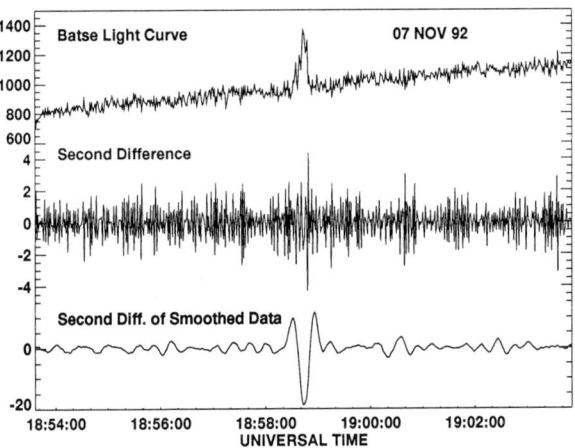

FIG. 3. Sample of raw BATSE data, its 2nd difference and 2nd difference smoothed with the parameters Z=4 and M=8 (2).

a possible candidate is found, its significance is compared to predefined limits based on a probability of 2.7×10^{-1} that the detection is random.

The second task in the search is locating events in space. This was accomplished in the solar study by testing the count distribution in the four most solar facing detector against a hypothesized burst from the sun using the mean count ratio, $\overline{R}$.

$$\overline{R} = \frac{\sum_{j=2}^{4} \frac{R_{j1}}{\sigma^2_{R_{j1}}}}{\sum_{j=2}^{4} \sigma^2_{R_{j1}}} \quad (1)$$

The R_j's are the ratio of the counts in detector j to the counts in the most solar facing detector. A mean ratio of one implies a solar event. Figure 4 shows the distribution of mean ratios for the solar study (1). The peak around unity corresponds to solar events and the finite peak width is a result of the finite angular resolution of BATSE. Following similar logic we wish to test the location of a γ-ray burst against a hypothetical location. However, because

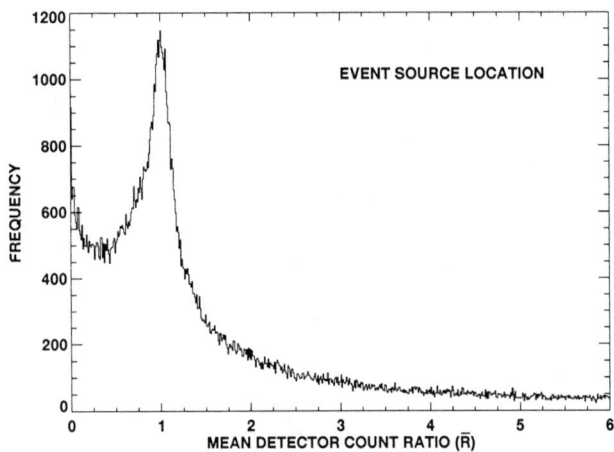

FIG. 4. Mean ratio distribution for the solar study (2).

γ-ray bursts are largely isotropic, there is no preferred origin for γ-ray bursts; or conversely no point in the sky is better than any other in terms of choosing a test source location. So, we chose the zenith as our test source location to suppress terrestrial events. We avoided the sun by considering only events where the sun was > 40 degrees from the zenith. Figure 5 shows the mean ratio distribution for the cosmic burst search. The mean ratios around unity correspond to events near the zenith but there is no peak in the distribution as the zenith is no more likely a source than any other point.

TRIGGER SELECTION

Our preliminary search covered 21 days of data. We searched the four most zenith facing detectors using the four energy channels ($25-50, 50-100, 100-300$ and > 300 keV). The background was fitted locally in time, where it can be treated as linear, then subtracted. Two significance criteria were used to select events. Events with a signal of more than 3.7σ above average in one detector/energy channel or more than 3.3σ above average in any two detector/energy channels were accepted. Events in a region centered around the zenith were selected by choosing those with mean ratios from 0.7 to 1.41. This selection range is the same as the solar study. Certain events such as particle events and instrument fluctuations can mimic acceptable mean ratios, even though their distribution of counts in four detectors reveals their true nature. This required that another parameter be used to eliminate these type of events. A chi-square was calculated for the mean ratio using the measured counts verses the expected counts. By considering events with good chi-squares (< 5 for 3 degrees of freedom), we are able to effectively remove

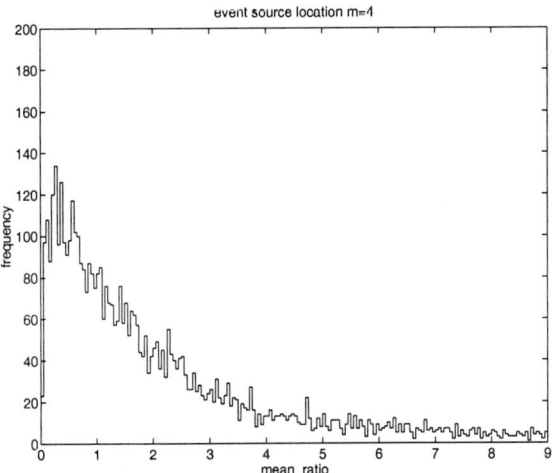

FIG. 5. Mean ratio distribution for the grb search.

detections of "spurious" events.

CONCLUSIONS

Biesecker's routine to search BATSE data for solar microflare triggers provides an automated algorithm with which we can search for triggers of GRBs. This routine allows us to search for bursts with rise times too slow or energies too soft to trigger the BATSE instrument. We eliminate terrestrial, solar, and spurious events, such as cosmic rays, and instrumental contributions, by limiting the field of view of the detectors, and by implementing stringent criteria for multi-detector detections. Information on the location of bursts is provided by the mean count ratios. We are enhancing the criteria for burst selection by including occultation edge detection and better accounting for Cyg X-1 fluctuations.

Our preliminary search detected all expected events (e.g., **3** 2B catalog events) as well as **2** new GRB candidates. These results remind us that there are biases in triggering which can significantly change parameters for global GRB characteristics and provide us with motivation to continue implementation of the search routine on the robust BATSE data set.

REFERENCES

1. D. A. Bicsccker, PhD Thesis, University of New Hampshire (1994).
2. M. A. Mariscotti, Nuclear Instruments and Methods **50**, 309 (1967).

AIP Conference Proceedings

	Title	L.C. Number	ISBN
No. 220	High Energy Gamma-Ray Astronomy (Ann Arbor, MI 1990)	91-70876	0-88318-812-0
No. 221	Particle Production Near Threshold (Nashville, IN 1990)	91-55134	0-88318-829-5
No. 222	After the First Three Minutes (College Park, MD 1990)	91-55214	0-88318-828-7
No. 223	Polarized Collider Workshop (University Park, PA 1990)	91-71303	0-88318-826-0
No. 224	LAMPF Workshop on (π, K) Physics (Los Alamos, NM 1990)	91-71304	0-88318-825-2
No. 225	Half Collision Resonance Phenomena in Molecules (Caracas, Venezuela 1990)	91-55210	0-88318-840-6
No. 226	The Living Cell in Four Dimensions (Gif sur Yvette, France 1990)	91-55209	0-88318-794-9
No. 227	Advanced Processing and Characterization Technologies (Clearwater, FL 1991)	91-55194	0-88318-910-0
No. 228	Anomalous Nuclear Effects in Deuterium/ Solid Systems (Provo, UT 1990)	91-55245	0-88318-833-3
No. 229	Accelerator Instrumentation (Batavia, IL 1990)	91-55347	0-88318-832-1
No. 230	Nonlinear Dynamics and Particle Acceleration (Tsukuba, Japan 1990)	91-55348	0-88318-824-4
No. 231	Boron-Rich Solids (Albuquerque, NM 1990)	91-53024	0-88318-793-4
No. 232	Gamma-Ray Line Astrophysics (Paris-Saclay, France 1990)	91-55492	0-88318-875-9
No. 233	Atomic Physics 12 (Ann Arbor, MI 1990)	91-55595	088318-811-2
No. 234	Amorphous Silicon Materials and Solar Cells (Denver, CO 1991)	91-55575	088318-831-7
No. 235	Physics and Chemistry of MCT and Novel IR Detector Materials (San Francisco, CA 1990)	91-55493	0-88318-931-3
No. 236	Vacuum Design of Synchrotron Light Sources (Argonne, IL 1990)	91-55527	0-88318-873-2
No. 237	Kent M. Terwilliger Memorial Symposium (Ann Arbor, MI 1989)	91-55576	0-88318-788-4
No. 238	Capture Gamma-Ray Spectroscopy (Pacific Grove, CA 1990)	91-57923	0-88318-830-9

	Title	L.C. Number	ISBN
No. 239	Advances in Biomolecular Simulations (Obernai, France 1991)	91-58106	0-88318-940-2
No. 240	Joint Soviet-American Workshop on the Physics of Semiconductor Lasers (Leningrad, USSR 1991)	91-58537	0-88318-936-4
No. 241	Scanned Probe Microscopy (Santa Barbara, CA 1991)	91-76758	0-88318-816-3
No. 242	Strong, Weak, and Electromagnetic Interactions in Nuclei, Atoms, and Astrophysics: A Workshop in Honor of Stewart D. Bloom's Retirement (Livermore, CA 1991)	91-76876	0-88318-943-7
No. 243	Intersections Between Particle and Nuclear Physics (Tucson, AZ 1991)	91-77580	0-88318-950-X
No. 244	Radio Frequency Power in Plasmas (Charleston, SC 1991)	91-77853	0-88318-937-2
No. 245	Basic Space Science (Bangalore, India 1991)	91-78379	0-88318-951-8
No. 246	Space Nuclear Power Systems (Albuquerque, NM 1992)	91-58793	1-56396-027-3 1-56396-026-5 (pbk.)
No. 247	Global Warming: Physics and Facts (Washington, DC 1991)	91-78423	0-88318-932-1
No. 248	Computer-Aided Statistical Physics (Taipei, Taiwan 1991)	91-78378	0-88318-942-9
No. 249	The Physics of Particle Accelerators (Upton, NY 1989, 1990)	92-52843	0-88318-789-2
No. 250	Towards a Unified Picture of Nuclear Dynamics (Nikko, Japan 1991)	92-70143	0-88318-951-8
No. 251	Superconductivity and its Applications (Buffalo, NY 1991)	92-52726	1-56396-016-8
No. 252	Accelerator Instrumentation (Newport News, VA 1991)	92-70356	0-88318-934-8
No. 253	High-Brightness Beams for Advanced Accelerator Applications (College Park, MD 1991)	92-52705	0-88318-947-X
No. 254	Testing the AGN Paradigm (College Park, MD 1991)	92-52780	1-56396-009-5
No. 255	Advanced Beam Dynamics Workshop on Effects of Errors in Accelerators, Their Diagnosis and Corrections (Corpus Christi, TX 1991)	92-52842	1-56396-006-0
No. 256	Slow Dynamics in Condensed Matter (Fukuoka, Japan 1991)	92-53120	0-88318-938-0

Title	L.C. Number	ISBN
No. 257 Atomic Processes in Plasmas (Portland, ME 1991)	91-08105	0-88318-939-9
No. 258 Synchrotron Radiation and Dynamic Phenomena (Grenoble, France 1991)	92-53790	1-56396-008-7
No. 259 Future Directions in Nuclear Physics with 4π Gamma Detection Systems of the New Generation (Strasbourg, France 1991)	92-53222	0-88318-952-6
No. 260 Computational Quantum Physics (Nashville, TN 1991)	92-71777	0-88318-933-X
No. 261 Rare and Exclusive B&K Decays and Novel Flavor Factories (Santa Monica, CA 1991)	92-71873	1-56396-055-9
No. 262 Molecular Electronics—Science and Technology (St. Thomas, Virgin Islands 1991)	92-72210	1-56396-041-9
No. 263 Stress-Induced Phenomena in Metallization: First International Workshop (Ithaca, NY 1991)	92-72292	1-56396-082-6
No. 264 Particle Acceleration in Cosmic Plasmas (Newark, DE 1991)	92-73316	0-88318-948-8
No. 265 Gamma-Ray Bursts (Huntsville, AL 1991)	92-73456	1-56396-018-4
No. 266 Group Theory in Physics (Cocoyoc, Morelos, Mexico 1991)	92-73457	1-56396-101-6
No. 267 Electromechanical Coupling of the Solar Atmosphere (Capri, Italy 1991)	92-82717	1-56396-110-5
No. 268 Photovoltaic Advanced Research & Development Project (Denver, CO 1992)	92-74159	1-56396-056-7
No. 269 CEBAF 1992 Summer Workshop (Newport News, VA 1992)	92-75403	1-56396-067-2
No. 270 Time Reversal—The Arthur Rich Memorial Symposium (Ann Arbor, MI 1991)	92-83852	1-56396-105-9
No. 271 Tenth Symposium Space Nuclear Power and Propulsion (Vols. I–III) (Albuquerque, NM 1993)	92-75162	1-56396-137-7 (set)
No. 272 Proceedings of the XXVI International Conference on High Energy Physics (Vols. I and II) (Dallas, TX 1992)	93-70412	1-56396-127-X (set)
No. 273 Superconductivity and Its Applications (Buffalo, NY 1992)	93-70502	1-56396-189-X

Title	L.C. Number	ISBN
No. 274 VIth International Conference on the Physics of Highly Charged Ions (Manhattan, KS 1992)	93-70577	1-56396-102-4
No. 275 Atomic Physics 13 (Munich, Germany 1992)	93-70826	1-56396-057-5
No. 276 Very High Energy Cosmic-Ray Interactions: VIIth International Symposium (Ann Arbor, MI 1992)	93-71342	1-56396-038-9
No. 277 The World at Risk: Natural Hazards and Climate Change (Cambridge, MA 1992)	93-71333	1-56396-066-4
No. 278 Back to the Galaxy (College Park, MD 1992)	93-71543	1-56396-227-6
No. 279 Advanced Accelerator Concepts (Port Jefferson, NY 1992)	93-71773	1-56396-191-1
No. 280 Compton Gamma-Ray Observatory (St. Louis, MO 1992)	93-71830	1-56396-104-0
No. 281 Accelerator Instrumentation Fourth Annual Workshop (Berkeley, CA 1992)	93-072110	1-56396-190-3
No. 282 Quantum 1/f Noise & Other Low Frequency Fluctuations in Electronic Devices (St. Louis, MO 1992)	93-072366	1-56396-252-7
No. 283 Earth and Space Science Information Systems (Pasadena, CA 1992)	93-072360	1-56396-094-X
No. 284 US-Japan Workshop on Ion Temperature Gradient-Driven Turbulent Transport (Austin, TX 1993)	93-72460	1-56396-221-7
No. 285 Noise in Physical Systems and 1/f Fluctuations (St. Louis, MO 1993)	93-72575	1-56396-270-5
No. 286 Ordering Disorder: Prospect and Retrospect in Condensed Matter Physics: Proceedings of the Indo-U.S. Workshop (Hyderabad, India 1993)	93-072549	1-56396-255-1
No. 287 Production and Neutralization of Negative Ions and Beams: Sixth International Symposium (Upton, NY 1992)	93-72821	1-56396-103-2
No. 288 Laser Ablation: Mechanismas and Applications-II: Second International Conference (Knoxville, TN 1993)	93-73040	1-56396-226-8
No. 289 Radio Frequency Power in Plasmas: Tenth Topical Conference (Boston, MA 1993)	93-72964	1-56396-264-0

	Title	L.C. Number	ISBN
No. 290	Laser Spectroscopy: XIth International Conference (Hot Springs, VA 1993)	93-73050	1-56396-262-4
No. 291	Prairie View Summer Science Academy (Prairie View, TX 1992)	93-73081	1-56396-133-4
No. 292	Stability of Particle Motion in Storage Rings (Upton, NY 1992)	93-73534	1-56396-225-X
No. 293	Polarized Ion Sources and Polarized Gas Targets (Madison, WI 1993)	93-74102	1-56396-220-9
No. 294	High-Energy Solar Phenomena: A New Era of Spacecraft Measurements (Waterville Valley, NH 1993)	93-74147	1-56396-291-8
No. 295	The Physics of Electronic and Atomic Collisions: XVIII International Conference (Aarhus, Denmark, 1993)	93-74103	1-56396-290-X
No. 296	The Chaos Paradigm: Developments an Applications in Engineering and Science (Mystic, CT 1993)	93-74146	1-56396-254-3
No. 297	Computational Accelerator Physics (Los Alamos, NM 1993)	93-74205	1-56396-222-5
No. 298	Ultrafast Reaction Dynamics and Solvent Effects (Royaumont, France 1993)	93-074354	1-56396-280-2
No. 299	Dense Z-Pinches: Third International Conference (London, 1993)	93-074569	1-56396-297-7
No. 300	Discovery of Weak Neutral Currents: The Weak Interaction Before and After (Santa Monica, CA 1993)	94-70515	1-56396-306-X
No. 301	Eleventh Symposium Space Nuclear Power and Propulsion (3 Vols.) (Albuquerque, NM 1994)	92-75162	1-56396-305-1 (Set) 156396-301-9 (pbk. set)
No. 302	Lepton and Photon Interactions/ XVI International Symposium (Ithaca, NY 1993)	94-70079	1-56396-106-7
No. 303	Slow Positron Beam Techniques for Solids and Surfaces Fifth International Workshop (Jackson Hole, WY 1992)	94-71036	1-56396-267-5
No. 304	The Second Compton Symposium (College Park, MD 1993)	94-70742	1-56396-261-6
No. 305	Stress-Induced Phenomena in Metallization Second International Workshop (Austin, TX 1993)	94-70650	1-56396-251-9

	Title	L.C. Number	ISBN
No. 306	12th NREL Photovoltaic Program Review (Denver, CO 1993)	94-70748	1-56396-315-9
No. 307	Gamma-Ray Bursts Second Workshop (Huntsville, AL 1993)	94-71317	1-56396-336-1
No. 308	The Evolution of X-Ray Binaries (College Park, MD 1993)	94-76853	1-56396-329-9
No. 309	High-Pressure Science and Technology—1993 (Colorado Springs, CO 1993)	93-72821	1-56396-219-5 (Set)
No. 310	Analysis of Interplanetary Dust (Houston, TX 1993)	94-71292	1-56396-341-8
No. 311	Physics of High Energy Particles in Toroidal Systems (Irvine, CA 1993)	94-72098	1-56396-364-7
No. 312	Molecules and Grains in Space (Mont Sainte-Odile, France 1993)	94-72615	1-56396-355-8
No. 313	The Soft X-Ray Cosmos ROSAT Science Symposium (College Park, MD 1993)	94-72499	1-56396-327-2
No. 314	Advances in Plasma Physics Thomas H. Stix Symposium (Princeton, NJ 1992)	94-72721	1-56396-372-8
No. 315	Orbit Correction and Analysis in Circular Accelerators (Upton, NY 1993)	94-72257	1-56396-373-6
No. 316	Thirteenth International Conference on Thermoelectrics (Kansas City, Missouri 1994)	95-75634	1-56396-444-9
No. 317	Fifth Mexican School of Particles and Fields (Guanajuato, Mexico 1992)	94-72720	1-56396-378-7
No. 318	Laser Interaction and Related Plasma Phenomena 11th International Workshop (Monterey, CA 1993)	94-78097	1-56396-324-8
No. 319	Beam Instrumentation Workshop (Santa Fe, NM 1993)	94-78279	1-56396-389-2
No. 320	Basic Space Science (Lagos, Nigeria 1993)	94-79350	1-56396-328-0
No. 321	The First NREL Conference on Thermophotovoltaic Generation of Electricity (Copper Mountain, CO 1994)	94-72792	1-56396-353-1
No. 322	Atomic Processes in Plasmas Ninth APS Topical Conference (San Antonio, TX)	94-72923	1-56396-411-2

	Title	L.C. Number	ISBN
No. 323	Atomic Physics 14 Fourteenth International Conference on Atomic Physics (Boulder, CO 1994)	94-73219	1-56396-348-5
No. 324	Twelfth Symposium on Space Nuclear Power and Propulsion (Albuquerque, NM 1995)	94-73603	1-56396-427-9
No. 325	Conference on NASA Centers for Commercial Development of Space (Albuquerque, NM 1995)	94-73604	1-56396-431-7
No. 326	Accelerator Physics at the Superconducting Super Collider (Dallas, TX 1992-1993)	94-73609	1-56396-354-X
No. 327	Nuclei in the Cosmos III Third International Symposium on Nuclear Astrophysics (Assergi, Italy 1994)	95-75492	1-56396-436-8
No. 328	Spectral Line Shapes, Volume 8 12th ICSLS (Toronto, Canada 1994)	94-74309	1-56396-326-4
No. 329	Resonance Ionization Spectroscopy 1994 Seventh International Symposium (Bernkastel-Kues, Germany 1994)	95-75077	1-56396-437-6
No. 330	E.C.C.C. 1 Computational Chemistry F.E.C.S. Conference (Nancy, France 1994)	95-75843	1-56396-457-0
No. 331	Non-Neutral Plasma Physics II (Berkeley, CA 1994)	95-79630	1-56396-441-4
No. 332	X-Ray Lasers 1994 Fourth International Colloquium (Williamsburg, VA 1994)	95-76067	1-56396-375-2
No. 333	Beam Instrumentation Workshop (Vancouver, B. C., Canada 1994)	95-79635	1-56396-352-3
No. 334	Few-Body Problems in Physics (Williamsburg, VA 1994)	95-76481	1-56396-325-6
No. 335	Advanced Accelerator Concepts (Fontana, WI 1994)	95-78225	1-56396-476-7 (Set) 1-56396-474-0 (Book) 1-56396-475-9 (CD-Rom)
No. 336	Dark Matter (College Park, MD 1994)	95-76538	1-56396-438-4
No. 337	Pulsed RF Sources for Linear Colliders (Montauk, NY 1994)	95-76814	1-56396-408-2

	Title	L.C. Number	ISBN
No. 338	Intersections Between Particle and Nuclear Physics 5th Conference (St. Petersburg, FL 1994)	95-77076	1-56396-335-3
No. 339	Polarization Phenomena in Nuclear Physics Eighth International Symposium (Bloomington, IN 1994)	95-77216	1-56396-482-1
No. 340	Strangeness in Hadronic Matter (Tucson, AZ 1995)	95-77477	1-56396-489-9
No. 341	Volatiles in the Earth and Solar System (Pasadena, CA 1994)	95-77911	1-56396-409-0
No. 342	CAM -94 Physics Meeting (Cacun, Mexico 1994)	95-77851	1-56396-491-0
No. 343	High Energy Spin Physics Eleventh International Symposium (Bloomington, IN 1994)	95-78431	1-56396-374-4
No. 344	Nonlinear Dynamics in Particle Accelerators: Theory and Experiments (Arcidosso, Italy 1994)	95-78135	1-56396-446-5
No. 345	International Conference on Plasma Physics ICPP 1994 (Foz do Iguaçu, Brazil 1994)	95-78438	1-56396-496-1
No. 346	International Conference on Accelerator-Driven Transmutation Technologies and Applications (Las Vegas, NV 1994)	95-78691	1-56396-505-4
No. 347	Atomic Collisions: A Symposium in Honor of Christopher Bottcher (1945-1993) (Oak Ridge, TN 1994)	95-78689	1-56396-322-1
No. 348	Unveiling the Cosmic Infrared Background (College Park, MD, 1995)	95-83477	1-56396-508-9
No. 349	Workshop on the Tau/Charm Factory (Argonne, IL, 1995)	95-81467	1-56396-523-2
No. 350	International Symposium on Vector Boson Self-Interactions (Los Angeles, CA 1995)	95-79865	1-56396-520-8
No. 351	The Physics of Beams Andrew Sessler Symposium (Los Angeles, CA 1993)	95-80479	1-56396-376-0
No. 352	Physics Potential and Development of $\mu^+\mu^-$ Colliders: Second Workshop (Sausalito, CA 1994)	95-81413	1-56396-506-2
No. 353	13th NREL Photovoltaic Program Review (Lakewood, CO 1995)	95-80662	1-56396-510-0
No. 354	Organic Coatings (Paris, France, 1995)	96-83019	1-56396-535-6

	Title	L.C. Number	ISBN
No. 355	Eleventh Topical Conference on Radio Frequency Power in Plasmas (Palm Springs, CA 1995)	95-80867	1-56396-536-4
No. 356	The Future of Accelerator Physics (Austin, TX 1994)	96-83292	1-56396-541-0
No. 357	10th Topical Workshop on Proton-Antiproton Collider Physics (Batavia, IL 1995)	95-83078	1-56396-543-7
No. 358	The Second NREL Conference on Thermophotovoltaic Generation of Electricity	95-83335	1-56396-509-7
No. 359	Workshops and Particles and Fields and Phenomenology of Fundamental Interactions (Puebla, Mexico 1995)	96-85996	1-56396-548-8
No. 360	The Physics of Electronic and Atomic Collisions XIX International Conference (Whistler, Canada, 1995)	95-83671	1-56396-440-6
No. 361	Space Technology and Applications International Forum (Albuquerque, NM 1996)	95-83440	1-56396-568-2
No. 362	Two-Center Effects in Ion-Atom Collisions (Lincoln, NE 1994)	96-83379	1-56396-342-6
No. 363	Phenomena in Ionized Gases XXII ICPIG (Hoboken, NJ, 1995)	96-83294	1-56396-550-X
No. 364	Fast Elementary Processes in Chemical and Biological Systems (Villeneuve d'Ascq, France, 1995)	96-83624	1-56396-564-X
No. 365	Latin-American School of Physics XXX ELAF Group Theory and Its Applications (México City, México, 1995)	96-83489	1-56396-567-4
No. 366	High Velocity Neutron Stars and Gamma-Ray Bursts (La Jolla, CA 1995)	96-84067	1-56396-593-3
No. 367	Micro Bunches Workshop (Upton, NY, 1995)	96-83482	1-56396-555-0
No. 368	Acoustic Particle Velocity Sensors: Design, Performance and Applications (Mystic, CT, 1995)	96-83548	1-56396-549-6
No. 369	Laser Interaction and Related Plasma Phenomena (Osaka, Japan 1995)	96-85009	1-56396-445-7
No. 370	Shock Compression of Condensed Matter-1995 (Seattle, WA 1995)	96-84595	1-56396-566-6

	Title	L.C. Number	ISBN
No. 371	Sixth Quantum 1/f Noise and Other Low Frequency Fluctuations in Electronic Devices Symposium (St. Louis, MO, 1994)	96-84200	1-56396-410-4
No. 372	Beam Dynamics and Technology Issues for + - Colliders 9th Advanced ICFA Beam Dynamics Workshop (Montauk, NY, 1995)	96-84189	1-56396-554-2
No. 373	Stress-Induced Phenomena in Metallization (Palo Alto, CA 1995)	96-84949	1-56396-439-2
No. 374	High Energy Solar Physics (Greenbelt, MD 1995)	96-84513	1-56396-542-9
No. 375	Chaotic, Fractal, and Nonlinear Signal Processing (Mystic, CT 1995)	96-85356	1-56396-443-0
No. 376	Chaos and the Changing Nature of Science and Medicine: An Introduction (Mobile, AL 1995)	96-85220	1-56396-442-2
No. 377	Space Charge Dominated Beams and Applications of High Brightness Beams (Bloomington, IN 1995)	96-85165	1-56396-625-7
No. 379	Physical Origin of Homochirality in Life (Santa Monica, CA 1995)	96-86631	1-56396-507-0
No. 378	Surfaces, Vacuum, and Their Applications (Cancun, Mexico 1994)	96-85594	1-56396-418-X
No. 380	Production and Neutralization of Negative Ions and Beams / Production and Application of Light Negative Ions (Upton, NY 1995)	96-86435	1-56396-565-8
No. 381	Atomic Processes in Plasmas (San Francisco, CA 1996)	96-86304	1-56396-552-6
No. 382	Solar Wind Eight (Dana Point, CA 1995)	96-86447	1-56396-551-8
No. 383	Workshop on the Earth's Trapped Particle Environment (Taos, NM 1994)	96-86619	1-56396-540-2
No. 384	Gamma-Ray Bursts (Huntsville, AL 1995)	96-79458	1-56396-685-9
No. 385	Robotic Exploration Close to the Sun: Scientific Basis (Marlboro, MA 1996)	96-79560	1-56396-618-2